Implement & Tractor

*Reflections on 100 Years
of Farm Equipment*

Implement & Tractor

Reflections on 100 Years
of Farm Equipment

Edited by Robert K. Mills

Photo:
Thomas Hart Benton's *"Threshing Wheat"*
The Sheldon Swope Art Gallery
Terre Haute, Indiana

INTERTEC®
PUBLISHING

CONTENTS

CONTENTS (CONT.)

PREFACE

Raising food was once this country's largest business (and is still the world's most important). America's history has been shaped by and has helped shape farming and its support industries. This book is about the last 100 years and the progress made toward effective and efficient production of food. These ads, articles and insights were gleaned from the hundreds of thousands of pages of the industry's leading "trade publications."

Suppliers have always had the problem of convincing retailers to sell their products to end users. Most companies have, through the ages, used traveling salesmen and public demonstrations to show the virtues of their products. However, articles and advertisements in industry "trade papers" have also been an effective way for the manufacturers to communicate with dealers.

The Implement Trade Journal began publishing a "trade paper" for the farm equipment retailers in 1886. Its primary business was helping retail dealers learn about new products. Industry changes including the development of tractors was reflected in the name change to Implement & Tractor magazine.

Implement & Tractor magazine has always been a liaison between the farm equipment dealers and industry suppliers. The success of the magazine was the result of serving both the supplier and the dealer. Manufacturers and suppliers could always count on Implement & Tractor to reach a large number of people who were interested in selling their product and to explain their product fairly. Dealers were also helped by articles explaining sound business practices and exposing frauds.

Dealers learned about new products and trends, often well before the public, by reading articles and advertisements in the trade magazines. Therefore, many products included in this book will seem to be introduced sooner than remembered. Some anticipated trends didn't develop as expected and other trends developed so quickly they were recognized only after being fully accepted.

To be successful, a dealer had to identify in advance the products his customers would want, decide how many of each item he could sell, then order quickly so that the product would be received before the market conditions changed. Implement & Tractor often showed several suppliers of similar products so dealers could select a supplier that would best meet their needs. A "hot" item was often impossible to get if the dealer hesitated. Change was so fast in the early days of the magazine that weekly publication was necessary to stay up with development of new equipment.

As Implement & Tractor celebrates 100 years of publishing a trade magazine for the farm equipment industry, this book looks back at the industry through the magazine's pages.

EDITOR'S NOTES

Seldom is an author offered the chance to "look back" at the past 100 years of an industry's growth through the pages of the leading trade paper. My travels back through time involved many articles, ads and thoughts that just could not be included, but I believe the illustrations and examples found here will help readers appreciate the people, companies and products that have contributed so much.

This effort was especially pleasant for me because of my long association with the magazine management and staff as well as others in the farm equipment industry. Many of the dealers who were profiled in articles were personal friends with whom I had worked. Educators and industry leaders that I have grown to admire and respect were often quoted or had contributed articles. Later, articles began to appear about retirements, deaths and dealership closings that affected people who were important to me. The tone of the magazine changed during difficult economic times and during times of war reflecting, I'm sure, how the country felt during these periods.

Neither Intertec Publishing Corp. nor I am responsible for the accuracy of articles reproduced from the original pages. As a matter of fact, much of the material was selected because of the flamboyant style. Every precaution has been taken to reproduce the facts as they were presented, but the original magazines contained errors too. Occasionally notifications of death were objected to in later issues by the people who were reported dead. Of course, offers reproduced from the original ads are probably no longer available and no effort has been made to verify any claim. Only a small part of this book is reproduced from the period after truth in advertising was anything but a consumers wish.

Ethnic, racial, sexual and religious slurs were far more numerous in the conservative trade press than I had believed. With no disrespect to any person or company intended, I have included samples to show the reader contemporary sentiment that was accepted at the time represented.

I especially wish to express gratitude to the original advertisers, editors, writers and other contributors for the success of Implement & Tractor magazine. Without their help the magazine and this "look back" could never have happened. Some by-lines have been included; more often they have not. The names of some companies have changed through the years from the original, but most are quickly recognized.

The Index of Articles and the List of Illustrations are provided to help find an illustration or article that you have seen or read previously and are not intended to research specific subjects. Company names, product names and the names of people have most often, but not always, been used for listing.

The Bibliography lists some books suggested for additional reading, but is certainly not complete. I constantly find books relating to the time and the equipment of rural America. It is important for us to remember the people who used the often difficult times and the sometimes crude equipment to overcome the always difficult task of feeding hungry people.

It is impossible to list everyone who was important to the production of this book, but the following is an incomplete list of some who affected the direction of this book:

Harold L. Adkins, Elmer Baker, Frank Buckingham, Paul Dumas, Howard Everett, Charles G. Ewing, Bill Fogarty, Jack Foland, Paul Hartsock, David E. Hertel, Robert E. Hertel, Wilber Higbie, Emmett P. Langan, Kenneth F. Long, Mel Long, Miguel J. Lozano, Henry Rinker, Dudley Rose, George H. Seferovich, Frank D. Smalley and Louis L. Tigner.

I hope that these 100 years are as enjoyable for you to reflect on through the pages of this book as its preparation has been to me.

Robert L. Smith

BEFORE 1886

THE BEGINNINGS

OH! TO BE A BOY AGAIN.

The Modern Miller
WEEKLY
FLOUR AND GRAIN TRADES REVIEW

KANSAS CITY AND ST. LOUIS.

Before 1886

Farming methods that had begun in England and Europe were brought unchanged to the "New World." Before 1800, most farm produce was used on the farm where it was grown. Only a few wealthy "gentlemen" experimented with ways to increase crop and livestock production.

Many companies still producing farm equipment can trace their roots to this period of American history. This was a time of many agricultural changes. The founders of these early companies usually built special equipment to help the local farmers, their friends and neighbors, solve the problems of plowing, sowing, reaping and transporting.

Charles Newbold patented an iron plow in 1797, but many thought the iron poisoned the soil and refused to use the new plow. Jethro Wood patented an improved plow design in 1814 which was similar, but it was made of several smaller castings that were bolted together. By about 1830, cast iron plows were widely used. The new plows could be pulled easier and faster than wooden plows and the result was that one man could farm more land than before. Horses proved better suited than oxen for drawing the new plows.

About 1800, the Lagona Agricultural Works was built on the bank of Buck Creek near Springfield, Ohio. That company would later incorporate as Wardner, Bushnell & Glessner and would produce the Champion line but began by making hoes, rakes and small hand tools, then later made plows, wagons and cultivators.

The frontier hero Andrew Jackson was elected president in 1828 and was reelected in 1832.

Sometime after 1800, a cradle was used to speed the cutting of grain, but harvesting was still done with hand tools. Cyrus McCormick patented a hillside plow of cast and wrought iron construction in 1831 and in late July of that same year, McCormick is said to have first demonstrated his mechanical reaper. Work on the reaper, that was to begin a new age in farming, continued until about 1840 when he sold two machines. The production of the new reapers was slow until 1847 when the McCormick Reaper Works was completed in Chicago, Illinois.

Obed Hussey, a Quaker from Ohio, operated and patented a mechanical reaper of his own design in 1833, several months before McCormick received a patent in 1834 for his reaper. The notoriety that Hussey's reaper received probably caused Cyrus McCormick to work harder and develop his design sooner than he would have without competition.

OBED HUSSY'S REAPER

Priority Claimed for the Invention of an Inventive Quaker—First Reaper Completed in 1833.

The following story from the Kansas City Star sets forth some alleged facts in relation to the mooted question, Who was the original inventor of the reaper? We pass it on for what it is worth:

Most of the inventions which have revolutionized industry and sped civilization on its way have brought fame to their creator—oftentimes a fame which puts his on the pedestal after death, but left him to starve while living. But fame and fortune both escaped Obed Hussey, the Quaker who made the first reaper, and by that invention made bread cheap, cutting much from the high cost of living.

The Harvester Trust for many years has held patents to various sorts of reapers and has enjoyed practically a monopoly of the manufacture. And the founder of this corporation, C. H. McCormick, dead for many years, has been regarded as the originator of this epoch-making agricultural implement. Yet his invention, if it can be called such, came a dozen years after this Quaker mechanic had put his machine in use. But his picture has been put on bank notes over the words, "Inventor of the reaper."

This is set forth in a book called "Obed Hussey," by Follett L. Greeno, who formerly lived in Kansas City. It is a collection of letters and pamphlets published heretofore by Hussey's friends and now collected and edited.

Hussey was born in Maine in 1792. After one or two whaling voyages he settled in Baltimore, where he became an expert draftsman and an incessant worker at different inventions. Among the devices he made were a steam plow, a machine for grinding out hooks and eyes, a mill for grinding corncobs, a husking machine run by horse power, a machine for crushing sugar cane and a machine for making artificial ice. In 1831 he was working in a Baltimore plow works on an invention for cutting grain.

His first reaper was completed at Cincinnati in 1833, and the principle was precisely that used today as regards the guards and knives. The scene of its trial was the farm of Judge Foster, a

SHOWING METHOD OF USING THE CRADLE.

FLAILING OUT THE WHEAT

THE McCORMICK SELF-RAKE REAPER.

few miles out of the city. Hussey, true to his modest and shrinking nature, wished to keep it secret, fearing jeers if it failed. But a crowd of farmers were on hand to scoff.

The machine was started, but some part gave way and it failed to work. One burly fellow picked up a cradle and, swinging it with an air of great exultation, exclaimed: "This is the machine to cut the wheat!"

The team was unhitched, and it looked like failure. But the merriment subsided when the inventor, assisted by some laborers, pulled the machine to the top of the hill. Then, alone, Hussey drew the queer-looking implement down the hill and through the standing grain, cutting every head.

The same machine was exhibited directly afterward before the Hamilton County Agricultural Society, where it was pronounced a success. By the side of a modern reaper Hussey's invention would look primitive, but it did the work of harvest faster than the best farmers could go by hand. Two men were required, one to drive the team and one to rake off the wheat as it fell on a platform.

In the years that followed the inventor traveled from state to state, often at the invitation of state agricultural societies. Owing to his limited means the reaper often was drawn from place to place by one horse, accompanied by the inventor, footsore and weary from walking hundreds of miles. This was in the days before the telegraph or the railroad, and news, as well as men, traveled slowly.

In such a humble way Hussey visited Ohio, Illinois, New York and Missouri, introducing his device. But scarcely a farmer could be found who was willing to adopt it, and it is a matter of history that the final adoption of the machine led to riots by farm hands, who feared there would no longer be work for them to do. He also introduced the reaper into England.

In 1835 Hussey brought two of his machines to Palmyra, Mo., and they were bought by William Muldow of Marion College for $150 each, a great deal of money in those days. One was used in a meadow near the settlement of Philadelphia and the other near Marion City. According to the Missouri Courier of that year they were a great success.

A letter from Hussey to his friend, Edward Stabler, in 1854, gives an insight of the hardships which the genius endured to make his dream come true. It was at the period when he was involved in defending the extension of his patents before Congress, a fight which he won. He wrote:

"I made no money during the existence of my patent, or I might say I made less than I would have made if I had held an underclerk's position in the patent office; I would have been better off at the end of the fourteen years if I had filled exactly such station as my foreman holds and got his pay, and would not have had half the hard work nor a hundredth part of the heart aching. I never experienced half the fatigue of rowing after a whale in the Pacific Ocean (which I have often done) as I have experienced for 18 years in the harvest field, I might say for 20 years, for I worked as hard in England as I do at home, for in the harvest, wherever I am, there is no rest for me. If I am guilty of no rascality why should I not be compensated for toiling to introduce an invention which I thought would be of so much advantage to the world? I know I was the first one who successfully accomplished the cutting of grain and grass by machinery. If others tried to do it before me it was not doing it. Being the first who ever tried it, why should I be obliged to suffer and toil most and get the least by it?

"No man knows how much I have suffered in body and mind since 1833 on account of this thing, the first year I operated in Balto. Three years after I cut the first crop I could not go to meeting for many weeks for want of a decent coat, while for economy I made my own coffee and ate and slept in my shop until I had sold machines enough to be able to do better."

The struggle of this old Quaker was ended when he was killed in a railroad accident near Boston in 1860.

On March 2, 1836, American settlers revolted against Mexico and declared Texas an independent state. To stop the rebellion, Santa Anna and his soldiers attacked and captured the Alamo on March 6. A short time later, Santa Anna and his army were defeated by General Sam Houston and other Texans. Martin Van Buren was elected President of the United States in 1836.

John Deere built his first steel plow in Grand Detour, Illinois, from an old saw blade in 1837. Later, on March 20, 1843,

John Deere and Major Andrus entered into a partnership and formed the Grand Detour Co. to build plows. Deere sold his interest to Andrus in 1847, then moved to Moline, Illinois, to begin Deere & Co. There was a hotly contested competition between the two companies over sales of plows both designed by John Deere. By 1850, more than 1600 of the Deere & Co. plows, made of special cast steel plates, were sold. Andrus moved the Grand de Tour Plow Co. to Dixon, Illinois, soon after purchasing the company.

In 1837, the Pitts Brothers (Hiram and John) patented the first successful threshing machine. The first machines were powered by horses or oxen on a treadmill of an inclined platform. Sweep powered units would soon follow and could use up to 14 horses walking around in a circle.

William Henry Harrison was elected president in 1840, but died soon after his inauguration and was succeeded by his vice president, John Tyler. Daniel C. Stover was born in Greencastle, Pennsylvania.

William Parlin began making plows at Canton, Illinois in 1842. Ten years later William J. Orendorff joined the W.H. Parlin & Co. Later the company was incorporated as the Parlin & Orendorff Co.

J.I. Case moved his company to Racine, Wisconsin, in 1844 and began to manufacture threshers. James K. Polk was elected president.

In 1845, the United States annexed Texas and crop failure in Ireland led to the Irish Potato Famine. Benjamin Franklin Avery started the Avery Plow Factory in Louisville, Kentucky.

Mormons settled in Utah during 1847. The arid land was irrigated with water diverted from City Creek near Salt Lake City and soon began growing a variety of crops abundantly.

In 1847, Edward P. Allis started the Reliance Works in Milwaukee, Wisconsin, the McCormick Reaper Works began in Chicago, Illinois, and Daniel Massey purchased equipment from R.F. Vaughn to begin manufacturing simple implements in New Castle, Ontario.

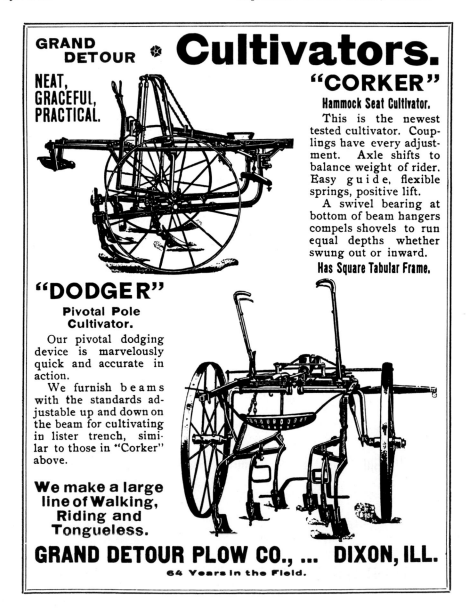

The area called the Mexican Cession (now parts of California, Arizona and Nevada) was purchased from Mexico for $15 million, GOLD was discovered at Sutters Mill, California, and Zachary Taylor was elected president in 1848. Bickford and Huffman began operation in Macedon, New York, and John Nichols opened a blacksmith shop in Battle Creek, Michigan. Nichols began building a threshing machine in the 1850's and David Shepard soon became partner.

Zachary Taylor died on July 9, 1850, and was succeeded by Millard Fillmore. In 1850, California was admitted as a "Free" state. Patterson & Bros. began in Woodstock, Ontario, and the J.O. Wiser, Son & Co. began in Brantford, Ontario. Edmund W. Quincy of Illinois obtained the first corn harvester patent in October 1950. Corn pickers were being developed in the 1850's, but the first practical models were available about 1895.

THE FIRST STUDEBAKER WAGON FACTORY—1852.

Franklin Pierce was elected president in 1852 and John H. Manny Co. began building the Manny reaper in Rockford, Illinois. The area known as the "Gadsden Purchase" (including a strip of Arizona and New Mexico) was purchased from Mexico for $10 million and John Studebaker occupied his first wagon factory.

The Rumley Co. was established at Laporte, Indiana, in 1853. The Furst & Bradley Co., later known for their Garden City Clipper plow, was started in Chicago, Illinois, by Conrad Furst and David Bradley. The company was later reorganized as David Bradley Mfg. Co., and the factory was relocated about 56 miles from Chicago in Bradley, Illinois.

It was 1854 before a railroad passed through Moline, Illinois. The year 1854 was also marked by the Congress passing the Kansas-Nebraska Act. The settling of Kansas and Nebraska was accomplished but was followed by much bloody fighting. Prairie fires, drought and dust storms during the years of 1854 through 1886 would also direct national attention to the Great Plains area. The area would be visited by dust storms frequently enough during the next century to keep the nation's attention.

Adriance, Platt & Co. began making haying and harvesting equipment in Poughkeepsie, New York, during 1855. The company of Buford & Tate began in Rock Island, Illinois, by Charles Bumford

and R.N. Tate. James Oliver purchased a quarter interest in a foundry located in South Bend, Indiana, and later in 1857, patented a method of "chilling" plowshears.

James Buchanan was elected president of the United States in 1856 and California farmers began exporting grain and flour to the country of Peru. The P.P. Mast Co. (Buckeye products) began in Springfield, Ohio, D.M. Osborne & Co. began in Auburn, New York, and the John H. Manny Co. was renamed Talcott-Emerson Co. after the death of John H. Manny.

In 1857 the A.B. Farquhar (Pennsylvania Agricultural Works) began in York, Pennsylvania, M. & J. Rumley Co. began building threshers in La Porte, Indiana, and W.H. Verity & Sons began operation in Exeter, Ontario. Inkpaduta and a small band of followers attacked the settlement of 40 at Spirit Lake, Iowa, on March 8. All but four women were massacred and two of the women who were captured were later killed. Mrs. Marple and Abbie Gardner were later purchased by friendly Indians and returned to St. Paul.

Marsh Brothers of DeKalb, Illinois, introduced a harvester in 1858 with a continuous canvas apron that swept the platform and elevated the grain over the drive wheel to a position on the harvester where two seated men could bind the cut stalks. William Deering and E.H. Gammon join the firm of Marsh, Stewart & Co. in about 1872.

THE OLD MARSH HARVESTER AT WORK

Oregon attained statehood in 1859.

Abraham Lincoln was elected president in 1860, then slightly more than a month after his election, South Carolina seceded from the Union. George W. Brown patented the two-row corn planter and the Van Brunt Co. began operation in Horicon, Wisconsin.

The United States was engaged in the Civil War, otherwise known as the War of the Rebellion, during 1861 through 1865. During this war, Charles Buford bought out R.N. Tate and formed Buford & Co.

Kansas became a state in 1861 and Daniel Massey purchased the Wood's Mower, manufacturer of a self-rake reaper. Harris soon introduced a reel-rake reaper. Stationary steam engines were added to the Rumley line.

The government was active in 1862. The Morrill Act was passed to create land grant colleges in each state, the Homestead Act was passed and President Abraham Lincoln signed a bill beginning the Department of Agriculture.

The Homestead law went into effect January 1, 1863, and the first homestead in the U.S. was in Gage County Nebraska and was taken by Daniel Freeman. The Superior Drill began during 1863 in Springfield, Ohio. The J.I. Case & Co., later renamed the J.I. Case Threshing Machine Co., was formed by J.I. Case, Massena B. Erskine, Robert Baker and Stephen Bull.

Lincoln was reelected president in 1864, but was assassinated on April 15, 1865. His vice president, Andrew Jackson, finished the term.

In 1865, Nicolaus August Otto and Eugen Langen formed N.A. Otto & Cie.

This was the first company to manufacture internal combustion engines as the primary product. Gottlieb Daimler became the production manager for N.A. Otto & Cie. in 1872. In 1876, the company built their first four stroke cycle engine. The first U.S. representatives were Jakob & Adolph Schleicher, nephews of E. Langen. Hermann Schumm, also related, joined later. The company's name in the United States was originally Schleicher, Schumm & Co., but the name was soon changed to Otto Gas Engine Works and was located in Philadelphia, Pennsylvania.

The bald eagle that would adorn Case equipment for many years was patterned in 1865 after "Old Abe," the mascot for Company "C" of the 8th Wisconsin Regiment during the American Civil War. The eagle perched on top of the globe was registered as the official trademark of the J.I. Case Threshing Machine Co. in 1894.

Before 1886

THE OVERLAND TRACTION ENGINE

In 1865, before construction of the Union Pacific railroad, a company called The Overland Traction Engine Co. was formed to provide rapid transit between the East and the Far West of the country. Jesse Fry, whose ideas were original and extreme, organized the company to build a traction engine of the desired dimensions. Everyone who has seen a farm traction engine has a general idea of the type conceived by Fry, except that his machine was to be ten times as large and a hundred times more powerful. The purposes of the company were to transport freight, haul immigrant trains or wagons and to carry passengers between the East and Far West. The first working engine was constructed at Paterson, New Jersey. The boiler was not like the type used on locomotives, though the lower half was similar. The principal difference was in the upper part and in the connections for exit of the gases. On top and parallel with the lower part was placed a long steel drum, which was connected to the lower section by necks in the shape of frustrums of cones having their smallest diameter at the drum. The latter was partially filled with tubes and the whole lower part entirely filled with tubes. The smoke and other products of combustion could pass directly out through the stack at the front end, as in ordinary locomotives, or if desired, they could be deflected so that they could pass backward through the tubes in the drum and out of an auxiliary stack in the rear. There were three separate water spaces, each supplied with its own feed connections and gauge cocks.

The arrangement proved to be defective, because circulation was poor. When the steam gauges showed 100 pounds pressure over the crown sheets, the third water space was cold. The traction locomotive had a large U-shaped tank,

between the upright sides of which the boiler was placed. This tank extended the length of the engine, which was about 20 feet long.

There were four driving wheels, each independent of the others. The main driving wheels were each 9 feet in diameter, had a tread surface 36 inches wide and were located at the rear of the machine. The drive axle shaft was 9 or 10 inches square with 2 inch holes bored in each end, 4 feet deep, plugged on the outside and six ½-inch holes drilled on top to lubricate the bearings. The front driving wheels were 6 feet in diameter and the tread was 30 inches wide. Each wheel was made of iron boiler plate half an inch thick, backed by 3 inch planking and the tread was filled with steel spikes projecting about 2 inches. These teeth were intended to ensure a sufficient grip. The forward wheels were arranged to be used for steering purposes.

The most peculiar thing about the machine was that each wheel had a complete pair of double (two-cylinder) vertical engines and a separate engine for steering, making a total of nine engines. There were also two steam pumps, one for feeding the boiler, the other (larger of the two) for drawing water from a river or creek on the journey for supplying the water tank.

The driving engines had 10 x 12-inch cylinders, and power was transmitted to the rear wheels by a steel link chain. The front wheels were driven by bevel gears. The axles were 14 feet long.

Before the machine was completed, the inventor naturally tried to devise some method of crossing the different rivers on the proposed route which the steam caravan (as the engine and train of wagons might properly be called) would take. At that time there were no boats on the Mississippi and other rivers capable of receiving and transporting a

whole train. The problem of how to cross these streams must have sorely vexed the promoter of the Overland Traction Company. One idea was as original as the rest of the invention, and this was to make the wheels 20 feet in diameter and of sufficient bouyancy to float the entire engine.

Mr. Fry must have feared that the engine and its train would be attacked by Indians, because the front of the engine was made of steel plate somewhat in the shape of a fortress, containing cannons in sufficient number to transform the motor into a veritable war engine if necessary.

When finished, it was decided to test the engine immediately on the street in front of the shop where it was constructed. This street was not paved and was comparatively level; however, the room between the railway track and the sidewalk was less than 30 feet. The engine was 20 feet long and 14 feet wide, so there was not overmuch space should any mishap occur. The results of this test were most extraordinary.

Several men were required to run the machine, because there were so many engines to be handled and whether the results of the experiment were due to a misunderstanding cannot now be ascertained. Suffice it to say that the telegraph poles were knocked down, the street was badly damaged and several other minor accidents occurred. The monster narrowly escaped running into a house. The steel teeth on the steering wheels gripped the earth and made it impossible for the small engines used for steering to properly control the direction. After making more working tests and repeated changes, the idea was abandoned.

The 120,000 pound machine had cost $45,000 to build and was sold for scrap.

1865 also marked the beginning of Candee, Swan and Co. started by Henry W. Candee and Robert K. Swan in Moline, Illinois. The company first made hay rakes, chain pumps, fanning mills and other items. Andrew Friberg (a plow maker) and George W. Stephens soon became partners in the firm. On April 6, 1870, the Candee, Swan and Co. of Moline, Illinois, incorporated as Moline Plow Co. S.W. Wheelock and another were added to the list of owners, bringing the total to six. Champion mowers cost the implement dealer about $85 and he sold them for as near $125 as he could. Interest was about 10%.

Edward Huber began production of a hay rake in Marion, Ohio, during 1865. Torre G. Mandt started a one man wheelwright shop known as Stoughton Wagon Co. in Stoughton, Wisconsin.

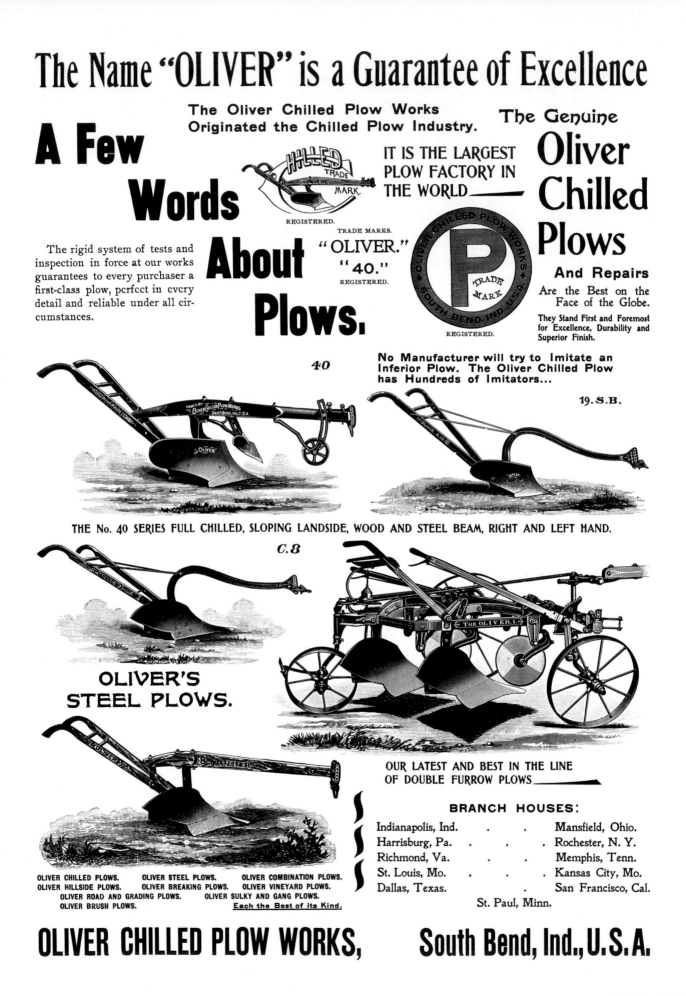
16

Before 1886

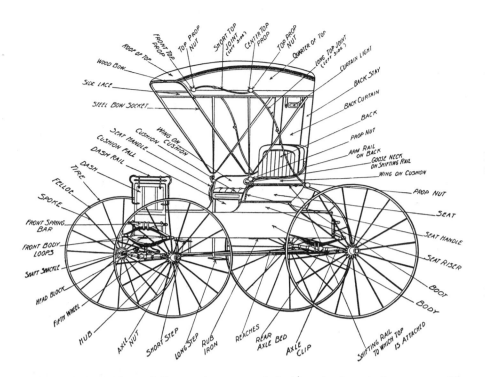

The Nichols & Shepard Co. was incorporated at Battle Creek, Michigan, in 1866 and was the state's first corporation. This was a reorganization of an ongoing partnership. The Sieberling combined reaper and mower, a dropper type, sold for $200. The Stover Experimental Works began at Freeport, Illinois, to build wireworking machines, but would eventually expand to include windmills, feed grinders, etc. The Moline Plow Co. was begun in 1866 at Moline, Illinois, and later, in 1884, distinguished itself by introducing the three-wheel riding plow.

Nebraska became a state in 1867 and the Joliet Manufacturing Co. was incorporated in Joliet, Illinois. The company had been manufacturing corn harvesting and shelling equipment since 1851.

In 1868, Ulysses S. Grant was elected president, John W. Henney began the J.W. Henney Co. in the town of Freeport, Illinois, the South Bend Iron Works began in South Bend, Indiana, and Deere & Co. of Moline, Illinois, organized from the John Deere Plow Co. The Champion hand rake harvester cost the dealer $175 and he sold it for $210. Late in 1871, the No. 4 Champion self-rake harvester cost the dealer $185 and was sold to the farmer for about $225.

James Oliver succeeded in "chilling" a moldboard plow in 1868. German or cast steel was introduced about 1850 in plow moldboards and proved satisfactory at first, but under repeated cultivation the soil became more difficult to plow. The plow makers had trouble adjusting the plows to correct complaints that their plows would not scour. Attention was directed to improving the moldboards. There was a limit to the temper that could be given to a moldboard of solid crucible iron, because a plow is subjected to severe use. The steel could be tempered enough to scour, but would break too easily. In 1862 William Morrison patented a moldboard that combined toughness and high temper, but it was difficult and expensive to manufacture. The front, or wearing plate, was cast steel, backed by a plate of soft iron, the two welded together. Since the iron and steel would not expand and contract together while heating and tempering, the moldboard would warp out of the

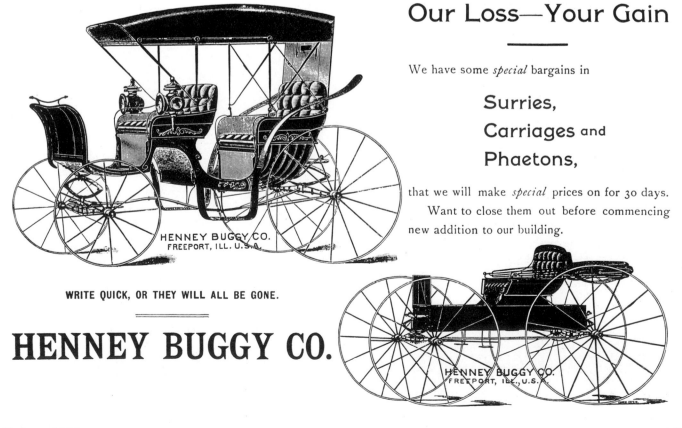

proper curvature. In 1868, John Lane, the superintendent of Hapgood, Young & Co. of Chicago, invented a moldboard that was a composite plate, consisting of a middle plate of soft iron and the front and back plates of crucible steel. The three plates were welded together and rolled out to make a solid moldboard. The steel front and back plates balanced each other while heating and tempering and permitted proper contours while giving the highest tempering. The soft iron plate in the middle made the plow tough enough to withstand hard usage. Singer, Nimick & Co. of Pittsburg made the first moldboard of this type to Lane's specifications, but soon the steel makers began making "soft center" steel for use by plow makers. Later, steel makers substituted mild steel for the iron center.

The Union Pacific railroad completed the link that would connect the two oceans by rail in 1869. Work was begun in 1867.

The spring tooth harrow was introduced in 1869. The disc harrow was introduced at about the same time, but didn't attract much attention until the 1890's. The disc harrow had the ability to cut and turn stubble (and trash) under, which was especially appreciated in the heavy clay soils of the corn belt.

The Philip Herzog fence factory began operation in 1869. Grain drills began general use in the 1870's. Dr. Gatling, the inventor of the famous Gatling gun, later claimed to have invented the first American wheat drill. He did not want to be remembered only as the maker of guns. A method called "checking" was used to plant certain types of seeds, like corn, at regular intervals. An early

STANDARD HARROWS

AND ⟩

Cultivating Implements.

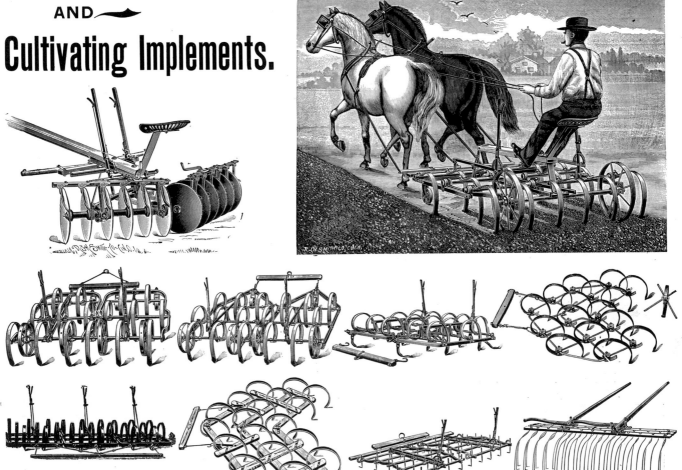

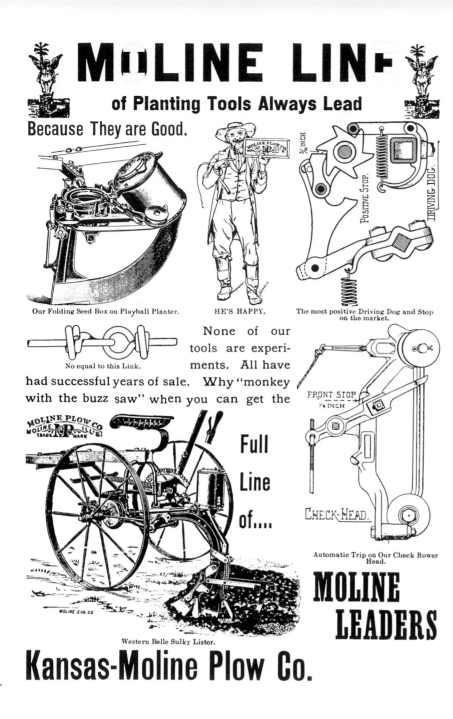
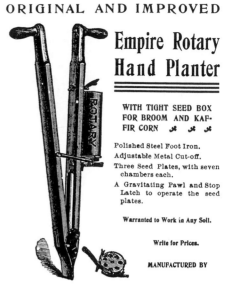
method of checking was by using a knotted cord that was stretched across the field. The planter was tripped by the knots and would plant a seed or several seeds. Later, a wire or special chain with "knots" at regular intervals would replace the knotted chord. Check rower attachments were sold for about $25 in 1877. Hand planting was still necessary where the seeds were blown or washed out and where the planter jammed. Portable steam engines gained popularity as power for threshers during this same period of time.

In 1870, Robert E. Lee died, the Marseilles Sheller Mfg. Co. began in Marseilles, Illinois, and the Massey Manufacturing Co. was incorporated. The Marsh harvester sold for $220. A 12-inch wood beam plow cost dealers about $16 and sold for about $20 to farmers. Plows with a steel beam were $2 more.

such as the "black waxy" lands of Texas. As a result of careful observations in field tests, the general design has been so perfected that in ordinary soil a plow will turn a true furrow without the driver touching his hands to the handles. A noticeable result of this balanced construction is that American plows have short convenient handles as compared to the awkard, ungainly, long handles considered necessary on the European plows of the same period.

In 1871, plans were set in motion as a result of the Morrill Act, that would be fulfilled in the opening of Texas A&M College about five years later. A fire destroyed about 3½ square miles of Chicago, including much of the McCormick Works in October of 1871. The Johnson Harvester Co. began in Batavia, New York, and a self-binding device was introduced for reapers. The sheaves were first bound with wire, but in 1874 or 1875 the twine knotter was invented for binders. Wire binders were not universally accepted, because some farmers believed the wire would be eaten by the livestock. Benjamin Ott, John F. Appleby and Hector Adams

AMONG the just measures that passed Congress during the late session was the direct tax bill which provides for the reimbursement of the states for money collected from them under the direct tax act of 1861. This tax was levied on the property of all the states alike for the purpose of raising funds for the support of the government in its efforts to put down the rebellion. Of this tax $15,227,682, most of it coming from the loyal states, was collected, but that portion of the tax due from the states in rebellion was only partially collected. Under this act $2,562,401 will be credited back to the states which did not pay their share, and the $15,227,682 paid into the government treasury by the several states will be refunded. The amount coming to Missouri amounts to $646,958; to Kansas, $60,981. Kansas was a small state then, her whole quota of the tax amounting to only $70,743. Utah was the only territory or state that paid no part of the tax; her quota of the levy was $26,982. Each state and territory will be entitled to its share when its legislature shall have accepted such sum in full satisfaction of all claims on account of the levy. Money appropriated to meet individual claims is to be held in trust by the state authorities, six years being allowed for the reception of these claims. Not a state or territory that will not receive some portion of this money, with the exception of Utah. The smallest amount goes to Dakota, $3,241, and the largest to New York, $2,213,330. It would seem that the distribution of this money among the states is eminently fair and ought to be thankfully received by the states, yet we notice that a majority of the Missouri delegation in Congress opposed the measure.

American steel plows, fitted with soft center steel moldboards, represented the highest quality of the plowmaker's art. The tempered wearing parts, in ordinary soil, will last for years. The curvature of the moldboard has been determined by years of practical plow building rather than by theoretical calculation. Different curvatures were

employed for sod and stubble plows and special moldboards with longer easier curves for exceptionally difficult soils,

Holmes were credited with inventing and doing most of the early development of the twine binding mechanism. By 1879 twine self-tying was perfected for use on grain binders and production of harvesting machines increased from 60,000 units in 1880 to 250,000 in 1885. These reaping and binding machines probably made a greater impact on farming than anything else up to this point. The twine mechanism, the transport mechanism and the mower made these harvesters the most complicated machine used by farmers up to this time. The passing of time and the gradual acceptance by the children growing up with these complicated machines would eventually reduce the time required by factory trained specialists to adjust and repair the harvesters.

EVOLUTION OF THE HARVESTER.

The Hired Man Must Let the Modern Machine Alone.

The original cave-dweller — dear child of the working scientist — harvested his wheat crop by going out to his fields and gnawing off the heads of the grain with his active jaws. The plan has its advantages and also its disadvantages — on the whole our able progenator longed for something better. Then there arose a thoughtful paleozic inventor who pointed out that the grain could be pulled up by the roots and the heads threshed out in the palm of the hand. This satisfied our esteemed ancestor, and matters ran along thus for a few hundred thousand years; indeed, I claim the working scientist's privilege to be vague as to years. Let us throw overboard the cave dwellers, for that matter, and come along down to modern times. Let us begin with the sickle, for instance.

You will still find old men who will tell you that they can remember when farmers in this country had nothing but the sickle with which to harvest their wheat and rye. A dozen men worked in single file, and cut the grain with one hand and gathered it on the other arm, stopping every "round" to drink earnestly out of a big jug of New England rum or Pennsylvania whisky. Then came the cradle, a scythe with "fingers" on it, which made the grain lie straight. Many farmers have a cradle yet for corners and odd nooks. With it one man cut down the grain and another bound it in sheaves. Then came a direct descend-

ent of the paleozoic genius, and invented a reaper drawn by two horses. This was in the '30s, say. A man drove and a small boy sat on a low seat and raked off the grain in gravels. He was practically the same small boy who used to pull the strings that worked the cut-off valve in the first steam engine. He soon lost his occupation in both instances — in the case of the reaper they invented a mechanical rake. It took five men to follow on foot and bind up what the reaper cut down. Still the farmer wasn't satisfied. So they made him the harvester. Two men besides the driver rode on this and bound the grain as it was brought up on an endless apron to where they stood. They had an awning over them and were very comfortably situated. This was in the '70s. Still the agriculturist fretted. Then he got the self-binder, which he has yet — though he is beginning to find fault with it and talk about electricity.

At first they tried to tie up the grain with wire, but it did not work very well, and the machines were abandoned and others using manilla or hemp twine were tried, with better results. The binder invented by a man named Appleby has, perhaps, been the most successful. The twine or cord is very strong, and a little larger than a round shoe string. It seldom breaks, and the sheaves are tied up firmer and better than by hand. The self-binder is somewhat complicated, but it seems simple when we consider what it does. It is the most intelligent machine used on the farm, if I may so express it. It would make the paleozoic man dizzy to watch it. All it asks is that the hired man shall keep his fingers out of it and furnish it plenty of grain to bind up. It does not tie a square or "hard" knot, nor yet a bow knot. Bring the two ends of a string together for two or three inches from their ends; then, considering the two strings as one, tie one single plain school-boy knot in it, and you have the knot made by a self-binder. It is the hardest knot in the world to untie, and it never "gives" a particle. In the machine it is made by a funny, crafty little thingumbob, which turns around half way, opens its mouth and seizes the cord, turns on around, and lets go sullenly, as if half a mind not to. A knife cuts the cord, another thingumbob holds the ends, two arms sweep the sheaf off on to the ground, and the binder waits for enough grain to accumulate for another sheaf, when it starts itself and repeats the operation. It works with the precision of a fine steam engine, if the hired man will only let it alone.

In 1871, dealers bought Buckeye Junior cider mills for $21 and sold them for $32. The larger and more popular Senior model cost $24 and sold for $40. Coquillard farm wagons cost $75 and sold for $90. McSherry wheat drills cost $85 and sold for about $100. Deere

12-inch wood beam plows cost $14 and dealers tried to sell them for $18. Steel beam plows cost $15.40 and were sold for $20. The war had affected the prices.

In 1872, Manny hand rake harvesters cost the dealer $194.80 and were sold on three-year time payments for $235.

Ulysses S. Grant was reelected in 1872. The Western Malleable & Gray Iron Manufacturing Co. began in Milwaukee, Wisconsin and the Furst & Bradley Manufacturing Co. was incorporated. A financial panic the following year (1873) resulted in a depression that would last for five years.

Robert H. Avery designed a cultivator in his head while he was held prisoner in Andersonville during the Civil War; then, with his younger brother Cyrus M. Avery began the Avery Planter Co. in Galesburg, Illinois, during 1874 to fulfill his dream. The Avery Planter Co. moved to Peoria, Illinois, in 1882. Robert

THE SENIOR CIDER MILL.

Avery died in 1893 and Cyrus Avery died September 15, 1905. The Fond du Lac Threshing Co. of Fond du Lac, Wisconsin, was also begun in 1874.

Joseph F. Glidden, a farmer of DeKalb, Illinois, received a patent on December 24, 1874, for barbed wire that he invented in 1872 to retain livestock. Also in 1874, Hector Adams Holmes whittled a wood model of a twine binder design. Later, in 1877, he began manufacturing the mechanism.

Farmers were not content with a walking plow, but demanded that the plow be placed on wheels so that the operator could ride. From 1850 to 1875 many patents were granted on "sulky" or gang plows having two supporting wheels. Soon many different types were sold, but two of the two wheeled or "sulky" types became famous and are recognized as the "pioneers," the "Gilpin" plow made by Deere & Co. (patented in 1875 by Gilpin Moore) and the "Casaday" plow introduced and sold by the Oliver Chilled Plow Works (patented by W.L. Casaday of South Bend, Indiana). The Gilpin plow had a regular bottom with a straight land side similar to a walking plow. The short curved beam was suspended from the axle or frame, which was carried on two wheels, one running in the furrow and the other on the "land." This plow enabled the driver to ride on a seat without materially affecting draft, but required straight driving to keep the furrow of even width.

The Casaday was radically different. The bottom was constructed without a land slide and was held against the turning furrow by the furrow wheel, which was set at an angle, leaning out at the top and in at the bottom, so that this wheel ran against the edge of the land at all times and carried the side pressure of the plow. The land wheel was straight and this one-sided construction gave the plow a very awkard appearance.

In Late July, 1874, the "grasshopper year" began in the Dakotas, Nebraska, Kansas, Texas and Missouri.

THE GRASSHOPPERS.

ARE we going to have another grasshopper scourge? Prof. Snow of Lawrence University says no, and we hope he is right. A vast army of locusts is working eastward along the line of the Kansas Pacific in Colorado, and is doing great damage to the crops. The pest has also appeared in dangerous numbers in South Dakota, and the farmers of the Northwest are becoming greatly alarmed. The worst case of "hopperdozing" ever known in this country was from 1873 to 1875, when crop after crop was destroyed by the rapacious pest. Colorado and Kansas were devastated first, and then Dakota and Minnesota suffered the terrible ordeal. The writer was living on a Minnesota farm at the time, and vividly recalls the bitter experi-

ence of the farmers. It seemed as if the scourge would never end. For three summers in succession the wheat and oat crops were swept away, and even corn and vegetables suffered. In the summer of 1874 the grasshoppers were so thick as to resemble a cloud above, while in the wagon tracks of the prairie roads shining lines of dead 'hoppers could be seen for weeks at a time; so numerous were they that they could not get out of the way of vehicles. Every method which the ingenuity of man could conceive was resorted to for their extermination, but with comparatively little effect. Every spring's plowing would show the ground to be full of eggs, which would hatch in spite of everything the farmers could do.

The greatest hardship was suffered, and despair had settled over the whole state, when suddenly in the spring of '75 (if our memory is correct) the 'hoppers disappeared. Nobody knew whither they went, nor did they care. Prayers of thanks went upward from every church and gospel hall—and it is a recorded fact that never before had there been so many tough old sinners brought to the religious altars as by the terror aroused over the proclamations of many industrious gospel promoters that the regularly ordained B. C. locusts had arrived and settled down for a *seventeen* years' siege !

Doubtless nothing could send a colder chill up the marrow of an old Kansas or Minnesota farmer than these reports from Colorado.

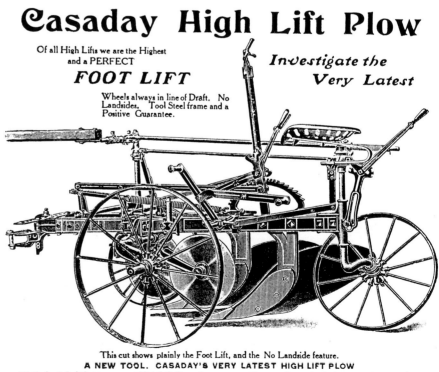

LIGHT RUNNING

PLANO

BINDERS, MOWERS, HEADERS,

We are the People.

OURS are the Machines.

UNRIVALED LIGHT-RUNNING PLANO BINDER.

UNEQUALED JONES STEEL HEADER—LIGHTEST WEIGHT AND LIGHTEST DRAFT HEADER EVER MADE.

They Sell at Sight.

Send for Catalogue

THE

Plano Mfg. Co.,

Chicago,

U. S. .A.

MATCHLESS JONES CHAIN MOWER.

Controversy over the returns and electoral votes was resolved by a congressional commission who awarded the disputed votes, and the election, to Rutherford B. Hayes in 1876. Texas A&M College was dedicated on October 4. The Case, Whiting & Co. was started by Jerome Increase Case. This company was never the same as J.I. Case Threshing Machine Co., but was renamed J.I. Case Plow Works in 1878. Also in 1876, the Fond du Lac Threshing Machine Co. reorganized as McDonald Manufacturing Co., Louis and Lorenz Mayer open a machine shop in Mankato, Minnesota, the Syracuse Chilled Plow Co. began in Syracuse, New York, the Deere & Mansur Co. began to manufacture corn planters in Moline, Illinois and General George Armstrong Custer was massacred with his troops at Little Big Horn.

In 1877 the Bland-Allison Act was passed, over the veto of president Hayes. This act required the government to purchase silver.

The B.F. Avery & Sons Co. was incorporated in Louisville, Kentucky. The initial capitalization of $1.5 million indicates the success of Benjamin Franklin Avery since establishment of the Avery Plow Factory. Avery started a plow factory at Clarksville, Virginia, on January 1, 1825, then moved the company to Louisville, Kentucky, in 1845.

The S. Morgan Smith Co. began in York, Pennsylvania, during 1877. Huber, Gunn & Co. began producing steam traction engines. In 1878, McCormick wire binders were costing the dealer $300. Wealthy farmers were the only ones who could afford the retail price of $375, but dealers were selling all the binders that they could get. Wire for the binders cost the dealer 11.5 center per pound and was sold for 15 cents. Old style wooden Shelby corn planters cost dealers $35 and sold for $42.50 cash or $45 on time. Nichols & Shepard horse drawn threshers cost the dealer $500 and were sold on three-year time payments for $655. Steam threshing rigs cost $1150 and sold to custom threshers for $1455.

During 1879 the Massey Manufacturing Co. moved to Toronto, the McCormick Harvesting Machine Co. began in Chicago, Illinois, Plano Mfg. Co. began in Plano, Illinois, Deering Harvester Co. began in Deering, Illinois, and the Stover Manufacturing Co. was formed. The Stover Experimental Works was a separate company and continued to operate. Twine binders began to be sold at about the same price as wire binders. Farmers were skeptical of the first twine binders, because they seemed so complicated, even magical. The twine binders often threw bundles off without being bound and were generally not as reliable as the wire type, but farmers ac-

cepted their faults because they felt that the wire type could injure the livestock.

The first "modern" refrigerator railroad cars to transport fruit and vegetables appeared in 1880. The cars had better insulation and special metal compartments to hold the ice. Special knotted wire began to replace the knotted cord used to trip the planter mechanism of check row planters. Self powered and self steering steam traction engines continued to gain popularity. Deering twine binders cost dealers $280 and were sold for $335. By 1882, the cost of self binders had fallen to $220 and were selling for $250 because of the stiff competition among dealers.

On March 26 and 27, 1880, possibly the worst dust storm of the 19th century began to sweep across the Great Plains from Las Cruces, New Mexico, to Iowa and eastern Missouri. James A. Garfield was elected president in 1880 and was shot in July of 1881. His vice president, Chester A. Arthur, was inaugurated

following Garfield's death on September 19, 1881.

In 1881 Kemp and Burpee began in Syracuse, New York, and the Massey Manufacturing Co. purchased the rival Toronto Reaper & Mower Co.

After a serious fire, Buford & Co. was reorganized as Rock Island Plow Co. in 1882. The Economy Baler Co. began in Ann Arbor, Michigan, and Philip Herzog was incorporated as the Herzog Mfg. Co. John L. Sullivan was the heavyweight boxing champion from 1882 until 1892.

In 1884, the Bullock Electric Mfg. Co. began operation and Furst & Bradley Mfg. Co. changed their name to David Bradley Mfg. Co. The close presidential election of 1884 was won by Grover Cleveland.

In 1884, the Moline Plow Co. introduced a new type of plow with three carrying wheels, appropriately named the "Flying Dutchman". This plow had one vertical wheel which ran on the land and

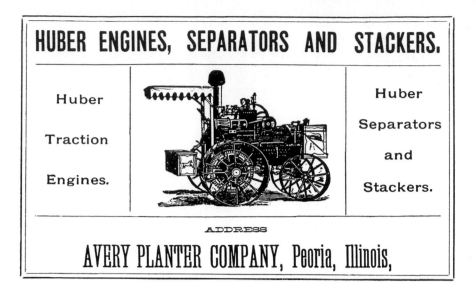

two wheels which ran in the furrow. The front furrow wheel ran in the forward or old furrow, opposite or a little in front of the moldboard. The other furrow wheel ran in the new furrow behind the plow almost in line with the land side. This construction cut a furrow of even width and relieved the furrow friction to some extent by carrying the downward strain of the turning furrow on the plow frame. All of the large plow manufacturers soon adopted this three wheeled construction as the standard type of riding plow.

Benjamin Franklin Avery, pioneer implement manufacturer, died at Louisville, Kentucky, in 1885. The C.L. Best Tractor Co. began in San Leandro, California, the J.W. Henney Co. was reorganized as the Henney Buggy Co., John Secor began experimenting with low grade fuel burning engines and the Kansas City Hay Press Co. began to manufacture the "Lightning" hay press. The price of self binders was down to $200, hay rakes to $28, wagons to $75, cultivators to $18.50 and sulky plows to $39. These were retail prices to the farmer.

1886-1895

SHAPING THE INDUSTRY

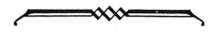

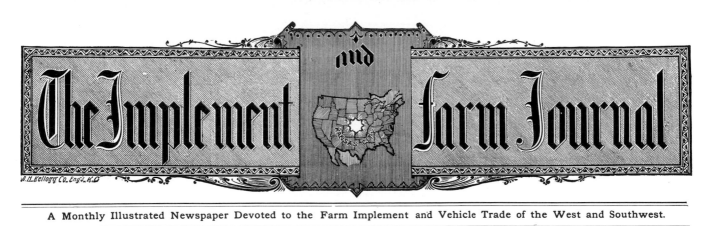

The Implement and Farm Journal

A Monthly Illustrated Newspaper Devoted to the Farm Implement and Vehicle Trade of the West and Southwest.

1886-1895

Horsepower was usually performed by horses before 1895. Dealer associations, trade magazines, industry shows and government support helped to fill the gaps that existed between the manufacturer, the dealer and the farmer. In 1886, 85 percent of the population in the United States lived on farms.

During 1886, the American Federation of Labor was organized, the Statue of Liberty was unveiled in New York harbor and, in August, the Apache Chief Geronimo was captured.

The city of Kansas City, Kansas, was formed on March 6, 1886, the *Kansas City Implement & Farm Journal* began publishing a monthly magazine for the farm equipment dealers in April and Coca-Cola was introduced in May. The Holt Manufacturing Co. began in Stockton, California. John Deere died at the age of 82 on May 18, 1886. Plymouth Iron Wind Mill Co. of Plymouth, Michigan began manufacturing an air gun in 1886 that was invented by Clarence Hamilton. The Plymouth Iron Wind Mill Company changed the name to Daisy Manufacturing Co. and stopped making wind mills in 1889.

The Hatch Act was passed on March 2, 1887, and created agricultural experiment stations as part of land-grant colleges. The McDonald Manufacturing Co. moved to Hopkins, Minnesota, and was reformed as the Minneapolis Threshing Machine Co. The Herzog Mfg. Co. was renamed the Gillette-Herzog Mfg. Co. and the M. Rumely Co. was incorporated.

Grover Cleveland lost his reelection bid to Benjamin Harrison in 1888, but was again elected in 1892 amid demands for free and unlimited coinage of silver.

A group of farm equipment dealers formed an association on February 5, 1889. This group was later called the Kansas Retail Implement Dealers Association, but changed the name to the Western Retail Implement Dealers Association in 1891.

At noon on April 22, 1889, about 2 million acres in Oklahoma were opened to homesteaders. Another 3,687,360 acres were opened to homesteading in 1890. The Stover Bicycle Co. began to build the Phoenix bicycle in Freeport, Illinois, the Twin City Iron Works began operation and the C.L. Best company was building three wheeled steam traction engines during 1889.

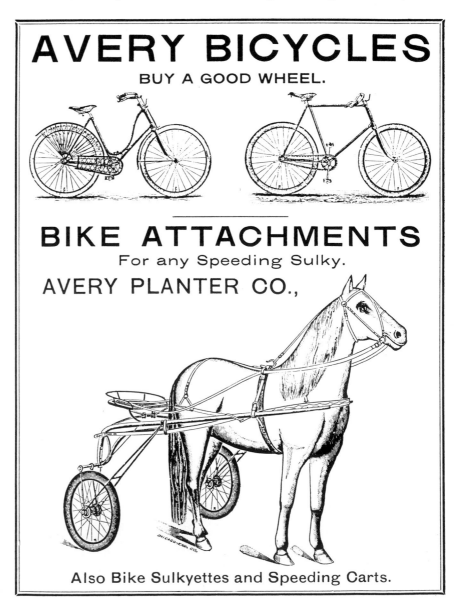

The 1890's introduced several new products including typewriters, Corn Husker-Shredders and Hay Presses (bailers). Bicycles in 1890 were very much like the type we have now, and they were being made by many different companies, including some really big names in farm equipment. By about 1896 the "Boom" had turned to "Bust," the prices fell and many bicycle companies failed.

What to Do With Dished Wheels.

On Saturday evening a two-wheeled road cart, drawn by a spirited bay horse, came down Woodward avenue. At Michigan avenue a coupe drove across the street directly in front of the rig, causing the driver of the latter to suddenly sheer off. As he did so the tire of his wheel caught on a paving block and the wheel was dished. The horse was caught by passers by and held, as the owner looked ruefully at his wheel, which was a new one.

Policeman Walpole, of the Broadway squad, came up and calling on two or three onlookers to assist him, succeeded in pulling the tire back into place and setting the wheel up as it had been. The owner, after thanking him, got in and drove off.

"That is something that more buggy owners ought to know. If a wheel is dished without breaking the spokes off it can easily be sprung into shape if three men will tackle it and, catching hold of the tire at various points, give it two or three good strong pulls. That makes five wheels that I have straightened that have been dished on this very corner."

A NEW STEAM HARVESTER.

Daniel Best, of San Leandro, Cal., has been giving agriculturists and manufacturers food for thought by his achievements in the production of steam plowing and harvesting machinery. His new steam harvester with Remington traction engine running together as a steam traction harvester has excited wonder and curiosity on the Pacific coast. It is said to be a practical success. Mr. Best claims that his machine will, under all tests, excel any other harvesting machine known to man. Its utility has been practically demonstrated in California wheat fields, and the general opinion of the farmers appears to be that it will revolutionize the harvesting of grain on large farms. Be this as it may, the machine is somewhat of a novelty which will excite the interest of our readers at this time. For several years past Mr. Best has been experimenting with horse-power harvesting machines, some of them having a cut of forty feet. Now he has substituted the traction engine for the horse-power, and how much wider swath he may be able to cut to advantage remains to be seen.

Mr. Best's Remington engine is said to be a great success. It is used for all sorts of purposes, and is much less ponderous than traction engines usually are. "It has," says Mr. Best, "pulled twelve 12-inch plows, running to the beam, at a speed that would soon exhaust 100 horses, and will, it is claimed, go into a field and reap, thresh, clean and prepare for the market more than 100 acres of grain in a single day. It moves in every direction, back and forth, in a circle and in all imaginable curves, and turns a square corner with the same ease that a railroad locomotive can move back and forth on the rails. It can climb as steep a hill as a team of horses with a loaded wagon, and it will climb over a larger log than a team of horses can be driven across. It plows and seeds the land, and then with the combined harvester with threshing and cleaning attachments, reaps, threshes, cleans and sacks the grain and hauls it to market. The cost of plowing with it is about 60 cents per acre, and the cost of harvesting the same or less. During the whole season the machine needed no repairs." If these statements be correct, this is undoubtedly a marvelous invention.

SOMEBODY up in Iowa has discovered that a good quality of binder twine can be made of plain, ordinary, every-day slough grass, and it is reported that a company has been organized at Cedar Rapids, with $300,000 capital to manufacture grass twine.

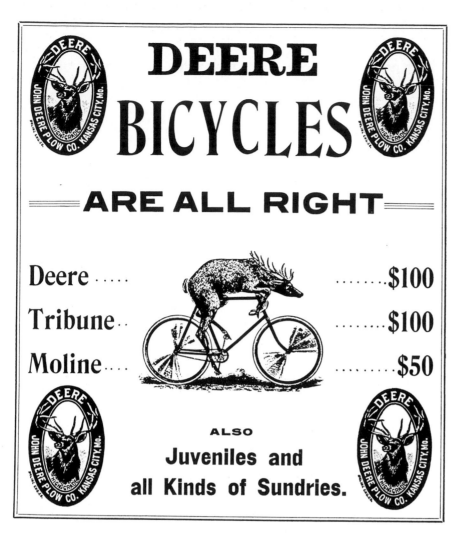
The Mormon church renounced polygamy in 1890, the Sherman Anti-Trust Act became law and the Second Morrill Act was passed. The Second Morrill Act offered more money than the first and described specifics in more detailed language, i.e. "shall be applied only to instruction in agriculture, the mechanical arts, the English language, and the various branches of mathematical, physical, natural, and economical sciences with special reference to their application in the industries of life." The Electrical Wheel Co. was incorporated in April at Quincy, Illinois, Sattley Manufacturing Co. began at Springfield, Illinois, the Stockton Wheel Co. of Stockton, California, began building steam track laying traction engines and the American Bridge Co. purchased the Gillette-Herzog Mfg. Co. The big news of the year was, however, that the American Harvester Co. was formed and then failed. Both farmers and dealers felt threatened by trusts that controlled the availability and price of products they needed to earn a living, such as binder twine and farm equipment. The founding of Sears Roebuck & Co. was also an event of major importance in 1890.

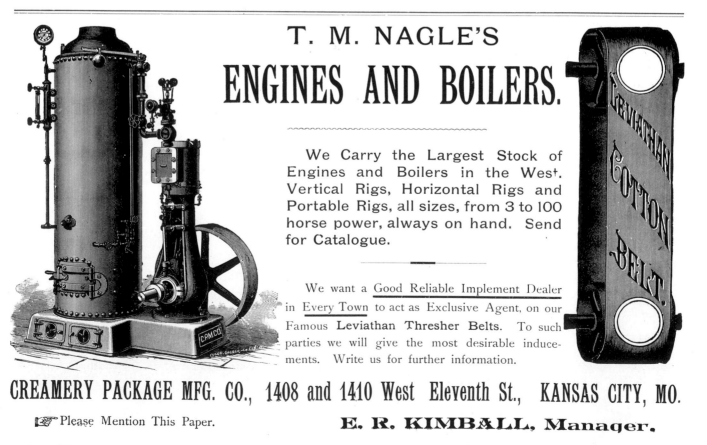

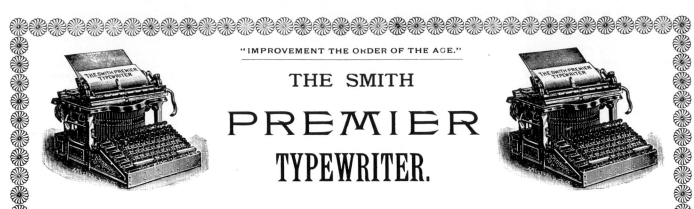
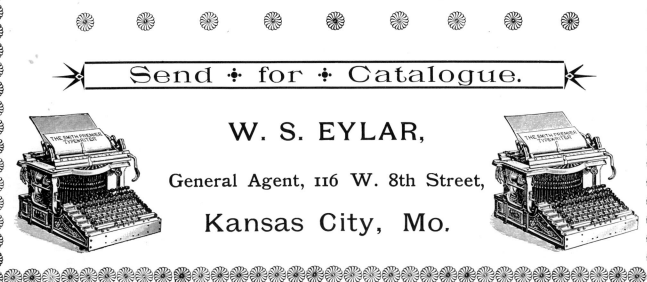

THE FREE COINAGE BILL.

THE substitution of a free coinage bill for the finance bill that had been so long before the Senate and its passage by a close vote (only two majority) in that body, created not a little surprise in all parts of the United States. Of course there is a wide difference of opinion as to its ultimate effects upon the business interests of the country, if it should become a law, but it has yet to pass the House and receive executive approval, and the indications are that it will be smothered in the house. and if this should be the fate of the free coinage bill we shall not get any further financial legislation from this Congress. We cannot help feeling that the friends of silver in the Senate rather over-did the matter, for in this case the more than half loaf the finance committee proposed to give them would have been infinitely better than no bread. Free coinage is sure to come sooner or later, probably not later than the next Congress. The bill is very brief and so plainly worded that anybody can understand just what it means. All the misunderstanding and disagreement relative to it arises from the difference in opinion as to its ultimate effect upon the financial welfare of the country. After defining the dollar as the unit of value of the United States coins of 412.5 grains of standard silver or 25.8 grains of gold, and as legal tender for all debts public or private, the bill provides that any owner of silver in gold bullion, whether a citizen of the United States or any other nation on earth, may deposit the same at any mint of the United States and have it converted into coin without charge. The right to refuse deposits of less than $100 value is reserved. Depositors of bullion may receive in payment certificates in place of coin, but the bullion received and paid for in certificates must subsequently be coined. The certificates are made receivable for all taxes and dues to the United States and a legal tender for the payment of all debts.

It is believed by men who have heretofore been regarded as our soundest financiers, that if the present bill becomes a law it will result in serious financial disturbance. They say the free and unlimited coinage of silver is certain to drive gold out of circulation and out of the country. This theory is based on the assumption that free coinage will make the disparity in the value of the gold and silver dollar greater than it is now. But this result need not obtain unless the people, or perhaps we should say bankers, choose to have it so. They can continue to do what they are doing now at the great banking centers, hoard the gold and refuse to let it out except at such a premium as shall equal the difference between the intrinsic value of the gold and silver dollar. Only those who buy in foreign markets must have gold. The people will have no special use for it, and so long as the white dollar goes just as far in paying debts contracted at home as the yellow dollar the people will be likely to say to the banker, "Keep your gold; silver certificates and silver dollars are good enough for us." Besides, silver men have theories as well as the goldbugs, and while they admit that the immediate effect of the free coinage of silver will be the retirement of gold to a greater or less extent, they claim that ultimately silver will appreciate in the markets of the world. One theory may be considered just as good as the other.

Why the American Harvester Trust Quit.

The following special dispatch from Chicago to the Pittsburg *Dispatch* contains what is claimed to be the true reasons why the Harvester Combine concluded to abandon the scheme before it was fully consummated :

When the great trust for the manufacture of agricultural implements was formed a few months ago, with a capital stock of $35,000,000, and put in operation with a corps of officers including some of the shrewdest business men in the country, and then suddenly dropped like a hot poker by all concerned, there was no little curiosity on the part of the public to know why it had been so suddenly abandoned. It was asserted by some that the outcry against it by the press was the chief cause of the downfall of the scheme, but, then, trusts usually pay little regard to the press. It was then said that certain large manufacturers like the McCormicks had finally refused to go into the scheme because they could not consent to give up their individuality and take trust stock for their immensely valuable properties.

The manufactures themselves explained that it was because of legal difficulties that the scheme had been given up, but very few people believed that, because trusts have too often met and overcome all the legal difficulties that could be put in their way. But the explanation last referred to is probably the correct one.

An attorney of one of the firms in the projected trust said this afternoon that the reason the scheme was not carried through was because the lawyers could not convince the capitalists engaged in it that the trust could be operated without running afoul of the act of Congress of last July in reference to trusts and conspiracies in restraint of trade. Pulling down a volume of session laws, he hastily read over the act, which, with the usual amount of verbiage, declares to be illegal " every contract, combination, in the form of a trust or otherwise, or conspiracy in restraint of trade or commerce between the several states or with foreign nations," and punishes everybody who enters into any such contract, combination or conspiracy with a fine not exceeding $5,000 or imprisonment not over one year, or both, in the discretion of the court, and imposes a like penalty on " every person who shall monopolize, or combine or conspire with any other person or persons to monopolize any part of the trade or commerce among the several states or with foreign nations ;" and as if that were not enough, provides further that any person who suffers injury in consequence of the formation of any such trust may recover triple damages, costs and attorney's fees from the parties engaged in the trust.

After running over the act of which this is the substance, the attorney said that when the attention of the firms had been called to its severe penalties they put the question to their counsel whether they were not violating that law by going into the trust. The latter replied that they were not. The trust, they said, was organized as a corporation under the laws of Illinois. Each party going into it would do so in good faith.

They took stock in it and paid for the stock its full value by turning into the incorporated trust their respective plants, and the trust would operate them in the same manner as an individual who might buy up and run half a dozen stores in various parts of the city. They cited the case of the United States Glass Co., a New York corporation located in Chicago, which operates about eighteen different glass factories, stretching from New York to St. Louis, which used to compete with each other, and said that that concern had never been interfered with.

But, the manufacturers urged, that company was formed before the act of Congress went into effect, and they could not be convinced that the Agricultural Implement Trust could be run without at least great danger of the prosecution of its members in the United States courts. " The upshot of it all was," said he, " that the scheme was abandoned. The advantages to be derived from it were not great enough to compensate for the risks."

Some of the manufacturers especially feared that section of the act giving to any person injured triple damages for injuries sustained through the trust, and the gentleman read the following :

Section 7. Any person who shall be injured in his business or property by any other person or corporation by reason of anything forbidden or declared to be unlawful by this act, may sue therefor in any circuit court in the United States, in the district court where the defendant resides or is found, without respect to the amount in controversy, and shall recover threefold the damages by him sustained and the costs of sui', including a reasonable attorneys' fee.

"The language of that section," he continued, "is very broad and sweeping, and the people engaged in that trust found that they would be at once swamped with troublesome litigation. That act is comparatively new and has been brought into the courts in only one instance, when proceedings were begun at Nashville, Tenn., to prevent the formation of a coal combine. The courts have not yet put an interpretation on this section of it, and, until they do, it will be hard to say just what it means."

"But there is no question that it might cause endless annoyance to a trust like that for the manufacture of agricultural implements. Suppose, for instance, that I am an extensive farmer, and must have many thousand dollars' worth of machinery to carry on my business. The trust is formed. All competition is put to an end and the price list doubled. I am, therefore, obliged to pay just twice as much for my machinery as I would have to pay if there were no trust, and I would be damaged just to that extent. Under that section of the monopolies act I could sue the trust, and if I made out my case I could recover three times the amount of damages suffered, and also my attorneys' fee. What an inducement there is in that to sue the trust! The cost of beginning and carrying on a suit is comparatively light, the profits of one, if successful, may be enormous."

"If that act prevents the formation of the Reaper Trust, why should it not also affect trusts in other articles—glass, for instance, or anthracite coal?" the attorney was asked.

"In time it probably will," was the reply. "Suppose the Window Glass Trust gained control of that article and forced the builders of all our big office buildings to make an extra expenditure of several thousand dollars. They could undoubtedly sue the trust for treble damages, and recover, too."

"But suppose the trust assumes the form of a corporation and takes the various plants as owner in return for stock regularly issued to their former owners, as in the case of the United Glass Co."

"That would make no difference except that it might make it more difficult and expensive for the injured party to make out his case. Of recent years the courts are coming more and more to disregard corporate forms, and I have no doubt that, if it could be clearly shown that the corporate form was resorted to only as a cover, the individuals composing the corporation would be held liable just as if they had acted as private individuals."

"Suppose, as in the case of the Patent Medicine Trust, the combine should refuse to sell goods to any person who would not agree to its terms?"

"In that case if a dealer lost money through the operations of the trust, I should say that he could recover damages under the provisions of the act. There is one thing to be noted however—the act of Congress applies only to trusts and combinations in restraint of trade between the several states and territories and the District of Columbia and foreign countries. Under its terms, therefore, it can not apply to a trust which restricts its operations within the limits of a particular state."

THE LATE AMERICAN HARVESTER CO.

Views of Dealers on the Causes of the Collapse.

Regarding the dissolution of the American Harvester Co., a reporter for the JOURNAL called on several of the leading agricultural implement dealers in Kansas City and asked their opinions on the subject. They are herewith given :

A. L Carson, of the Moline Plow Co., said: "It was nothing more nor less than I expected, but I must say that I was surprised at the movement holding out as long as it did. I did not look for a dissolution, as is claimed was the case, based on the menace of the laws to the combination, but I did expect that personal jealousies and the inclination of some of the members to have things their own way would disrupt the combine. While I am not in a position to know the true inside workings of the mammoth corporations, yet I have a pretty good idea as to how they intended to carry on business. If it had been possible for the combine to prove a success and it had continued in operation during the coming year, I believe prices would have been raised on all harvesters and machines manufactured by the company. Competition has been so sharp among the various firms during the past two or three years that it was quite natural that all should get together and form a combine that would forever put a stop to any cutting of prices. All these firms have made big money, even if they did shave prices a little, but there was not one of them but would like to make all the money possible out of the business, and this would have been done if they could have been made secure against a further cutting of prices. It would not surprise me in the least to see a war on prices. It would be a mighty poor policy for the manufacturers to pursue, it is true, but mighty fortunate for the farmers. There is no denying that there is lots of bad blood existing among the big harvester men, and it is likely to take a decidedly war-like hue.

Harvey Rhodes, manager for Aultman, Miller & Co.: "I was considerably surprised when I heard that the gigantic combine had floundered, just as it had fairly set sail and was apparently in deep water and away from the breakers. The breakers were there, however, in the persons of McCormick, Whitman & Barnes, and Warder, Bushnell & Glessner. I don't know what provisions were made by other managers in Kansas City whose firms were interested in the combine, but I bare deeply in mind the fact that only one man could be manager of all companies at this point, and hence I was preparing to go, and I did not intend to stand on ceremony, but to offer my resignation as early as I considered would be proper. I intended to engage in the transfer and storage business, handling implements and machines for various firms that make Kansas City a distributing point. I also expected to devote considerable time to the manufacture of my knife grinders, a patent I hold and consider valuable. I shall, however, engage in the manufacture of the grinders, but probably not as soon as I would like. I think probably the farmers and agricultural classes are the gainers by the dissolution of the trust, for implements certainly can be purchased cheaper when there is plenty of competition than where there is none whatever. And then, again, there would have been thousands of mechanics and expert workmen thrown out of employment had the trust been a success."

George W. Fuller, manager of the John Deere Plow Co.: "I think the failure of the American Harvester Company trust to hold together was a mighty good thing for the country in general and the agriculturalists in particular. I am opposed to trusts of every kind for the simple reason that trusts are hurtful to every community. The sole object in forming the American harvest trust was to keep up prices. Last year, in fact, for the past three or four years, there has been considerable cutting in prices, and the combine appeared to many the only reasonable solution to a restoration of old time figures and consequent big profits. Whether there will be serious cutting the coming season as a natural result of the failure of the combine can only be inferred. I have already heard some threatening talk, but I am sure I cannot say whether it will ever amount to anything. Whenever there is more of a supply than a demand there will always be a shaving of prices. Whether this will be the case the coming season no one knows. There are really too many manufacturers of farm machinery in business. More money could be made by a consolidation of some of the smaller concerns. This would have a tendency to steady the market by regulating the supply."

George Wilkins, manager for Walter A. Wood Reaper & Mower Co.: "When Cyrus H. McCormick made overtures to other harvester men looking to the promotion of a trust I did not think that as soon as the combine was fairly organized he would be one of the first kickers and use means which, in common parlance, would be considered dishonorable. It appears to me that he utilized every move for his own personal agrandizement and advantage. I may be mistaken in this, but anyone who can read between the lines ought to be able to discern as much. In the first place, after the organization, Mr. Butler, Mr. McCormick's head man, was made general manager of the American Harvester Co. This was a master stroke. With possibly one exception, I am told that McCormick men were given the best positions under the company. This exception was at Omaha. At Lincoln, St. Joseph, Indianapolis, Columbus, Kansas City and other points McCormick men were either already installed as managers or it was understood they would be. If, after a time, the trust had been in business and the members had become dissatisfied with the way things were running, Mr. McCormick could easily have invited them to step down and out, while he would have remained master of the whole situation. With McCormick men in every important position he would have been able to dictate any policy that he might choose. It has been given out by members of the combine that the dissolution was the natural result of the conflicting laws, but had it not been for the 'laws' it is pretty hard to tell what would have been the result of the combine. Our people will make no effort to cut prices. We are enjoying a good healthy trade and are satisfied with it."

J. M. Patterson, manager of the Keystone Implement Co.: "Not handling the implements and machinery which the trust would control, I hardly think I am a suitable person to deliver an opinion. I think the doom of all embryo trusts, of which I believe there were many, is now forever sealed. I presume the attorneys employed by the American Harvester Co. went to the uttermost depth and exercised the best of judgment in searching out and delivering their opinions on the legislation against trusts in various states effecting the company."

W. H. Town, manager for McCormick: "It was stated that I went to Chicago for the purpose of accepting a position with the American harvester Co.. This is a mistake. I was assigned to look after the Chicago house, and was not connected with the trust. The dissolution occurred for the simple reason that there were grave legal obstacles in the way of a consummation, and the $15,000,000 in bonds could not be floated. However, there was some trouble before this matter was thoroughly considered. Mr. McCormick's idea was to equalize, if not to reduce, the price of machinery. This could be done by a curtailment of expenses. He was met by a strong opposition, and a stormy scene ensued. It was the object of some members of the trust to raise prices, and when they became "sot" in their opinions there was a crash. We shall operate the business in Kansas City as we have been doing for years, and we expect to do as large if not a bigger business the coming year than we have ever done during any season in the history of the house. It is not our purpose to cut prices in any manner, and I do not believe other firms have an idea of making any serious cuts."

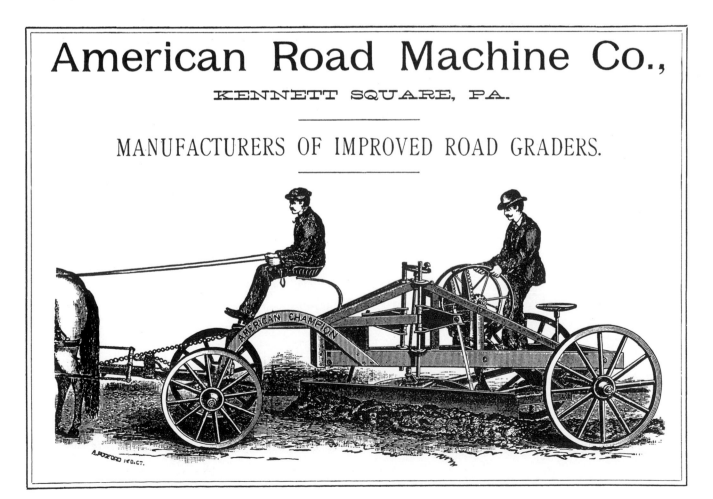

M. H. Losee, manager of the Sandwich Mfg. Co.: "It is well understood that the object in forming the American Harvester trust was to keep up prices. Had the scheme been successful it no doubt would have been millions in the pockets of the companies interested. I believe the principal cause for dissolution lay in the fact that the scheme conflicted with state and national laws. There were of course some hot words passed between certain members of the trust which only precipitated the dissolution. The success of the trust certainly boded no good to the farmers and agricultural classes. Disaster, on the other hand, means agricultural implements to the farmers at the lowest figure. I look for cutting in prices, but I cannot say just now how serious the fight will be."

W. J. Kirk, Rock Island Plow Co.: "I am innately opposed to monopolies on general principles, and to a monopoly of any branch of the implement business in particular, therefore I have no tears to shed over the collapse of the harvester combine. While I believe it would have been a good thing if the several harvester manufacturing concerns could have held together in an honest effort to prevent reckless competition, yet if they could have held together at all it is doubtful if they could have resisted the temptation to put up prices."

THE COLLAPSE OF THE COMBINE.

WE devoted a good deal of space in last month's issue of the JOURNAL to the causes leading to the formation of the mammoth combine of the reaper and harvester manufacturers and its probable effect upon the trade in this class of farm implements. At the date of going to press with our December number the American Harvester Co., with its enormous capital of $35,000,000, was supposed to have fully completed its organization and to have been ready for business. To the farmers generally it appeared to be a monster of frightful mien, for it said to him, "hereafter there will be no competition in the sale of reapers and mowers; while as heretofore you may be allowed your choice of machines, there will be no bantering on prices or time of making payments," and they were disposed to take very little stock in the statement that they were hereafter to get their machines for less money than they have been

obliged to pay in years past; so when the report got abroad that the company had decided to advance the price of harvesters $35, the Northwest, especially, blazed up with indignation, and straightway there was a calling together of the Alliances, and hot speeches were made, and resolutions passed pledging the members of the body not to purchase a dollar's worth of farm machinery of any kind from the "monster combine." There was to be war to the knife and knife to the hilt. The farmers, who are in a majority in the legislatures of most of the Northwestern states, went up to their capitols with their pockets full of ammunition in the shape of anti-trust and anti-combine bills to fire at their common enemy. But in the midst of all this preparation for war came the announcement that the trust had dissolved, that the American Harvester Co. had collapsed, dropped to pieces of its own weight, as it were. The following dispach from Fargo, N. D., indicates the feeling among all classes not only in Dakota, but throughout the West:

The collapse of the American Harvester trust at Chicago is hailed with delight by users of farm machinery in this section. A visit to machinery warehouses elicits the fact that the order has been received to re engage employes of the company the same as last year. When the new company was organized the services of something over 100 men were dispensed with. A majority of these men were about leaving the city in search of employment. Business men express the greatest satisfaction over the news and consider it a great victory for the Northwest.

Elsewhere in this paper we give the opinions of leading jobbers in Kansas City as to the cause of the sudden and unexpected collapse. It will be noticed that the opinions given are varied and all do not agree with the following statement of President McCormick:

The dissolution was decided upon only after we were assured that we could not carry out the objects of the association. Able lawyers were consulted, and they were unanimously of the opinion that under the laws of the states the company could not continue in the form in which it had been started. Therefore, like sensible men we concluded to abandon the whole affair. The dissolution of the company will have no effect on the different companies or firms which were merged in it. No change had been made in the business arrangement of any of them, and hence the action of the company would not affect them. There will be no effort to reorganize the defunct company. It was only after long deliberation that the gentlemen composing the harvester trust decided to abandon it. For more than a week the board of directors had been holding daily sessions in Chicago. The whole field of trust possibilities was thoroughly canvassed and fully discussed, and the decision was the abandonment of the present form of organization. The hardest rocks against which the trust ran were decisions in the supreme courts of Illinois and New York, the former in the gas trust and the latter in the sugar trust case, that one corporation may not legally hold stock in another corporation for the purpose of controlling it. The interests involved were so great that it was agreed that $50,000,000 capital would be necessary to put the combination upon its feet. Of this sum $35,000,000, representing the capital of the concern, was to be apportioned among these several constituent institutions and $15,000,000 in bonds wers to be issued for working capital. When the company attempted to float these bonds the United States Trust Co., of New York, through which the negotiations were conducted, instituted a careful inquiry into the statutes of the American Harvester Co., and declined to make loans upon the ground that the company had been organized in such a way as to violate the statutes prohibiting trusts and similar combines of capital to limit production and to control trade. Other financial institutions were appealed to, but with the same result. Then it was proposed to raise the necessary funds within the company itself, but the various concerns in the deal becoming frightened at the legal aspect of affairs, refused to tie up their individual establishments where they would be handicapped by the heaviest legal and financial liabilities. Practically this has already been accomplished but a good deal of red tape formality must be gone through in winding up the last earthly affairs of the concern.

The JOURNAL last month undertook to show that there were good business reasons for a combination of the reaper and mower manufacturing interests. It was plain enough to be seen that by it the profits of the manufacturers would be greatly increased without the slightest necessity of raising prices. How much truth there was in the report that some of the concerns in the combination proposed to advance prices on some of the machines, we do not know, but if there was any probability of such a course being taken, it is a good thing that the combine went under. The spirit of our institutions is against all such schemes, and the JOURNAL is with the people in its detestation of money extorting monopolies.

The large number of agents and traveling experts who would have been thrown out of employment had the combination succeeded, are feeling very happy over the outcome, and well they may. The farmers are also exceeding jubilant over what appears to them to be a prospect of a reckless price cutting campaign, but in this we sincerely hope they will be disappointed.

THE HUBER THRESHER AND ENGINE.

The Avery Planter Co. have been handling the Huber Mfg. Co.'s threshers and engines at this point for the past four years. Each year has marked an increase in their sales, and this season they expect to sell more outfits than ever before. They have already more orders than were booked up to June 1 of last year.

The Huber engine has features which are worthy the special attention of threshermen. Steel wheels have been substituted for cast iron, the fire-box and other parts are done away with, making it much lighter than other engines. The entire weight of the engines is carried on springs. The front end of the boiler also rests on springs. The main gear of the drive wheel is supplied with cushion springs, which relieves the traction gear from sudden strain when starting its load. A cab is furnished for the protection of the engineer. A steel water tank is attached to the front of the engine, which holds seventy-five gallons, and is supplied with a steam jet pump and hose to fill the same.

The boiler is of the Scotch marine return tubular type, and is made from the best open-hearth flanged marine steel with 60,000 pounds tensile strength, and every boiler is tested both under steam and hydrostatic pressure to 150 pounds before leaving the factory. The fire-box is a large tube in the center of the boiler. The flues or tubes are above the fire-box, therefore the boiler is non-explosive. There are no leaky flues, no stay bolts, no reversible link, only one eccentric, one eccentric strap and one eccentric rod.

The new Huber separator does its work thoroughly. The separation begins at the cylinder, and the grain and straw do not again come in contact after leaving it, the grain going to the measure and the straw to the stack. Only one belt besides the main drive belt is used By a separation in the tailings spout the good grain is put behind the cylinder, while the small seeds, cheat, etc., go to the ground. It has other good points which will be appreciated when seen.

ENSILAGE AND FODDER CUTTER.

The Belle City Mfg. Co., of Racine, Wis., have now on the market an ensilage and fodder cutter that is claiming the attention of all progressive dealers. This company has long had the reputation of manufacturing reliable goods, so that it is not surprising that this late success should command so much attention. In the manufacture of cutting machinery the Belle City Mfg. Co. has always ranked high, and the opinion expressed by competent judges is that their latest machine is decidedly the best yet produced. Herewith is presented an illustration of a complete Belle City cutter at work. They are made in different sizes, requiring different styles of powers. Looking at the illustration, it will be seen that the chain feeder does away with the help of one or more men, as the man on the load can put the ensilage into the running feed box, which takes it on without any further attention from the operator. This attachment is furnished only by this company. Their carriers are driven principally by rope instead of belting, although belting is furnished if preferred. Rope has proved much more satisfactory, as it runs in grooved pulleys, and in this way does not run off, even if the carriers are not set perfectly true. The carriers can be swung several feet without interfering with the running arrangement; adjustable tighteners are used, taking up all the slack. Any style of knife can be

BELLE CITY MFG. CO'S ENSILAGE AND FODDER CUTTER.

furnished with the Belle City—the convex and concave, also the straight knife, as on a cylinder cutter. These are one-fourth of an inch thick and are preferred by some operators. This company will be pleased to answer any inquiries regarding their machines, and will also send upon application of interested parties their latest publication on ensilage, which contains much valuable information on that subject. Their illustrated catalogue and price list may be had by writing them.

THAT HARVESTER DEAL.

The Walter A. Wood Co. Now Controls the Appleby Reaper Patents.

The Walter A. Wood Mowing and Reaping Machine Co. is now, to a large extent, a western enterprise. Re-

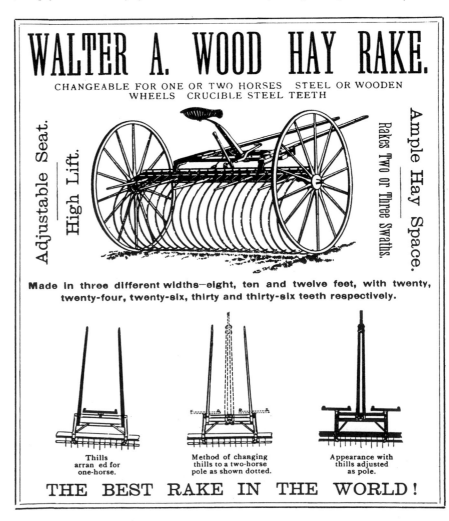

cently a contract was made by that company by which it gained control of the works formerly operated by the Minneapolis Harvester Co., and as a result the machines coming from that factory this season will bear the Walter A. Wood brand. In getting control of the Minneapolis Harvester Co.'s plants and patents, the Walter A. Wood Co. has completed a plan which will permit it to build in St. Paul an immense branch manufacturing establishment. A company is now being organized in St. Paul to have a capital stock of $2,500,000. This company, when its factory is completed, will manufacture the Walter A. Wood reapers, improved with all the Appleby patents owned by the Minneapolis Harvester Co., many of which Inventor Appleby did not dispose of to any but that company. It is not yet settled whether the Walter A. Wood Co. will take the mower product of the Minneapolis works this season or not, and the fact that this is not fixed upon leads to the belief that the Minneapolis company may continue to manufacture mowers only. That part of the plans of the two companies has, however, not yet been made public. Owing to the fact that the Minneapolis plant has only just commenced operation, its product this season will not be as heavy as usual.

Latest Flying Machine Reports.

The experimental flying machine of Mr. Hiram J. Maxim, built at his works near Kent, England, says the New York *Engineering News*, "promises good results." As Mr. Maxim is the inventor of the celebrated gun bearing his name he is entitled to a patient hearing. "Mr. Maxim, in a late interview, describes the machine as an inclined plane, 13 feet by 4 feet. This plane is balanced on a revolving arm 30 feet long, and so arranged that it can rise and fall. It is propelled by a wooden screw revolving at the rate of 1,000 to 2,500 revolutions per minute. When the machine travels at the rate of 30 miles per hour it remains on the same plane; at 35 miles it begins to rise, and at 90 miles it broke the guy wire and showed great lifting power. Mr. Maxim says he has already expended $45,000 on these tests, and is now at work on a large machine of silk and steel, with a plane 110 feet by 40 feet, with two wooden screws 18 feet in diameter. A petroleum condensing engine will furnish the power. In his previous experiments he found that one horse-power would carry 133 pounds 75 miles per hour. It was also proved that the screw would lift forty times as much on the propelled plane as it could push. A motor has been built, weighing 1,800 pounds, which pushes 1,000 pounds, and will consequently lift 40,000 pounds. The estimated weight of his engines, generator, condensor, water supply (2 gallon), petroleum (40 pounds per hour) and two men, is about 5,000 pounds. Mr. Maxim is in earnest and is very confident of ultimate success."

A NEW SPINDLE SURREY.

Who has not heard of the Bush road carts? No handler of vehicles in this country, we dare say, who has been in business for a single year. The fame of these vehicles has gone abroad throughout other countries than our own, which, however surfeited it may be with road carts, still retains its respect for the name of Bush, while thousands of dealers harbor sweet remembrances of the palmy days when to reap profits from the sale of Bush carts was like plucking the rosy peach from the bending branch eager to be rid of its tender burden. With the prestige of its success with the one line, the company, in the course of its business progress, took up others, and its four-wheeled vehicles are now offered to the trade with a confidence in their capabilities of winning public favor that is born of a consciousness of honest effort well rewarded in the past.

This enterprising Michigan concern takes special pride in calling the attention of dealers to its new spindle surrey, which it offers as a vehicle within the reach of the ordinary mortal who cannot afford to pay a fancy price for a family carriage. It is strong, light, durable and handsome, and not too heavy for one horse. It is finished either in oil or painted and furnished with canopy top and curtains or open. An inspection of this job is sufficient to satisfy any experienced dealer that it is one of unusual merit for the price asked.

John Deere Lister Plows.

DEERE DISC LISTER AND DRILL.

They have moldboard and share made of the very highest grade steel, and of such shape and temper that they will scour where others fail. They are provided with either disc coverers or with shovel coverers.

The sub-soil attachment is so made that the ground is sub-soiled, and the corn is, in consequence, dropped on loose ground, thus enabling it to take root.

COMBINED LISTER AND DRILL, 1891 PATTERN.

This Lister is adapted especially for use in gumbo soil. Made only by **Deere & Co.,** Moline, Ill.

IS THE LISTER SUPERFLUOUS?

A writer in the *Farm Implement News* has this to say about the lister: "The history of the rise and use of the lister is an interesting one, save that the concluding chapters can not be written. In the minds of many the lister is indispensable, while others would regulate it to the limbo of abortive experiments. The present condition governing the sale of the rival implements is one more of compromise than peace.

BUSH ROAD CART CO.'S NEW SPINDLE SURREY.

The demand for either is amenable to the whim or style—whatever it may be called—of the user. The planter and the lister revolve in concentric circles about a common point—now one and now the other in ascendency. The periods of demand and decadence are cyclical. The year 1891 by right belongs to the planter; but prevailing conditions, climate and otherwise, determined differently. The first orders indicated a largely increased demand for planters and check rowers. This tendency was arrested, however, by early spring rains and the lateness of the season. The result was that the lister has regained the lost ground and more—a larger number, probably, than ever before having been sold. It, too, has invaded territory which ever before knew it not. Our northern country has been peculiarly the property of the planter, but this season has seen the general introduction of the lister therein. It is impossible to say that one is the greater favorite than the other, but there are times and seasons when the advantages of the lister preponderate.

"From the jobber's standpoint, the lister is an unfortunate invention, and not one but would be glad to see its total extinction.

"From the farmer's standpoint, it is an implement strongly commending itself on the score of economy, if no other. It takes the place of a plow, a harrow, a planter and a check rower. It enables him to plant in a wet season when the old method is impracticable, and it insures a fair return almost in spite of drouth or extreme damp. To the debt-burdened farmer of the West it has proved a blessing.

"On the other hand, the use of the lister necessitates the manufacture and use of specially prepared harrows and cultivators for the growing plants. The difficulty of cultivation is increased, and it is questionable if the average of yield is above that obtained under the old method.

"Were I to express an opinion as to its probable position in the trade, I should say that the need of such a machine is doubtful, but that it has a hold upon the popular fancy which makes it a necessity in the trade. We who are on the battle ground watch the contest with interest and base our preference on the fancy of the individual buyer."

SEVERAL professors in agricultural colleges are trying to steal the laurels of Chancellor Snow, of the Kansas State University, as the discover of the minute insect that preys upon the chinch-bug. This won't do. Chancellor Snow is entitled to all the glory of that discovery, and the method of exterminating the wheat pest.

The year 1891 was marked by the deaths of several pioneers of the farm equipment industry, including: Jerome Increase Case, William Parlin, John Nichols, and Nicolaus August Otto.

At noon on Tuesday, September 22, 1891, about 1,282,434 acres of the Oklahoma Territory was opened to settlers. On April 19, 1892, about 4.25 million acres of Indian land were opened to settlement. The following year, on September 16, six million acres of land opened in the Cherokee strip (or Cherokee Outlet) was settled in a land rush that lasted about one day. Additional land in Oklahoma was opened for homesteading in 1895, 1896 and 1901. Most of the Oklahoma land was settled by farmers relocating from eastern states that were becoming "too crowded."

Also in 1891, the Populist Political Party was begun. The Avery Planter Co. began building steam traction engines and threshers, the Minneapolis Threshing Machine Co. began building Minneapolis steam engines, the Scott Hay Press Co. began manufacturing the "O.K." hay press in Kansas City, Missouri, and the American Wheel Company was formed. The Walter A. Wood Co. purchased the Minneapolis Harvester Co. and thus gained control

of many patents granted to John F. Appleby. John Lauson bought part of a business in 1891 that specialized in building boilers, tanks, smoke stacks, etc., and repairing traction engines. The other partners were his uncle, George Lauson, and a third partner J.H. Optenburg. John's brother Henry joined the firm in 1895. Brothers John and Henry Lauson and H.N. Lauson built their first gas engine in 1898.

UP in Manitoba, farmers are paying harvest hands $30 and $40 a month, which farmers in the Eastern provinces of the Dominion consider remarkably high wages. So they are, for "Canucks," and it is not a wonder that the liberality of the Manitoba farmers has created the surprise of their Eastern countrymen; but in the United States a man who offered a harvest hand less than $2 a day would not only be considered stingy, but would not find many willing to accept it As a rule, harvest hands in the United States get $2.50 and sometimes $3 a day during the season, and nobody is surprised, either.

It is sometimes well to look back over the road, take bearings, and measure progress. When we find farmers and laborers politically in arms against alleged oppression and hard times, we naturally turn to view the conditions of "the good old times" when everybody was prosperous and contented.

The farmer of the day makes two special complaints: high prices for what he buys, and low prices for what he sells. How was it with our former fathers?

Take the staple dry goods article, calico. In 1790 it cost 58 cents a yard; in 1830, 29 cents; in 1860, 11 cents; in 1891, 5 cents.

Take the staple grocery article, sugar. In 1790 it cost 18½ cents for cheap brown grades; in 1830, 15 cents; in 1860, 10 cents; in 1891, 5½ cents for granulated.

For what the farmer sells, take the staple dairy product, butter. The prices in Massachusetts are as follows: In 1790, 11 cents; in 1830, 18 cents; in 1860, 26 cents; in 1891, 30 to 35 cents.

The staple meat product, dressed beef, in Massachusetts sold in 1790 at only 3½ cents; in 1830, 7½ cents; in 1860, 12 cents; in 1891, 12 to 18 cents.

Our farmer fathers of revolutionary days whistled among the stones and pumpkin vines of sterile New England, attired in cotton jeans and shirtings that cost 50 cents a yard, slept on ticking at 90 cents, and if rich enough wiped the perspiration from their brows with handkerchiefs that cost 70 cents. Their wives, if unusually stylish, paraded in muslin at 75 cents, gingham at 55 cents, and cambric at $1. Pins were 15 cents a paper; for matches everybody borrowed fire; and farm implements and machinery, the hoe and scythe were as all-important as to-day's sulky plow and self-binder.

Those were the "good old days" when there were no debts nor Donnellys, no machinery nor mortgages, and few products, Peffers, or people's parties. Yet what the farmer sold brought not much more than one-half what it brings to-day, and what the farmer bought cost more than double what it costs to-day.

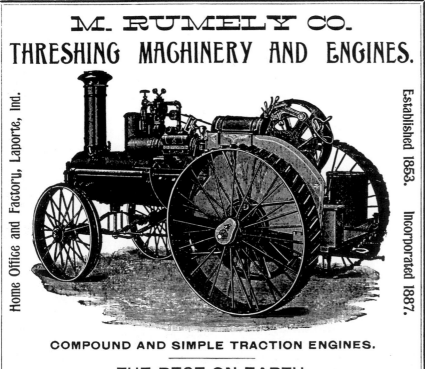

On July 22, 1891, the Massey-Harris Co. was formed by the merger of:

Massey Manufacturing Co. of Toronto, begun by Daniel Massey of New Castle, Ontario, Canada.

Alanson Harris, Son & Co. begun by John Harris of Brantford, Ontario, Canada.

Later in 1891, the Massey-Harris Co. bought Patterson & Bros. of Woodstock, Ontario, and the J.O. Wisner, Son & Co. of Brantford, Ontario. The Massey-Harris Co. became affiliated with W.H. Verity & Sons of Exeter, Ontario, Canada, in 1891 and 1892 purchased exclusive marketing rights to Verity Plow Co. Much later on May 14, 1986, Massey Ferguson Ltd. changed its corporate identity to the Verity Corporation.

THE ANNEXATION OF CUBA.

THE annexation of Cuba is being talked about again. For more than seventy years American statesmen and writers have given some attention to this subject, generally agreeing that the island would be a beneficial acquisition to the United States. John Quincy Adams, as secretary of state in the Monroe administration, wrote that within half a century the annexation of Cuba to the United States would be regarded as "indispensible to the continuance and integrity of the union." Thomas Jefferson advocated the annexation as early as 1823. Secretary of State Everett was another advocate of annexation, all arguing that the United States required the protection that possession of the island of Cuba would afford.

General Thomas Jordan, in a recent number of the *Forum*, agrees with Jefferson, Adams, Everett and all others who have argued in favor of annexation from a protection standpoint, and then proceeds to show that the possession of Cuba by the United States would be a rich acquisition from a commercial view. He advances the belief that Cubans would be delighted with such a change of government and makes plain his argument that the trade relations of Cuba and America are such that a union is the most desirable course to pursue. Calling attention to the agricultural resources of Cuba, he says that in 1883 that country produced 50,000,000 pounds of coffee. Coffee is no longer cultivated there because of Spanish commercial restrictions which give Brazil a monopoly of the United States coffee trade. He says, too, that only 5,400,000 acres of the 23,040,000 comprising the area of tillable lands on the island are under cultivation and that the agricultural possibilities are immense. Mineral, also, is no unimportant factor in estimating the wealth of that country to this country.

It seems now, as it has for almost a century, that Spain does not want the United States to possess Cuba, even though a good price were paid for it. The benefits to be gained by such a possession are unquestionably great, but are not worth going to war about. If it can be bought, the United States could produce her own coffee, and sugar would no doubt be lowered even below the free trade prices of to-day.

CUBAN RECIPROCITY.

The treaty recently concluded with Spain under the reciprocity clause of the McKinley tariff bill, will be of great benefit to the American farmer and manufacturer. The first schedule takes effect September 1, 1891, and gives American farmers a free market in Cuba for their bacon, ham, lard, tallow, oats, rye, barley, straw, fruits, vegetables and farm products of all kinds. On January 1, 1892, American flour goes in at $1 per barrel, which means the restoration of the flour trade of Cuba with the United States and a market for at least 400,000 barrels of flour per annum. The Boston *Journal of Commerce* in commenting upon the subject says: "The treaty was the price of peace in Cuba. Without it, those who know the situation best, say that it would have been impossible for the Spanish cabinet to have much longer resisted the rising storm of discontent. The development of beet-sugar culture in Germany had excluded Cuba from her European market, and had reduced the price of sugar from 6 to 3 cents a pound. The markets of the United States were alone left to the Cuban planters, and an ominous provision of the McKinley act made it possible to close the doors of the United States to Cuban sugars. To have rejected the offer of the United States would have been to sign the commercial death warrant of Cuba. There were, however, ominous signs in the island, of hostility to the mother country, and the talk of annexation to the United States is said to have been stronger than was known in this country. The problem before the Cuban planters was a simple one. It was loyalty to Spain, with bankruptcy, or reciprocity with the United States and prosperity."

DEVELOPMENTS show that Russians, though stricken with famine, are among the most greedy speculators in the world. The millers of Warsaw had shipped most of the flour at their command, and the railroads leading from Russia to Germany were blockaded with grain, when the czar's ukase prohibiting the exportation of breadstuffs went into effect. Russian shippers and grainholders didn't care if their nearest of kin or neighbors starved to death if they were enabled to add to their fortunes by the influence of calamity. In this instance the czar has shown that he is as much entitled to the respect of honest men as some of his subjects are.

TWO GOOD IMPLEMENTS.

The Patent Office at Washington is filled with ideas in wood or other material, many of which are the ideas of "wooden heads," while others have been the means of blessing humanity the world over. The records of the office show that no pressing need in the line of invention has been without a champion, who believed he could supply it, and his crystalized idea of the means of so doing has its place among the models which pose before the American people as representatives, not of the nude in art, but of the crude in invention. The simplest of these inventions have oftentimes proven to be the most valuable, while some have been but mere suggestions, which some fertile brain has seized upon, and by developing it has made a future for himself, and also conferred a lasting benefit upon mankind. In the line of agricultural implements, the nineteenth century has seen some wonderful developments, and machines have multiplied until Solomon's language might be paraphrased, and we might almost say, "Of making many *machines* there is no end."

One of the crying wants of these latter days has been something that would take the place of cutting corn by hand, and the "long felt want" is being supplied by corn harvesters. One of these, to which we call especial attention, is the Moline Corn Harvester, manufactured by that old and well known house, the Moline Plow Co., and illustrated on this page. How well the inventor has done his work in securing the simplest machine that would do *its* work perfectly, and at the smallest cost, can be seen by noticing some of the especially commendable features of it. The blades are adjustable, and it is the only machine of its kind in which the cutting surface may be given any desired angle. The wings fold under the platform also, and a hill can be missed

MOLINE CORN HARVESTER.

when it is desired to form a gallows for a shock, or they can be put entirely out of the way when on the road. The caster wheel under the front center is a feature peculiar to the Moline alone, and by pressing upon the long lever shown in the cut, it can be turned in either direction without having to lift the machine around. In all other regards it will be found to be just what is wanted.

The "Flying Dutchman, jr.," which is

FLYING DUTCHMAN, JR.

claimed to be the first 3-wheeled sulky plow made, is another valuable implement, which is most favorably and widely known. Among the claims made for it are that it is the most successful, the most popular, the lightest draft, the

easiest operated, does the best work, is the best made, and the only sulky plow which lifts the plow out of the ground, raises the rolling coulter and levels the frame simultaneously with one motion of the lever, and the only one on earth which can plow across corn rows or ridges and maintain an equal depth of furrow. It is hardly necessary to attempt a review of the grand record made by this plow. For a dozen years its name and fame have been known throughout the land. To new men in the business, we would simply say, by all means post yourselves on the merits of the Flying Dutchman, jr. The Moline Plow Co., of Kansas City, will be pleased to answer all inquiries.

A QUARRY OF PETRIFIED GRAIN.

A absolute quarry of petrified grain, either wheat or barley has been unearthed near Talmage, Nemaha county, Nebraska, and is attracting considerable attention. The kernels of grain are perfect in form but have become as hardened as solid rock and are well matted together. It is with considerable difficulty that one or more of the petrified grains can be separated from the body of a chunk. It is a pretty stone or composition, or whatever it may be called, to look at, and with a few finishing touches would make a very ornamental material to be used on a building. J. H. Thompson, of the Chicago Lumber Co., says that the people in the vicinity of the quarry are using the stone for all sorts of building purposes. Specimens of this discovery will be placed on exhibition at the World's Fair.

A SHORT distance out of Buena Vista, Cal., is a cave literally swarming with spiders of a curious species of enormous size, some having legs four inches in length and a body as large as a canary bird. The cave resounds with a buzzing which the spiders emit as they spin and weave.

IMPROVEMENTS AT MOLINE.

How the Great Plow and Vehicle Factories Are Growing—
The Biggest Business in Their History Being Done.

MOLINE, ILL., Aug. 15.—To feel the pulse of the agricultural implement trade, go to Moline, Ill. This oft repeated assertion is as *apropos* to-day as at any time in the history of this famous manufacturing city. From no other point can a more accurate knowledge be obtained of the trend of the implement business for the entire country than here. Located as it is in the heart of the agricultural West, it is a sort of industrial hub from which radiate the spokes of an immense business wheel whose scope embraces every part of Uncle Sam's domain. Of the thirty and more manufactories, five have become world-famous for their immense annual output of farm implements and vehicles, and when this year's work has been completed the greatest sales in their history will have been recorded.

The writer is pretty familiar with the growth of Moline's industries, having resided here for a number of years, and from personal observation can say that never before were the signs of their prosperity more abundant and impressive than now. The business of the great plow works of Deere & Co. and the Moline Plow Co. have outgrown their capacity, and large additions to the plants are being made. The latter concern is building a 5-story stock and store-room 25x110 feet in dimensions, and adding a fourth story to a building 300x50 feet. It is putting in three new Sioux City Corliss engines with a nominal rating of 400-horse power, which is in addition to the 1,000-horse power furnished by the Mississippi river. Besides these improvements a double 50-inch Sturtevant exhaust fan of 35-horse power and a new steam economizer are being added. This fall a complete electric system will be put in. It is estimated that the Moline Plow Co.'s

business this year will reach nearly to $2,500,000, or 20 per cent more than for any previous year. The improvements being made will enable it to increase the output about 25 per cent. The buildings now occupied number twenty-six, and cover nearly four acres of ground. More than 500 men have been employed, and the number is now steadily being increased.

BIRDSEYE VIEW OF MOLINE, ILL.

When it is considered that this company commenced business in 1864 with a capital of $20,000, and a capacity of 1,000 plows annually, the astonishing strides it has made may be better appreciated. It now makes 300 different

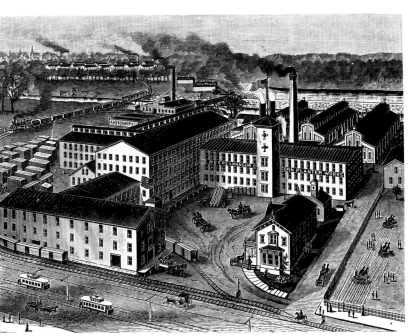

WORKS OF DEERE & MANSUR CO.

sizes and styles of steel walking plows, adapted for all purposes and conditions of soil, besides sulky and gang plows, cultivators, corn and cotton planters, check rowers, sulky hay rakes, scrapers, harrows, etc. How is this for a record of twenty-seven years,

The Deere & Mansur Co., which was established a little less than thirteen years ago on a small scale, and now employs over 200 hands in turning out nearly $600,000 worth of goods annually, is laying the foundation for a new 66x208 feet 4-story addition to their main shop, the same to be completed within ninety days. Their present buildings are nine in number, and occupy 500x800 feet of ground. None of the Moline factories have had a more rapid growth than this. Its production this year of corn planters, seeders, cultivators, drills, disc harrows, listers, stalk cutters, check rowers, cotton planters, and hay loaders has far exceeded that of any previous seven months, and the pressure of orders is now so great that they find it necessary to increase capacity. The manager informed the writer that the John Deere Plow Co., of Kansas City, had sold more of some lines of their goods this season than any other jobbing house in the country, going considerably ahead of their St. Louis house. This is especially noteworthy as indicating the growth of the implement business at Kansas City, for Mansur & Tebbits have heretofore lead all other jobbers in the amount of sales of this company's goods. The reception given their new Deere hay loader this year has been a source of great satisfaction to the company. It has fairly bounded into favor, and a great many more orders for it were received than could possibly be filled. Next season they will have ample facilities for supplying the big demand for it that is inevitable. The cut of the works given herewith gives a view of the addition to the main building as it will appear when completed. The new structure joins the main building with tower on the left at the end, forming a T.

Great changes are about to take place at Deere & Co.'s. For several years they have been cramped for office room, and are now preparing to tear down the front end of their main building and erect in its place a 3-story structure with basement. It will be of brick, with iron columns and glass front on the first

WORKS OF THE MOLINE WAGON CO.

story, brick pilasters between the windows on the second and third stories. It will have a frontage of forty-two feet. The offices of the company will be on the first floor, where also there will be a large reception room. The first floor will be used for sample rooms, and the third for storage purposes. The company also contemplate the erection of extensive additions to their mechanical departments in the near future. It need hardly be said that Deere & Co.'s business is growing rapidly. They now employ 800 hands, their shops and warehouses to the number of thirty, have a floor space of over thirteen acres, and their annual product amounts to 125,000 complete implements. To give the reader a proper idea of the magnitude of this industry, I may say that in the construction of their implements there are used annually 2,000,000 feet of oak and ash lumber, 6,000 tons of coal and coke, 5,000 tons of iron, 2,000 tons of steel, 1,500 tons of grindstones, 200 tons of oil and varnish, 30 tons of emery, and other material in proportion. Their goods are shipped to England, France, Germany, Russia, Australia, Cuba, South America, the Sandwich Islands, and other foreign countries. I am told that the government of Japan is now negotiating with the company for plows.

About twelve years ago the Moline Wagon Co. erected one of the largest farm and spring wagon factories in the country, to which they have made additions from time to time, until now they have a capacity for the employment of 400 men in producing 100 complete wagons per day. Their buildings and lumber sheds cover about six acres of ground. A fair idea

of the extent of these immense works may be obtained by looking at the cut. The handsomest business offices in the city are those of the Moline Wagon Co. Rich in the display of a variety of beautiful woods, the walls and ceilings present an appearance that charms the eye by its artistic appropriateness. The great success of this concern, in the face of the fierce competition in wagon making that has existed, is remarkable, and proves that thorough practical experience and honest work must be its basis, for the Moline wagons long ago became standard. They are handled to-day in all the imple-

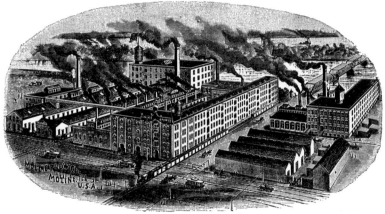

WORKS OF THE MOLINE PLOW CO.

ment jobbing centers of the country. The company's output will be larger this year than ever before.

The usual midsummer repose reposeth not at the D. M. Sechler Carriage Co.'s. At least the appearance of it is not plain to the eye of the visitor. The manufacture of buggies and carts goes steadily on regardless of 98° in the shade. This season there has not been such a howling demand for roadcarts as in years past, but the market for "B. B." carts has been good, and more than the usual number have

been made. The output of buggies and road wagons has been very large. This is comparatively a new concern, having been established less than three years, but it has already worked up a wide reputation for its goods. The magical name of Moline seems to bring success to every industry planted there. The works of

WORKS OF THE D. M. SECHLER CARRIAGE CO.

this company compare favorably in size with some of the older establishments. Their main building, of brick, is in the form of an L, 194 feet front by 207 feet deep, four stories high in front, and running back to five stories in the rear. Adjoining this is a 3-story brick building 70x70 feet on the ground. These form the buggy works, aside from which is a large cart factory. The capacity of the buggy works is 12,000 vehicles a year, and of the cart factory, 20,000 a year.

The Moline Buggy Co. is another concern of large proportions, which is enjoying a good trade. They are widely extending their business this year.

The signal prosperity of its manufacturing establish-

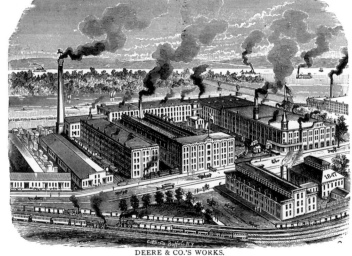

DEERE & CO.'S WORKS.

ments has given an impetus to Moline's growth and quite a lively little real estate boom is now on here. C. F. H.

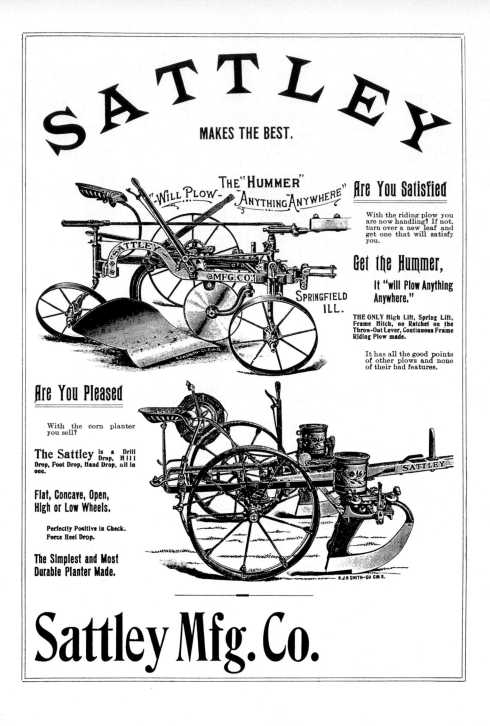

The Sattley Manufacturing Co. introduced the high spring lift plow. The beam was suspended on two bails so that the plow could be raised out of the ground or lowered into the ground without raising or lowering the frame on the wheels. This movement was made easier by a lever aided by a powerful spring. Raising or lowering and leveling could be accomplished using levers which adjusted the depth at which the plow ran. This construction was soon adopted by nearly all manufacturers and has also been applied to "gang" plows.

On January 5, 1892, A.S. Peck of Geneva, Illinois, patented principles used for nearly all corn harvesters. The Moline Plow Co. purchased the Mandt Wagon Co. of Stoughton, Wisconsin, the Stockton Wheel Co. changed their name to Holt Manufacturing, the Stover Heater Co., began in Freeport, Illinois, and the Waterloo Gasoline Traction Engine Co. was formed in Waterloo, Iowa, by John Froehlich. About 1892, the Sandwich Mfg. Co. of Sandwich, Illinois, added the "Southwick" hay press to their line of corn shellers, feed mills and other equipment.

Some claimed the financial panic in 1893 was caused by the severe gold drain. Thousands of banks and companies failed before the general recovery began in 1897.

The Columbian Exposition (World's Fair) took place in Chicago during 1893, the Stover Engine Works began operation in Freeport, Illinois, the Wood Brothers Thresher Co. began and the Massey-Harris Co. bought Corbin Disc Harrow Co. of Prescott, Ontario. The price of binders continued to fall. Champion binders were sold to dealers for $125 in 1893, then reduced the price to $105 in 1894. Mowers cost the dealer $37 to $39 and margins were very small on all equipment. Almost all large equipment had to be sold on time and collections for equipment sold earlier took up much of the dealer's time, effort and money (profit). One bad crop meant that the farmer couldn't pay.

The History of the Hay Press.

BY O. V. DODGE, VICE PRESIDENT KANSAS CITY HAY PRESS COMPANY.

Early in the present century the manufacture of hay presses was begun by Mr. Dederick, the father of P. K. Dederick, who is known throughout the civilized world as the pioneer hay press manufacturer, and who now owns, perhaps, more hay press patents than any other manufacturer. For years Mr. Dederick sold but few presses outside of his native state. The press made by him at that time was what is known, to all those familiar with the trade in the early days, as the "Beater press." It was a large, upright press, and worked something on the principle of a piledriver, the baling being done by raising the beater to a certain height, where it tripped, descending with great force into the baling chamber. These presses made a bale about 24x42x42 inches, weighing from 300 to 400 pounds. The bale was fastened with hoop poles, baling wire being unknown in those days. On account of their size and weight, making them so cumbersome, it was difficult to move one of these presses, the great weight making high freights, which, together with meager transportation facilities, and the necessary high price, accounts for so few being used, especially in the West. Later, the growing demand for baling presses led Mr. Dederick to make improvements in the way of portable presses that made smaller bales, but, even yet, inventors and manufacturers did not foresee the great field lying open to this industry, that is now such an important factor in America, and Mr. Dederick proceeded, unmolested, for years without a

competitor. He enlarged his works and extended his business until, here and there, hay presses were found in nearly every state.

It was not long, however, until others began to realize that there was a rapidly increasing demand for this class of machinery, and George Ertel commenced manufacturing baling presses in Quincy, Ill. His product also found ready sale, and Mr. Ertel and his presses are now known throughout the world.

The Whitman Bros., forming the Whitman Agricultural Co. of St. Louis, also put in the market an improvement in presses.

The success attained by Mr. Ertel had a stimulating effect on other enterprising men

of Quincy, and soon Quincy was known as a hay press manufacturing center, the firm deserving special mention being the Famous Manufacturing Co., headed by Andrew Wickey and his son, E. W. Wickey. Being men of push and enterprise, their presses also became widely known and this company, feeling the necessity of enlarging its borders, sought a location affording better facilities for transportation and in 1890 moved their plant to Chicago, where they have largely increased their output and are the only manufacturers of hay presses in the city of Chicago.

The Collins Plow Co. of Quincy also entered the race, placing in the market a press which they evidently intended to "get there," for they christened it "Eli."

As yet, there was not a hay press factory west of the Mississippi until in 1885, when the Kansas City Hay Press Co. was formed, with E. C. Sooy at its head, and thus was the

foundation formed for a new era in the hay press business. In this year (1885) the Kansas City Hay Press Co. began the manufacture of the "Lightning" full circle steel press, being the first full circle steel press placed on the market. Until this time, every press that had been placed on the market was constructed largely of wood.

The Scott Hay Press Co. of this city was formed in 1891. The product of this company has the significant title of "O. K.," which

the manufacturers say means "all right," which it undoubtedly is.

In (about) 1892 the Sandwich Mfg. Co. of Sandwich, Ill., the well known manufacturers of corn shellers, feed mills and kindred implements, in addition to their regular line, began the manufacture of hay presses, making what is known as the "Southwick" press, the invention of Mr. Southwick, and they also have succeeded in establishing a good trade in this line.

It is remarkable to note the very important place the haying industry now occupies in this country, the hay crop of 1893 exceeding in value that of any other agricultural product. The vast prairies of the West now yield millions of dollars' worth of hay, whereas, but a few years ago, except what was utilized for grazing purposes, was simply allowed to stand and in the fall was consumed by the great prairie fires that swept over the country. Many railroad companies now build cars intended especially for the transportation of hay. Thousands of men throughout the West give their entire time and attention to the hay business; a great many stations marked on the maps of the various railroad companies passing through the Indian territory are simply hay camps. The uplands of Kansas and the Indian territory produce the finest grade of prairie hay in the market. This hay sells in advance of any other wild hay, being much finer than that produced in Iowa, Nebraska, Minnesota and the Northwest. Many farmers now sow large tracts of land in timothy, which they raise for the market. The hay crop for 1894 has been estimated at 60,250,000 tons, about 5,000,000 tons short of the crop for 1893.

Some idea of the magnitude of the hay business may be had when we consider the fact that the hay crop of this country amounts to twice as much as the wheat crop, the crop of 1893 being valued at $590,000,000.

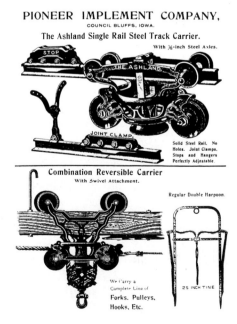

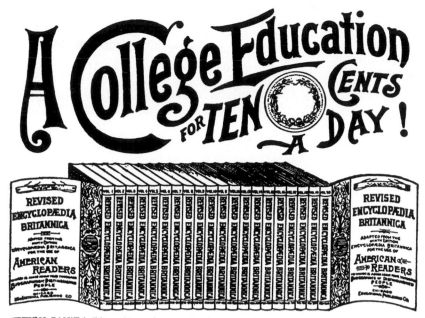

A College Education for Ten Cents a Day!

YOU CAN'T build anything unless you first have a foundation, whether it's a ten-story hotel or a success. Neither a house nor a man can reach any great height unless it has something to stand on. You build a foundation for your home in order that your family may be protected. Now, is it not equally wise to build a foundation for a success for your children in order that their future, too, may be protected?

Education is the only sure foundation which will elevate character and brains to the height where their possessor can reach the fruit which nature intended should be his. It is not the teaching a boy receives, but the learning, that counts in the race of life. You know it is the knowledge that you acquire by puzzling out your problems yourself, not that which was given you off-hand by a perfunctory teacher, that has stood you in good stead.

The **REVISED ENCYCLOPEDIA BRITANNICA** is the learning of the world concentrated. It has all there is to a college education except the college buildings and the "larks of the students." Is it not worth securing? Ten cents a day saved and this priceless work is yours for all time. We even furnish you the bank to save up the dimes.

The edition we offer is not a reprint, but a new edition. It is published in 20 volumes of over 7,000 pages, 14,000 columns, and 8,000,000 words. It contains 96 maps printed in colors, showing every country of the world, and separate maps of every State in the Union. It contains every topic in the original Edinburgh edition, and biographies of over 4,000 noted people, living and dead, not contained in any other edition.

READ OUR PROPOSITION:

On receipt of only **One Dollar** we will forward to you, charges prepaid, the entire set of 20 volumes, the remaining $9.00 to be paid at the rate of **10 cents a day** (to be remitted monthly). A beautiful dime savings bank will be sent with the books, in which the dime may be deposited each day. This edition is printed from new, large type on a fine quality of paper, and is strongly bound in heavy manilla paper covers, which with proper care will last for years. Bear in mind that the entire 20 volumes are delivered to your address, with all charges paid to any part of the United States.

The illustration shows a traction engine drawing six wagons of wheat as compared to the team that was able to draw two wagons to Belle Center, Ohio. This was another example of the superiority of the new machines and a precursor of grain transports to come in the future.

MINNESOTA THRESHER MFG. CO.,
STILLWATER, MINN.,

Makers of First-Class——————————

THRESHING ∴ MACHINERY.

Stillwater Traction Engine

The Aultman & Taylor Machinery Company.

WE DESIRE TO CALL YOUR ATTENTION TO OUR NEW——————

STACKERS,
SAW MILLS,
FEED MILLS.

Columbia Separator.

Shown in cut, sizes 33x50, 36x56 and 40 x60. Also our new

SELF-FEEDER,

unequaled for large jobs.

Our **DIXIE MACHINE** is the best ever offered for moderate sized jobs and flax, and other seeds, etc.
Also write for circulars on our **CLOVER HULLER.** We have bought all the rights, etc., of the Ashland (Ohio) Clover Huller. Our 1894
TRACTION ENGINES will be equipped with **FRICTION CLUTCH,** when so ordered.

In 1894, Adriance, Platt & Co. purchased D.S. Morgan & Co. and the Stover Novelty Works began (and probably replaced the Stover Experimental Works). Products made by the Stover Novelty Works included drill presses, power hack saws, spring making machinery, etc. The Waterloo Gasoline Engine Co. dropped traction from its name at about the same time John Froehlich left the company. The Duryea Manufacturing Co. was incorporated at Peoria, Illinois, with a capital stock of $50,000 by D.S. Brown, H.H. Faynestock and Charles E. Duryea.

In 1895, the Massey-Harris Co. purchased Bain Wagon Co. and the Stroughton Wagon Co. changed its name to T.G. Mandt Vehicle Co. Mr. T.G. Mandt died on March 1, 1902. Ball and roller bearings became less expensive. Bearings became inexpensive enough to be used in a wide variety of applications including bicycle wheels and gears of farm equipment. The reduced friction made it possible to use a smaller engine or fewer horses. Mass production methods of 1895 permitted one machine to make about 11 balls per minute. A machinist was paid about $2.50 per day to make the bearings.

VICTOR COB MILL.

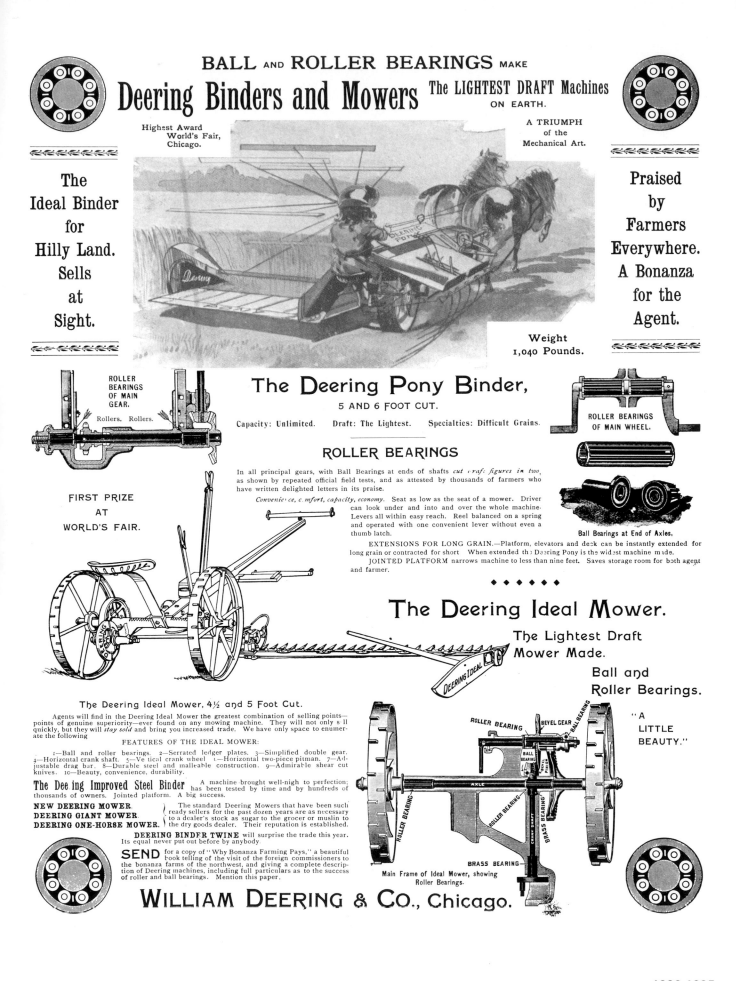
50

1886-1895

What is Claimed FOR THE ROCK ISLAND LOADER

Diagram Showing
Motion of Rakes on Rock Island,
1,700 Strokes Per Mile.

HOW Other Loaders DO IT.

Strokes on others varying from 3,500 on some to 4,500 on others, which simply means in our case that little wear is taking place, and that the work is being done with the least effort, while with the others high speed is required to do the same work, taking much more power and causing a rapid wearing of parts.

THAT in furnishing means for automatically attaching to and detaching Loader from wagon, we have deprived our would-be competitors of the principal claim of advantage which has been urged for other machines.

That the oblong strokes of the rakes makes the "Rock Island" run easier, last longer, rake cleaner, break the hay less, and gather the least dirt with it than any other loader on the market.

That it draws lighter and will do better work and more of it under any and every condition, and with less breakage and annoyance than any other loader.

That in hay yielding three tons per acre, it will rake and load a ton in ten minutes.

That, while it is not sold as a windrow loader, it may be used in gathering a light windrow, and will take up heavy ones by driving diagonally or squarely across them, or if the rake loads are left thus - - - - - - - - the loader will do much better work, and a moderately heavy windrow may be gathered.

That it will not only work in all kinds of grass, but under every and any condition of meadow will successfully rake from the swath, elevate the hay onto the wagon, and leave the ground cleaner than if raked with a spring tooth horse-rake.

That if properly put together and attached to wagon, a good team and one good man will handle it with ease on level ground. (It must not be expected that a machine elevating a ton of hay to the top of a wagon every 10 minutes is going to run itself and help push the wagon.)

That it works equally well on hillsides as on level ground.

That it is provided with the best means of raising and lowering the rakes at will.

That it is necessary to have such provision to adapt the loader for work in different kinds of hay and varied conditions of meadow.

That it has a successful ADJUSTABLE DELIVERY, by which, in beginning a load, the hay may be delivered near the bottom of the rack and raised higher as the load grows larger.

That with the new "automatic coupler" it is easily attached to and detached from the wagon.

That as the hay is "pushed" or "rolled" upon the load, it becomes bound together, and does not fall off as if pitched on or taken from the delivery of an endless chain loader.

That if kept steadily at work during haying season, it will more than pay for itself in the actual saving made over any other known process of making hay and delivering it on the wagons.

That no one can afford to use a hay-rake if they can borrow, to say nothing of buying one, when a "Rock Island" Loader can be bought for so little money.

That it makes the farmer almost entirely independent of hired help, and at the same time saves money and makes the house-work lighter for the wife.

Ask for "Songs of Praise," that will be gladly furnished on application.

Hardware Department of J. P. Burtis' Store, 1895.

1896-1907

COMPETITION BETWEEN THE COMPANIES

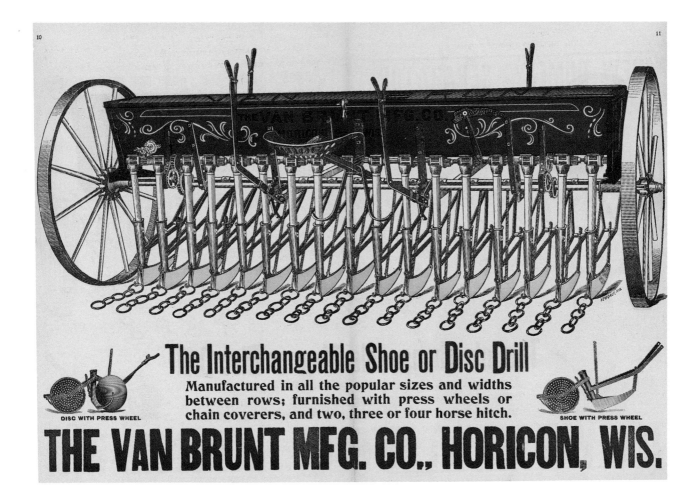

The Interchangeable Shoe or Disc Drill

Manufactured in all the popular sizes and widths between rows; furnished with press wheels or chain coverers, and two, three or four horse hitch.

DISC WITH PRESS WHEEL

SHOE WITH PRESS WHEEL

THE VAN BRUNT MFG. CO., HORICON, WIS.

THE IMPLEMENT TRADE JOURNAL.
A WEEKLY NEWSPAPER.

1896—1907

A depression occurred during 1896-1897 followed by a subsequent recovery. Difficult times, death of more industry pioneers and "Harvester Wars" led the way toward consolidation of the companies. Sales and distribution networks were developing to smooth the movement of farm equipment to the farm.

NEW FAST CALIFORNIA TRAIN.

On October 29th, the Santa Fe Route inaugurated new and strictly limited first-class service to Southern California. The California Limited leaves Kansas City at 9:10 a. m. daily, reaching Los Angeles and San Diego in two and one-half days and San Francisco in three days, thus reducing the time half a day. Equipment will consist of superb new vestibuled Pullman palace and compartment sleepers, chair car and dining car, through from Kansas City to Los Angeles without change. Entire train lighted by Pinsch gas. This will be the fastest and most luxurious train via any line to California. The present train leaving Kansas City at 2:00 p. m. will be continued, carrying through palace sleeper and tourist sleeper to San Francisco and tourist sleeper to Los Angeles. Full particular can be obtained by addressing, GEO. W. HAGENBUCH, P. & T. A., northeast corner Tenth and Main streets, Kansas City, Mo.

GAS ENGINE POSSIBILITIES.

While the gas engine has been known and used for many years in a small way, and with remarkably good results as far as economy goes, it is only since a short time that its merits have been fully appreciated. It is now, with the introduction of new methods of gas production, by the use of by-product-saving appliances that go far toward paying the original cost of fuel, and thus reducing the cost of the fuel gas to a very low figure, doing much to solve the problem of cheap and effective power.

As these gases are low in illuminating qualities, they are very much better suited to give the highest efficiency in the gas engine. Another gas that has recently been discovered has remarkable qualities under compression, and can be reduced in volume 400 times at 800 pounds, and when expanded will burn with twenty times its volume of air, requiring only 0.4 of a pound of it when compressed to develop one-horse power per hour. Each cubic foot of it at this pressure weighs thirty pounds, and, therefore, contains seventy-five H.-P. hours, being the greatest storage of energy ever known for a given weight. This opens a wonderful field for the development of power for motors for tram cars and other classes of motor vehicles, as well as pleasure boats. Gas engines, working with this new fuel gas, are likely to have a very large use in all stationary work and for propelling boats, and it may not be beyond the bounds of possibility to drive ocean steamers and locomotives of the future by gas engines.—Cassier's Magazine for August.

THE DAVIS GASOLINE ENGINE.

THE Davis Gasoline Engine Co. of Waterloo, Iowa, is sending out to customers and those making inquiries concerning the Davis gasoline engine, or requesting information on power questions, a budget of useful information. In perusing this printed matter we found much to interest and instruct and have thought it well to call the attention of readers to this engine by publishing a cut of the same, and some description of its workings.

This engine is furnished with water tank, gasoline tank, glass sight feed oil cups, muffler for exhaust pipe, pipe cut and fit for water tank, sufficient exhaust and gasoline pipe for all ordinary places and connections necessary to put an engine in running shape.

The manufacturers assure all users that it is so simple to operate that "all classes of people, irrespective of trades or professions, manage it without the slightest difficulty." It is in no sense dangerous; being equipped with every safety and economical device known, and these have been proved to be reliable by the most exacting experience.

The manufacturers claim superiority for this engine and offer the appended eight reasons—quoted from a recent circular to the trade—why:

First—The only gasoline engine having a throttle by which it is started, stopped and run slow without stopping, just the same as the steam engine is handled by the throttle.

Second—The only gasoline engine feeding the gasoline direct through a throttle and having a governor valve operated upon by a fly-wheel governor, by which each charge is proportioned to suit the work being done, thus keeping a regular speed, a uniform temperature of cylinder, saving a large per cent of what is usually lost through radiation of the old method of alternately heating and cooling the walls of the cylinder, and at the same time consuming more air in proportion to gasoline used than any other engine. Hence our guarantee of economy.

Third—The only gasoline engine in which all sizes is provided with an automatic crank oiler, which is very important, as many engines have proven a failure through nothing but improper lubrication of the crank bearing.

Fourth—The only gasoline engine having a shield over the crank, preventing the engine from throwing oil, and guarding against the possibility of anyone getting caught, as everyone knows an uncovered crank is dangerous.

Fifth—The only gasoline engine having an oilway around the base, thus enabling the operator to catch all the oil, filter and re-use.

Sixth—The Davis is the only gasoline engine in which there is no odd constructed movements and no sensitive or variable valves or other apparatus to regulate.

Seventh—The only gasoline engine that is entirely free from cams, rollers, triggers or anything that will get out of shape by wear.

Eighth—The only gasoline engine of which every bearing or wearing surface can be adjusted to take up wear.

Rural Mail Delivery began during 1896, William Jennings Bryan, the Populist presidential candidate, was defeated and Utah was admitted as a state. The Marseilles Mfg. Co. entered receivership on January 27, but was able to reorganize and continue normal operations in April.

The gold rush began in the Klondike in 1897. The Hart-Parr Gasoline Engine Co. began in Madison, Wisconsin, while both Charles W. Hart and Charles H. Parr were engineering students of the University of Wisconsin. Also during 1897 the *Kansas City Implement &*

Farm Journal was incorporated as Implement Trade Journal Company and the magazine was renamed the *Weekly Implement Trade Journal.* O.B. Kinnard and A. Haines began building Flower City gasoline traction engines in Minneapolis, Minnesota.

Rural Free Delivery.

 FEW weeks ago a representative of the JOURNAL called upon a manufacturer of vehicles in a town in Indiana and was invited to make a tour of his factory. Much to his surprise, he found that the labor of the entire working force was concentrated upon the production of rural mail delivery wagons. The proprietor imparted the information that he had "struck a snap," and had ceased to manufacture a general line of vehicles. There was every indication of "snap" about the place, to say the least. A small army of men was busily engaged in the construction of a variety of styles and sizes of delivery wagons, and in the finishing room the last touches of the painters revealed the fact that the United States comprised this manufacturer's trade territory, and suggested that for some time to come "U. S." would be the talismanic letters with which he would command prosperity.

Not the least of the benefits to be extracted from the creation and extension of the free rural mail service is the profitable employment of labor in the manufacture of the delivery wagons. Thus again do the interests of the farmer and the manufacturer converge to their mutual advantage under a wise governmental dispensation. As the writer understands, the Government does not contract for the wagons; the mail carriers must do that for themselves; hence competition is open to all, and the manufacturer who hustles the hardest and offers good wagons at the most attractive prices is the one who will secure the orders in this as in other branches of trade.

The story of the evolution of the rural free delivery service, as told by Charles H. Greathouse, M. A., in the Yearbook of the Department of Agriculture, is very interesting. It dates back to the "village delivery" which Postmaster General Wanamaker recommended in 1890, but that was simply an extension of the city delivery system by carriers on foot in towns with a less population than 10,000, or less gross postal receipts than $10,000, the limit at which city delivery stops under existing law. After a brief experimental existence of a little more than two years, the movement for free delivery on a broader basis was not suspended but grew in intensity. The new agitation took the form, not of a request for free delivery in villages where none of the patrons lived more than a mile or so from their village postoffice, but of a movement to give the country delivery to farmers who lived from two to twelve miles from any

postoffice, and who in consequence had to waste the better part of a day whenever they wished to mail a letter or expected to receive one, or desired to obtain a newspaper or magazine for which they had subscribed.

It was not until 1897 that Congress appropriated a sufficient sum of money to warrant an undertaking in this direction by the Postoffice Department. With the assurance of having $40,000 at his disposal, Postmaster General William L. Wilson, in the fall of 1896, inaugurated the experiment by the selection of forty-four routes in widely differing localities in twenty-nine states. Fifteen routes were going in October of that year, fifteen in November, eight in December, three in January, 1897, and one each in February and April following. The very first routes established were at Halltown, Uvilla and Charlestown, . Va.

It was midwinter before the work got fairly under way; the officials who were intrusted with the inauguration of the service were often dissatisfied and unfavorably disposed toward the work, because they had to be detailed from their regular work in such a way as to hinder their probable promotion, and therefore some of the first reports received were quite discouraging.

RURAL MAIL DELIVERY WAGONS READY TO START AT HOUSTON, TEX.

MAIL WAGONS AND BOXES AT DUNKARD CHURCH,
NEAR DEFIANCE, O.

These carriers are also authorized to receipt for applications for money orders, and while they can not yet issue the orders, they can save the farmer the trip to the postoffice by acting as his agent. It has been found possible in the coldest or hottest weather in any part of the country to deliver the mails with very little interruption. When a heavy snow blocks the way for the rural carrier, it is customary for the farmers to turn out and break the roads, and this is done several days earlier than would be the case ordinarily. This way communication throughout neighborhoods and with the outside world is opened up promptly. In consequence the farmer is able to take advantage of good markets, and the townspeople are not cut off from the supply of fresh country produce, as often has happened in cases of severe storms. Also cases of distress in isolated farm houses are sooner reached and relieved.

In 1898 the number of routes was increased to 128, and by Nov. 1, 1899, it had jumped to 634. These radiated from 383 distributing points and served a population of 452,735 persons. On June 30, 1900, a little more than six months later, the number of routes had grown to 1,214 for a population of 879,127, and in the next four months the system had again more than doubled its proportions, showing on Nov. 1, 1900, 2,551 routes for 1,801,524 persons; and there were also at that date 2,158 applications for the establishment of other routes. The applications for new routes by March 1, 1901, had reached 4,517, and the number has since greatly increased.

Rural carriers are now authorized to receive and deliver registered mail. As the law requires such matter to be delivered personally, the carriers are obliged to go to the houses instead of dropping the letters or packages in the farm box.

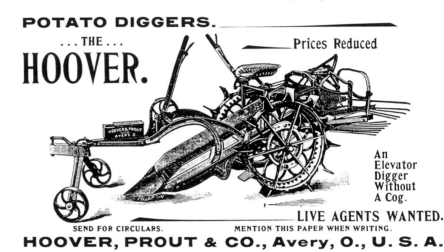
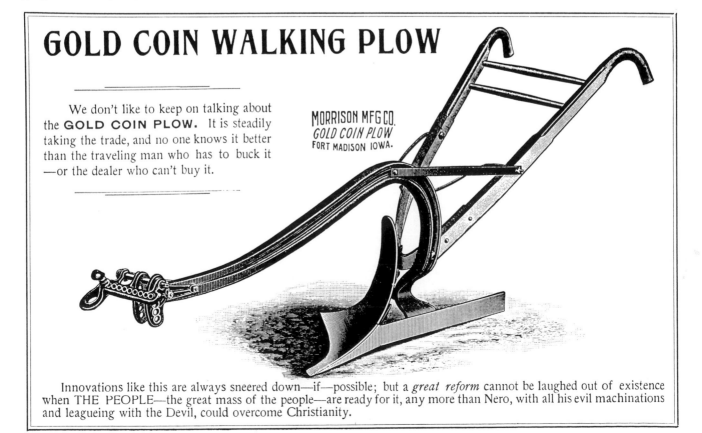

THE HOOVER POTATO DIGGER AT WORK ON THE FARM OF A. J. HULETTE, OLD FRANKLIN, MO.

On February 15, 1898, the U.S. battleship Maine was blown up in Havana Harbor, killing 264. President McKinley demanded, in a communication dated April 20, that Spain withdraw from Cuba, then blockaded the Cuban ports. War was declared on April 24, but was made retroactive to April 20. After several months of fighting, an armistice was signed on August 12 which gave Cuba and Puerto Rico to the United States. The U.S. bought the Phillippines from Spain for $20 million and thus ended the Spanish American War. The U.S. also annexed Hawaii during 1898.

Alvah Mansur, an early partner to John Deere, died on January 8, 1898.

All through the years, there has been strong opposition to the sale of farm equipment through catalog houses. Attempts to stop the sale of equipment by nonstocking dealers continued, but so did the catalog houses. The Rural Mail Delivery and Parcel Post made strong

THE BUCKEYE.

The Buckeye is the only Binder successfully built with Header Attachment, thus saving the farmer the expense of buying a header when the grain is short. The GENUINE BUCKEYE MOWER, built exclusively by ourselves, is the only mower that has stood the test for thirty-five years without change in plan of construction and is now being copied by pretended rivals.

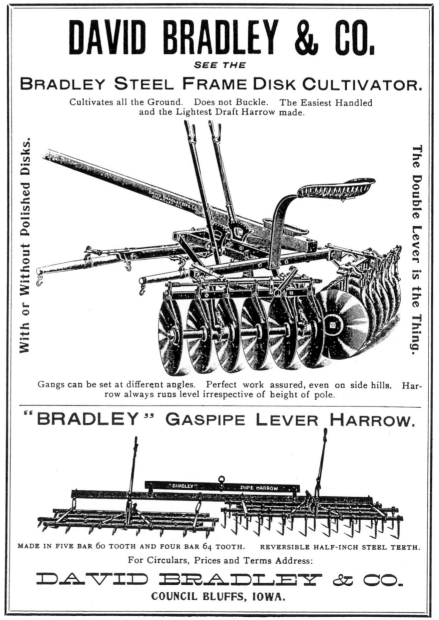

AULTMAN, MILLER & CO.

contributions to the success of mail order catalog companies. The Huber Mfg. Co. purchased Van Duzen of Cincinnati, Ohio, in 1898 and, on September 15, the Weir Plow Co. plant in East Moline, Illinois, was bought by Kingman & Co. of Peoria, Illinois, at a sale for $35,000. The first Lauson engine was made during the year 1898.

COMING OF THE HORSELESS CARRIAGE.

A little reflection will convince anyone that the use of motocycles, or, in other words, horseless carriages, will improve the roads. General Morin of France is authority for the statement that the deterioration of common roads, except that which is caused by the weather, is two-thirds due to the wear of horses' feet, and one-third due to the wheels of vehicles. This being the case, if the same amount as usual continue to be laid out upon the roads, and the continual damage decrease two-thirds, then the amount spent will go to increased and permanent improvement, and the roads will be "as smooth as a barn floor."

There are many questions to be solved, many difficulties to be surmounted, before the unexceptional vehicle appears. It was a long time before the difficulties of making sewing machines, revolvers, repeating rifles, typewriters and typesetters were overcome. Yet, examine them! It is all plain and simple, and not at all marvelous now, and we can hardly imagine how any mechanic could spend years of time studying over such easy problems. So it will be with motocycles. The mountains of difficulty will sink into molehills, and the ingenuity displayed will be found to take the form of judicious application of ordinary mechanical appliances, approved by the final umpire, the common sense of mankind.

Those who build automobiles must not permit themselves to think that they were born with all the carriage makers' lore inherent in them. A man may be a first-class theoretical and practical mechanic and not be able to make a good vehicle to run on wheels. The perfect carriage, as we know it to-day, is the aggregate of the years of exhaustive trial and experiment and the improvements on that experience, made by a thousand men of genius.

If the carriage builders bestow upon the new carriage all the art acquired in building the old, and the motocycle men learn the reasons of the conventionalities of the trade and adapt their improvements to them, with reference of the opinions of those who are not prejudiced against innovation, they will both work together in harmony, and with one purpose, and, so united, they will make rapid progress in the development of the inevitable vehicle of the future.

Lewis Miller, a leading engineer and partner of Aultman, Miller & Co., died on February 17, 1899, David Bradley, founder of David Bradley Mfg. Co., died on February 19 and Titus G. Fish, president of Fish Bros. Mfg. Co. died on December 31.

Many manufacturers added a foot lift to riding plows during 1899. New Idea Co. was started in Coldwater, Ohio, and the Foos Mfg. Co. of Springfield, Ohio, built a new foundry and pattern shop. R.D. Scott & Co., carriage builders, were incorporated at Pontiac, Michigan, and the Challange Wind Mill Co. was incorporated at Dallas, Texas. The Avery Planter Co. of Peoria, Illinois, changed its name to Avery Manufacturing Co. to recognize their expanding product lines. In August 1899, Deere & Co. became controlling owners of their branches and distributing houses.

John M. Gaar, president of Gaar, Scott & Co., died on August 9, 1900. Self-feeding and wind stacker (blower) attachments were available for threshers in 1900 and the American Bridge Co. formed the Minnesota Malleable Iron Co. Later in 1902, the Minnesota Malleable Iron Co. became part of the Minneapolis Steel and Machinery Co. and the American Bridge Co. was sold to the U.S. Steel Corp.

SELLING BINDERS IN THE EARLY DAYS.

BY COL. JAS. H. SPRAGUE, NORWALK, O.

[The Implement Trade Journal takes great pleasure in presenting to its readers the following reminiscences from the pen of Col. Jas. H. Sprague of Norwalk, Ohio. Colonel Sprague is a well-known author, song and music writer, and an inventor of vehicle canopies, etc. He was formerly attorney for D. M. Osborne & Co., and was at one time general agent for the Plano Mfg. Co. for Ohio. He speaks from experience, and although president and general manager and principal owner of the well known Sprague Umbrella Co., he manages to find time to write books, stories, songs, and to run about all of the lodges in the city in which he resides. Being the leading spirit in the Knights Templar, Odd Fellows, Elks, G. A. R., Royal Arcanum and many other societies one might wonder how he got time for business, but everything around his mammoth establishment goes on like clock-work, where a large force of well-paid and competent employes help the colonel make a success of the business.]

Editor of the Implement Trade Journal: You ask if I will write about selling binders in the early days if I have the time to do so, and this is my answer:

A man who rightly divides his time can do almost anything for his friends. Some people handle their time to a great disadvantage and others make their business too serious a thing. It is easy to work between meals if you are cheerful and take it cool. There's always luck in well-aimed leisure. The man who worries about his business cannot be a success. If he boils his pot night and day he will soon run out of steam and burn out his flues. 'Way back in the eighties, poverty compelled me to accept a position as attorney for D. M. Osborne & Co., one of the best concerns that anyone ever had the privilege of working for. When I say poverty compelled me to work for them, I mean it in its literal sense. Very few of us work for the pure love of it; we need the money, and so we exchange our time, on which we are long, for some one else's money, on which we are sadly short. It is not an even exchange always, but poverty compels it, and we grab at the chance.

The life of the binder man in those days was one of uncertainty and toil, and much more strenuous than it is today. The farmers viewed the "newfangled riggin'" with distrust; they were fresh from the embraces of the factious lightning-rod man, who by the way, was gener-

The EMERSON Foot=Lift
Sulky and Gang Plows

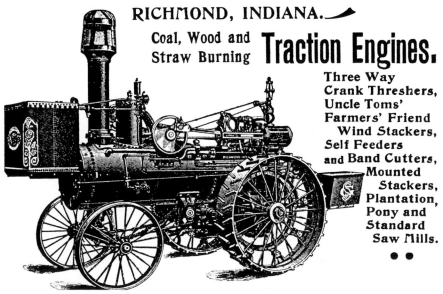

GAAR, SCOTT & COMPANY

RICHMOND, INDIANA.

Coal, Wood and Straw Burning **Traction Engines.**

Three Way Crank Threshers, Uncle Toms' Farmers' Friend Wind Stackers, Self Feeders and Band Cutters, Mounted Stackers, Plantation, Pony and Standard Saw Mills.

ally a disappointment to them. They had the dropper and knew they could get along with it, and they didn't believe it was possible to cut grain and bind it all with one machine. They

had seen the Marsh harvester, but it was only a dropper with a platform, and it was hard expert work to accomplish much with it, so when they were offered the binder they were in doubt, and it was the business of the binder agents to remove the doubt. You old ones all know what a hard job that was. There was no binder at that time that could use twine; several of them were quite successful with wire, and there were stories started that the wire would choke cattle, and that lightning would be liable to strike the field and burn up the crop if it was tied with wire—so much steel would invite electricty, etc. But these dreadful forebodings never materialized, and our men succeeded in "forcing" a goodly number of wire binders on to the nervous farmers, and they were, as a general rule, well pleased and glad they had bought them. The writer went one time with one of our agents to Sebawing, Mich., to see our local agent, who was a German, and kept a general store, to which was attached a beer annex. The old fellow had sold ten binders for three hundred dollars each, and we got them in shape and started them, and he got "fur baar geld" and cashed his contract and arranged for the next year. The price was necessarily high, as it was a new thing, and we were also new but expensive. After a year or two it became necessary, in order to supply the popular demand, to get out a twine binder, and there was a great strife between the large concerns as to which would get one that would work. Appleby had succeeded in getting a twine binder attachment that did very fair work, but neither the Osborne nor Champion would use it then. D. M. Osborne, one of the grandest men who ever trod the earth, was made of too stern stuff to accept another's invention for his machine. Wm. Whitely, of the Champion, was also too great an inventor to use any machine not conceived by him. The Osborne tried the wire and twine; it worked well as a wire binder, but countless semi-wet nurses were needed to keep it in the field when the twine attachment was used. We followed it night and day, camped on its trail, greeted it with our morning "smile," and cursed it with our evening breath. We sat on the fence anticipating the farmer's call for help and rushed to his rescue in time, if possible, to save the sale. One minute we were ready to swear it was a grand success, and the next we were hanging by the neck in the slough of despond. We were not alone; "there were others." The corners of the wheat field fences were strewn with the flotsam and jetsam of disappointed hopes and binder experiments. It required tact in those days for a binder agent to get out of a farming community alive. We grew expert and eloquent in explanations of why it wouldn't work. " 'Twas a new thing, you know, and you couldn't expect too much from it." But they did expect it to work some, and it generally didn't, and the genial farmer who had fed us with so much care before the binder was started was ready to kick the meal out of us after the machine had gone half around the field. He was disappointed, and so were we-uns. He thought we were frauds, and we half suspected as much ourselves. He swore

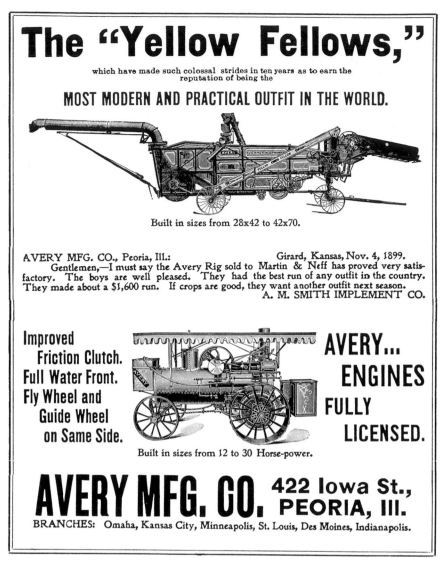
he wouldn't settle for the machine, but we doubted him, and at the moment the poor old binder behaved a little and the dear granger had a lucid interval, we flashed a note in his face, and with the assurance that the embarrassment of the binder was only temporary and not chronic, we closed the incident and tried to get away before the pesky thing got on another tantrum and the dear farmer had another bad spell. We were almost sure to see him that evening, however, at the local agent's, bearing the news that "The Only" was on a strike again, and the genial scientist who did our expert work was billed for his farm at the dawning of the day morning. Then we all remember how kind we used to be to the farmer. How we used to lie awake nights for fear he would buy some other machine than ours, and be hornswoggled out of his hard earnings by the "lying agents" of the other machines. This used to worry us a whole lot, and we admonished him to sign an order for "The Only" and be safe from the wicked wiles of the adversary; and the other fellows did all they could to match us, and even went so far as to get the farmers to afterwards sign for their machines and repudiate our contract, a contract that for strong legal terms made a death warrant pale into insignificance. Yet he waived it all and bought of the other fellow, and then we tried to hunt up some one who had bought the other fellow's machine, with the hope of persuading him to do the same shady thing. It was Greek meet Greek, if not dog eat dog.

But the great machine companies were

hard at work. The steel binder with its dead sure knotter was on the way, and the lumbering horse killers were relegated to the rear. The skies became brighter, and no longer was the genial expert required. No longer now do the crowds of machine experts swarm the hotels; no longer do they board with the long-suffering granger, waiting and watching for the success or failure of a doubtful aggregation of cog-wheels and useless cast-iron; no longer do those animated scrap-piles worry the manufacturer. The splendid steel structures elegantly finished, each a plaything for a good team, sure workers moving across the almost countless acres of the inhabited globe, carrying the names and fame of those grand pioneers, Osborne, Deering, McCormick and others, to the uttermost parts of the world; spreading civilization and happiness wherever their wheels roll around, until we can almost say they are distributed in the places old Bill Jones declared he would send the gospel to, "where the foot of man never trod, and the eye of God never see."

Those avant couriers of the modern harvester and binder toiled for years through evil and good report, through failure and success. Small were their beginnings, humble was their origin, small shops and discouragements the rule. Yet these grand men, working with an unfaltering courage, builded better than they knew. The vast factories and fortunes with which their successors are endowed were made possible only by the sweat of the brows of those splendid old giants, who often risked all they had in doubtful experiments, and whose genuine grit and perseverance made it possible for these United States to lead the world in grain cutting implements. Then all honor to the names and fame of McCormick, Manny, Osborne, Deering, Appleby, and all of the giants of the early days of the binder. Their monuments are more enduring than marble or brass; their names will be household words so long as grain shall be grown.

NICARAGUA CANAL.

From the report of President Theodore Search to the National Association of Manufacturers at the fourth annual convention held in Cincinnati last week we take the following:

"Interest in the construction of a Nicaraguan canal has been stimulated powerfully by the incidents of the late war with Spain, and by the commercial expansion of the United States which has resulted from the brief conflict. Arguments heretofore advanced in behalf of the various interoceanic canal projects have been more essentially of a commercial character; but while the importance of a canal from a purely mercantile standpoint has in no way diminished, the political and strategic value of a waterway across the central American isthmus has been increased enormously during the past few months. We have heard much of the importance of a shorter route from the Pacific coast to the Atlantic seaboard, based upon the theoretical necessities of a possible war, but when our magnificent 'Oregon' steamed from San Francisco to Key West last spring, traversing a distance of over 15,000 miles in an actual sailing time of 59 days, then it became clear to every mind not hopelessly obscured by preconceived opinions how immeasurably important would have been a channel of communication between the two oceans capable of saving one-half this time, and avoiding the constant danger of destruction by a hidden foe in South American waters. Experience has added to theory an argument which cannot be overthrown.

"The whole tendency of the Nicaragua Canal agitation of late has been more and more strongly in the direction of the United States, and, in fact, under the actual ownership of our nation. In the minds of the people this subject is becoming more and more a public enterprise in which private interests are figuring only in the promotion and preliminary work.

"The commission created under the Act of June 4, 1897, for the purpose of investigating and considering the entire field of canal possiblilities in Nicaragua, has finished its field-work during the past year, and the results of its labors will give to this Government a more accurate and complete basis upon which to consider legislation touching this project. The preliminary report of this commission roughly figures the cost of constructing a canal upon either of the two best known routes at about $125,000,000, this estimate providing for a waterway of larger dimensions and greater capacity than any heretofore proposed.

"Legislative consideration of the Nicaragua Canal project centers now upon Bill S. 4792, which provides for the amendment of the original act of incorporation of the Maritime Canal Company by virtually making the United States the sole owner and proprietor of the enterprise. Several propositions for accomplishing this end are now under discussion, and the whole consideration of the matter now is directed towards determining to what extent and in what manner this nation shall take part in the construction of this great waterway.

"This whole problem has been complicated somewhat by the granting of a new concession by the Republic of Nicaragua to Messrs. Eyre and Cragin, representatives of an American syndicate, of whose executive committee William R. Grace is chairman. This new concession covers the same route now embraced by the concession under which the Maritime Canal Company is operating, and is to become effective upon the expiration of the present contract, October 9, 1899, or sooner if agreement to that effect can be obtained. The claim that the concession held by the Maritime Canal Company will be forfeited upon expiration is subject to dispute, however, and the conflicting interests of the holders of the two concessions have added an unfortunate element of confusion to the problem. We are not called upon to determine the rights of the parties in this controversy, nor are we asked to pass upon the technicalities of route, cost and engineering. It seems highly proper, however, for us, as business men, to give clear expression for our belief in the commercial necessity of a canal across Nicaragua, and to our conviction that whether constructed by a private corporation or with funds supplied by our Government, the canal should be first, last and always an American enterprise and under the control of the United States. With our new possessions in the Far East we need now as never before a way of quick communication by water between our Atlantic and Pacific coasts. In the Orient are the largest undeveloped markets for our products, and our growing trade there demands better and cheaper means of transportation than our transcontinental railway lines afford. The millions which a Nicaraguan canal would cost would be well spent if devoted to such an end by our Federal Government."

During 1900 and 1901, the Hart-Parr Co. moved to Charles City, Iowa. The first tractor was built in 1901, but production of the first fifteen began in 1903.

In 1901, the South Bend Iron Works changed their name to the Oliver Chilled Plow Works, William Deering retired at the age of seventy-five and Allis Chalmers was formed by a merger of:

Edward P. Allis Co.
Faser and Chalmers Co.
Gates Iron Works
Dickson Mfg. Co.

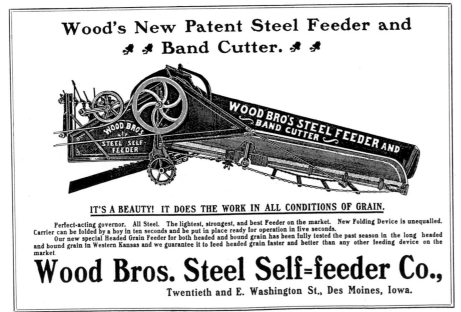

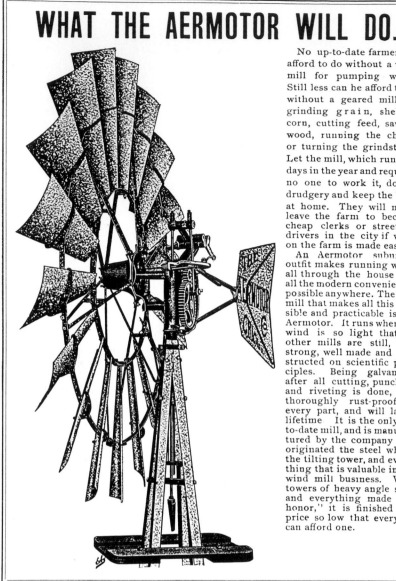
The Secretary of Agriculture Wilson ventured these opinions in 1901.

"The prediction of Mr. Edison that electricity will come to the rescue of the farmer during the next fifty years, is likely to prove true, but not in the way of heavy machinery. It will come about, in my opinion, through the use of electricity in transportation, mining and manufacturing. It is just a trifle improbable that anything will ever be invented to take the place of a team of horses for farm work. Automobiles run smoothly on a level road – but not in the mud. But electric railways are going out into the country, radiating from every town and city in America. Every one of these benefits the farmer. City people move out, build houses, beautify grounds, and come into healthful contact with mother nature. The farmer, not to be behind, brightens up his own place a bit, uses the trolley himself, enlarges his horizon and his market."

The United States government passed the National Reclamation Act in 1902.

By 1906, the reclamation service had irrigated 13,600 acres by constructing 241 miles of main canals, 116 miles of distribution system, 388 miles of ditches and 5.5 miles of tunnels, including 2.5 miles of the Gunnison, Colorado, project tunnel.

THE FIRST GOVERNMENT RESERVOIR.

C. W. Talley, who represents the Parlin & Orendorff Co. in Colorado, writes that the first reservoir which the agents of the Government have recommended is to be well located at Pawnee Lakes, fourteen miles northwest of Sterling. It is to have a capacity of twelve billion gallons, and with it and the Platte river, it is expected to irrigate a million acres of land. Mr. Talley also writes that the new ditch west of Ft. Morgan will have an abundance of water next year, as an immense reservoir is being built at the head of the ditch, which will supply all the water needed so badly in that section in the months of July and August. He says the crops in the San Luis valley are completely ruined for lack of water this season, which will make the conditions serious for the farmers for at least eight or ten months. The alfalfa and wild grass in the Platte river valley are doing finely. They have had rain about every ten days since the first of May from La Salle east on the Platte river, and the nearer to Nebraska and Kansas the better are the crops.

In the Mancos and Dolores valleys, Mr. Talley says, there is prospect of three-fourths of a crop of alfalfa. About the same result is looked for in the neighborhood of Montrose and Delta. The fruit crops will be large about Grand Junction and Fruita. Water has been short at De Beque and crops will be poor. Around Rifle, New Castle and Carbondale the conditions are more favorable. At Canon City they have had plenty of water and will secure a fine fruit crop. At Florence there has been complaint of dry weather, but the oil, coal and mining interests there offset the shortage in farm products. The mining interests of the state are quiet, excepting in the San Juan country around Durango, where there has been a rich strike. At Ft. Collins the crops are in splendid condition, as well as at Loveland and Berthoud. There will not be full crops about Longmont. The entire eastern part of the state was in straits for water when Mr. Tally wrote, but since then some heavy rains have fallen.

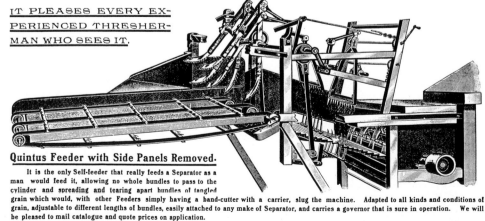

WE WISH YOU ALL

A Merry Christmas

STYLE

STRENGTH

FINISH

NO. 1 PORTLAND

NO. 2 PORTLAND

KINGMAN 2-KNEE BOB

STYLE

STRENGTH

FINISH

A light flurry of snow during the Holidays calls for Sleighs.
A Wise Dealer will be prepared to catch that Trade.
An Order now, while Our Stock is Complete, will be shipped promptly.

KINGMAN-MOORE IMPLEMENT CO.

OKLAHOMA CITY   KANSAS CITY

THE NEW CHAMPION RAKE.

There are several practical features on the New Champion rake which mark an advance in rake construction over anything heretofore offered to the trade. These features meet needs of farmers and of dealers not met by other rakes. The first and perhaps the most important of these is the lock lever for holding down the teeth of the self-dump rake, either in raking or bunching. It has been a common practice to lock down the teeth of hand dump rakes for this purpose, but a positive lock for holding down the teeth of the modern self dump has not heretofore been put on the market. The hold-down lever on the New Champion is so arranged that by the changing of a single bolt it will lock or not lock as may be desired.

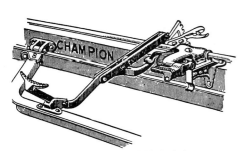

Lock Lever Set to hold Teeth down.

In ordinary raking the teeth of the Champion will remain down of themselves, conforming to any irregularity or unevenness of the ground. There is a convenient foot lever and treadle by means of which the teeth may be easily held to position. On no other rake are the teeth held down more easily. Nevertheless when raking heavy hay or bunching, especially if a small boy or girl is driving, it is desirable to lock the rake, so that no effort will be needed in holding the teeth down while it is being filled. An adjustment for this purpose is quickly and easily made on the New Champion.

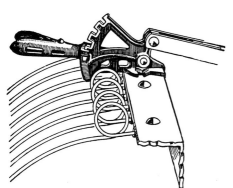

Lever and Hand Latch for Adjusting Angle of Teeth.

In the first days of the mowing machine when the driver wanted to change the angle of the cutters he raised or lowered the pole by shortening or lengthening the breast straps. Something of the same primitive plan has been used on hay rakes up to the present time. When the driver wished to change the position of the teeth to suit the condition of the ground over which he was raking, he had to get off of the seat and remove and replace bolts or keys. On the New Champion, this primitive method is replaced by a convenient hand lever and latch, by means of which the driver may change the position of the teeth as easily as he can change the tilt of the guards on his mowing machine.

Tripping Mechanism Rachet Disengaged.

Tripping Mechanism, Hooks and Ratchets Engaged.

An improvement of the ratchet engagement of the New Champion consists of a locking device on the trip, by means of which when the trip is made the dumping hooks are held in positive engagements with the ratchets until the discharge is completed. On other rakes the hooks are held into position by contact merely and are liable to slip off the ratchets, especially when slightly worn.

One-Piece Dump, Rod Support and Reversible Ratchets

The wheels and ratchets on hay rakes have heretofore been made rights and lefts, and this has been a source of much annoyance both to dealers and farmers as well as to manufacturers. In shipping and in loading on to farmers' wagons it is no uncommon occurrence to find two rights or two lefts with a rake instead of one right and one left, causing annoyance and delay, and other additional expense. On the New Champion rake there are no rights or lefts,— any two wheels complete the rake. This construction also doubles the life of the ratchets, as they may be used on both sides. The ratchets are also removable from the hub.

For fine or short grass it is necessary to have the teeth on rakes closer together than ordinarily used. It has been the common practice of manufacturers to build their rakes with entirely different heads for this purpose. If a dealer or

Teeth so Fastened as to get full benefit of Coil Spring.

farmer purchased a rake with the ordinary spacing between the teeth, and found afterwards he needed a close teeth rake, he had to exchange the rake head as well as the tooth holders.

On the New Champion rake the same head is used on both styles. If a dealer or farmer wishes to change his rake from an ordinary spacing to a close spacing between the teeth, all he need do is to put on new tooth holders and the additional teeth. The old holders may be used again so that the only expense to make the change is the additional teeth. The holders are so constructed that the teeth bear upwardly against the steel rake head to prevent breakage of the holders.

In the features above described the New Champion is in advance of other rakes. In other respects it is built according to the best construction found on other rakes. The head is supported by a strong truss rod so that the teeth are always in proper position on the ground. The pole and thills are of the combination pattern. The axles are reversible, giving them double life. The hubs of the wheels are removable, so that it is not necessary to buy new wheels when the hubs are worn. The snub block, which causes the disengagement of the dump rods after the discharge is made, is adjustable to accommodate the height of lift to different conditions of crop, and to regulate the time of the return to suit the speed and size of the horse or team. The hand lever is placed to come to the driver's hand in bunching without bending his body in any direction.

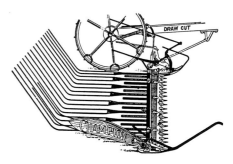

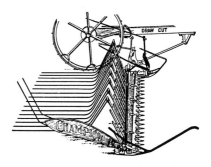

NEW CLOVER GATHERING ATTACHMENT.

The Warder, Bushnell & Glessner Co. is building a new attachment for its mower. It is a carrier and buncher for gathering clover which is cut for seed. It is certainly an ingenious device. The long upturned fingers carry along the clover as it is cut. There is no raking of the ground or gathering of trash and dirt. The clover is not rolled over and over. The heads stay up away from the ground so there is no shelling of the seed. The carrier is discharged by a foot lever which places the fingers in the position shown in the second picture. The bunch is carried off immediately by the stubble which projects between the fingers. A complete separation is made so there is no straggling swath left after the bunch.

The New Champion Self-Dump Hay Rake.

(Front View)

The best Hay Rake is not lowest in price.

The New Champion Hay Rake is the best.

Self-Dump and Hand-Dump; Made in 8, 9, 10 and 12-ft. Sizes.

(Rear View)

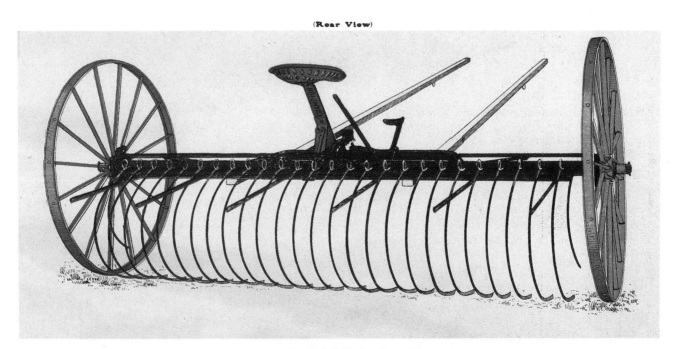

THE WARDER, BUSHNELL & GLESSNER CO.

THE STUDEBAKER EXCURSION.

On Sunday evening, January 22d, at 6:20 o'clock, a special excursion train of two Pullman sleepers left the Kansas City Union Depot over the Santa Fé route. Special trains and excursions have gone out from the same depot times almost without number, but this particular excursion differed from all others. The party constituting this excursion was made up of forty-four implement and vehicle dealers; J. S. Welch, manager of the Kansas City branch of the Studebaker Bros. Mfg. Co.; his assistant, S. B. Robertson; Chas. Woolverton, manager of the Springfield, Mo., branch; O. S. Webb and C. E. Miller, travelers; and a representative of the Implement Trade Journal—making a total of fifty in the party.

It was a free show. Mr. Welch and his assistants, ever anxious to please their customers, conceived the idea of giving an excursion to the factory at South Bend, Ind. The plan once decided on, preparations were begun to carry it out in a manner in keeping with the Studebaker motto of having nothing but the best. Sixty customers were invited, but sixteen were prevented from accepting on account of sickness or other hindrances which made it impossible for them to leave home. The latest improved Pullman cars were provided for the guests, everything was free; not a cent of expense was permitted on the part of those invited.

The excursion was attached to the regular Chicago train of the Santa Fé, and as soon as the train had started the guests were provided with tickets which entitled them to supper in the dining-cars, where they proceeded to satisfy their appetites, which had been sharpened by a tallyho ride over the hills of Kansas City for several hours before starting (also at the expense of the Studebaker Bros. Mfg. Co.). There were no restrictions, every man ordered according to his appetite. The evening was spent in getting better acquainted, and by bed-time everybody knew everybody else, and a jollier and at the same time more civil crowd never went on an excursion.

Soon after breakfast, Monday morning, Streator, Ill., was reached where the special cars were transferred to the I. I. & I. R. R., reaching South Bend at noon. A reception committee from the home office, headed by Mr. C. A. Carlisle, met the visitors with two special trolley cars, and they were at once taken to Johnson's dining hall, where dinner was prepared and where they ate their meals during their stay in South Bend. After dinner the guests formed in line and marched by twos to the office of the Studebaker Bros. Mfg. Co. After registering their names on the visitors' registers, each man passed into the private office of the Hon.

Clem. Studebaker, president of the company, and was given a formal introduction. To have an opportunity to shake hands and speak face to face with this venerable man, who has done so much for his town, State and country, and for humanity, whose name is linked with that of philanthropy, as well as progress, was a favor highly appreciated by every visitor.

A general reception was then held in the traveling men's room, after which the party was given a trip over the Chicago & South Bend R. R. This road is the property of the Studebaker Bros. Mfg. Co., and embraces ten miles of track in the lumber yards and among the factory buildings, and is connected with the railroads entering South Bend, thus enabling the company to promptly handle all cars in and out. The lumber yards of this company contain over 50,000,000 feet of lumber, and are said to be the largest private lumber yards in the world. From the lumber yards the guests entered the spring-making department. This department, including the stock-room, covers an area of 25,000 square feet. Much interest was manifested by the dealers, who gave close attention to the various processes of the manufacture of this very important part of a vehicle. Perhaps the most interesting feature in this department was the testing of completed springs. Every spring is placed in this testing machine and tested above the carrying capacity for which it is intended, so that it is known absolutely what weight they will carry without breaking.

From the spring department the guests were shown through the steel skein department, where the various operations in the manufacture of steel skeins were gone through. Here also are finished the cast truss skeins for which the Studebaker wagons are celebrated.

From the skein department the foundry was visited. A heat was just being taken off, which proved an interesting sight to those who had never witnessed this operation on a large scale. The welding and finishing department was next on the route, and the manner of welding and setting tires, boring out and setting boxes, and finishing wheels gave the dealers a good opportunity to see how the work was done. Welding by electricity was a novel sight to most of the visitors. It is employed largely in the Studebaker works. This department alone covers 32,000 square feet of surface, and has a capacity of 300 complete sets of wheels per day.

The next department visited was the blacksmith shop of the wagon department. A blacksmith shop containing an area of 120,000 square feet and employing 218 men is a sight not soon forgotten. The various departments were gone through from the wood-working machines used for cutting up the stock, through the bending rooms, body-making rooms, wheel rooms, to where the finishing

touches are put on. Dealers were enabled to see for themselves the grade of stock that enters into the construction of the Studebaker vehicles, and the manner in which it is put together and finished, and they were one and all satisfied that they could go home and say to their customers that they knew the materials used in the Studebaker are the best that money can buy.

Every department is in charge of competent men who thoroughly understand their business, and they took especial pains to enlighten the visitor in regard to anything about which they inquired.

The second day at the factory was devoted to inspecting the finished work in the repository. Every style of vehicle was found on the various floors of the repository, from a pony-cart to the heaviest mountain wagon, and from a one-man buggy to the most capacious coach. Among the novelties on exhibition in the repository was the famous aluminum wagon which was made by the Studebaker Company and exhibited at the World's Fair. With the exception of the tires, the metal parts of the wagon are all made of aluminum. The box is made of rosewood, inlaid with a border of holly, and 35 medals awarded to the company at various expositions since 1852. The total cost of this wagon, including the aggregate labor employed, which was 424½ days, was $2,110.68. Less handsome, but equally attractive, is the State carriage used by President Lincoln, which is now owned by the Studebaker Company. They also have on exhibition the carriage used by Gen. La Fayette when he visited the United States in 1824.

guests. The only regret expressed by anyone was that the only other surviving brother, J. M. Studebaker, was not at home. Hon. Clem. Studebaker presented to each visitor a handsome china souvenir on which was a picture of his residence, "Tippecanoe Place."

The "Western Yell," adopted by the crowd, was frequently rendered, and attracted a great deal of attention among the Hoosiers. It was as follows:

"Rock! Chalk! Jay Hawk!
We are the men who make the talk.
What about? The Studebaker; it's all right.
What 's all right? The Studebaker."

C. B. Wells, of Sedalia, left the train at a little station over in Illinois to chase a man who was driving a Studebaker wagon, and give him a cigar. He nearly missed the train by his daring. The crowd cheered so lustily as to almost cause the man's team to run away.

There was something of a rivalry between Seery, of Topeka, and Wallace, of Stillwater, as to which was the most popular with the dining-room girls.

H. T. Black gives promise of becoming a champion at tenpins.

Behind the Gun", "Steel Winged Beater"
and "Beating Shakers." Several farmers
near Wichita, Kansas, contracted with
threshers to separate their wheat first
by paying the premium price of 10 cents
per bushel. The usual pay had been 7
cents. The Wallis "Bear" tractor was announced and the Butler Manufacturing Co. began making steel tanks in Kansas
City, Missouri. The Moline Plow Co. contracted for the total production of the
D.C. Stover Co.—Henney Buggy Co.
from November 1, 1902, to July 1, 1903.
Later, on July 20, 1903, the Moline Plow
Co. purchased the Henney Buggy Co.
from the D.C. Stover Co. On April 24,
1902, the Minneapolis Steel and
Machinery Co. (manufacturer of the
Twin City tractor) was begun in Minneapolis, Minnesota, by a merger of:

Minnesota Malleable Iron Co.
Twin City Iron Works

The biggest news of the year waited
until near the close of 1902 when the International Harvester Co. of Chicago, Illinois, was formed by the merger of:

McCormick Harvesting Machine Co.
  of Chicago, Illinois.
Deering Harvester Co. of Chicago,
  Illinois.
Plano Mfg. Co. of Plano, Illinois.
Milwaukee Harvester Co. of
  Milwaukee, Wisconsin.
Wardner, Bushnell & Glessner (the
  Champion line) of Springfield,
  Illinois.

Competition between harvester
manufacturers was so fierce before the
merger that something had to be done.
William H. Jones, manager of the Plano
Works when it became part of the
merger, said that dividends had not been
paid "for ten or twelve years before
1902, and the only profit we made was
on goods sold in the foreign markets."

T.G. Mandt died on March 1, 1902, at
the age of fifty-six, Henry G. Olds died
on May 1, George Stephens died on July
12 and Hector Adams Holmes died on
October 26.

The Nichols & Shepard Co. introduced
the "Red River Special" thresher in
1902. The "Four Threshermen" were
features of Nichols & Shepard Co. and
included "The Big Cylinder", "Man

This photograph was taken in front of the store of O. Gossard of Oswego, Kansas, on delivery day shortly before the harvest of 1902 and is additional evidence of how one good, live dealer can stir up the people and get up a crowd. There was the brass band accompaniment, with a free lunch for the farmers and their families. It kept the people talking about Gossard and Milwaukee binders for weeks and was a good advertisement for both. No less than 78 binders were delivered on this day. Mr. Gossard was a Kingman-Moore Implement Co. carload customer and had already ordered several carloads of drills, wagons, haytools and vehicles this season.

PLANO HUSKERS BY THE TRAINLOAD.

Every day for the past month or more long trainloads of Plano huskers and shredders have left the great Plano works for different sections of the country, while the Plano factory has kept humming both night and day to supply the demand. There are several makes of huskers in the field this season, but probably none has attracted more attention by distinctly novel and at the same time practical features than the 10-roll Plano—"the only husker with a duplex husking belt"—a machine we have before briefly described in these columns. Evidence of this is shown in the fact that often during the state fairs recently held large numbers of threshermen and engine owners left the ground and crowded to the Plano agent's quarters to see this unique machine in operation. The practically perfect work it does on all occasions, seemingly regardless of conditions, compels universal admiration, and has resulted in a big shredder business for the Plano Division. In fact, the company writes that during fair week at Indianapolis, Ind., 42 orders for shredders were taken by the Plano representatives, six men being employed at one time writing orders; a record that has been seldom equalled in selling a machine of this size and cost. Every few days large delegations of farmers and shredder men have visited the Plano factory in Chicago to obtain full information regarding this machine and see it work. Here the sight of hundreds of these big machines in process of construction, the perfect system prevailing in their manufacture and the wonderful equipment of the works proves as interesting to visitors as it is novel. No one can witness the care and precision with which each part of the work is executed without feeling increasing confidence in the construction of modern farm machinery as now built by a concern of this size and equipment.

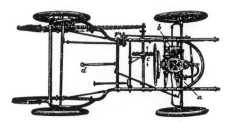

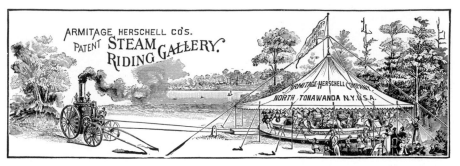

A GERMAN LEADS.

It may not be particularly pleasing to Americans to be told that a German has led them in an important invention, but the information furnished by Consul Monaghan of Cheminitz, Germany, is to the effect that one Ludwig Maurer, after thirteen years of practical experimentation, has produced an automobile that is simple, practical and durable. As shown in the illustration, the drive wheel, which is also the fly wheel, transfers its power directly to the friction wheel (c). The latter can be slid both ways upon the shaft by means of a forked clutch worked by the driver. On the left end of the shaft containing the friction wheel is a sprocket wheel, over which runs a chain which drives both back wheels.

It will be seen that the further the friction wheel is removed from the center of the fly wheel, the higher will be its speed, and the nearer it is moved toward the center the slower will be its speed. When the friction wheel is moved across the center of the drive wheel, the motion is reversed and the automobile runs backward. When the friction wheels runs close to the center, at the point of slowest motion, the automobile can climb ascents up to 30 percent, as determined by the gear arrangements. At the point of greatest speed, on the circumference of the drive wheel, a velocity of 30 to 60 kilometers (19 to 38 miles) per hour is attainable, varying with the power of the motor employed.

The contact between the drive wheel and the friction wheel is effected by moving the lever (e) and forcing the drive wheel backward against the friction wheel (c). Reversing this motion forces the drive wheel forward and away from the friction wheel. This simple mechanism obviates complicated structures, reduces friction to a minimum, and leaves few parts exposed to dust and grease.

Because of the possibility of attaining any speed desirable, the motor is used to its fullest extent, whether running on level or ascending ground, no matter how frequent the change in elevation. As a result, a four-horsepower machine equipped with the Maurer system is said to have a traveling capacity equivalent to that of other motors of six or eight horse power.

Another advantage consists in the ease with which the driver can get at all of the movable parts of the machine without besmearing himself in the act. The simple construction reduces the weight of the machine, decreases the wear on the pneumatic tires, and effects a pronounced economy in the consumption of gasoline, not only because of the lessened weight to be carried, but because of the gain in horse-power.

The simplicity of the machine enables a ready understanding of its parts and thus enlarges its usefulness for different persons, whether familiar with mechanics or not. When running, the operator is required to manipulate but one lever with each hand, thereby reducing the possibility of mistakes.

The machine, though readily manipulated by means of the lever (e) which causes the friction contact, can also be regulated by an appliance cutting off the supply of gas. The gas-feeding apparatus can be so operated as to permit of the running of the motor at a uniform speed of 400 revolutions per minute, the friction wheel being set at the circumference or point of greatest velocity. On uphill ground, the motor is supplied with more gas, thus increasing the power of the motor and carrying the machine up

the ascent without a change of speed, through the manipulation of the friction-wheel lever. Only on unusual inclines need speed adjustments be made by means of the friction wheel. In this way the wear upon the motor is reduced by virtue of its uniform speed, and the gas consumption is regulated by the changing requirements of the road traversed.

AN AUTOMOBILE LAWN MOWER.

An automobile lawn mower has been set to work on the greensward of Capitol Hill at Washington. The new machine resembles both a steam road roller and a steamboat whistle. A huge brass dome, surmounted with a brass smokestack is mounted on a pair of rollers. Ahead of the front roller is fixed a lawn mower mechanism, over which sits the chaffeur. The little machine puffs up the hills and swings around and goes down again, while the grass flies up in front like a green mountain. It cuts about three times the amount of grass that can be laid low by the old one-horse machine, which it superseded. The advantage which is claimed for the mower is that the roller passing over the ground after the grass has been cut, crushes down the weeds and kills them.

THE DEMPSTER PICNIC.

The Dempster Mill Mfg. Co.'s picnic at Beatrice, Neb., became an annual feature several years ago, and on the **second** occurred the fifth of these occasions which the employes of the company, and especially their families, look forward to with such pleasure. The picnic this year was pronounced the most enjoyable of all. The day was fair and the program an interesting one. An hour was spent in singing, recitations and music, and after dinner sports of all kinds, including races and ball games, were indulged in. Over eleven hundred people participated in the picnic, from President Dempster to the office boys. Much interest was aroused in the game contests between different departments of the shops. The picnic was held at the Chatauqua grounds.

THE GATLING PLOW AGAIN.

There is no question but that the time is rapidly approaching when the farmers will have a corner on the aristocracy of the country. A few more inventions only are necessary to provide the farmer with appliances and automatic machinery that will indeed make farming a pastime. The Gatling motor disc plow, for instance, which has recently been brought before the public again, will effectually solve the problem of plowing and seeding, of which the St. Louis Globe Democrat says:

One man, it is claimed, operating the Gatling plow can accomplish as much in a day as thirty or forty men using from sixty to eighty horses operating the old-fashioned plows. Dr. Gatling's invention is operated by steam. It will cost about $6 per day if coal, wood or oil be used for fuel, or $2 per day with gasoline. A comparative statement showing the cost of plowing by the acre will best show the advantages the owners of the Gatling plow will have over the farmers who adhere to old-fashioned methods. Such a statement shows three items in the cost of plowing at present:

1—Plowing of land by horse power (per acre).	$1 50
2—Rolling or pulverizing the soil	30
Harrowing	30
Total	$2 10
Plowing, rolling and harrowing with Gatling plow	50
Saving	1 60

A wheat drill may also be attached to the Gatling plow and the grain sown as the soil is turned, thus effecting a further saving of 40 cents per acre. Then the motor can be separated from the plow and used for all kinds of heavy hauling, now done by horse power, and which, it is estimated, costs the farmers of the United States about $900,000,000 annually. Several gentlemen are forming a syndicate for having some of Dr. Gatling's plows made and practically demonstrating what can be accomplished with them in farm work.

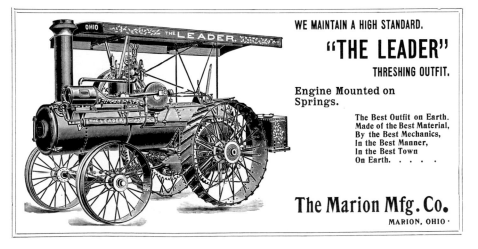

THE GLIDE MOBILE.

One of the attractions at the Chicago Automobile show was a new medium-weight gasoline machine of the runabout type made by The Bartholomew Co. of Peoria, Ill. It is called the "Glide Mobile," and is the product of the genius of Mr. J. B. Bartholomew, president of the company and also vice-president of the Avery Mfg. Co. A view of the machine is given herewith. It is propelled by a single cylinder four-cycle motor of five inches bore and six inches stroke. The cylinder and valve chest are water-cooled and the valve cams and levers are encased. The speed of the engine is controlled by throttling the charge by means of a foot lever and by adjusting the time of the start. The speed may be varied between the limits of 100 and 1,200 revolutions a minute. The water jacket around the cylinder is separate and is packed at the front with metallic packing. The cylinder hood is attached with six 9-16 inch bolts and the packing is said to give no trouble. The motor is started by means of a detachable starting crank inserted through an opening at the side of the frame. Ignition is effected by means of the jump spark, with current supplied by a dynamo, the latter being driven by friction off the flywheel. In addition to the dynamo a battery of four dry cells is carried and a switch permits the operator to use either the battery or dynamo as desired. The spark plug furnished with the machine is one insulated with mica. The carbureter is of the float-feed type, and is provided with a throttle valve in the air passage. While the vehicle is standing the vapor that arises from the jet mixes with the air so that the engine may be started without any preliminary operation of the carbureler. The transmission gear is of the sun and planet type and gives two speeds ahead and one reverse, all these changes being obtained with only nine gear wheels. The body is mounted on a spring frame independent of the engine frame, and may be easily removed without disturbing any of the machinery. The wheels are of the steel suspension type, 32 inches in diameter, and fitted with 3-inch pneumatic tires. The rear axle runs on four hardened steel roller bearings, one bearing near either end of the axle, and one on each side of the compensating gear. The front axle is provided with ball bearings. It is a lively, powerful, easily-managed runabout, one suitable for country roads as well as for city streets. The price of this machine is $750.

THE GLIDE RUNABOUT.

FAVORS NOISY AUTOMOBILES.

Chas. E. Duryea, of automobile fame, in an article in the Cycle and Automobile Trade Journal, talks in favor of motor vehicles that are noisy. He says:

"The oddity of the noise emanating from a motor vehicle at first attracted considerable attention and received adverse criticism, with the result that preference was universally given to those vehicles which were silent. Now that the automobile is more common a reverse sentiment seems probable, and some buyers express a preference for vehicles that make some noise. One experienced driver recently stated this preference, and added that he had tried both and found the silent vehicle kept him busy looking out for other people, whereas the noisy one gave warning of its approach. There is unquestionably much sound truth in this position. One of the arguments against motor vehicles has been the fact that they do not make as much noise as horses' hoofs and therefore do not automatically warn pedestrians, and such warning devices as lamps and signals have been required. No device of this kind, however, is constantly acting, as is the noise of the vehicle, and expert drivers do not object to the constant murmur of the exhaust, but on the contrary rather like to know what their vehicle is doing by the sound of the exhaust. It is therefore a question whether the silencing of automobiles should be carried to an extreme point. The writer's opinion is that the noise should be reduced to a point permitting conversation among the occupants of the carriage without undue effort, leaving a perceptible amount of noise which acts as a certain warning of the approach of the vehicle. It will readily be seen that the danger of the railroad trains and trolley cars and the like would be much greater if they made no noise whatever, and motor vehicles are no exception.

"Of course, such noises as clatter of mechanical parts and roaring of gears are objectionable, and these should be eliminated as far as possible, but the puff of the exhaust indicates a saving of power rather than a loss through an over small muffler, and against this slight noise of the exhaust no argument can be raised. The use of locomotives without mufflers of any kind would seem to bear out this position."

A NEW COTTON PICKER.

A Memphis paper reports that a machine that will pick cotton from the fields has been invented by G. H. Zempter of that city. The construction of the machine is very simple. It is mounted on high trucks and has nothing which will injure the standing stalks of the cotton. The high stalks which grow in the rich bottom lands easily pass under the machine. It is drawn by two mules that walk in the middle of the rows. The tongue between them is divided and has an arch between its two sections which allows the stalks to pass there without injury. Thus, it is safe to go into the fields with the machine when the cotton is only half open and gather the cotton which is ready without injuring that which remains. The machine, which is set upon a frame, is composed of a gasoline engine, a fan that creates the suction, pipes through which the cotton is gathered, a cleaner through which it is passed and a final receptacle for the picked cotton. None of the parts is complicated. Five men are required to operate the machine, though four can do the work. It is stated by Mr. Zempter that the cost of the gas for use in the engine will not exceed 50 cents per day. The machine is exhibited in a warehouse in Memphis.

GASOLINE ENGINES.

It has been estimated, says a writer in the American Miller, that about a hundred new makes of gasoline engines are put on the market each year. Many of these are all right and, especially if backed by men of successful business qualifications and with sufficient capital, remain on the market and eventually become known as among the standard makes. Others of these new engines are good enough in themselves, but fail to stay on the market because of improper backing as respects both the bank account and the business activity of their promoters. Still others of these engines fail because they cannot be built in competition with standard engines and still leave any profit. In other words, they are too expensive to make. Still another reason for failure of some of these engines is that inherent defects, either in design or construction, are present. Perhaps the igniter is of imperfect construction. Possibly the valves are defective. These and many other things, well known to gas engine men, cause many an experimental engine to be taken off the market. As a usual thing these engines do not stay on the market more than a year. In some cases the defects are eventually corrected.

But that the gasoline engine has come to stay there is plenty of evidence on every hand. In 1900 over 18,000 internal combustion engines were built, their aggregate horse-power being 164,662, and their value $5,000,000. Great improvements are being made in these motors every year, and they are bound to grow in importance as power producers.

ABOUT MOTOR VEHICLES

NEW MODELS OF GASOLINE AUTOMOBILES.

The two-passenger runabout, 1902 model, shown by the Haynes-Apperson Co. at the Chicagoe xposition, will embody the latest improvements incorporated in their machines. These include direct gearing, water circulation by means of a radiator and pump, a new design of steel wheel rims of greater strength than used in earlier machines, and improvements in the carbureter, clutch and a new pump feed·lubricator. The motor is a double-cylinder engine with cylinders arranged horizontally on opposite sides of the shaft—an arrangement that gets rid of troublesome vibration. The sparking device is of the make and break positive contact type, which the company claim is not affected by wet weather and muddy roads. The particular model shown weighs 1,250 pounds; the motor is of 6 horse power, wheels are 32 inches in diameter, and the carriage is handsomely finished with leather upholstery. The machine has three speeds forward and one reversed, all controlled by a single lever. The Haynes-Apperson people entered two machines in the New York and Buffalo endurance contest, and they finished second and third out of a field of 89 that started. All of the Haynes-Apperson carriages are fitted with wood wheels, which the company claim to have found equal to the most exacting requirements of the road. A feature on which much emphasis is laid is the fact that they are larger in diameter than those customarily in use, and that consequently there is considerably less jar in traveling over rough roads than there would be with smaller wheels.

The Gasmobile embodies all the features of the best French machines, and the motor and machinery are so situated that they can easily be inspected.

The 25 h. p. four-cylinder engine of the Gasmobile stanhope, vertically disposed within the framework of the car, produces, when in motion, but a slightly perceptible vibration. The system of lubrication is most thoroughly carried out. Besides a small quantity of oil in the crank-case, automatic oiling devices are also provided. The water circulation is established by a rotary force pump. Both the inlet and exhaust valves are thoroughly water-jacketed. The Gasmobile vaporizer is highly efficient, simple in construction and never-failing. Its throttling feature, together with the quickly adjustable timing of the sparking device, make it possible to vary the speed of the motor over a range more than ample to secure the greatest variation in speeds desirable. Gasmobile engines are readily started notwithstanding their high compression. They are not, strictly speaking, high-speed motors. They are of slightly greater weight and larger proportions than other so-called high-speed motors, and therefore while quite as powerful as the latter, are more durable, since they do not run so fast.

The expanding friction clutch, which transmits the power from the motor to the driving-gear, ranks high among devices employed for this purpose. It is positive and substantial and will take hold of the clutch casing attached to the flywheel gradually, thus causing the car to start smoothly and without jerks.

At the right of the chauffeur three levers, forged from tough steel and operating on a double notched sector, serve for starting, stopping and reversing, and changing from the first to the fourth speed, as well as for operating the main brake.

The Gasmobile stanhope has a 71-inch wheel base and the regular standard tread of 56½ inches. All four wheels are 32 inches in diameter, and fitted with clincher tires. Three forward speeds and one reverse are furnished, and the 12 horse power motor will drive the machine as high as 30 miles an hour.

The Fischer Motor Vehicle Co. have for the past five years been perfecting a combination system, which possesses all the good qualities of

THE HAYNES-APPERSON GASOLINE RUNABOUT.

the electric and gasoline combined, while the disadvantages inherent in each alone are practically eliminated by the combination.

The system consists of a combined gasoline engine and dynamo, one motor for each (rear) drive wheel, and small storage battery and controller. It will be noticed that there is no mechanical connection between the engine shaft and vehicle drive wheels, therefore the dynamo is free to run at a practically even speed, producing a constant supply of electricity. The electric circuit is so arranged that when running the vehicle under normal conditions (loaded on the level) the current goes directly from the dynamo (through the controller) to the motors; but when coasting down grade, slowing up or in general when less power is needed than that furnished by the engine, the current is automatically taken up by the battery, which is connected to the wiring at a point between the dynamo and controller. Again, when extra power is needed, as in ascending steep grades or starting heavy loads, the battery promptly furnishes the deficiency. This action—the carrying the peak of the load—does not have to be watched by the operator, being entirely automatic. As the output of the engine does not vary, no governor is required, and the gas and air mixtures can be set permanently for perfect combustion. This insures great saving in fuel and prevents the usual bad odor. As the speed of the engine is almost constant, the balance is nearly perfect, thus preventing vibrations. Another very important and convenient feature is the starting of the engine, which is accomplished by simply throwing in a switch controlled by the driver.

Fig. 1 shows an 18-passenger omnibus recently completed. Fig. 2 is a photograph of one of the standard running gears, which consists of an angle steel underframe to which various parts of the machine are attached. The front portion carries the gasoline engine, dynamo, controller and steering gear. The front axle, instead of being made of the usual heavy forging, is trussed somewhat on the principle of a bridge, and carries extra long and flexible platform springs, bolted to the brackets on the frame. The rear axle with wheels, springs, two motors and reduction gears forms a complete driving unit. All parts subject to wear are entirely incased so as to be properly lubricated and at the same time protected against dust and moisture. No reach is used, the half-elliptic springs conveying the power from the rear axle to the frame. The water cooler is suspended under the frame be-

THE GASMOBILE STANHOPE.

tween the front and rear axle. The illustration, Fig. 2, shows the running gear complete in every respect, and Fig. 3 the body in place, making a finished bus.

The power equipment consists of a 10 horse power, 3-cylinder, 4-cycle gasoline engine running 600 revolutions per minute. Directly on the engine shaft is placed the armature of a 5-kilowatt 110-volt dynamo. The motors are of special "twin" type (built together), 5 horse power each, and will stand an overload of 100 per cent for a half hour, or 200 per cent for 5 minutes without sparking at brushing. The controller is of the series parallel type of five forward speeds of from 2½ to 10 miles per hour, and two reverse speeds of 2½ and 5 miles per hour, all controlled from one lever. The batteries consist of 50 cells of 90 ampere hours' capacity. The front wheels are 38 inches in diameter and the rear 46, all equipped with 4-inch Calumet solid rubber tires.

The company has orders for a number of heavy trucks, some of them being of the beer truck type, built to haul seven tons.

THE FISCHER GASOLINE-ELECTRIC BUS.

THE FISCHER BUS WITH BODY REMOVED.

The Peerless motor car has a long wheel base and low center of gravity, with motor situated

THE PEERLESS GASOLINE TONNEAU

in front. The frame is built of channel iron after the style of a locomotive, thus carrying out the idea that the motor car is a road locomotive rather than a horseless carriage. Both front and rear wheels are pitched inward, after the manner so long in use with all vehicles not run on prepared tracks. The pitch of the rear wheels is made possible by the flexible driving axle, which also obviates all loss of power by excessive friction when strains of the road tend to throw the rear wheels out of alignment.

The motor consists of two vertical 4x4⅜ inch cylinders, with cranks inclosed in tight aluminum cases and running in oil, an arrangement which automatically lubricates the cylinders and all bearings. Owing to the vertical position of the cylinder it is thoroughly lubricated, since the piston rings wipe uniformly the entire circumference, and thus prevent any oil getting into the firing chamber, which does away with obnoxious odors and keeps the spark plugs clean. Both grease cups and force feed lubricators are used throughout the machine, and all are situated on the dashboard before the driver. The mufflers used produce very little back pressure and yet almost eliminate the noise of the exhaust.

Ignition is by the jump spark system, the make and break of the circuit being accomplished by means of a mechanically operated vibrator of unique and entirely original design, which requires no adjustment for months. Heavy insulated cable is used in all the wiring.

A circulation of water through all the engine jackets is obtained by means of a centrifugal pump operated by a friction disk against the flywheel. The water is pumped through a very effective system of radiating coils at the front of the car, and only two or three gallons are used.

An atomizing float feed carbureter of improved design, requiring absolutely no adjustment to the varying speeds of the motor, is used to furnish gas for the latter. The motor is started by a half turn of the crank, which is placed at the front of the car. The speed gear is connected with the driving gear by a flexible shaft and with the motor by a universal coupling, which protects the bearings, gears and clutch from any strain due to an inequality of the road. The gears are inclosed in an aluminum case and run in an oil bath which automatically lubricates all bearings. The speed changes are obtained by means of a single lever at the right, which gives three speeds forward and one reverse, while the speed of the motor is regulated by varying the time of the spark.

The changes of speed are made by friction clutches that go in without clatter or vibration, and the gears operate without noise. A powerful band brake on each rear wheel is operated by a lever at the right and held by a ratchet until released. A foot brake operates on a drum on the change gear shaft between the motor and the compensating gear.

All two-passenger cars have a rear platform which may be used for luggage, rumble seat, or two-passenger tonneau. The driver's seat is either double or divided into individual seats. The cars are geared to make 30 miles an hour at a speed of 1,200 R. P. M. of the engine, but are capable of being speeded up to 40 miles an hour. They are equipped with two kerosene side-lights and a very powerful acetylene headlight, or with two side-lights and two acetylene headlights having a combined power about equal to the single headlight which is offered as an option. The mudguards are of aluminum with front guards flared out, protecting both occupants and the car from mud when the wheels are at an angle.

Comments on the Harvester Merger.

From O. V. Eckert, president Iowa Implement Dealers' Association, Northwood, Ia.: "We consider it now too early to form any definite opinion of the new binder organization, as we have yet very little information in regard to the plans of the company for the distribution of its products. It is evident that the movement is made in the interests of economy, primarily for the manufacturers, as self-interest is the first law of nature. This of itself should not militate against the best interests of the retail dealer, unless arbitrary and unyielding rules prevail to interfere with our ideas and methods of business, which is quite unreasonable to expect. We are in hopes to arrange for a complete illumination of this subject before the annual convention of the Iowa Implement and Vehicle Dealers' Association, to be held in Des Moines December 2, 3 and 4. To bring about such a result, we have made an urgent request of the International Harvester Co. to have a representative there who will give full and complete details in regard to its methods of work and its expectations from the dealers who will be associated with it in the distribution of goods, and are glad to note that our request will receive favorable consideration by the company. As this has now become a question of general and vital interest, we trust that it will excite new inspiration on the part of all dealers in the state to attend the next meeting, and prompt those who have not already joined our organization to send the membership fee of $2 at once to Secretary D. M. Grove, Nevada, Ia., who will be pleased to return a certificate of membership which will entitle the holder to all the privileges of the next convention. By doing this much profitable information may be easily and quickly obtained, not only upon the question referred to, but upon many others of equal importance that otherwise could be had only by slow and tedious effort.

"We find by sad experience that striving alone without the expressed experience and observation of others in the same line of business, that the best part of our lives are spent in discovering glaring and fatal mistakes in our methods of business and judgment, which could have been avoided had we have heard and grasped the more practical ideas of the eminently successful men in our trade. These men, we have good reason to believe, can be met in convention and by proper effort be constrained to give up some of their business secrets. They are approachable and have their eyes and ears open to see and hear the wants of their fellows and always have an encouraging word for them.

"Permit me to close this somewhat lengthy epistle by an urgent appeal to all retail dealers of the state to promptly resolve that they will make the next convention glad because of their presence and manifest interest in its deliberations. The program is intentionally made flexible so that time may be given for full consideration of what will seem to be the most important matters at the time of our meeting.

"I thank you for this opportunity and hope you will assist us in arousing a great interest in the convention."

From the Nation Hardware Co., Fairland, I. T.: "Replying to your favor of the 3rd, would say, there is no doubt in our minds that the combination of harvesting machine companies can be a help to the dealers and the trade in general, but the question is, will it? We are agreed that in some hands, not all, competition is the life of trade. Will this so-called binder trust hold the prices of machines down regardless of the rise and fall of manufactured products, or is it only a sly way of getting control of the business in order to raise the prices to its heart's content? The combination gives it the power to do as it sees fit. We are in favor of putting such a price on a machine that our neighbor can not afford to cut that price. It is very discouraging to know that you have only the name of selling a machine and no profit out of the deal. We understand there is to be a fixed price on machines under the new management, which we certainly hope is true and heartily endorse."

From the Bethany Hardware Co., Bethany, Mo.: "It is a hard matter to say whether the merging of the harvesting machine companies will be a benefit or a detriment to the trade, as it is not yet known what course the combine will pursue. If it does what it says it will, it can not possibly harm the dealer, but our experience so far is that trusts are a soulless set of fellows, organized for the express purpose of getting more for their goods than they are worth. All we can do is to wait and see what it will do."

From J. H. Hill, Russell, Kan.: "In regard to merger of harvesting machine companies, I believe it will be a benefit to the retailer, provided they quit the commission business and sell direct to the retailer. Let him buy them the same as he does wagons, plows, etc., and it will shut out these crossroads agents who have no money to buy with, but take any agency and turn over farmers' notes in settlement for what they sell, while the company carries over all unsold machines. These are the fellows we dealers want to shut out. They make no money themselves and we must meet their prices and do business for almost no profit at all. I believe that as a result of the merger prices will be higher and the farmer will have to pay more for his machinery, but it should do away with some of the evils in the harvesting machine trade. If not, it will not be long before the companies will have to establish their own agencies and sell direct to consumers, as dealers are tired of the way it now is, and many have quit that line of business and do not handle harvesting machines at all. I think one of the principal topics for discussion at our next convention should be this harvester question, and the dealers certainly realize that fact by this time. Another one is credits and collections. This might be discussed through the columns of the journal. The credit business is certainly growing and dealers are obliged to sell on longer time than a few years ago. Farmers seem to prefer buying on time, even at a higher price, than for cash. Possibly this is not so all over, but in certain localities it is especially so in farm machinery.

From S. W. Engler, Clay Center, Kan.: "Regarding the recent merger of the harvesting machine companies, I firmly believe it will be a blessing to each and every business man, and particularly the implement men, for, in my judgment, the present methods employed by machine companies are demoralizing and detrimental to the best interests of all concerned. Simply because any 2x4 crossroads agent can take an agency, the machine company will furnish lightning-rod peddlers, or worse, to assist them in doing "dirty work," so that a legitimate dealer is compelled to do likewise or get little or no business. The expense incurred by these peddlers, paid by the company, is not less than $12 per harvester and $8 to $10 to the agent; hence, if the International Harvester Co. will give the dealer one-half the amount paid to peddlers and retain the other one-half, we will all be better off. It is true, the dealers may not sell quite so many harvesters, but they will get pay for just as many as they do now; besides, the expense of collectors and commissions to "jackleg" lawyers will be done away with. There is no reason on earth why harvesting machines cannot and should not be sold the same as other goods, and no reason why farmers should not buy them the same as other goods. The sooner this is accomplished the sooner the harvesting machine business will be on a business basis, and the foundation of nearly every evil attached to the implement and vehicle business destroyed. A dealer who is not competent to handle a binder without the aid of a so-called expert should not be classed as a dealer. These views are based on facts by one who has served 17 years in the retail and wholesale binder business."

From John Q. Adams, Stockton, Kan.: "Replying to your favor of Sept. 3, which has reference to the International Harvester Co., would say that in my judgment this can not help but work out good for the retail implement dealer. I can not think of any policy that they may adopt that would make the business any worse than it now is, and they can certainly get rid of a great many other errors, which would be to their advantage as well as the dealer. I hope to see this new company adopt the policy of selling its goods outright to the dealer, compelling him to settle for all goods sold. This would get rid of the canvasser and the farmers' paper deal, which is the source from which most all of the trouble comes.

"I think that the principal topic for discussion at the next dealers' convention would be for dealers to get right themselves. The retail implement dealer should get down to business principles and recognize that every one engaged in the trade is entitled to his share and that no dealer can do business unless he can make a profit on the goods which he handles. If the retail implement dealers will get right themselves, it will not only kill off other errors, but will set an example that will have an influence on the jobbers and the manufacturers, and we would have no use for conventions and associations except to get together and have a good time. The things we complain of nearly all accrue from the unbusiness like manner in which the retail implement dealer has conducted his business."

From E. R. Moses Mercantile Co., Great Bend, Kan.: "The merging of the five harvesting machine companies into one will be of more benefit to them than any one else. I sometimes think that they merged to relieve themselves of the great burdens which afflict all companies who want to do a large business and find that the expenses are eating them up. It is a wonder they did not get all of the companies into this. Possibly they were satisfied to form a trust of five companies and relieve themselves of the burdens and let go if they found that it did not pay them. We do not think they have yet formulated any plans as to how they will run the business, and, of course, do not know what their future operations will be, but we think they will get things somewhat mixed up before they get down to a systematic and harmonious basis of running it. I sometimes think that we should discuss how best we could combine our forces in buying goods, so that we could get them at the very lowest prices and meet competition from any source whatever. In fact, if all the institutions enter combines, we will have to begin this business ourselves, so as to protect ourselves from the encroachments and ravages that are being made on our business."

From the Arcadia Hardware and Lumber Co., Arcadia, Kan.: "In regard to the recent merger of the five leading harvester companies, will say that we are opposed to any and all so-called trusts or mergers. The companies, to a certain extent, lose their individuality, and after such a consolidation the "big" company becomes very arbitrary; at least such has been the case up to date. There is no such thing as concession when they are all one, and a great many times a dealer is entitled to such. Of course, on the other hand, it will have a tendency to stop price-cutting, and it may be a help to the dealer, providing they don't cut out all the commissions. This is a hard subject to discuss, as we don't know yet what policy Bro. Morgan will pursue. We swell up and say we are like Russell Sage, and believe that all these combinations will work to the harm of the country."

From George Boensel, Chandlerville, Ill.: "I think the formation of the International Harvester

Co. will be beneficial to the dealers in some ways and detrimental in others. If the manufacturers will do away with the so-called agents handling their goods in blacksmith shops and livery stables, at every crossroads, and place them in the hands of reliable dealers, I think it will do away with the present evil of cutting and slashing prices the dealers have to contend with, and place the manufacturers in position to have uniform prices and get more money for their goods. The principal topic for discussion at the next dealers' conventions should be the forming of some plan, advantageous to both the manufacturer of harvesting machines and the dealer, whereby the profits would justify the dealer in pushing the trade."

From Jenkins & Wells, Sedalia, Mo.: "In regard to the merger of the harvester companies, would say, we believe it will be beneficial to the retail dealer if managed as combinations of like nature in other lines are. However, if it is the intention of the combine to still keep the canvassers out, with no more restrictions in the matter of price to the farmer than have prevailed in the past, there will be no benefits derived by the dealers. We are waiting with some curiosity and, we might say, anxiety, to see what methods will be adopted by the combine in handling the trade next spring, and trust some move will be made to eradicate some of the evils existing in the past which have made it unprofitable for the best trade to handle binders."

From E. I. Holley, Erie, Kan.: "The recent combination of five of the leading harvester companies will, in my opinion, work against the interests of both the farmer and dealer. It may be that this change will do away with the companies' canvassers, which would be of great benefit to the dealer, but there are several reasons, which I will not now take time and space to give, why this gigantic combination of capital will, in the end, work more for its own benefit than for the interests of the dealer or farmer.

"One of the worst things the dealer has to contend with is the catalogue houses and this should be the principal topic for discussion at the next dealers' convention."

From M. J. Irvin, Fulton, Kan.: "As to my opinion on the harvesting machine companies that have formed a trust, it is too early yet to know just what it will amount to, as no one seems to know just what they are going to do. I think that before we talk too much we had better wait and get acquainted with their scheme, see their contracts, etc. We can then form an opinion as to whether the trust is to our advantage or not. I have an idea, though, that when we see the contracts they will all be in favor of the harvester companies with the dealer left on the outside."

From L. J. Bowers, David City, Neb.: "I have no way of telling by what method the binders will be put on the market, but if it is any worse than in the past, all the best dealers in the country will stop handling binders. My opinion is that conditions will be better for the local dealer and no worse for the farmer, and will gradually grow better for both. I have confidence in the ability and fairness of the manufacturers to do right."

From George Messelheiser, Alexander, Ia.: "I think no harm will result from the recent merger of the leading harvesting machine companies, as it will do away with the several thousand local canvassers that bored the dealers nearly to death. There should be a large saving to the companies, as they can economize in many ways. The competition hereafter will be among the local dealers, and I think it will be better all around."

From H. M. Stewart, Medford, O. T.: "Discuss plans for dealers to eradicate canvassing evil. This is the worst evil I have to contend with at present. I believe the merger to be a good thing if it will insure a uniform price and confine agencies to legitimate dealers."

From Henry Lubker, Columbus, Neb.: "It is hard at this time to form an opinion of what effect the merger of the five leading harvesting machine companies will have on the retail implement trade, as they have not outlined their policy clearly enough for anyone to form an opinion; but it seems to me that they now have the opportunity to shape the machine trade so that it will be of a great benefit to the retail machine dealer, and I believe that the local agents can look forward to better conditions in this line of their business than have heretofore existed for many years.

"I think that the principal topic for discussion at dealers' meetings should be the harmful competition between dealers. It seems to be a hard lesson for dealers to learn that each can secure no large amount of gain by reducing the profit on what they sell, in order to prevent their competitors from making sales. There is at this time no one firm that can have very much of an advantage over its competitors, and if it has it ought to use such an advantage to earn larger profits, rather than try to destroy its competitor.

From J. C. Foster, Knob Noster, Mo.: "I hardly feel qualified to discourse on the abstract questions pertaining to the general business world. My horizon of observation is necessarily limited, and such views as I have are therefore likely to be based upon ex parte investigations. From a cursory examination of the situation, as it pertains to my own business, I am at a loss to see how the retail dealer can suffer either loss or inconvenience as a result of the combine. It ought to be possible, with the resultant saving in expenses, for the combine to place machines on the market at greatly reduced prices, which, in turn, ought to stimulate buyers and increase sales. I can discover no reason why the dealer should not obtain the same profit under the present regime as under the old dispensation. I have never been much imbued with this trust opprobrium. I do not recall any instance where a trust has injured me. I am led to believe that the high-price nightmare is a pipe dream. Most of the articles controlled by trusts are cheaper than before they became subject to trust domination. I presume that all dealers have their troubles; I know I have mine; but I find that hard work is an excellent remedy. If dealers will go hard after business, my experience teaches me that they generally get it."

The International Harvester Co., formed the previous year, purchased the D.M. Osborne & Co. of Auburn, New York, in 1902. The Mayer Bros. Co. was incorporated in Mankato, Minnesota, to build gasoline and steam engines, boilers and other machinery. The Minneapolis Steel & Machinery Co. began building Corliss steam engines and the New Holland Co. began in New Holland, Pennsylvania. The usefulness of the disc plow continued to invite arguments from both sides.

A 1903 article explained the experimental use of the automobile for plowing, then for mowing. A walking plow was tied to the rear of an automobile when horses couldn't be found and a furrow was necessary to stop a grass fire near a railroad. The speed necessary for the auto quickly exhausted the man who had to walk with the plow and the plow had a tendency to skim the ground, but the results were encouraging. The men made the comment that "it would be necessary to gear down the machine to a slower rate of speed." Experimental mowing was soon tried, attaching a horsedrawn mowing machine to the auto proved to be more successful. An acre and a half of grass was mowed in one hour when, it was noted, a horse would have required three hours.

S. H. VELIE, JR.'S RIVERSIDE CART

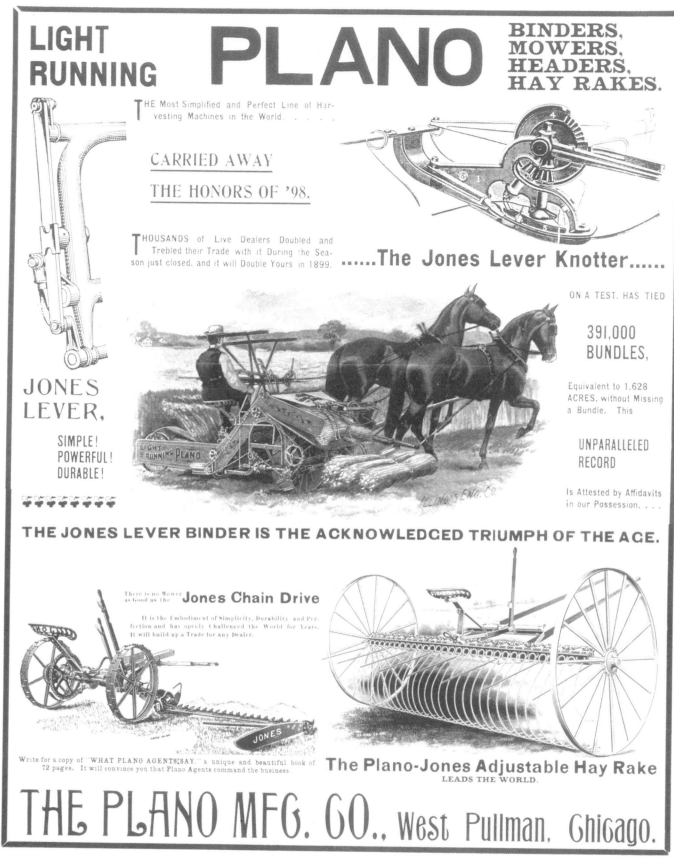

LIGHT RUNNING PLANO

BINDERS, MOWERS, HEADERS, HAY RAKES.

THE Most Simplified and Perfect Line of Harvesting Machines in the World. . . . .

CARRIED AWAY

THE HONORS OF '98.

THOUSANDS of Live Dealers Doubled and Trebled their Trade with it During the Season just closed, and it will Double Yours in 1899.

......The Jones Lever Knotter......

ON A TEST. HAS TIED

391,000 BUNDLES,

Equivalent to 1.628 ACRES. without Missing a Bundle. This

UNPARALLELED RECORD

Is Attested by Affidavits in our Possession. . . .

JONES LEVER,

SIMPLE! POWERFUL! DURABLE!

LIGHT RUNNING PLANO

ILLINOIS ENG. CO.

THE JONES LEVER BINDER IS THE ACKNOWLEDGED TRIUMPH OF THE AGE.

There is no Mower as Good as the **Jones Chain Drive**

It is the Embodiment of Simplicity, Durability and Perfection and has openly Challenged the World for Years. It will build up a Trade for any Dealer.

JONES

Write for a copy of "WHAT PLANO AGENTS SAY." a unique and beautiful book of 72 pages. It will convince you that Plano Agents command the business.

The Plano-Jones Adjustable Hay Rake
LEADS THE WORLD.

THE PLANO MFG. CO., West Pullman, Chicago.

In 1903, Andrew Carnegie collected $1 million each month from U.S. Steel as payment on a note which he held. His wife was paid $83,333 as similar payment. It was noted that the total stock of the Pennsylvania Railroad was worth about $55 million more than the huge U.S. Steel Corp.

On April 21, 1903, patent number 725,860 was granted to S.S. Morton of York, Pennsylvania, for the traction truck. The Ohio Manufacturing Co. of Upper Sandusky, Ohio, was organized to manufacture this "Morton Traction Truck" or "Ohio Patented Traction Truck." The company folded by about 1923, but formed the beginnings of many great tractor companies by providing a chassis on which to install a normally stationary engine.

IKE other inventions which have changed customary methods and introduced new principles, the disc plow was in the minds of inventors and had received much attention, as shown by the files of the Patent Office, before it attained commercial success.

The first tree fork drawn through the soil introduced the sliding wedge action and all modern mould-board plows are but modifications and refinements of that principle.

The disc plow brings to bear a different action, using a rotary or revolving wedge in turning the soil, and in certain sections has now been in use a sufficient time to demonstrate its utility and advantages. The older patents which are here illustrated and mentioned are but a few of the most notable ones on file, and are chosen for the reason that they seem to best show the state of the art, and to represent the efforts of the most practical inventors—men whom it is clear have applied themselves closely, and who have very probably failed only for the reason that they were not in localities where the action of the disc would show to the best advantage in comparison with the mould-board.

Like most useful modern inventions in the implement line in America, the history of disc plow development is linked closely with the West. One of the first patents for disc plows was granted to M. A. & I. M. Cravath, of Bloomington, Ill., (No. 66,802) and this plow was the first to handle the soil with discs which produced a plowing action. While crude compared with modern plows, as shown by the cut, yet it will plow when constructed exactly as shown in the patents.

Following the Cravaths comes a long list of inventors, many of whom devoted much time and great energy to the development of this type of plow, and it is to be noted that the methods employed were in most respects similar to the devices used on the most modern successful tool. Two, three and four carrying wheels were used, both straight and staggered. The caster wheel

No. 66,802.—M. A. AND I. M. CRAVATH, July 16, 1867

was arranged to run in the furrow behind the last disc to guide the plow, and was given a sharp edge to enable it to hold its position more firmly. Various forms of scrapers were used, both mould-board and revolving. The front carrying wheels were pivoted and attached to a pole, and it is evident that several of the inventors experimented with discs set at different angles in order to properly penetrate and handle the soil. Several of them made workable plows and took out many patents, but all failed to arrive at the proper relations and sizes of the various parts, and so missed perfection and commercial success.

Among the most interesting of these inventions are those of J. K. Underwood, Cedar Rapids, Ia., (No. 169,499) who took out several patents and evidently studied the tool closely and experimented much for several years. Some five

or six patents for different styles of disc plows were granted to him. As shown by reference to the cut from one of his patents, he was first to arrive at a three-wheel construction with a pivoted caster wheel behind the plow, which caster wheel was arranged to control the plow and hold it in line with the furrow by a stop limiting

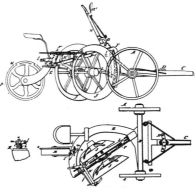

No. 177,668.—J. K. UNDERWOOD, May 23, 1876.

its action, and to increase its holding power he gave the caster wheel a sharp edge. It is evident that Mr. Underwood made a plow which worked, to an extent, and which was closely along the correct lines. In his first patents he used very large discs cut away in center, but a subsequent patent shows the use of much smaller discs (No. 177,668).

It would seem that many of the first inventors were hampered by being unable to get solid discs struck from one plate of steel, and were forced to build up a suitable disc from parts and plates, and several patents were granted for discs so constructed. Underwood, in his later patents (No. 233,455), attached the beam of the plow to a two-wheel truck which was controlled by a pole and also arranged to tilt the caster wheel by suitable levers to raise the plow for transportation and to adjust it to its work.

D. H. Lane, of Anoka, Minn., (No. 208,246) also received several patents for disc plows, and his work is notable in that he placed a wheel in

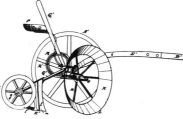

No. 208,246.—D. H. LANE, Sept. 24, 1878.

the furrow behind the disc to control the action of the plow, and also arranged a knife behind the disc to cut under the wall of the furrow and so destroy any ridge left by the disc and making the turning of the soil easier for the disc on the following round.

J. Austin, Chicago, Ill., (No. 244,367) was granted several patents and constructed a number of different styles of plows. In order to overcome the sidewise pressure of the discs, he resorted to staggering the carrying wheels of the plow (No. 291,127), which is the first instance shown of the use of this principle on disc plows. Evidently Mr. Austin used his plows in the field sufficiently to disclose weakness in the bearing of the disc, for he subsequently shows them mounted on a shaft running through to the frame of the disc next ahead, making a very long and durable bearing.

S. S. Gardiner (No. 285,809) also received several patents, and all of his plows show evidence of careful work and experience in the field. Mr. Gardiner's efforts are notable in that he is first to claim an adjustable disc whereby the penetration of the disc in the soil could be varied by adjusting the angle.

S. C. Baucum, Waco, Tex., (No. 257,914) devoted some time to working on a disc plow, but stopped short of commercial success.

The first disc plow to be placed on the market by any of the well-known plow companies was manufactured under the Kirk patents by the Weir Plow Co., of Monmouth, Ill. It contained none of the principles of a disc plow, but simply substituted a disc for the mould-board in the endeavor to make a tool which would scour in Texas black land. The landside and share of the ordinary mould-board sulky plow were combined with a disc mould-board rotated by a chain driven by the landside wheel of the plow. But

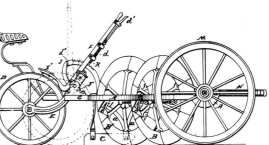

No. 244,367.—J. AUSTIN, July 19, 1881.

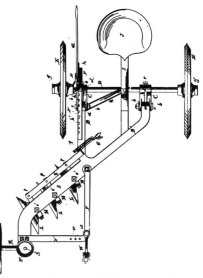

No. 285,809.—S. S. GARDNER, Oct. 2, 1883.

few of these plows were placed on the market, and it did not meet with the success expected.

The foregoing patents have been mentioned in about their acknowledged order and simply to show the various good methods and devices which are suggested by them. In none of them, however, does there appear that proper arrangement and balancing of parts which are now known to be necessary in a successful tool. A study of modern disc plows shows the following broad resemblances:

The furrow wheels are staggered and usually have edges adapted to prevent their sliding sidewise, and are generally about 24 inches in diameter. A caster wheel, to enable the plow to turn at corners, is located in the furrow behind the disc and provided with a stop to limit its action and enable it better to resist the side pressure of the discs and keep the plow in line. So, with the landside wheel, a three-wheel construction is formed which is but slightly different from the ordinary sulky plow frame which seems to be the best type of frame for riding plows. All use a 24 or 25-inch disc and seem to have concluded that this size gives best results.

These general resemblances are seen in all, and it is safe to conclude that these points are typical and necessary to the successful plow, but it is not until the patents granted M. T. Hancock appear that this proper arrangement begins to be indicated. In his plow first appears a medium-size solid steel disc mounted on substantial bearings and secured to a frame constructed with carrying wheels and necessary levers for operating, and furrow and caster wheels properly disposed; in fact all the parts combined practically as is now shown in the most modern plow. Rights were disposed of under these patents, and in certain sections plows were sold which were constructed practically as shown in his patents, and it is fair to say that his invention contained the necessary combinations on which to construct a successful plow, although these combinations were strongly foreshadowed by former inventions.

Mr. Hancock's plow, as constructed by the Texas Disc Plow Co., after certain modifications became very popular in Texas, owing to the difficult soil in that state, and is still manufactured by that concern and largely sold. This company was the first to demonstrate that a single disc could be so set as to turn a full 12 to 14-inch furrow, as claimed in other patents controlled by them. Patent No. 556,972 (C. A. Hardy, March 24, 1896,) is the single disc plow, first sold by the Texas Disc Plow Co., and it has been a pronounced commercial success. The success of this plow in Texas soon attracted the attention of other companies, and many plows were brought out in response to the demand, and from that time it has been a well-recognized fact that the disc plow would become a very considerable factor in the plow trade, and that the territory in which it would prove popular was not confined solely to the "black waxy" of Texas, but included all that portion of the West and South where the soil became too hard after harvest to allow the mouldboard plow to work well, while the disc seemed especially adapted to those conditions, pulverizing hard soil well in place of breaking it into large clods.

Those best acquainted with the work of the disc plow claim that dry ground plowed with it requires far less harrowing than with other plows. For several years it was asserted that the only reason the disc plow penetrated hard soil was because of extreme weight. While the principle of its construction and work requires more weight in the frame, to give the necessary strength, than a mould-board plow needs, yet a disc has certain angles which if found enable it to enter the soil easily. This is best shown by the action of a disc harrow, which, when set straight, will not penetrate or work, but which will immediately cut deeply into the soil and throw small furrows when set by the levers to take the ground. Manufacturers have learned this, and the modern disc plow closely approximates the

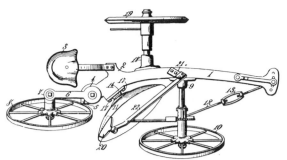

No. 556,972.—C. A. HARDY, March 24, 1896.

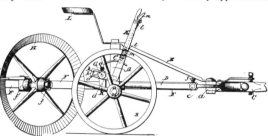

No. 257,914.—S. C. BAUCUM, May 16, 1882.

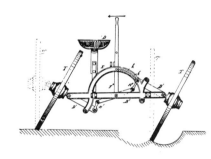

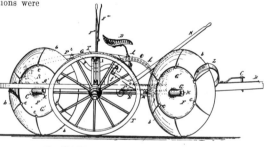

No. 291,127.—J. AUSTIN, Jan. 1, 1884.

weight of the mould-board sulky plow. It is claimed that this saving of weight can only be accomplished by a careful placing of the disc, regulation of the draft and proper location of the furrow and land wheels.

Its enemies have claimed that the disc plow has excessive draft, but careful dynamometer tests show marked advantage for the disc plow per cubic inch of dirt turned, and that advantage is especially noticeable in very hard, dry soil. Careful investigation and test also shows that the bottom of the furrow is practically level, no ridge being left between furrows when the plow is properly operated and set to plow the proper depth. It does not perform so satisfactorily in shallow plowing as when set to plow six inches deep or more.

As a sod plow the manufacturers unite in making no strong claims for the disc, but users agree that its work in certain soil on clover and even prairie sod is good and satisfactory.

In its simpler forms there is really nothing to do but hitch on to a disc plow and drive ahead, there being but one place to hitch and but one lever to operate, and in this particular form it would seem destined to find a wonderful trade in those vast plains of Siberia and South America where the laborer demands a simple tool which will work without adjustment and which will continue to work until worn out without much attention. The field known to be favorable to the disc plow is growing larger each year and more attention is now given it than ever, and while few new principles are brought forth, yet more attractive forms of construction are continually being tested and each season finds several new plows on the market. Each has its particular advantages, but in general it will be seen that they have already commenced to approach that similarity of construction which now marks the leading mould-board plows. The broad reason for this is that in every mechanical device for performing a given action there are certain vital principles which must be contained in every successful machine, and consequently, in a term of years, the constant tendency of manufacturers is toward a type which contains these successful principles and which eliminates all useless construction, thus creating a distinct resemblance.

J.S. Gibson's Present Place of Business.

How To Run Disc Plows.

By C. A. HARDY.

AS the trade on disc plows develops naturally an increasing number of the dealers and their men become acquainted with the proper methods of operating them. However, a few words regarding their limitations and possibilities may assist those who have found conditions which seemed beyond the ability of the plow to meet.

A disc plow will work in its own way and cannot be forced to do more than the underlying principles inherent in the tool will permit. If it is not handled under each condition in the proper manner, some fault will develop which would not occur if the tool were adjusted for that particular condition. In the first place, it is evident, from the fact that the disc is a round, concave furrow-opener, that it will not exercise the same turning action on the soil if run shallow as though run deeper. Right here many have found a stumbling block in the operating of disc plows. They have tried to run the plow cutting a furrow twelve to fourteen inches wide and less than five inches deep with each disc. A little reasoning along this line will show that the wider a disc of any certain diameter is asked to cut and turn, the deeper it will of necessity have to run. Many makers of disc plows are careful to state that the disc will not do good work unless run from five to six inches deep. Generally these makers set their discs to cut ten or twelve inches wide. If run as directed they will do good work, but if the attempt is made to run shallower, then troubles arise, such as lack of capacity to cover and cut trash, leaving the bottom of the furrow irregular and in a rounding shape, and with ridges. The same plow set to cut deeper will generally obviate most of these troubles.

As a general proposition, the shallower it is wished to plow the narrower should be the furrow, and the wider the furrow the deeper it becomes necessary to run the plow. Increasing the size of the discs does not remedy these troubles, owing to the fact that at any given distance from the edge of a disc there is so little difference in the amount of disc covered by the soil that increasing the disc within any reasonable size gives no more width to act on the soil, as it is only the bottom six inches or so of the disc which is cutting at any time, and then only that portion of it next to the land. Then, as is well known, the larger the discs the less penetration they have. Dealers have all learned this from the disc harrow business, as it is now a settled fact that a sixteen-inch disc harrow will penetrate better than a twenty-inch. The same principle applies to disc plows.

The varying conditions of soil and the varying conditions to be met in the field are almost innumerable. Frequently several different conditions are met in the same field, and these conditions also change from time to time during the plowing season, each requiring certain changes in the handling and adjustment of the disc plow in order to do the best possible work.

Frequently a field which will plow perfectly when dry and hard with a disc plow, will fail to handle or the plow will refuse to work well after a light shower. This is especially true of soils known as "gumbo." When dry the plow works nicely, but as soon as wet the rear wheels lift out and the plow draws sidewise and refuses to stay to its work. Under these conditions the narrowing of the cut for each disc generally makes the plow work well, and frequently it is necessary to reduce the cut to eight inches to the disc.

In stubble fields which have become "soddy" from the growth of "foxtail" or "crab grass," it is frequently found impossible to get the disc plow to cut through the heavy growth; the wheels lift and ragged work is the result. This condition requires more judgment in handling than any other condition the plow is asked to meet. In the first place, the plow must be run deep enough to go below the sod, and get loose dirt enough to properly cover the grass and sod, and allow the discs to get a proper turning action on the soil before it begins to break, so that the sod will not be turned part one side up and part the other. If running the plow deep does not properly effect this, then the cut of each disc must be reduced until the proper quality of work is obtained.

Naturally, where there is a large amount of such growth on the stubble, it is more difficult for the disc to cut through it, and the disc frequently shows a disposition to stop and only pushes along. When running in this manner it will not cut at all. In such instances, setting the plow to run about two inches deeper will generally cause the disc to cut clean and turn well, for as soon as the disc is run deeper the added friction of the soil on its face gives far more power to its cutting edge as it keeps revolving and gives the drawing cut so necessary to a clean cutting of trash.

Some complaint is generally made that at corners it is difficult to keep a good even depth. This is easily obviated if the operator will drop the plow about two inches just before reaching the corner, and after turning raising it again to the old level. It is also well to run slowly at the corner and not allow the plow to be pulled forward when turning. Good work can also be done by making a round corner and swinging around without allowing the team to stop.

In order to produce nice, even work through the entire field, it is absolutely necessary to so regulate the plow that both discs are cutting the same width of furrow, since if one is cutting ten inches and another thirteen, the entire field will be thrown in ridges, and if there is any loose trash it will not be nearly so nicely covered as when all the discs are so regulated that they are cutting, not only the same depth, but the same width.

Disc bearings should be carefully lubricated with a heavy oil or axle grease, and the scraper should be so adjusted that it does not bear too heavily on the disc. If one scraper bears heavily and one lightly, then a difference in the throw of dirt from the discs will be seen; the one which is held back by the scraper not showing so neat a turn as the free one. The same condition arises if one bearing is well lubricated and one is running nearly dry.

To sum up the points:

Do not try to cut a wide furrow without giving the plow depth enough to properly cover the trash and stubble and leave the bottom of the furrow level.

If the plow labors and tries to dodge out of the furrow, run it deeper; and if the same trouble shows then, the width of cut for each disc should be reduced.

Plow deeper in soddy ground than in clean stubble, if you wish to cut and cover the trash cleanly.

Adjust your scrapers carefully, and keep the disc bearings well lubricated.

Adjust the plow carefully, so that both discs are cutting the same width of furrow and the same depth, if you want even plowing.

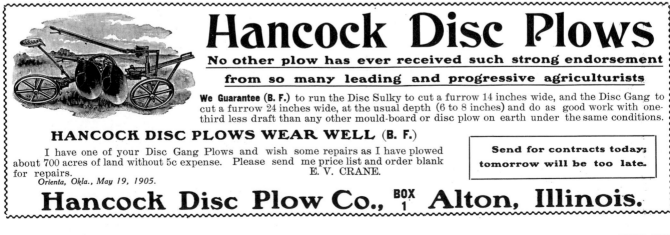
1896-1907

REX BEAUTIES

ARE YOU WITH US?

REX BUGGY CO.

CONNERSVILLE, IND.

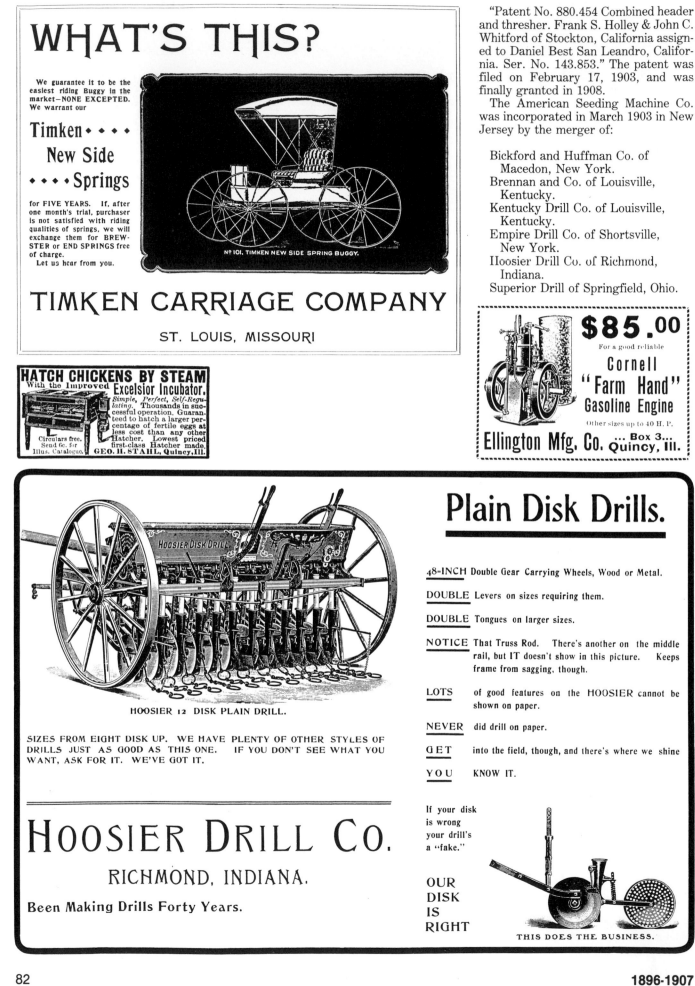
"Patent No. 880.454 Combined header and thresher. Frank S. Holley & John C. Whitford of Stockton, California assigned to Daniel Best San Leandro, California. Ser. No. 143.853." The patent was filed on February 17, 1903, and was finally granted in 1908.

The American Seeding Machine Co. was incorporated in March 1903 in New Jersey by the merger of:

Bickford and Huffman Co. of Macedon, New York.
Brennan and Co. of Louisville, Kentucky.
Kentucky Drill Co. of Louisville, Kentucky.
Empire Drill Co. of Shortsville, New York.
Hoosier Drill Co. of Richmond, Indiana.
Superior Drill of Springfield, Ohio.

TO THE IMPLEMENT, HARD-
WARE AND BUGGY TRADE.

Cadillac at Factory
with Single Tube
Tires **$850**

GENTLEMEN:—We want to call your attention to the Cadillac Automobile as a means of facilitating your business and increasing your income. Very few persons who have not had experience with the Cadillac in a business way have any idea of what a great seller it is.

In fact Western dealers have but a vague idea of how fast the Automobile is coming upon them. Some of them say: "Why, you couldn't sell an Automobile in our town," and the very next day two or three machines come to town, having been ordered direct from the factory by some of the dealers' best customers, much to their surprise and chagrin. The way to get into the Automobile business is TO GET INTO IT.

Buy a sample machine, run it yourself, and you will find customers for Automobiles where you least suspected it. No other line offers such quick returns on the investment. No other line will grow big so fast. There is a regular cyclone of Automobile enthusiasm spreading over the country. The bicycle craze was slow and weakly in comparison with it. But Automobiles have come to stay, and the business will be permanent. Now is the time to act. Start at once, and start RIGHT, by securing the Cadillac agency. Only one man in each town or territory can have it. Be that man.

The Cadillac is the only successful high class runabout and light touring car in the world. It is built by the Leland & Falconer Co., of Detroit, Mich., a concern whose reputation for fine mechanical work is second to none in the United States. For forty years they have been the largest builders of fine tool-makers' machinery in America. For five years they have been preparing for the production of the Cadillac, during which time thousands of dollars have been invested in fine special tools, jigs, templates, gauges, etc. New buildings were erected and power plants enlarged. When they had perfected the Cadillac and the tools for making it, the first lot of 3000 machines were brought through—three thousand Automobiles at one run, each part like the other; each part interchangeable, one with another; each part true to the gauge of one thousandth part of an inch. No other Automobile in the world is built in this way. No other Automobile has the advantage of forty years' fine shop practice behind it. No other Automobile is built by such a high class of skilled mechanics. No other Automobile has the one great advantage of being built in such a factory.

Is it any wonder the Cadillac is equal, if not superior, to any Automobile that sells for twice as much money?

The Cadillac, as a runabout with big detachable double tube tires, sells in Kansas City for $815.00, and as a touring car for four or five passengers for $915.00, and it takes the hills and grades of this city better than any car built in America costing less than $3,500.00; and it will wear longer than the $3,500 car.

Is it any wonder it is popular? Isn't this the kind of a machine to have the agency for? Isn't it a pleasure to sell only the best, and to know that your competitor can't find a machine in the world that is even a fair comparison to your own line?

Then get the Cadillac agency, and get it now. Our new branch agency just opened for Kansas City and the West will take care of your requirements. Catalogues and further information for the asking. Let us hear from you. Write today. Don't lay this paper down and forget it—do it now.

Yours for business,

Cadillac with
Single Tube Tires
at Factory . . . **$750**

Cadillac Automobile
Company

Care of Richter Bros. 17th and McGee Streets,

Kansas City, - - - - - - Missouri

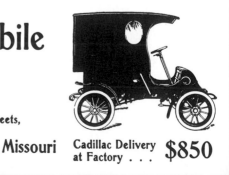

Cadillac Delivery
at Factory . . . **$850**

What Irrigation is Doing for the West.

NO subject that engages the minds of Western people today is of so much importance as that of irrigation; compared with it, the free silver agitation and the opening of the Kiowa country to settlement were questions of far less moment to the inhabitants of the western half of our national domain. While the silver question affected, in one sense, the nation, it was simply a political issue which, settled in either way, meant nothing in the nature of providing homes for the homeless. The opening of the Indian reservations to settlement meant more, but comparatively few of the many who entered the race for a home there were successful. Although as old as the Pharoahs, irrigation in the United States is comparatively new. The movement, which was started by the passage at the last session of Congress of the National Irrigation bill, means, ultimately, homes for many times the number who were successful in the Government's grand Indian land lottery a few months since.

As was stated in National Irrigation a short time prior to the action of Congress on the irrigation question, "the key-note to the national irrigation movement is *home-building*—the creation of millions of prosperous and happy homes in places in the land which are now waste and desolate, the creation of a dense population in the West, which will be a source of permanent and constantly increasing prosperity to the commercial and manufacturing interests of the country. What the West wants is school-yards full of merry children as well as corrals full of fat cattle and sheep. It wants homes, small farms, towns, villages, cities, rather than great ranches and a few lonesome sheepherders and cowboys. It wants roads rather than trails, fenced fields rather than fence riders, and county fairs rather than yearly roundups."

Thanks to Congress and our progressive President, the movement that shall culminate in this happy state of affairs is well under way, and the fund already available for the purpose of building reservoirs amounts to several millions of dollars. What this irrigation movement means to the West, and to the country at large, can only be appreciated by those who have seen the semi-arid districts in their natural state and the transformation wrought in the portions of those same districts where water has been systematically and scientifically applied. To those who live where the rainfall is more than sufficient for the crops, even making cultivation at times impossible, it is difficult to imagine vast areas parched and brown the greater part of the year, and when such an one has an opportunity to make a trans-continental journey and is compelled to ride for hours through a land barren of trees and green fields; where prairie dogs, owls and rattlesnakes seem to be the only inhabitants, and buffalo grass, cactus and sage-brush the only vegetation, he is ready to say: "This land is worthless"; and until he has seen the magic influence of water on the same kind of soil, he is ready to admit that it is not worth an attempt at reclamation.

While, as stated, irrigation in the United States is comparatively new, it is no longer an experiment. In certain portions of the western country irrigation has been carried on for years, and most successfully. It is true that millions of dollars have been invested in irrigation schemes that have never returned a dollar

for the investors. This, however, was because of lack of knowledge, of poor judgment, of attempting to cut too large a garment from a limited supply of material. There is in California, for instance, a dry canal which cost thousands of dollars, and it is today dry because the work was not carried far enough. It was planned to irrigate about 156,000 acres, and is estimated to have cost $750,000. Land owners became frightened and fought the project, and the canal was involved in litigation from the beginning. After $575,000 had been expended the work was stopped, and unless the canal is completed this expenditure will be a total loss. In speaking of this instance, Elwood Mead, who was the expert in charge of irrigation investigations for the U. S. Department of Agriculture in California in 1901, says: "If, instead of this immense canal, work had been begun in a small way with a ditch large enough

AGRICULTURAL SCENES in CALIFORNIA.

LIVE STOCK NEAR WOODLAND.

HARVESTER AND THRESHER, SAN JOAQUIN VALLEY.

HOME OF GENERAL BIDWELL, LEADER OF FIRST OVERLAND PARTY OF EMIGRANTS TO CALIFORNIA.

OSTRICH FARM NEAR LOS ANGELES.

FARM SCENE, NORTHERN CALIFORNIA.

CORNFIELD IN SACRAMENTO VALLEY.

to water 5,000 acres, and its extension deferred until the profits of irrigation had been demonstrated, the central district (where the canal is located) would, in my opinion, have been a success."

It has been demonstrated that irrigation is a success in California. "In order to realize what irrigation has accomplished in California," says Prof. Mead, "one must go to the southern part of the state, where land, not worth $5 an acre in its original condition, has sold, when irrigated and planted to orange trees, for $1,700 an acre; where valleys, which were originally deserts of sand and cactus, producing nothing more valuable than stunted grass, and where a whole township would not keep a settler and his family from starving to death if compelled to cultivate it in its natural state, have been transformed into the highest priced and most productive agricultural lands in this country; where water, which formerly ran unused to the ocean, is worth for irrigation alone ten cents per 1,000 gallons; where $3.50 an inch, and forty cents an inch extra for its carriage, was paid last year (1900) for a twenty-four hours' flow."

To show the advantages that will accrue to the country with small, irrigated farms over large farms or ranches irrigated, nothing more forcible could be cited than a portion of Prof. Mead's report of investigations in California in which he tells of a trip through a portion of the Sacramento valley, covering a distance of about thirty-five miles, between the towns of Chico and Willows. He says, in part: "In the suburbs of both Chico and Willows there were seen attractive homes surrounded by orchards and gardens, but five miles would cover the distance required to get beyond the town limits of either place. In the remaining thirty miles only six houses were passed, and surrounding these were neither orchards nor gardens. The distressing effects of a two-months' drouth and the absence of water to mitigate its influence were only too manifest. These homes were a signal illustration of the truth that a world without turf is a dreary desert. Instead of the refreshing green of an irrigated district, or of a country where there are summer rains, everything was parched, dusty and lifeless. Practically all of the land was being prepared for small grain. Less than one hundred acres of alfalfa were seen and this looked as though it was prepared to surrender the unequal struggle. One-third of the land had been summer fallowed, but much of it was in no condition to be benefited, as the clods had never been pulverized and the fertility of the soil was being burned out by the heat and dryness of the summer sun. The region visited was one of bonanza farms, the road traversed crossing one of 40,000 acres. A mortgage was the most important re-

sult of wheat growing in recent years, and the land is now being sold to satisfy the debts thus created. The boundaries of other large estates were pointed out whose owners were historic figures in the early days of California. In nearly every instance their estates have passed out of their hands and out of the possession of their descendants, and are now owned by banks or capitalists in San Francisco, having been taken in payment of loans made to meet the losses incurred in growing small grain. Nor has the change of ownership affected the general results. The present owners of these properties will not rent them to tenants who cannot give a satisfactory bank reference, experience having shown this to be a needed precaution. Although equipped with teams and machinery and understanding the California climate and California agriculture, many of these tenants have at the end of the year walked out of the valley, leaving both crop and equipment to pay the debts created by their failure.

In the thirty-five miles traversed there were only two school-houses. Attending these schools was only one child whose parents owned the land on which they lived. The other pupils were the children of foremen and tenants. The county superintendent told me that at these two schools there were only fifteen children. These conditions of alien landlordism, tenant farming, unoccupied homes and scanty population, in a country so rich in possibilities, show a vital economic defect in methods. The situation here was in such striking contrast with what had been seen in traveling through an irrigated valley in Utah the month before that the difference seems worthy of statement. In a distance of fifteen miles along Cottonwood creek, Utah, there was not a farm of over thirty acres. The houses and barns on these little farms indicated more comfort and thrift than those of the Sacramento valley, where the farms are ten times as large. The average population of the Utah district was over three hundred people to the square mile; the district traversed in California has less than ten people to the square mile. The Utah lands range in value from $50 to $150 an acre, the lands of the Glenn estate in the Sacramento valley are being offered for sale for from $10 to $40 an acre. Every natural advantage is in favor of California, but the Utah district is irrigated, the other is not."

In much of California, however, irrigation is employed, not because it is necessary to the production of a crop, but because it is expedient; experience having shown that crops are benefited by irrigation during a dry period. It is being employed with beneficial results in the central West and in the Eastern states. There are millions of acres of land in Western Kansas and Nebraska and Eastern Colorado, as well as

in Utah, Nevada, Wyoming, Arizona and New Mexico, where it is impossible to raise crops without the aid of irrigation. It is in these states that the work of the Government will be undertaken; the work that will eventually transform what is now little more than a barren desert into thousands of small, well-tilled farms, each one occupied by, not a tenant, but the actual owner of the land.

Enough has already been done in Western Kansas and Eastern Colorado to demonstrate that water is all that is needed to make of that wilderness a veritable garden. There are irrigated farms along the Arkansas valley that cannot be bought today for $200 an acre, the land of which is no better than thousands of acres lying a few miles distant from the irrigating canals that go begging at one dollar an acre. In that region may be seen land producing the most luxuriant vegetation, alfalfa making four crops a year, and the most delicious fruits and melons grown anywhere in the United States, because of irrigation, while across the road may be land through which no irrigating ditches have been run and which produce scarcely enough vegetation to hide the rattlesnakes that infest it.

The only thing necessary to make the vast arid plains of the West valuable farming lands is the creation of huge storage reservoirs to catch the surplus water in times of floods and freshets and preserve it against the time it is needed. The natural streams of the West do not carry enough water in the dry seasons to supply the irrigating canals, but enough water goes to waste each year between the Rocky Mountains and the rain belt to thoroughly water every acre of arid land in that territory if it was saved and utilized for that purpose. This is the work that the National Government has begun and which it is hoped will be pushed as rapidly as the funds available will permit. Then, indeed, "the desert shall rejoice and blossom as the rose; it shall blossom abundantly and parched ground shall become a pool, and the rejoice even with joy and singing; and the thirsty land springs of water." Isaiah did not say that such a condition would bring about an increased demand for agricultural implements and vehicles, but what the irrigating of the now arid lands means to this branch of business can scarcely be comprehended. This, in turn, will mean more activity in mines and factories, increasing the markets for the products of the soil, making an endless chain of benefits in which all classes shall participate. No single movement ever inaugurated in the West can compare in value to the people with the irrigation movement, which is now to receive a strong impetus as the result of recent Congressional action.

THIS PHOTOGRAPH REPRESENTS A SCENE ON THE VAN NUYS RANCH, NEAR LOS ANGELES, CAL.
Thirty thousand acres are cultivated each year in wheat and barley, while twenty thousand on the hillsides are used for grazing purposes. It requires eighty eight-mule teams to plow in this crop and harvest same each year.
Sixty Hancock six-disc Gangs have been ordered to take the place of the eighty Stockton Gangs, which will put in the crop in the same limited time. Mr. Van Nuys is owner of the Van Nuys Hotel, which is considered the best on the Pacific Coast by Eastern tourists.

IRRIGATION CANAL.

BREAKING LAND.

PUMPING STATION.

DRILLING.

HARVESTING.

THRESHING.

RICE MILL.

BALING STRAW.

MARKETING — BAY CITY.

RICE FARMING FROM START TO FINISH.—Farm and Ranch, Dallas.

The Rise of Rice.

IT is small wonder that there is a heavy immigration to the United States from foreign countries. Poets may sing of Italian skies and orators may paint beautiful word pictures of the grandeur of old world scenery, but nowhere in the wide world can be found a country so well adapted to the habitation of mankind. From developments of the past few years it would seem that man was only beginning to learn of the resources of our country. From an experiment a few years ago, rice growing in Louisiana and Texas has developed into an industry that promises great things for the future.

The Burton D. Hurd Co. of Kansas City recently issued a very handsome booklet on the subject of rice culture and to this company we are indebted for the illustrations produced herewith. As stated in the booklet referred to, few people of the United States stop to think of rice as a domestic product further than to inquire the price when they see a barrel of it on display at their grocer's, yet the most remarkable phenomenon of industrial development in the United States at the present time is presented in the already extensive and rapidly increasing cultivation of this cereal, which is the recognized standard foodstuff of three-fourths of the world's millions. The reason for this apparent lack of interest is probably due to the fact that the area in which it has been demonstrated that rice can be profitably grown in the United States is not large—that is, not large as compared with the great wheat and corn growing areas of the Central West and Northwest—yet when values are considered, the comparison is more favorable.

Rice has been raised in the United States for a hundred years in a primitive fashion, but the great development of the industry did not come until after modern methods of cultivation were introduced, which change began to be effected less than ten years ago. At that time the greater portion of the rice raised in the United States was "Providence rice"—that is, it was rice raised without artificial watering and was wholly dependent upon Providence for weather and rainfall suitable to mature the crop. In this primitive method it was necessary to utilize the lowlands for rice culture in order to retain the season's rainfall, and from this fact the general impression seems to have gone out that rice can only be cultivated in swamps or on swampy land. Such is far from the case, as we shall see later.

As is well known, every available foot of land in India, China and Japan is devoted to the cultivation of rice, a large part of which is consumed at home, though there is a large export trade from those countries. Cuba and Porto Rico, with their tropical climate and soil, can only produce about one-fourth, and the Philippine Islands about three-fourths of the amount they consume. These islands now being a part of the United States come in for attention from domestic rice growers and to them are shipped annually a large amount of our poorer grades. Reports for 1901 show that the exportation of rice from the United States to Cuba and Porto Rico during October was about 240,000 bbls.; during November, 300,000 bbls., and a varying amount each month of the entire year, which amount ranges from 100,000 to 300,000 bbls. To offset this exportation, which, as stated, is of the poorer grades, the United States imports head rice or fancy grades for home consumption.

As to home consumption. The rate of increase in the consumption of rice in the United States is approximately 20,000,000 lbs. per year, and while the production is increasing at a very rapid rate, the domestic growers can only produce about 52 percent of the total amount consumed. From this fact it is readily seen that with the demand so far in excess of the production, there is practically no danger of there ever being a material drop in the price or a lessening of the profits derived from its cultivation. The truth of this statement is borne out by two facts. First, the available area adapted to rice cultivation in the United States, with the exception of a few very small tracts on the Atlantic coast, is confined to a strip of territory bordering on the gulf coast of Louisiana and Texas, about 25 miles wide and not to exceed 300 miles in length, and comprising a total area of about 4,000,000 acres, which is designated as the Great Southern Rice Belt. After deducting from this acreage all swampy, marshy, timber and waste land, we have an available acreage for rice culture of not to exceed 2,500,000 acres. Of this acreage not more than one-half is susceptible to profitable development because of a lack of sufficient water supply; thus we have 1,250,000 acres of actual rice land of which it is estimated there is now under cultivation and in process of development, approximately 850,000 acres. This statement should be further explained by saying that of the 850,000 acres, nearly 400,000 acres are actually cultivated, while the remainder, 450,000 acres, has been made available for cultivation by the construction of rice canals, and a large portion of it, it is expected, will be put in rice within a year. Second, the comparatively low cost per acre for production and the growing demand for and the rapidly multiplying uses to which rice and its by-products are being put, have a tendency to not only uphold but to advance prices.

One fact must not be lost sight of before getting too far away from the question of available acreage, and it is this: There are in the Southern rice belt thousands upon thousands of acres in which the soil is suitable and the climate right, yet the land cannot be cultivated in rice because the main condition, the principal factor is missing, namely, the water supply or means of getting to it, and again because the land is rough or "bumpy," cannot be flooded evenly or cannot be properly drained. All these things are of vast importance and must be considered in selecting a rice farm. This lack of proper conditions cause an immense reduction in the acreage adapted to the industry as already spoken of.

From a few "Providence rice" growers of a decade ago, the industry has grown to a point where there is now invested the vast sum of $25,000,000 in lands, canals, mills, machinery and other equipment. The aggregate mileage of the canals in operations is not far from 1,500 and the number of rice farmers is probably 30,000, and their combined product amounts annually to, approximately, 4,000,000 bbls. of 162 lbs. each, and requiring 15,000 cars to transport it to market. Four million barrels, or 648,000,000 lbs., represents, at a very low estimate, $12,000,000 in cash for the crop, or practically 50 percent on the total capital invested. So much for the industry in general; now for some of the details, for every great business is made up of details.

Rice is planted, harvested and cared for in practically the same manner as the cereals of the North, the main difference being in the preparation of the farm by the construction of levees and drainage ditches. To do this levels are taken on every foot of the farm by a civil engineer, and all that part of the farm, be it large or small, that is of equal level, or not varying more than six inches, is enclosed by a levee, which is made by plowing four furrows together and then pushing and tamping them into a compact bank or ridge. These levees follow the contour of the land and run irregularly across the farm, always following the line of the level. When the levees are once made they will do duty for two or three years with very little repairing, and no further surveys need be made. When, after a few years, the elements have beaten them down to too low a level, they are again plowed up and reconstructed. The levees being constructed or put in shape for the season, the next step is the preparation of the ground to receive the seed. The ground is plowed, harrowed and often "floated" or rolled to get it into prime condition and the seed rice is sown by means of a press seed drill in exactly the same manner as oats or wheat are planted by the Northern farmer.

The plowing may be done, and frequently is done, late in the fall, in November, December, January, February, and up to planting time; the planting months being April, May and June; the early plowing is the best, as it produces an early crop and an early crop usually commands just a little better price on the market than does the later crop. Seed time over, the farmer has no further trouble or labor on the crop save the labor necessary to keep his levees in good shape, stop up leaks and regulate the water supply, for water is the necessary thing and to be without water means to be without rice, and the season's labor is lost. Under the old "Providence" method, complete losses were not uncommon, because of drouths, though now they are very rare.

When the rice has been sprouted and begins to paint the field a delicate green, the farmer turns on the water and begins to flood the growing grain, at first lightly, and as it grows the flood is increased until when the rice begins to "stool" or throw out its spur roots, it is completely flooded and by the system of levees just described the deepest water on any tract enclosed by levees can only be six inches, and when the water reaches that depth at the levee, the highest point in the enclosure is also covered, thus forming a series of artificial lakes, each one of which is on a six-inch lower level than its neighbor, the steps grading downward from the canal or source of water supply to the remotest corner of the field.

Assuming that the water supply has been abundant and the farmer attentive to his duties, then in from 70 to 90 days from the planting, the rice kernel is perfectly formed and has only to ripen to be ready for the harvest. Now the water is drained off the farm, the source of water supply shut off, and in from ten days to two weeks the farmer goes in with his "binder," for he uses a harvesting machine very similar to those used in the wheatfields, and cuts and shocks his crop. In this semi-tropical country, the evaporation is so rapid that from the time the water is drained off the field until the harvest begins, the ground becomes very dry and firm and no difficulty is experienced because of the "late flood."

Next the threshing, which is done from the shock (very little of the grain being stacked) by a threshing machine especially adapted to handling rice. From the thresher the grain is hauled to the farm warehouse where it is tested, graded and bought by the rice mill men, and the farmer is ready to begin work for the next season.

THE "GUS" ENGINES.

The Carl Anderson Co., of Chicago, seems to be fully alive to the possibilities of the gasoline engine and is making them in all sizes and styles, adapted to all classes of work. The "Gus" engine, as it calls its product, possesses many good features, among which is a quick opening and closing of the exhaust valve, which is accomplished by the use of a special cam, keyed to the crank shaft, in connection with a simple, although very ingenious arrangement for locking the valve every alternate revolution. This does away with all gears, eccentrics, etc., and apparently makes a very simple engine. As the opening of the exhaust valve is regulated entirely by the amount of explosions in the clyinder, it will not open until such explosions occur, and then opens instantly, closing just as soon as cylinder is cleared.

This engine will start on either cycle, regardless of where it may have stopped, and in starting it is not necessary to pay any attention to this matter whatever. The manufacturers claim that, in taking this engine apart and putting same together again, there is no possible chance for getting same out of time.

Gasoline is taken direct from tank by means of a pump, which is operated by the same cam that operates the exhaust valve, and is pumped into a small reservoir directly underneath the inlet valve, from which all surplus flows back into tank, through an extra large overflow pipe. A small pipe leads directly to air pipe, where the gasoline mixes with the inrushing air, which is drawn through the inlet valve into explosion chamber, where it is exploded at the proper time by a hot tube or an electric spark, such as the case may be.

The governor on this engine is of the graduation type, and regulates the gasoline and air according to the amount of work which is being done. It is located around the inner hub of the fly wheel.

The Panama Canal.

THE latest and most comprehensive authoritive treatise on the subject of the Isthmian Canal is found in the concluding chapters of "Ancient and Modern Engineering, and the Isthmian Canal," by Professor Wm. H. Burr, professor of civil engineering at the Columbian University, and a member of the Isthmian Canal Commission. The following summary was compiled for Dun's Review, to which publication we are indebted for the use of the cuts illustrating same:

The route adopted by the Commission is that of the New Panama Canal Company. Starting from the harbor of Colon it traverses low, marshy ground to the Mindi river, and thence to the Chagres river; which it meets at a point about six miles from Colon. It then follows the general course of this river to Obispo, which is about 30 miles from the Atlantic terminus. Here it leaves the Chagres and follows a small stream called the Camancho for five miles until the continental divide is reached, at the great Culebra cut, the distance from Colon being 36 miles and from Panama 13 miles. After traversing the cut the canal route follows the course of the Rio Grande River to its mouth at Panama Bay. This route presents several marked advantages over the Nicaragua route. Its total length is 49.09 miles, as compared with 183.66 miles for the other. The time for traversing it is estimated at twelve hours, while 33 hours would be required for the longer route. This means that vessels can generally get through by daylight, while in the other case it would always

THE PORT OF PANAMA.
The Pacific Terminals of the Panama Canal.

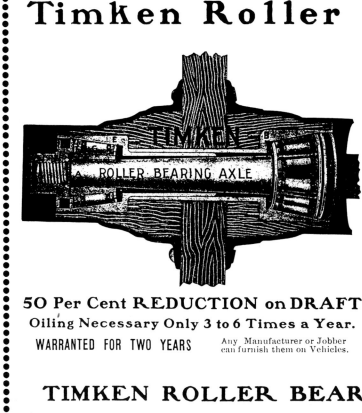

THE CITY OF PANAMA.
Street Scene in the Best Quarters.

THE CULEBRA CUT.
Showing French Method of Excavation.

be necessary to slow down or stop at night, especially in the case of large or heavily loaded vessels. There are 29 curves on the Panama route, of which only one has a radius of less than 6,000 feet. This is at the entrance to the inner harbor of Colon, where the width of the channel is 800 feet. On the Nicaragua route there are 56 curves, of which no less than 33 have a radius of less than 6,000 feet. This matter of curvature is of little importance for small vessels, but it is at all times difficult for large steamships to navigate a tortuous channel. The estimates of the Commission involved the expenditure of large sums to avoid curves on the Nicaragua route wherever practicable, but in so difficult a canal country they were necessarily very numerous and, as indicated, frequently quite sharp. The Panama curves are, with the single exception noted, of large radius and will offer little difficulty to navigators.

The plans of the Commission contemplate a channel 500 feet wide and a minimum depth of 35 feet at low water across the harbor of Colon, with a width of 800 feet on the sharpest curve. This channel will be two miles long. Across the Bay of Panama a similar channel four miles in length will have to be excavated. From Colon to the River Chagres and thence for a considerable distance no engineering difficulties are encountered. Here and throughout the route the American canal will be a marked improvement upon that contemplated by the French engineers. The original De Lesseps company planned for a bottom width of only 72 feet, and the new company of 98 feet, with a depth of 29.5 feet in each case. As the largest ships now afloat draw 32 feet in salt water and would draw at least 33 feet in the fresh water portions of the canal, the Commission determined to increase the depth to a point that might at least anticipate naval construction for some time, and fixed that dimension at 35 feet throughout. The ships having the greatest beam at the present time are naval vessels, the maximum now being 77 feet. As it is essential that the canal should accommodate these the bottom width was enlarged to 150 feet. The locks planned are all to have a usable length of 740 feet and a clear width of 84 feet. As stated, no obstacles are encountered in the marshes above Colon and the lower reaches of the River Chagres. At Bohio, a point on this river about 14 miles from Colon, one of the greatest engineering feats in connection with the canal will be undertaken. The River Chagres is subject to sudden and violent floods, the water rising frequently to a point 30 feet above low water, and in one instance 39.3 feet. These floods proved troublesome to the first French company and presented a problem that the second company found most difficult of solution. The Commission, adopting the plan finally evolved by the French engineers but improving upon it in several important respects, proposes to erect a dam perhaps 2,000 feet long and rising to 100 feet above sea level. This will cut off the entire Chagres valley and back the river up into an artificial lake 25 miles long and having a surface area of about 40 square

miles. At a point three miles from this dam a second dam is to be erected having an overflow weir 2,000 feet in length, which will serve as a spillway or wasteweir in time of flood. At low water the Commission estimated the flow of the Chagres as reaching a minimum 600 feet per second, while at flood it was assumed that there might be a possible discharge of about 136,000 feet per second. This spillway is designed to dispose effectively of 140,000 cubic feet per second, the water running over its crest and thence through low marshes again to the Chagres and then to the sea. As the canal runs in proximity to the marshes which would thus be flooded it will be necessary to protect it to some extent by embankments or levees. As the water in this artificial (Bohio) lake may be 90 feet above sea level, two great locks will be constructed near Bohio, each of 45 feet lift. Twin locks are to be built at each point, making really four locks for the two flights—an arrangement that is to be followed throughout. They will all be constructed of masonry and concrete, and that at Bohio will be located about 1,000 feet from the Bohio dam.

At Obispo, a point about 14 miles from Bohio, the canal leaves the lake and shortly beyond strikes the continental divide. Here the greatest open cut in the world will have to be made. The entire section is only 7.9 miles in length, but the cost of the canal at this point will be nearly one-third of the total estimated cost, the Commission estimating it at about $44,400,000. At its highest point the depth of cut to the bottom of the canal will be 286 feet. The material is for the most part hard clay, and the sides must therefore slope. Considerable work has already been done on this section by the French companies, but their machinery is now antiquated and the Commission believes that much will be saved by the employment of entirely new and modern plants. American invention in this field has made great progress in recent years, notably in connection with the Chicago drainage canal, and it is possible that the cost of completing this section may fall materially below the Commission's estimates. At Pedro Miguel, just beyond the Culebra cut, is another flight of two twin locks, each with a lift of 30 feet. These bring the canal level down sixty feet nearer to that of the Pacific. A little farther on is a single flight of twin locks at Miraflores, having a maximum lift of 40 and a minimum of 20 feet, to allow for variations in the tide level. This brings the canal down to the Pacific, after traversing five locks in all. From this point to Panama is 8.54 miles, the construction of the canal offering no difficulties. In connection with the locks another advantage of the Panama route over Nicaragua may be noted. There are eight locks on the latter, and the principal danger from volcanic disturbance being of injury to the locks it is evident that this is a factor of some importance. Volcanoes are somewhat numerous in Nicaragua, while the nearest one to the Panama route is 175 miles away.

The total cost of the Nicaragua canal was estimated at $189,864,062. That of the Panama Canal, including $40,000,000 to be paid to the

New Panama Canal Company, is $184,233,358. The annual cost of maintenance and operation is estimated at $3,300,000 for Nicaragua and $2,000,000 for Panama. In the hygienic conditions along the two routes there seems to be little to choose, although in this respect Nicaragua may be superior. The Commission estimated that eight years would be required for completing the Nicaragua canal and ten years for the Panama, but there is excellent authority for the belief that these figures may be safely reversed, since there are great opportunities for concentration of work upon the Bohio dam and the Culebra cut. The Nicaragua route would have been 378 nautical miles shorter than the Panama route as between the Atlantic ports and San Francisco, and 580 miles shorter for the trip from New Orleans, but, on the other hand, the time thus saved would be wholly lost in the increased time required for traversing the longer and more tortuous canal.

The commercial effect of the canal will be in line with that of most great modern public improvements, a saving of labor and time. The distances between the principal Atlantic ports of Europe and North America and the ports of western South America, the East Indies and Asia will be in most instances immensely reduced. This will produce results of the utmost importance in the way of altering existing commercial routes, and may materially affect the trade of certain cities and possibly that of several countries. Such changes may, however, be anticipated and will in any event be gradual, so that the readjustment of commercial relations caused by the opening of the Panama Canal need occasion no such economic disturbances as those that were attributed to the opening of the Suez route in 1869.

The changes in trade routes constitute a phase of the subject too large for complete consideration in the space available in this issue and may therefore best be postponed to a later number. For the present it will suffice to indicate briefly the saving in the distance between leading ports that the Panama route will make possible. This may best be shown by the following table, showing the present distance between the ports named by existing steamship routes and those by the Panama route:

From	Via Existing Route.	Via Panama.	Distance Saved.
N. Y. to San Francisco ..	13,174	5,299	7,875
" " Melbourne........	12,860	10,427	2,433
" " Auckland, N. Z..	11,599	8,892	2,707
" " Honolulu, H. I..	13,290	6,795	4,495
" " Valparaiso.......	8,440	4,630	3,910
" " Callao...........	9,640	3,359	6,261
" " Guayaquil.......	10,300	2,864	7,436
" " Yokohoma.......	15,217	9,835	5,382
Liverpool to San Fr'isco.	13,494	8,038	5,452
" " Guayaquil..	10,620	5,603	5,017
" " Callao....	9,960	6,098	3,862
" " Valparaiso..	8,760	7,369	1,391

Distances from Hamburg, Bordeaux, Antwerp and other leading European ports to the foregoing ports on the Pacific are likewise reduced from 20 percent to nearly 100 percent. To ports in the Orient the Suez Canal will continue to be preferable for ships sailing from European ports, but the Panama route will in many instances be better from American ports. The whole commercial world will benefit by the new canal, and its influence will be especially marked in stimulating the growth of trade between Europe and America and the countries bordering upon the Pacific.

THE CANAL, COLON SIDE.
Showing Methods of Excavating in Lowland Section.

An article from Scientific American was quoted in a 1903 issue of the Implement Trade Journal describing how an engine could be cooled with oil in much the same way as Oil Pull tractor engines were cooled in 1910.

A record 4,071 bushels of wheat were threshed by a single crew and machine in one day on September 19, 1903. The Avery 42x70 separator driven by an Avery 30 horsepower traction engine was operated by its owner F.A. Krhut on the farm of J.W. Groachmer in Trego county near Salina, Kansas. The wheat was in 33 different stacks, requiring the rig to be moved and a 45 minute breakdown was caused by a five pound ball of twine accidentally fed into the machine. Work began at sunrise and continued until 45 minutes after sundown, when the last bundle was finished. Krhut claimed that the rig had threshed 71,433 bushels of wheat between October 18 and September 19 for a record daily average of 2,304 bushels for a 31 day period.

The New Side Crank
SPRING MOUNTED ENGINE

Here It Is!   There are No Others!

The Agitator Separator

NOW, AS ALWAYS, SPEAKS FOR ITSELF.

Case Weighers, $50.   Case Loaders, $35.

J. I. CASE THRESHING MACHINE CO., Racine, Wis.

The J.I. Case Threshing Machine Co. introduced the first all steel thresher in 1904. The company was producing more steam engines and threshing machines for farms than any other company. Lauson "Frost King" engines were introduced. The International Harvester Co. expanded further in 1904 by purchasing the Aultman-Miller Co. of Akron, Ohio, the Keystone Co. of Rock Falls, Illinois, and the Weber Wagon Co. of Chicago, Illinois. The Massey-Harris Co. purchased Kemp Manure Spreader Co. of Stratford, Ontario, and the Moline Plow Co. purchased Acme Steel Co. of Chicago, Illinois. The purchase of the Acme Steel Co. permitted the Moline Plow Co. to make their own plow shears.

Russia and Japan were engaged in war during 1904 and 1905, which af-fected the export of farm equipment, especially plows. In the U.S. steam plowing was the rage in 1905. Standard wheel heights set by the National Wagon Mfg. Assoc. were:

Front	Rear
44 inches	52 inches
40 inches	48 inches
40 inches	44 inches
36 inches	44 inches

1896-1907

Benjamin Ott of LaCrosse, Wisconsin, said to be one of the inventors of the twine binder, died at the age of 74 on October 2, 1905.

In 1905, Allis Chalmers purchased Bullock Electric Mfg. Co. and the Minneapolis Steel & Machinery Co. began building Muenzel gas engines under license of G. Luther Co. of Braunscheig, Germany. Deere-Clark Motor Co. began at Moline, Illinois. W.E. Clark was to design the automobile. W.H. Williams became sales manager for Hart-Parr Co. in 1905 and claimed, in 1927, to have coined the word "tractor" first in some of the company's 1907 advertising. Williams also claimed to have been the first to add the word "farm tractor" in 1912.

On April 18, 1906, an earthquake and fire destroyed four squares miles of San Francisco, California. Ten percent of all business failures in 1906 were directly traceable to fraud by employees or principals of the company. More than $21

million was involved in frauds during 1906. During Theodore Roosevelt's presidency investigation of International Harvester was begun by the Federal Bureau of Corporations.

Joseph H. Glidden, the inventor of barbed wire, died on October 9, 1906, and Col. F.W. Blees died on September 8.

International Harvester Co. purchased the Waterloo, Iowa, based manure spreader division of the J.S. Kemp Manufacturing Co. of Newark Valley, New York. International Harvester also began tractor production in 1906 by installing their own one-cylinder engine on tractor frame and drive gear manufactured by the Morton Manufacturing Co. The Moline Plow Co. purchased the Mandt Wagon Co. The American Seeding Machine Co. purchased the A.C. Evans Manufacturing Co. of Springfield, Ohio, and the American Seeding Machine Co. was reincorporated on April 1, 1906, in Springfield, Ohio.

The Relation of the Agricultural College to the Implement Business

More and more the public is coming to realize and acknowledge the power and influence of the agricultural college in its relation to agriculture.

The farmer has always held the throne of autocracy in his relation to the rest of the world, for all of us are dependent upon him for what we eat, drink and wear.

The farmer has been slow to acknowledge the influence of the agricultural college, or to realize that "book farming" may be as practical as theoretical, and that the man who combines both will not only have higher ideals, moral, intellectual and social, but will be able, with even less labor, to secure not only better crops and fatter flocks and herds, but a better price for them. But of recent years these facts have been demonstrated so frequently and so forcibly that the average farmer has begun to yield, with more or less grace, to the inevitable, and to be ready to accept suggestions from these very "book farmers" whom formerly he professed to despise.

But recently the agricultural colleges have seen a light, and a number of them have come to sit at the feet of the implement and farm machinery masters. Doubtless there has heretofore been no disposition whatever to ignore the mighty part which these articles play in the cultivation of the soil and harvesting of crops, but it is only recently that it has been deemed advisable or necessary to create a department in the agricultural colleges to train the students in the science of farm mechanics.

Among the first to adopt the new idea was the Iowa agricultural college at Ames—a mighty force in its particular field, and one which for many years has wielded a strong influence on the agriculture of Iowa, and which undoubtedly has been a large factor in helping that state to take such gigantic strides toward the head of the column.

The following article from the pen of Prof. J. B. Davidson, professor of Agricultural Engineering at Ames, is here reproduced from a recent issue of Wallace's Farmer, and will serve to throw additional light on the work of this department:

Agricultural Engineering.

It is largely due to the introduction of improved agricultural machinery that agriculture has become one of the most desirable of all vocations. As long as hand methods prevailed the life of the farmer was anything but to be envied. The application of power through machines has relieved the farmer of much physical toil but has demanded from him greater mental activity.

Practically all of the change from hand to machine methods has taken place during the last half of the past century. However, it is doubtful if the development of agricultural machinery was ever more rapid than at the present time. Machines and appliances used five years ago are out of date today.

Through the aid of modern agricultural machinery the American farmer is not only relieved of hard labor to the extent he is no longer "the man with the hoe," but he is able to compete in the markets of the world with cheap labor abroad and at the same time be well paid for his services.

WAGON WITH RECORDING DYNAMOMETER ATTACHED FOR A DRAFT TEST

From this we must conclude that modern machinery is from necessity quite complex and complicated, and that it is an important factor in modern farming operations. Not only is the farmer of today blessed with machinery to aid him with his work, but city conveniences have been brought to the farm. The farm water supply and the telephone have added much to the pleasures of country life. Owing to the advance in the price of farm lands the farmer finds that he is often confronted with the problem of draining land which may be made productive or more productive. The farmer of today gives thought to the planning of his home and farm buildings in order to secure the maximum comfort and pleasure. The improvement of the rural roads is a problem which will for all time depend largely upon the farmer to work out.

To operate and care for the machinery of the farms, to secure for himself the greatest returns in the way of salable products and comforts, the successful farmer of today must be trained along mechanical and engineering lines.

To supply this training is the function of the department of agricultural engineering at the Iowa Agricultural College. This branch of agricultural instruction is perhaps the newest of all branches, but nevertheless it is believed to be very important and practical to the agricultural student. It has been but a few years since the first department of this kind was organized in an agricultural college. Before this time the only work offered to agricultural students along these lines consisted in certain courses of shop work. At the present time there is a general awakening along these lines and many colleges throughout the country have established, or are preparing to establish, a department of agricultural engineering, or farm mechanics, a name often applied to this work.

The lines of study offered by the department of agricultural engineering at the Iowa Agricultural College comprises drainage,

THE SIGN

THAT DRAWS THE TRADE

An Engine that doesn't require an Engineer

If you want over your door a sign that draws the trade. get a 1906 CONTRACT for

I·H·C· GAS OR GASOLINE ENGINES
AND DRAW THE TRADE YOUR WAY.

*Write to-day to our nearest general agency
or see blockman for terms and territory*

All dealers handling I. H. C. engines will, upon application to their general agency, be furnished an attractive "cut-out" metal sign lithographed in the natural colors of the engine, similar to that shown in this advertisement.

INTERNATIONAL HARVESTER CO. OF AMERICA
(INCORPORATED)
CHICAGO, U. S. A.

road construction, farm machinery, farm motors, rural architecture, and shop practice in blacksmithing and carpentry. The object of the department is to make the work as practical as possible and to take up such lines of work as will prove useful and helpful to the student when he turns his attention to the practical affairs of life.

The department of agricultural engineering at the Iowa Agricultural College is provided, without doubt, with the best and most complete equipment of the kind in existence. A substantial fireproof building is provided, with suitable rooms for the class, laboratory, and shop work. This building cost, with the equipment, about $70,000. The department also has in its care over $20,000 worth of agricultural machinery to be used for instructional purposes.

The work of the department is principally that of giving instruction to those who intend to make the farm the object of their life work. However, the demand for men as instructors who are trained along these lines requires that the department offer special work to meet this demand. There is no other place where so desirable conditions are to be had for special study along these lines. Furthermore, it is hoped that the department will also be able to furnish men who will be useful in the various phases of the farm machinery industry.

J. B. DAVIDSON.
Iowa Agricultural College.

———

The students are required to make a careful study of each type of implement or machine and are required to write their observations, upon which they are graded. Some shop work is also required, including blacksmithing and carpentering, so as to prepare the students to set up machinery or to ascertain what may be wrong with it in case of damage or faulty setting up, and to construct necessary farm buildings, etc. A short course in road and drainage engineering is also included and is found to be very practical. The course includes lectures as well as laboratory work.

Prof. Davidson recently has been experimenting, in connection with his work, on alcohol as a fuel, an illuminant and as motive power, and informs the writer that as either fuel or illuminant it requires about 30 percent greater volume than gasoline to accomplish a given result, although it is much safer, being non-explosive, and the flame can be put out with water. These tests were made with 95 percent alcohol. The proposed commercial article will be of only 90 percent grade and proportionately lower heat unit capacity. Used in the motor, it is necessary to start the engine with gasoline to heat the cylinder, else the spray of alcohol will condense on the cold sides of the iron cylinder. Once started, however, its work is satisfactory and it does not smut and clog the cylinder as gasoline is apt to do. The experimentation has not yet progressed far enough to determine the proportionate power per gallon as compared with gasoline.

Combined Harvesters in Oregon.

Combined harvesters are becoming popular with the big wheat growers in the Umatilla country. One dealer, E. L. Smith of Pendleton, Ore., sold thirty-four of the Holt combined harvesters in Umatilla county this year, besides several others in Gilliam county. These big machines harvest and thresh the grain at one operation, each machine having an average capacity of a thousand acres per season, and it is predicted by those who claim to know that the old style of harvesters and threshers will soon become obsolete in that section. The combined machines have a cut of 14, 16 or 20 feet and make short work of a large field of grain.

Scene on a Washington Wheat Ranch.

The illustration shown herewith represents a familiar scene on the great wheat ranches of Washington and other Pacific coast states. It is that of one of the mammoth "combined harvesters," which cut the wheat, thresh, weigh and sack it at one operation. The harvester in the picture is propelled by a thirty-three horse team, something after the style of the big teams that formerly were the feature of the old-time circuses and which never failed to attract the ardent admiration of the country youth from miles around. Naturally these machines are very expensive and are praticable only for use on the great ranches where horses are plentiful and men are scarce, and where they like to do everything on a big scale. Large as the machine is and apparently unwieldy, as the illustration shows it can be used successfully even on steep side-hills.

These machines are used under different conditions than prevail in other sections where the binders are used. The wheat must be dead ripe and entirely dry, otherwise it could not be threshed immediately after being cut. The dry summer climate makes this possible where it would not be elsewhere. However, this fact is responsible for a large proportion of the wheat scattering and being lost, the waste amounting to more, in all probability, than the saving in time and labor. Then, again, the

process serves to scatter weed seeds widely and so to make the wheat foul and to make it almost if not quite impossible to eradicate the weeds from the soil.

While within the past two seasons quite a number of these combined machines have been shipped into Washington and the Inland Empire, it is claimed by many that as a whole they are going out of vogue and are being replaced by the binder. This will undoubtedly be more and more the case as the big ranches are cut up into smaller farms, as is the tendency nowadays, as land becomes scarcer and more valuable, and the necessity more apparent of intensive farming in order to produce larger profits from the soil. In the light of this information the following recent newspaper dispatch from Lodi, Cal., is both interesting and instructive, and putting two and two together may serve to throw some light on the recent reports that the manufacturers of these big machines are trying to sell out to the International Harvester Co.:

M. M. Laufenberg, residing a few miles south of Lodi, is a close observer of the wheat-growing business in this state, and agrees with the majority of experts that the reason that California wheat has lost its gluten and is unfit when used alone for flour is that the grain is left standing in the fields long after it ought to be cut and bound and stacked and put into a sweat.

The state and national agricultural experts have for the past two years been investigating the cause of the deterioration of California wheat as a millstuff product, and they have concluded that when the stalk of the grain is ripened above the

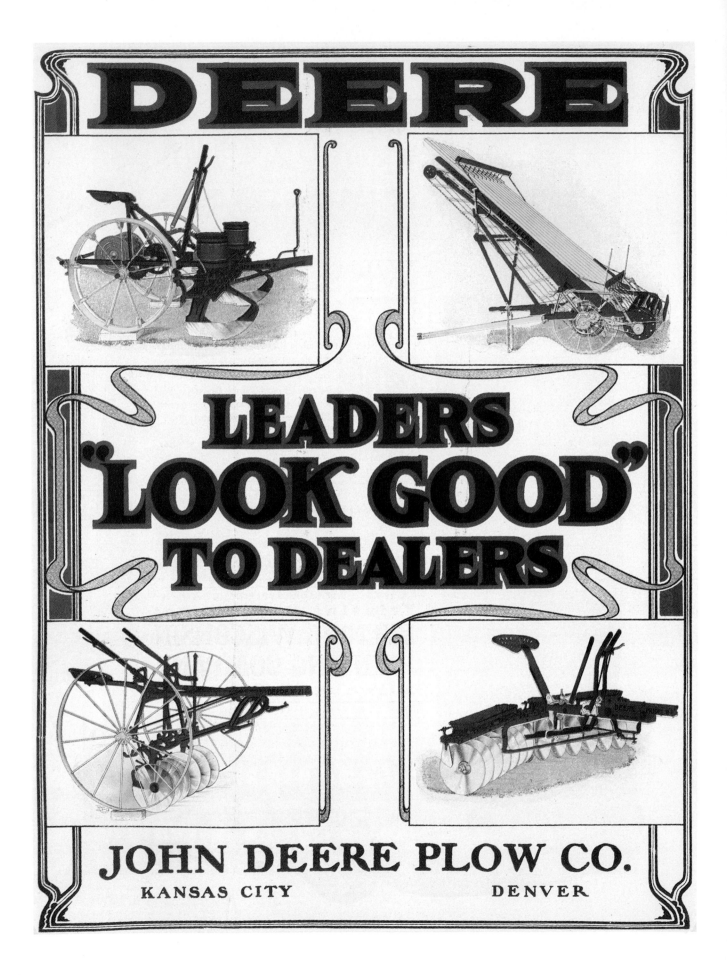

root, all the substance remaining in the stalk should go into the berry. When the grain is left standing exposed to wind and sun this substance is vitiated into the atmosphere instead of going into the berry, and thus the berry loses its substance, or the gluten which is necessary for the making of good flour—of flour that "stands up," as the bakers say, when put into dough.

Mr. Laufenberg agrees with this theory, and states that in the great wheat-growing country in the Chico locality self-binders are taking the place of the big harvesters, and the wheat is much better. In the harvester country the native wheat has to be reinforced with about 66 percent or Eastern wheat to make first-class flour.

It is probable that in many parts of the state that wheat growers will discard the big harvesters and go back to the old Eastern way of gathering crops. This will be easy when the farms are cut up, as they will have to be when California has a population that will fill up the state as it should be.

Cheap Automobiles.

Ever since it was first announced a year or so ago, that the International Harvester Co. had undertaken to experiment with the construction of a new cheap automobile for farmers' and general use, there has been a lively interest in the proposition. The price of automobiles heretofore has been prohibitive to the average man, although at the admittedly inflated prices asked nearly every automobile factory in the country has been badly rushed and then unable to fill orders within a reasonable time. But the concern that succeeds in putting out a dependable machine at a reasonable price is sure to find a ready and eager market. That the International's machine has passed the experimental stage and by another year will be ready for the dealers' hands seems to be indicated from the following article republished from a recent issue of the Sterling (Ill.) Gazette:

The International Harvester Co. of Rock Falls had the new automobile out on the streets on a trial trip Wednesday afternoon and the latest test has proved very satisfactory. Edward Johnson, the head inventor of the Rock Falls plant, has been working on the machine for some months perfecting the mechanism. The car is equipped with an air cooled, two cylinder engine and does away with the differential gear used on most of the motor cars. It is constructed in form similar to that of the common top buggy and is capable of high speed. The machine is equipped with heavy buggy wheels and solid rubber tires. The motor power is connected direct with the rear wheels by three drive chains and is also equipped with a four-way speed chain. The machine was run up an eight foot twenty-five per cent incline as one of the test made early in the week and performed the feat with much ease. The engine is of eight horse power. Many of the deficiencies in other cars are overcome in the International, and the motive power is so directed that the life of the engine is about twice as great in this

machine as any other. In turning corners with the auto, both drive chains act the same, thus centering the wear on both sides instead of one as the larger percentage of the machines manufactured. Twenty-five machines are now under construction, as the auto has passed the experimental stage and is said to be a winner. A man in the employ of the company, says the machine could be sold at a cost of from $250 to $300. The cars under construction, will be finished within a short time and if they work as successfully on the road as expected, the company will go into the business on a large scale.

It is to be hoped that the price suggested by the employe is approximately correct, although it is considerably lower than local representatives of the company have been led to expect, they having understood that the probable retail price would be $500. At the latter price the runabout would, however, probably have a rather slow sale, while at the former figure—if the machine proves to be all that is claimed for it—it would sell like the proverbial hot cakes. However, it seems scarcely probable that the machine could be built and sold for $250, although undoubtedly some time that may become a popular price.

That Farmers' Automobile.

Mention was made a few weeks ago in these columns of the fact that a number of experimental automobiles are now being built in the International Harvester Co.'s Keystone shops at Sterling, Ill., with a view to creating and supplying a demand among the farmers for a low-priced machine that shall be practical for farm and country road use. These machines, it is understood, are being built of a type similar to the ordinary runabouts for use with horses. They are built with high wheels, steel shod, and are of sufficient power for ordinary roads and to perform such service as might be expected and demanded for country use.

That there can easily be created such a demand goes without controversy. Farmers nowadays are progressive and, being prosperous beyond the dreams of a few years ago, are ready to buy anything that strikes them as being practical and desirable. Many uses can be made of such a machine that will effect a saving in time, trouble and cleanliness, and leave the horses free for the ordinary work of the farm. In fact, it will not surprise us to see, in the course of a few years, the use of the automobile, in some form or other, almost universal in country and town. There seems to be no reason why some enterprising manufacturer should not before now have taken up the manufacture of a cheap, plain, but practical automobile for village and country use, except the fact every manufacturer of automobiles has more than had his hands full trying to supply the city trade with higher priced machines. But the cheap and yet practical automobile is bound to come, and the implement and vehicle dealer is the logical man to sell it. If he has his ear to the ground he will have already taken notice and will be getting ready to take advantage of the first opportunity to extend his business.

For Gasoline Engine Men.

The following is said to be an excellent anti-freezing solution for gas engine jackets: Four pounds of chloride of lime to each gallon of water will form a solution that will not freeze at 17 degrees below zero, 3½ pounds to the gallon at 8 degrees below zero and 3 pounds to the gallon at one degree below zero. This solution has a very slight, if any, effect upon the walls of the cylinder, pipes or tank, and can be carried in solution from one season to another. It doesn't form a sediment in the tank. The chloride of lime is sold in hard, stony cakes, chunks or crystals, and may be purchased at about one cent per pound.

The Cartograph for Autoists.

American automobilists will soon be crying for the cartograph, an almost human invention which is being shown here, if it comes up to the claims made for it. Think of an attachment resembling the contrivance by which self-playing pianos are made by the unskilled to produce masterpieces. The cartograph, instead of being a perforated music roll, is a map of the roads to be traversed by the motor car, unrolling a panel in front of the chauffer, so that he can tell at a glance where he is and which turning to take. The speed of the machine governs the motion of the map, so that it always indicates—or should—the exact point where the traveler is. Moreover, the cartograph is provided with perforations just ahead of where the short turns and corners are, and these perforations ring a bell to warn the motorist in time. Even on the darkest night, by means of this device, it is asserted, a wholly unknown route can be covered without danger of being lost or ditched. The next logical step would be a contrivance to attach the cartograph to mechanical means of controlling the steering gear and levers so that the motorist can set it going and look for the machinery to do the rest.

To Abolish Bright Headlights.

A movement is said to be on foot in New York City to abolish the glaring acetylene gas headlights of automobiles. They have proven themselves to be a great nuisance and to be liable to cause unnecessary accidents by reason of their dazzling intensity, which startle not only pedestrians and drivers of ordinary vehicles but also drivers of other automobiles, bewildering them and causing them to lose their bearings. For unknown and dangerous roads they are all right, but there seems to be no good excuse for them in the crowded and well lighted paved streets of a city. And along with this movement, which is said to have the active support of real automobilists, there should by all means be a crusade against the prevailing reckless and insane craze for high speed, which is responsible for so many accidents.

Motor Problem Solved (?)

Word comes from New York that Thomas A. Edison has at last solved the problem of cheap power by producing a storage battery that will last indefinitely, at a minimum cost. It is said that two large factories are being built for the purpose of manufacturing motor batteries and that in a few months they will be placed on the market. Mr. Edison is quoted as saying that in fifteen years the horse will be a curiosity.

The death knell of the faithful horse has been sounded many times during the past fifteen or twenty years, but, somehow, good horses are in as great demand as ever and bring good prices.

While it isn't at all likely that the horse will become a curiosity and be exhibited in side-shows, as Mr. Edison says, yet it is quite certain that whenever a cheap motor power can be placed on the market there will be a big demand for it. Such a motor as is now credited to Mr. Edison, which can be sold for $200, and which is practically indestructible, would be purchased by thousands of persons who cannot afford to buy an automobile at prevailing prices, and could not afford to maintain one if it were given to them; persons who do not now, and perhaps never could, own a horse. There will still be a place for the horse, and there is also a demand for a motor vehicle at a moderate price, and a fortune awaits the individual or corporation that will produce it.

History of the Cream Separator

**By E.W. Curtis, from the Implement
Trade Journal, 1906**

When the Cream Separator is mentioned nowadays the best usage concedes that the Centrifugal cream separator is meant. A history of this most valuable farm implement is really a history in dairying, because no dairy progress was made until the invention of the centrifugal separator.

Since the beginning of time it was the habit of those engaged in animal husbandry, to set the milk of goats and kine away in vessels to await the rising of cream. This required from twelve to thirty-six hours, according to conditions. Only a part of the cream could be secured by this method, and both skim milk and cream were frequently sour. This injured the quality of butter, and the skim milk was less valuable for young stock.

The discovery that centrifugal force would separate cream from skim milk

belongs to an obscure German chemist, in the late 1850's. He found that he could fill a small glass tube with milk, then by whirling this tube in a revolving mechanism the skim milk would go to the bottom and the cream to the top. He was not particularly interested and did not pursue his investigations further. It was fifteen years before even a crude way of applying the discovery was invented, and the old German chemist died without knowing that his discovery would revolutionize dairy methods.

The next step was by a Danish horse doctor by the name of Jensen. He rigged up a crude contrivance consisting of a vertical wooden pole with a cross-piece at the top. On the ends of the cross-pieces he hung two milk pails. By attaching a belt he was able to revolve this upright pole. His method of operation was to fill the two milk pails with fresh milk and attach to the cross-pieces. After whirling twenty to thirty minutes

the machine was stopped. A thick cream would be found on top of the milk.

A German engineer, Lefeldt, worked on this idea about 1874, and continued to design a practical machine. In 1877, he introduced a new machine of greatly increased efficiency. The skimming part of the machine consisted of a metallic drum about the size and very much the shape of a common wash tub except that the top was slightly drawn in. This was not a continuous skimming machine, and required a great deal of power to drive. The operator would fill the drum with milk. In about ten minutes a speed of 800 revolutions could be attained. The cream would be separated in from fifteen to twenty minutes and the machine would run down in about twenty-five minutes. It must be understood that these machines were driven by steam engines—hand separators not having been thought of until many years later. Lefeldt sold his separators at the following prices:

No. 0, 110 lbs. of milk per hour, $200.
No. 1, 220 lbs. of milk per hour, $300.
No. 2, 440 lbs. of milk per hour, $600.
No. 3, 660 lbs. of milk per hour, $750.

This brings us down to the invention, several years later, of the continuous separators. The first one patented and sold was the Danish-Weston by Neilsen & Petersen. Thousands were sold, and indeed they were so well and substantially built that their sale continued until recent years. A great many of these old Danish-Weston separators are in use today (1906) in some of the older dairy districts of the United States and Europe.

Following the Danish-Weston came the DeLaval, a continuous separation Lefeldt, the Fesca, and others too numerous to mention—all of them European inventions. These machines were largely imported to this country. The duty, however, was so high that most of the old country manufacturers made arrangements in the early eighties with machine shops in the United States to build the cruder parts of the machine. European manufacturers believed that it was impossible to secure the right grade of steel for cream separators bowls in this country and that our mechanics were not sufficiently skilled to properly balance the bowls or to construct the various other parts. The owners of some of these same machine shops afterwards engaged in the manufacture of cream separators, on their own account, and have become the largest builders of cream separators in the world.

About 1883, several cream separator factories believed that a hand or dairy size cream separator would be prac-

*"And the gentle cows at her winsome call,
Trudge homeward o'er the lea;
Her voice, full of laughter, quite charms them all,
E'en as her eyes charm me."*

WANTED--Agents and Dealers...

To Handle, Represent and Sell the

DE LAVAL
Cream Separators

in every county in which such machines are not already represented.

The De Laval Centrifugal Cream Separator is superior in all respects to any other machine or system in use for the separation of cream from milk. The Separator will effect a saving of fully $10 per cow per year, over and above any other system, in actual cash returns, aside from its many conveniences. It is guaranteed as represented in all respects, and satisfaction to the buyer is an absolute condition of sale.

Address for Catalogue and any desired particulars:

THE DE LAVAL SEPARATOR CO.,

ticable, and a number of different makes were placed on the market. Very few were made, however, until in 1896 when Mr. Irv. Moody of Nashua, Iowa, began to investigate the possibilities of introducing hand separators in a general way among his creamery patrons. For many years it had been the practice of the dairy farmer to haul the milk to the local factory or creamery, where it would be run through a large factory cream separator and the cream extracted. The cost of manufacturing a pound of butter has been reduced at least 40 per cent wherever the hand separator system has been introduced.

The system of the farmer hauling milk to the factory was wasteful for several reasons: First, the skim milk returned from the creamery was frequently sour and unfit to be fed to young stock. Second, it was necessary to haul the heavy load of milk every day.

Mr. Moody advised his patrons to buy hand cream separators, and pointed out to them that they could skim the milk while it was warm and fresh. The expense and trouble of hauling the cream, was much less than the bulky whole milk, and it was not necessary to haul it so frequently. The idea proved popular and several hundred dairy farmers in the vicinity of Nashua bought hand cream separators. The dairy press sent representatives there to investigate, and their reports as to the success of the movement were so glowing that many creamerymen also investigated the Moody system, as it was called.

The sale of hand separators has been heaviest in Kansas, Nebraska and Iowa. Probably 100,000 separators were sold in these three states in 1904 and 1905.

It is doubtful if any tools used on the farm will make their owner more money, and the statement often made that they will save their cost in one year is no exaggeration. A modern machine properly cared for will last a life time.

Cultivate Their Confidence.

Instead of complaining about the competition of mail-order houses, suppose you put in the same time it would take to make your complaint, says Stoves and Hardware Reporter, in doing some good strong thinking along lines which would help you to be as thorough a merchant as the mail-order house is.

Too many retail merchants are today holding wrong impressions of their own customers' intelligence. They figure that it is good business to get rid of some old out-of-date goods to a customer who is not likely to know that they are out of date, and make him believe that they are the very latest thing, and in best demand. This may be classed as good salesmanship by a good many people, and it may account for a good many failures, incidentally. The safest way to compete with mail-order houses is to always sell your goods for exactly what they are. You may profit more temporarily, by pushing off something which is out of date, and claiming that it is the very latest, but as soon as your customer shows the purchase to a more enlightened neighbor, he will hear about that neighbor buying identically the same thing several years before, or he will know of someone else who made such a purchase, or will know that the brand has been an off one for some time. Do you fool yourself into the idea that your customers feel good towards you after such an experience? If he had bought with an express understanding as to actual conditions, then the remarks of his neighbor would not hurt you in the least, for he feels that you were honest with him. No matter how small and mean your customer may be, he will not approve of such action on your part, and your best trade bringer will be the absolute confidence of all the people who deal with you.

Every merchant makes mistakes in buying, and if he does not keep close track of his stock, such mistakes will soon accumulate and make much dead stock. The thing to do is to locate mistakes in buying as soon as possible, and then get rid of them without delay. Every day you carry them will make them poorer property, and to try to sell them under false representations, either expressed or implied, will be sure to cause you much damage.

If you will think the matter over quietly, you will see that people all around you are very careful how they trade with people who are considered "tricky." When these same people find a retailer misrepresenting his stock, they put him on the "tricky" list, and after that he is watched carefully, and is forever laboring under a disadvantage. The mail-order house has an advantage over him, even if the prices are the same; such a retailer has a very hard fight ahead of him, for he must restore confidence in himself, and that is a long struggle. The retailer who now has the confidence of his trade, and is only tempted to take up such methods, thinking they will help him fight the big battle, will do well to hold the confidence of his people, regardless of all other considerations. Confidence is an asset which no merchant can afford to lose, and it is one which is very easily lost. It is hard to keep your trade feeling right under the best of circumstances, for people are always suspicious of each other, but you can strictly adhere to the rule of selling your goods for exactly the quality they are sold to you for, and thus avoid any possibility of being charged with selling out-of-date goods as brand new, or cheap goods as a better quality.

While the mail-order house is tricky, and is occasionally caught at it by a customer, they have a splendid excuse which does not apply to you. They write a fine letter and beg the pardon of the customer, explaining that they cannot personally supervise all shipments and that their help sometimes makes mistakes. Your stock is before the people, and you cannot lay your "tricks" on the "help."

Cheaper Liquid Air.

Reduced cost of liquid-air production is indicated by an article in the London Times. Recent experiments in England of an invention by Mr. Knudsen, a Dane, furnished liquid air at one-sixth of the present market price, and give promise of an ultimate low price of a fraction over 2 cents per gallon. The result is secured by purely mechanical means, without an atom of added chemicals. Atmospheric air is first purified and then compressed by stages to 2,500 pounds to the square inch. It is finally reduced to 125 pounds to the square inch, which then cools and liquefies the high-pressure air.

The oxygen gas produced by separating the nitrogen from the liquid air is claimed to be purer than by the old method, and can be supplied in the liquid as well as the gaseous form. One gallon of liquid air equals approximately 128 cubic feet of oxygen gas, which retails at 6 cents per cubic foot. The new price is 1 cent. Liquid air has been successfully used in coal mines as an explosive, being quite safe where fire damp and other explosive gasses exist. Liquid oxygen is also used for welding steel pipes, boiler shells, and plates for ship-building instead of riveting.

That oxygen and nitrogen can be separated from liquid air and sold retail at $1.20 per gallon shows great commercial possibilities. The use of nitrogen for agricultural purposes opens yet another field. The maturing of liquors will be helped by liquor air, as also the preservation and purification of milk. As a motive power its use is considered to be quite practicable for small powers. The British government is already carrying out a number of experiments with a view to the utilization of liquid air for various purposes.—Government Report.

If subsequent developments shall demonstrate the truth of these claims, even approximately, the Journal forsees gratifying practical possibilities both in the way of refrigeration and power, in this recently discovered substance. It already has been demonstrated that it can be used to furnish enormous power for steam engines, and may yet be used for almost all the purposes for which motive power is required. Further developments will be awaited with interest.

Marshall Sattley died on January 1, 1907, and the inventor of the Stevens rifle, Joshua Stevens, died on January 21. Samuel L. Avery, eldest son of B.F. Avery and former president of B.F. Avery & Sons, died on Feburary 27 at age 60. Charles H. Deere died on October 29 at 70 years of age and Dester M. Ferry, head of the largest seed company in America, died on November 11.

Platinum was in very short supply because the largest deposits, near the Ural mountains in Russia, were becoming depleted and there were fewer laborers because of the recent war. Two short pieces of platinum wire were used in every incandescent light bulb and the rapid acceptance of electric lights had increased the use of the metal. Nothing else could be found to successfully replace platinum for this application in 1907.

The International Harvester Co. began work on the "Mogul" tractor line in 1907. International Harvester Co. also began selling the Belle City thresher made by the Belle City Mfg. Co. of Racine, Wisconsin. Edward Johnston was the chief engineer of the tractor project and success was so great that a separate assembly line opened in 1908. Since McCormick and Deering each had established their own dealer organizations before the merger, separate tractor lines were used, the "Mogul" for McCormick; "Titan" for Deering dealers.

Traction engines, threshers (or separators) and steam plows usually cost from $1000 to $1500 and were usually sold on time payments directly to the operators by the manufacturers. Only a few were sold by local dealers.

Because of the transient nature of their work, laws such as the mechanics' lien were enacted to encourage the custom threshermen to pay for repairs to their equipment. Thresher operators often ran out on small bills.

Tools specifically for the fasteners of the new mechanical equipment like automobiles, traction engines, gasoline engines, threshers, etc., started becoming available. Tools for blacksmithing, cabinet making, harness making and other trades wouldn't work well on the threaded fasteners used to attach the metal parts together on "modern" farm equipment of 1907. Operators of gasoline engines discovered that they would run equally well on kerosene if first heated by running on gasoline for a short time. Farm equipment dealers were advised to look into the possibility of selling automobiles. Dealers were told that because of the demand for automobiles, most manufacturers refused consignments and could make sales direct if they so desired, but there was opportunity for those who became automobile dealers. Automobiles were popular. Even in Salt Lake City, Utah, 280 were registered with more coming as fast as possible. American farmers used 50,000 miles of wire fence each month of 1907 to enclose their land and to repair or replace old fence.

A New Toggle Bolt.

If wood screws held firmly in hollow tiling, contractors doing work in modern buildings would find their work considerably simplified. But this is not the case, so the next best plan is to punch a hole in the tile and use a so-called "toggle-bolt," i. e., a bolt with a headpiece which will catch inside the tile. Bolts for this purpose have for years been made with an anchoring bar pivoted to the head; then when the bolt is tightened, the nut and surplus of thread are left exposed. The extra projecting thread is usually cut off, but this still leaves a rough bolt end and a plain nut, neither of which make a finished-looking job. At the same time, the increasing use of tiles for ceilings and partitions has created an increased demand for such bolts to be used where a nice-looking finish is essential.

To meet this demand the Ajax Line Material Co. of 12 South Jefferson street, Chicago, is putting out the new type shown in the accompanying cuts, which is offered in eight sizes as the Ajax toggle-bolt. Unlike the old forms, it has the locking bar or headpiece riveted to the nut of the bolt, this

"ELGIN" WRENCH BOXED SETS.

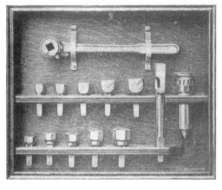

FIG. 2—NO. 0 WRENCH AND ATTACHMENTS ON OAK BOARD.

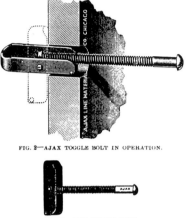

FIG. 2—AJAX TOGGLE BOLT IN OPERATION.

FIG. 1—AJAX TOGGLE BOLT.

headpiece being a V-shaped strap with an ingenious arrangement for keeping it parallel to the shank of the bolt while the latter is being inserted; then, when inserted, a half turn of the bolt drops the headpiece squarely across the hole, and screwing up the bolt hides the surplus of thread within the wall, leaving merely the slotted head of the bolt exposed. It therefore requires no cutting off of the surplus thread, nor covering of the nut to get a finished job, and hence will lend itself to a large variety of work for which the old style toggle-bolts have been poohpoohed. The latter have been found a good source of profit for hardware stores in the larger cities, and the new type (which also holds well between laths on an ordinary wall) may be worthy of a still wider field. It is said to be useful also for fastening to marble and wire lathing.

Gas Mantles Only Ashes.

The gas mantle is nothing but ashes, and it is wonderful how science makes its particles cling together as long as they do.

The gas mantle was invented by a chemist of Vienna. He noticed the intense heat given out by a small quantity of thorium thrown into a stove in his laboratory. He realized the importance of the discovery and in 1880 began a series of experiments to utilize this remarkable quality of the element in intensifying light. He found that pure thorium would not cohere well enough to be of use, and he then began searching for a combination of elements that would answer.

In 1887 he produced his first mantles, but they were so delicate that they could not be transported and were delivered by hand. A boy was trusted to carry two, one in each hand, for delivery about the streets of Vienna. In the early '90s he found that a good mantle could be made from a combination of the two substances, thorium oxide and cerium oxide, and that a coating of collodion would give a sufficient firmness to allow it to

"THE SUCCESS" RATCHET CYLINDER WRENCH MANUFACTURED BY THE PARSONS BAND CUTTER & SELF FEEDER CO. NEWTON, IOWA.

be transported. Since then gas mantles have gone into general use in nearly all parts of the world and thousands of factories are producing them.

Oklahoma became a state in 1907 and the prepaid McKinley postal cards began to be sold. Government regulations governed the size, shape and material of the cards. A postal employee had cut his finger handling a picture card and got blood poisoning, so metal, glass, sand and tinsel cards were barred from the mails. The Hunter Arms Co. of Fulton, New York manufactured a 16 gauge shotgun for the first time in 1907. Large areas of Texas that were previously grazed began to be opened by the plow for planting wheat, cotton and other crops. A group of Moline, Illinois, men purchased the machinery and material of the Deere-Clark Motor Car Co.

The Pope Manufacturing Co., including its subsidary the Pope Motor Car Co., failed in August of 1907 as a result of a 6 month long strike and improper use of capital resources. The Pope Mfg. Co. was the successor to the American Bicycle Co. The Moon Motor Car Co. was organized at St. Louis, Missouri, in October. The new company was a new direction for the Joseph W. Moon Buggy Co. Buggy companies were changing in an effort to stay alive in the changing times of the gasoline auto.

Steam powered equipment was also fighting to justify itself by emphasizing its high horsepower and economy of operation. Gasoline engine advocates stressed low initial cost and ease of operation. An operator of a steam boiler had to be trained and still accidents were common. A big argument about the need to license steam engine operators resulted because of the boilers blowing up and killing innocent people working near them. The arguments persisted until gasoline engines provided almost all power used on farms.

It was reported in 1907 that a German invention would revolutionize the refrigerator. "The device is an extremely simple and inexpensive affair costing perhaps no more than $1 to manufac-

ture. It consists of a double-wall tin vessel with a capacity of 5 gallons or more. There is a hollow space between the two walls, or inner and outer vessels. This space completely surrounds the inner compartment and is about an inch in width. By the graduated admission of carbonic acid to this surrounding chamber at the bottom of the vessel, and from this surrounding chamber into the vessel proper at the top through a cross-armed tube, the contents of the vessel are frozen quickly and completely. Water is changed into ice in the space of sixty seconds. Meats, fruits, bottled beverages, such as beer, champagnes, wines, etc., may be chilled or frozen in a few seconds. This effect is produced by the sudden great reduction of temperature caused by the rapid expansion of the carbonic acid, which is admitted from an ordinary carbonic acid reservoir. The invention is at present designed for hotels, restaurants, hospitals (particularly field hospitals), and the ordinary household."

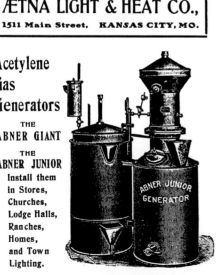

This is our 200-egg incubator. It is the most popular size we sell. We are proud of its construction. It is simple and easily operated. There are no useless frills. We have perfected it by long and careful experimentation. When you buy a Victor machine you are not getting an experiment—you get one that can be counted on to hatch every fertile egg. You get a machine made from carefully selected and seasoned lumber. You get fourteen-ounce copper in the boiler and tanks. You don't get green lumber that the maker hasn't had time to properly cure. You don't get some cheap substitute for copper. You don't get a machine so poorly made that you would be ashamed to show it to your neighbors. In fact, "V-I-C-T-O-R" spells incubator satisfaction.

GEO. ERTEL CO.,
QUINCY, ILL.

1908-1913

PRE-WAR GROWTH

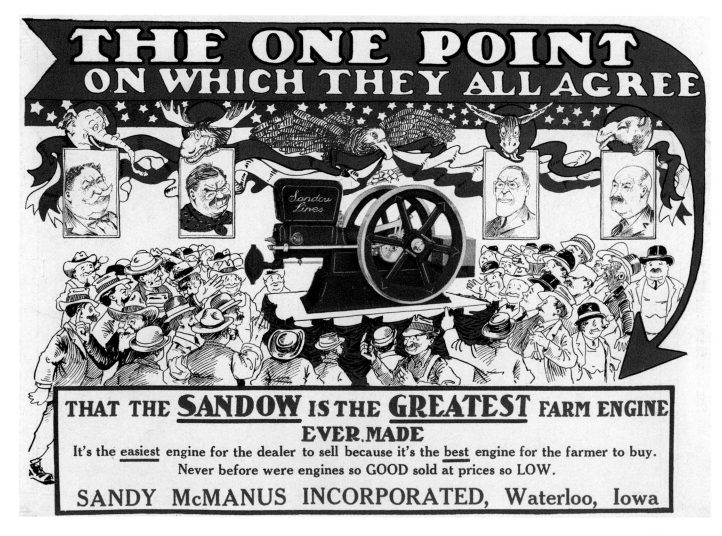

Farm work was beginning to be mechanized, but it was evident that training in operation and maintenance was necessary. Differences obvious to the engineers and designers weren't noticed by the farmers who were used to dealing with the needs of horses.

About 1908, it was apparent to some that education was necessary for people, accustomed to animals, to properly operate and maintain the new machines of farming. Schools began to open for the purpose of teaching operators how to run equipment. The Theo. Audel & Co. at 63 Fifth Ave., New York, began selling their 469 page Gas Engine Manual. The book traced the history of internal combustion engines beginning with Lenoir's engine of 1860 and explained the laws of permanent gases,

A Tandem Spinner.

The angling season will soon be here. In view of this fact the illustration here given of Pflueger's luminous tandem spinner will be especially interesting. This reversible "spinner," when in motion, is said to so strongly resemble the sure enough wiggles

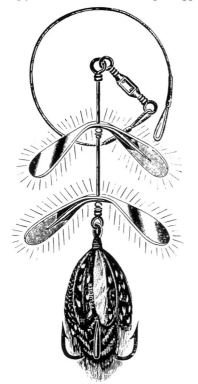

TANDEM SPINNER.

of a live minnow as to absolutely deceive the wily bass. The cut shown here is exact size of the No. 2, which is the largest size. Two smaller sizes are made. They are made by the Enterprise Mfg. Co. of Akron, O., manufacturers of luminous baits of all kinds.

theoretical working principles, actual working cycles and much more. During the winter of 1908-1909, the Hart-Parr Co. began to conduct schools at each of its branch offices to teach operators how to use their equipment. A similar correspondence course was added in 1911. The Kansas City Automobile School opened in 1909 to teach repair and operation of "any make of machine." The classes lasted 6 weeks.

Boy's Corn Clubs were organized in Jack County, Texas. About 1912, Girl's Tomato Clubs and Canning Clubs were formed. These clubs were forerunners of the 4-H Clubs.

The deficit of the Postoffice Department was said to be $16,910,279 which was the largest known in 1908. The Postmaster General assured all that the condition was a result of unprecedented use of the mails by catalog houses and would be quickly adjusted.

In 1908, supplies of tinned meats, arms and Gatling guns were being sent to "European governments preparing for an emergency of some kind in the near future." In 1909 it was noted that Great Britain was importing an abnormally high quantity of acetone. The principle uses include smokeless powder, gun cotton and mine explosives, but acetone is also used for making chloroform, preparing photographic plates and for certain dyes.

An article from April 1908 explained that tools introduced for sale in Germany would be closely examined, then copied or perhaps improved by a German company. Exporting tools to Germany was not considered a worthy enterprise. Germany tried to improve trade relations with Canada during 1908 and 1909. Canada didn't buy much from the United States, because of trade restrictions. Canada bought most of its equipment from France.

In 1908, William Howard Taft was elected president, the country was afraid of the "approaching shortage in the supply of hardwoods" and an important trans-continental railway crossed

Guatemala, linking the Pacific and Atlantic oceans.

Some still believed that rabies was an imaginary disease. Rabies first gained attention of the medical and veterinary professions in 1892, when a resident of the District of Columbia died of a confirmed case of rabies. "Madstones" were still being rented in 1908 as a cure for those bitten.

Texas retailers were charged 50% tax on the sale of hand guns so there were some really long term rental contracts to get around the tax. The Colt .25 caliber 6 shot hammerless and Savage .32 caliber 10 shot pistols were two of the popular new automatic hand guns.

Roller skating increased in popularity around the world due partly to the recent use of ball bearings in the wheels.

The Unique Vest Pocket Pistol.

The accompanying cut illustrates the Unique vest pocket pistol, which is a radical departure from the usual form of pocket arm, and is far superior for self-defense. This pistol takes up no more room in the pocket than a watch, as it is very small, the 22-caliber weighing but six ounces. It can be carried in the hand without detection and fired without raising the hand, or can be fired from the pocket or hand bag, should the

UNIQUE VEST POCKET PISTOL.

situation require it. It cannot be knocked out of the hand or discharged accidentally, as it requires the full sweep of the trigger to explode the cartridge. It holds four cartridges, is double acting and can be fired very rapidly; the four shots can be fired in one second. With 22-short cartridge, black powder will penetrate one inch of pine. The recoil when fired is scarcely noticeable.

This pistol is manufactured by the C. S. Shattuck Arms Co. of Hatfield, Mass. The list price for 22-caliber is $5, and for the 25 and 32-caliber $5.50. It is said to be a very strong and durable arm.

The Electrical Suction Sweeper Co. of New Berlin, Ohio, introduced an electrically powered brush and fan to suction dirt into an attached bag. Attachments were available at extra cost for cleaning beneath radiators, upholstery and other places where the machine could not work. Another helpful household article invented about 1908 was the dish-pan drainer. The upright position of the dishes allowed boiling water to be poured upon them to rinse. The Lionel Manufacturing Co. Inc. began making its New Departure motor.

It was thought that the newly discovered mercury rectifier would solve the biggest problem connected with operation of electric vehicles. The new rectifier would permit economical battery charging from normal house (alternating) current. Thomas Edison and others were constantly working to improve the quality of batteries and

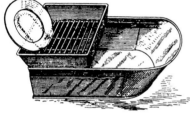

related equipment so that electrical automobiles would be practical. In 1910, Edison announced his new storage battery that "a man can take his wife out to the country in his electric car and have

no fear about getting back. On such roads as those between here (Detroit) and Philadelphia 150 miles ought to be made without recharging. On hilly roads

the machines ought to run about 125 miles. I feel confident that there is a great future for the storage battery in the automobile business, for there is

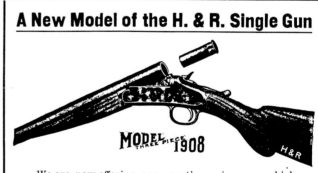

such less wear on the tires and the whole machine than with the gasoline engine."

George O. Erskine, one of the founders of the J.I. Case Plow Works, died on January 16, 1908, at age 80. Daniel C. Stover, founder of one of the early gasoline engine companies, died two days later at the age of 68. James Oliver died on March 2 at age 82. George J. Cram, secretary and manager of the Pontiac Buggy Co. and member of the board of directors for the Oakland Motor Car Co. died of cancer on November 7 at 38 years of age. George W. Corbin, of the Corbin lock companies, died of cancer on December 1.

A gasoline engine built in Los Angeles, California, had the cylinders revolve while the shaft remained sta-

A REVOLVING GAS ENGINE.

tionary. The engine was in daily use in an automobile that was especially adapted to its use. Planned experimental application of the low vibration, light weight motor was for "Aeroplanes."

A Wonderful Electrical Automobile Equipment.

The Witherbee Igniter Co. of New York has just finished for Mr. M. R. Hutchinson, the well-known inventor, the most complete electrical installation ever put on an automobile.

This electrical equipment consists of two Witherbee No. 86 batteries connected in series, located under the rear seat of the machine. The batteries are charged by a dynamo situated under the front floor boards. A wood split pulley, attached to the shaft between the clutch and the gear box, together with the belt, drives the dynamo and the air compressor, which inflates the tires and operates the pneumatic jacks for raising the machine. In addition to the dynamo there is a Wico charging device located on the running board, which enables the driver to charge the storage battery from any electric light socket.

The headlights are equipped with stereopticon incandescent lights, which can be turned off or on at will. The side and rear lights are fitted with small incandescents.

By each of the side doors of the tonneau there is a lamp turned on automatically when either door is opened, lighting the way into the tonneau, where another lamp operates simultaneously by the same means, illuminating the interior.

By raising the bonnet of the engine four lamps on either side of the motor are automatically turned on. There is also a lamp in the pan under the engine, and also lamps beneath the chassis which are turned on from the switchboard.

On the dashboard are five lamps, illuminating the speedometer, ammeter, voltmeter, pressure gauge, oil feed drips and clock.

Attached to each wheel rim is a device which indicates when the air pressure in the tires is below 60 pounds, by sounding one of the electric horns attached to the dash and an indicator on the dash instantly locates the trouble. The same horn blows, and the same indicator operates when any bearings are hot, the water in the radiator gets low, the oil in oil box gets half empty, or when the gasoline gets down to five gallons.

Putting on either of the foot brakes or the emergency blows an electric horn attached to the rear of the car, and drops a sign "Stop."

As soon as the brake is released the horn stops and the sign disappears.

There are four lights in the folding top, which are turned on when desired. On each end of the rear seat, and on the back of the front seat, there is an electric cigar lighter. Situated near each of the wheels are extension lamps, for use about or under the chassis.

In the rear of the car is a box with a celluloid front, through which the license number for whatever state the car is in is visible. These numbers are painted on a curtain and can be turned at will.

The signaling equipment consists of three Klaxon horns. A chain attached to the steering column operates these electric horns, and when all three are going, they can be heard for a mile distant. If the occupant of the tonneau wishes to speak to the driver, a special telephone transmitter is used, and an electric horn on the dash proceeds to talk in a loud tone. If the driver wishes to say a few things to a teamster, it is not necessary to waste strength shouting. He simply speaks into his transmitter and the electric horn on the mud guard repeats his words loud enough to be heard several blocks.

Three-in-One Oil.

A prominent Eastern paper speaking of Three-in-One oil says:

This oil is made by a secret process at Rahway, N. J., where a large plant is kept in busy, continuous operation. The manufacturers—G. W. Cole Co.—have met with pronounced success in marketing "Three-in-One" through its reliability and superiority to all other oils whatsoever, and lastly, its adaptability to a multitude of uses. Originally introduced as a bicycle oil its peculiar properties were soon realized and today, it is safe to say, there is no more widely known lubricant in the market than "Three-in-One." Unexcelled as a lubricant it also cleans and polishes all surfaces, prevents rust and can be used in any case where oil is employed. It is particularly valuable for cleaning and polishing furniture, having been found superior for these purposes to any other oil in existence. "Three-in-One" will not eradicate, but it will prevent rust. Used judiciously on any surface or machinery, etc., exposed to the elements or which are commonly subject to rust, "Three-in-One" will completely eliminate this annoying trouble. For fire-arms, bicycles, typewriters, sewing machines, talking machines, etc., ad infinitum, "Three-in-One" is a lubricant and a tonic whose value can only be appreciated after use. To catalog all of its advantages would prove rather voluminous, but it is an oil that will take the place of any other oil used for any purpose, do that oil's work more satisfactorily than it has ever been done before and in short, "Three-in-One" is indispensable. Those not already familiar with its advantages should investigate it at once. The manufacturers, the G. W. Cole Co., will gladly send a liberal sample to any address upon receipt of the interested party's name. The headquarters of the company are at No. 42 Broadway, New York.

In 1908 it was noted that the custom of yearly automobile model changes was not proving as popular as it once was. "The simple fact of depreciation in the value of the old models adds very greatly to the final cost of an 'up-to-the-minute, car." It was also noted that it was difficult to handle the "second-hand" cars. The automobile industry affected wages of all other industries. In 1908, automobile factory workers were paid an average of $752.11 per year, while regular foundry and machine shop workers were paid about $150 less. The automobile industry soon began to exhaust the supply of rubber for tires and the price was quickly increased.

American cars dominated the Vanderbilt cup race on Long Island in 1908. A 120 hp Locomobile won, but foreign made automobiles had difficulty finishing. This was one of the first times that the European cars didn't sweep all of the top spots.

A twelve cylinder 180 hp Maxwell car was built to attempt to break the world speed record for self-propelled vehicles. The racer was driven by Arthur Lee and came within 3 seconds of the record at Altantic City, going 1 mile in 31 1/5 seconds or 116 miles per hour. The world's fastest enclosed automobile race course was built at Indianapolis, Indiana, in 1909. All stunts involving automobiles were not speed events, though most were. Hill climbing was a challenge for most cars of 1909 and just completing a reasonably long distance, with or without repairs, was difficult.

The Automobile Club of Council Bluffs, Iowa, conducted evaluations of oiled road surfaces in 1908. Some of the best maintained natural road surfaces were located in this area and were used to compare with the modern oiling methods. It was decided in favor of the oiled surface in quality of the surface as well as ease of maintenance.

The 1908 Buicks used a new timer (distributor) that was mounted vertically so that it could be checked and adjusted easily. It also operated at half the speed of most other timers.

A new Chalmers-Detroit "30" automobile was tested for 20,000 miles by driving the car between Detroit and Pontiac, Michigan, four times each day. The actual mileage was 208 for each day for 100 days for a total of 20,800 miles.

In August 1907, a company known as the Pontiac Buggy Co. was formed in Pontiac, Michigan, by such notables as E.M. Murphy, Alanson P. Brush and Frank E. Kirby to design and produce a 20 hp automobile in the $1000-$1500 price range. The Company absorbed the Dunlap Vehicle Co. and established the Oakland Motor Car Co. in 1908. Brush had designed the Cadillac engine.

The Velie Motor Vehicle Co. made their first automobile in November 1908. The rakish 30 hp model A touring machine was able to attain 45 mph. Delivery to dealers was handled by Deere & Co. distributors. It seemed that newly introduced cars were both better and cheaper than models by established manufacturers. The cost of change was high and many had expended much of the engineering talent and operating budget to design the first model.

Oxy-Acetylene welding process, new in 1908, became a valuable repair tool for mechanics servicing automobiles of that time. An axle could be straightened without removing it from the car.

A successful test of wireless telephony was conducted at sea between the British cruiser *Furious* and the schoolship *Vernon*, both at full speed and separated by about 50 nautical miles. The DeForest system was used and Mrs. DeForest received the message, which was mostly stock quotations. Out of 154 figures, there were only two mistakes. In the later part of 1909, the Great Northern steamship *Minnesota* sailed from Seattle, Washington, to Yokohama, Japan, and was able to communicate by wireless with one city or the other at all times.

The Winnipeg Motor Competition was held in July 1908. The soft and muddy fields really tested some of the traction engines. Field tests to evaluate equipment began to be conducted at different parts of the country. The Winnipeg Motor Contests were held each year through 1913.

Wet weather during the harvest of 1908 resulted in several successful experiments with gasoline power for binders. The additional 400 pounds (about) was compensated on some units by moving the bull wheel approximately 4 inches to the rear. Since the wheel no longer provided power for cutting and binding, it was necessary to use only two horses to pull the unit. Drive attachments were soon available as either a tag-along or mounted unit.

William Butler of Carthage, Missouri, invented an auto thresher that he called "New Process." If was self-propelled and powered by a gasoline engine.

The Acme Harvesting Machine Co. of Peoria, Illinois, was reincorporated. The Emerson-Brantingham Co. was reorganized again in 1908. The company was renamed and reorganized twice earlier, first called the Talcott-Emerson Co. until Talcott died, then the Emerson Manufacturing Co. The Moline Plow Co. purchased the Freeport Carriage Co. of Freeport, Illinois. Dr. Edward A. Rumely and John Secor began designing the Oil Pull Tractor. Dr. Rumely was the head of the Interlaken school at La Porte, Indiana, and M. Rumely & Co.

In 1908 and 1909, farmers were readily accepting the automobile, but because of the difference in city and country roads, they were not yet interested in the large, high powered models. Country roads required light weight models with tall wheels and narrow tires. It was predicted that the "car of the future" would be small, trouble free and inexpensive. The B.F. Goodrich Co. of Akron, Ohio, began testing a truck rated at 5 tons, driven by a 50 horsepower, 4-cylinder engine.

The 1909 Ramblers were fitted with a spare tire mounted on a wheel that was complete except for the hub center. This revolutionary idea required only three minutes to remove the nuts from the six bolts and remove the complete tire and wheel. The spare tire was of course a handy item, but it was thought that the spare wheel could also be valuable in case of trouble. It was noted that the same spare fit on either the front or the rear. Another feature of the Rambler was the interchangeability of parts. "A duplicate of every part ever made for any Rambler car is kept in stock and orders are filled within twenty-four hours." In 1908, the company claimed "Fourteen thousand Ramblers have been built and sold and every single one of them is still running today."

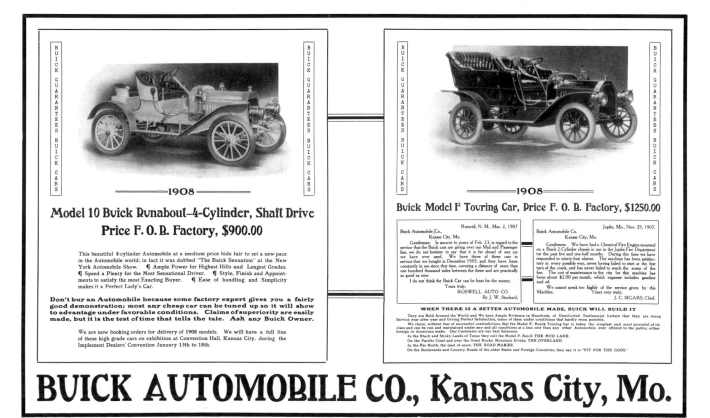

Everyone was aware that repairs were a problem and some noticed that it was really difficult to identify what part to order from the factory. Parts were ordered by name in 1909, but soon parts would be numbered to help ordering, shipping and receiving the one that was needed.

Theodore Roosevelt was still much admired, especially regarding the intelligent use of our resources. President Taft's speech about the same time at Winona, Minnesota, in defense of the Aldrich-Payne tariff bill was a real blow to farm related interests. The Gunnison, Colorado, Tunnel was completed on September 23, 1909 and the gates were formally opened by President Taft. The cost of construction was $3 million and 15 lives. About September 1909, gold was discovered near Salida, Colorado and "The Farmers Gold Mine" was explained to be his manure pile. Agricultural experimental stations devoted considerable attention to the subject of fertilizers. Some of the products sold as fertilizer were analyzed and found to be of little or no value. Dealers were called to assist in stopping the fraud of some factories.

A Minnesota firm introduced a simple hitch to pull a binder with a tractor instead of a horse. The American Seeding Machine Co. purchased the P.P. Mast Co. (makers of "Buckeye" farm equipment) of Springfield, Ohio.

THE MANURE SPREADER IS A BOON TO FARMER AND DEALER ALIKE.

In 1909, the Avery Co. of Peoria, Illinois, secured exclusive rights to the Cockshutt engine gang plow, the John Deere Plow Co. purchased the Marseilles Mfg. Co. and the Moline Plow Co. purchased the Monitor Drill Co. of St. Louis Park, Minnesota. The Western Malleable & Gray Iron Co. moved to Port Washington, Wisconsin. The J.I. Case Plow Works began to sell the Ohio automobile in November. The Common Sense Harrow & Mfg. Co. incorporated at Windsor, Missouri and the Harley-Davidson Motor Co. increased their capital from $35,000 to $100,000. The W.H. Kiblinger Co. changed its name to W.H. McIntyre on the first of January at Auburn, Indiana. Senator F.L. Maytag bought and incorporated Maytag-Mason Automobile Co. in Jasper county, Iowa, on December 5. George White Buggy Co. was formed in Rock Island, Illinois, to begin making automobiles and the Stratton Carriage Mfg. Co. of Muncie, Indiana, began manufacturing automobiles.

Rumors began in 1909 that Henry Ford was inventing a new automobile for farms that would combine all of the essential details of crop work, such as plowing, harrowing, seeding and rolling into one operation. A four gang (bottom) plow was claimed to be perfected for the Ford, but it was said that Henry wanted to perfect a six gang before introduction. A suit against the Ford Motor Car Co. by George B. Selden was tried in the courts in 1910. Selden won in 1910 and got automobile manufacturers worried, but the decision was overruled in 1911. Selden held the patent for the gasoline powered automobile. Henry Ford and the Ford Motor Co. were very visible and were always in the news for something.

Infringements of Selden Automobile Patents.

After nine years of litigation, a decision has been rendered in the celebrated Selden patent case. The decision of the United States Circuit Court sustains the claims made by George Selden that automobile manufacturers have been infringing his patents. Manufacturers of the Ford machine and of the Panchard French car were the main defendants. It was held that the Selden patent was a pioneer one, and had precedence over all machines using the features of the patent. The Selden patent is so far-reaching that it covers almost every modern car, and the decision affects the entire automobile industry. It is needless to say that the decision is disappointing to automobile manufacturers, for the decision was vital to the industry. The patent will expire in Nov. 1912, and its validity may never be finally settled before that time, if an appeal is taken.

Corning Dealer Poisoned.

A. T. Wheeler, a prominent implement and lumber dealer of Corning, Ia., was poisoned October 23d, by drinking from a bottle of cough medicine, sent to him by some person in Chicago. His life was saved by prompt work of physicians, who worked with him that afternoon and all night.

The bottle, labeled Foley's Honey and Tar, came in the same mail with a letter urging him to try the medicine. It was found to contain a solution of strychnine, in sufficient quantity to kill several men.

Mr. Wheeler thinks the sender of the poison will be found, although that has not been accomplished thus far.

Electric Light Bulbs

Drawn Wire Tungsten

Clear Bowl, Standard Edison Base, Can be used on Either Direct or Alternating Current, and Burned at Any Angle.

No.	15110	20110	25110	40110	60110
Watts	15	20	25	40	60
Volts	110	110	110	110	110
Candle Power	12	16	20	32	50
Size, inches	2⅛x4½	2⅜x5¼	2⅜x5¼	2⅝x5⅜	3 1/16x7⅛
No. in Package	100	100	100	100	50
Weight per Package, lbs.	45	45	45	60	50

These New Drawn Wire Tungsten Lamps produce a pure white light which so clearly resembles natural daylight that colors and shades can be accurately matched. The saving of Electric Current and its long life makes it a more economical lamp than the ordinary Carbon Lamp; it uses less current, produces more light and lasts longer than any other lamp. The 25 and 40 Watt Lamps are the most satisfactory for residence use. The 15 and 20 watt for use where not so much light is required as on the Regular Lamps.

In 1909, the Colby Combined Harvester & Thresher Co. began in Walla Walla, Washington. Combines began in the Dayton, Washington, area about 1899, then around 1901, they were discarded and considered impractical, because they were horse killers. It was necessary to have two teams of 26 to 32 horses to harvest and even then, the horses were worn out quickly. About 1905, improved lighter combines started gaining popularity, especially in Washington and northern California. About 5 men were required to operate a combined harvester; Driver, Separator tender, Header man, Sack man and Sack sewer. A combine with a 20 foot header could harvest about 35-40 acres per day, working both men and horses very hard.

The nickle was considered the most important coin in the United States and white oak wood is in short supply. General Frederick D. Grant recommended to the War Department that laws should be enacted that would oblige private owners to turn over to the government on demand their automobiles at the first cost of the machine in case of war. The Goodyear Tire & Rubber Co. of Akron, Ohio, increased its capitalization from $2 million to $6 million.

H. Luke Owen, a young mechanic of York, Pennsylvania, was reported to have sold his patents to an invention of a fibrous asbestos fireproof covering for wire, for the sum of $100,000 to the General Electric Co. of Schenectady, New York. The company hired the inventor at a large salary to superintend the manufacture of the new product.

Motor Car Laws.

It is wise to be prepared beforehand when taking a trip in a motor anywhere away from home. In fact, a good many persons owning cars apparently do not give a thought to legal regulations until they happen to run up against some zealous official who proceeds to make it warm—and expensive—for them. The gist of the laws relating to motor cars in some of the Western states is given below, condensed from a summary of the laws of the various states, in Motor:

California—Speed limit 10 miles per hour in built-up sections of municipalities, 15 miles elsewhere; 20 miles outside of municipalities; 4 miles on approaching bridge, dam, sharp curve or steep descent; must stop on signal from driver of restive horse.

Iowa—Substantially the same.

Kansas—Practically the same.

Minnesota—Outside speed limit 25 miles, one mile in 8 minutes on curves and approaches and steep descents; "joy riding" a misdemeanor.

Missouri—Eight miles in business portions of municipality, 10 miles in other portions, 6 miles at curves and corners; elsewhere, 15 miles. Must stop on signal.

Montana—Substantially the same, with outside limit of 20 miles.

Nebraska—Practically the same.

North Dakota—Eight miles within any village, town or city, 25 miles outside.

Oregon—Eight miles within thickly settled or business portions of villages or cities, 8 miles in country when within 100 yards of any horse-drawn vehicle, 4 miles per hour at crossings; elsewhere outside of villages and cities 24 miles.

South Dakota—Ten, 15 and 20-mile limits.

Texas—Eight to 18 miles per hour. Racing on highways is prohibited.

Utah—Speed limits 10, 15 and 20 miles, with 6-mile speed on approaching crossing, descent, curve or street intersection. Speed rates cannot be lowered by local authorities.

Washington—Twelve and 24 miles per hour allowed, with rate of 4 miles in business portions of municipalities or over crossings. Must turn to right in passing or overtaking vehicles or persons.

Penalties for violations range up to $500, or in some cases, as in Missouri, include imprisonment. All provide for registration fee and displaying of license number, with temporary exemptions for cars showing license tags from other states.

Auto Carries Telephone.

Something new in telephone work during an automobile run was developed the other day in Texas. One of the numerous reliability runs was conducted between San Antonio and Dallas. Among the entrants was D. A. Walker, the president of the telephone company, with a big Rambler car carrying a portable telephone. By means of a long fishing pole with a hook at the end Mr. Walker was enabled at any time to ring up any connecting point along the lines without getting out of his car. The usefulness of this member of the touring committee can be appreciated by some of the things which Mr. Walker did while en route. One day, while many miles from any station, he made arrangements with the governor of Texas to be entertained by the auto men at dinner where the night stop was made. If a car broke down the wire along the roadside was tapped and messages for relief were promptly sent. Points ahead were kept well informed of the progress of the cars from time to time, and the usefulness of the plan was demonstrated many times.

Here's a Tongue-Twister.

Grenville Kleiser, instructor of the Public Speaking Club of America, and formerly a member of the faculty at Yale, whose headquarters is at the West side Y. M. C. A. in New York, defies anybody to repeat accurately from memory the following tongue-twister: "Esau Wood sawed wood. Esau Wood would saw wood. All the wood Esau Wood saw Esau Wood would saw. In other words, all the wood Esau saw to saw Esau sought to saw. Oh, the wood Wood would saw! And, Oh, the wood saw with which Wood would saw wood! But one day Wood's wood saw would saw no wood, and thus the wood Wood sawed was not the wood Wood would saw if Wood's saw would saw wood. Now, Wood would saw wood with a wood saw that would saw wood, so Esau sought a saw that would saw wood. One day Esau saw a saw saw wood as no other wood-saw Wood saw. In fact, of all the wood-saws Wood ever saw saw wood, Wood never saw a wood-saw that would saw wood as the wood-saw Wood saw saw wood would saw wood, and I never saw a wood-saw that would saw wood as the wood-saw Wood saw would saw until I saw Esau Wood saw wood with the wood-saw Wood saw saw wood. Now Wood saws wood with the wood-saw Wood saw saw wood."

A STUNT WITH A McINTYRE MOTOR VEHICLE.

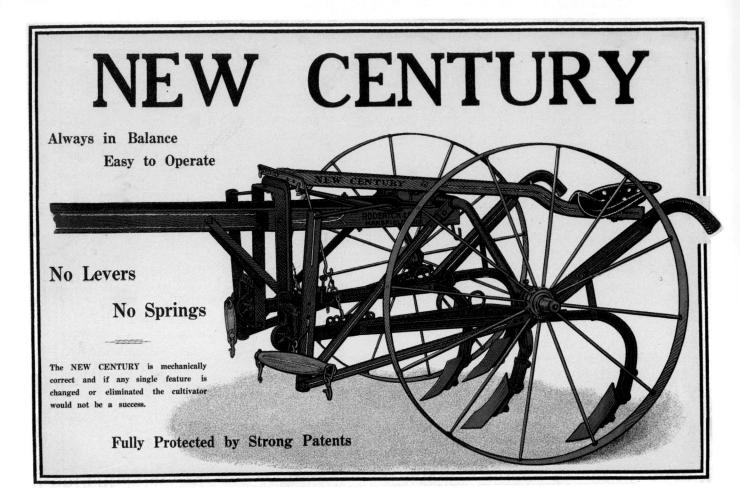

An Anti-Flicker Device.

The peculiar flicker of the moving picture has given rise. it is said, to an entirely new disease of the eye among habitues of the ten-cent celluloid theater. To overcome this objectionable scintillation Mr. Thomas A. Edison has recently taken out a patent for an "anti-flicker device for viewing moving pictures." His invention consists of leaves or members provided with small eyeholes about the size of the pupil of the eye, which may be held in front of the eye and through which the moving pictures are viewed. Mr. Edison states that such a device reduces considerably the objectionable flicker defect, when the device is held a short distance in front of the eyes. He suggests that the great beneficial effect may result in some measure from the cutting down of the supply of light entering the eye when the shutter is open, thus reducing the baneful shock caused by the impact of light on the retina when the shutter is fully opened.—Scientific American.

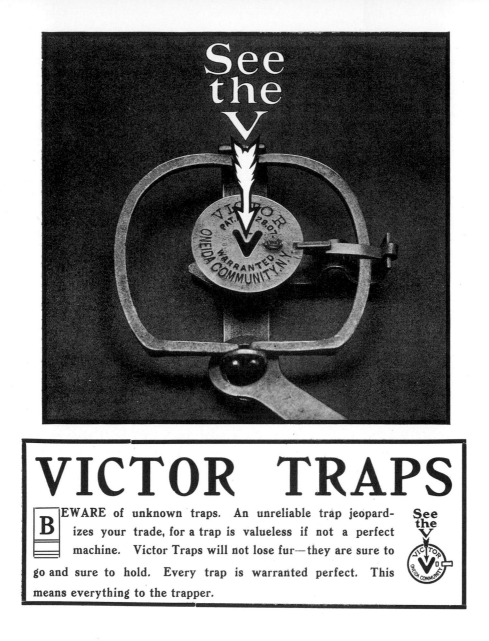

VICTOR TRAPS

BEWARE of unknown traps. An unreliable trap jeopard-
izes your trade, for a trap is valueless if not a perfect
machine. Victor Traps will not lose fur—they are sure to
go and sure to hold. Every trap is warranted perfect. This
means everything to the trapper.

775J—7 -inch front, each	$2.80
776J—8¾-inch front, each	2.90
777J—9½-inch front, each	3.40

The two most important accessories to the **AUTOMOBILE.** Dealers make big
money by handling Automobile Supplies. Send for Special Booklet No. 16,
and see what low prices are now quoted. These Booklets sent to dealers only.
ALL GOODS Guaranteed. The Pioneer Automobile Supply House of Kansas City.

Winslow's Skates
Are Best for
Men, Women and Children

Dictating Letters by Telephone.

In many busy offices the phonograph has entirely displaced the stenographer as the intermediary between the dictator and the typewriting machine. The time of the typist is thus economized, and the dictation may be recorded when any at any rate desired.

A further improvements has recently been developed, which is thus described in the Scientific American: The phonograph has been entirely eliminated from the dictator's office. In its stead a desk telephone is used, and the words spoken into the transmitter are conducted to the typist's room, where they are automatically recorded on one of a battery of wax cylinders, ready to be reproduced and transcribed on the typewriter. Each desk in an office may have its own recording cylinder, so that any number of persons can dictate simultaneously. Thus, instead of having separate phonographs, the apparatus is all centralized in a single multiple recorder, which also includes a reproducer used by the typist in transcribing the dictation on the typewriter. One of the accompanying illustrations shows a recorder cabinet provided with five recorders and one reproducer.

CABINET WITH FIVE RECORDS.

The desk telephone is not furnished with a receiver, as all communication between the dictator and operator regarding the operation of the recorder is carried on by automatic signals. Wherever one wishes to dictate a letter, he picks up the transmitter on his desk, and, in doing so, unconsciously grips and depresses a button in the standard of the instrument. This closes the circuit of a magnetic clutch in the recorder cabinet, whereby the drum which carries the wax cylinder operated upon by this particular transmitter is set in motion. The same button closes the circuit of a signal lamp at the top of the telephone standard, and as long as this glows, the dictator is aware that the recorder is operating properly.

The cylinder will take an eight-minute dictation before it needs to be replaced. At any time, if the dictation is interrupted, the cylinder may be stopped by releasing the button on the transmitter. Half a minute before the cylinder is filled, a switch is thrown automatically which extinguishes the signal lamp, giving the dictator warning that he must stop at the end of the next sentence

OPERATING MECHANISM OF A RECORDER.

or paragraph. At the same time a buzzer is sounded in the cabinet which notifies the operator that a cylinder must be replaced, and a lamp opposite the cylinder glows and indicates which one needs replacing. As soon as a new cylinder is placed on the drum, and the recorder is ready for further dictation, the fact is signaled to the dictator by the relighting of the lamp on his transmitter. In the meantime, while one or more persons are dictating, the typist may be transcribing the dictation of previously filled cylinders. The reproducing instrument may be controlled by pressing a button with the foot, so that the typist may take the dictation sentence by sentence as desired, and has the instrument under control, while her hands are free for typewriting.

Efforts to record the vibration of a telephone receiver on a wax cylinder have heretofore proved unsuccessful. The main difficulty has been to produce a mechanism which would be sufficiently sensitive, and which at the same time would follow the eccentricities of the wax cylinder. This difficulty has now been overcome very cleverly. Pictured in one of the illustrations is a recorder inverted to show the operating mechanism. It comprises a pair of electro-magnets A, which acts upon an armature of soft iron mounted on a strap of thin metal B, which serves as a diaphragm. Instead of connecting the diaphragm directly with the recording stylus, an arm C is hinged immediately below it. Fulcrumed in this arm

is a lever D, which at one end is connected by a link to the diaphragm B, and at the other end carries the recording stylus E. The arm C is provided with a weight at its free end which is sufficient to hold the stylus against the cylinder. Any eccentricity of the wax cylinder is taken up by the motion of the arm C, which carries the fulcrum of the lever D up and down without interrupting the vibrations communicated from diaphragm B to the stylus through the link. The shaft on which the recorder is mounted is hollow, and within it is the screw which feeds the stylus axially along the cylinder. The shaft is slotted to admit a nut which is carried by the recorder. When resetting the recorder, a cam lever is operated which throws this nut out of mesh with the screw, and breaks the circuit of the magnet A, as well as of the dictator's signal lamp.

The automatic system of signals between the telephone and the cabinet is very complete, and so simple as to be understood by any one. The business man has at his beck and call a mechanical ear, in which he can dictate his letters without the distraction of adjusting or regulating a machine. The device is ready to record his words in or out of office hours. It might be so arranged as to permit him to dictate letters from his home. For long dictation an automatic relay may be used to connect the transmitter with a new cylinder as soon as the first is filled. The dictation may thus be continued without interruption, indefinitely.

The Holt Mfg. Co. of Stockton, California, secured control of the Best Mfg. Co. of San Leandro, California, and the Northwestern Harvester Co. of Spokane, Washington, in 1909. The Spokane Harvester Co. had been reorganized as the Northwestern Harvester Co. Pliny E. Holt had organized the company to build the "Caterpiller" that he had invented, but by 1909 he controlled or owned every known patent on combined harvesting and threshing machines. The Holt companies made Daniel Best models at San Leandro, California; Haines-Houser, Holt and Holley Jr. at Stockton, California; McRae models at Spokane, Washington. "Caterpiller" traction engines, which had grown popular on the sandy soils of the Pacific coast, began to be made at the old Colean plant at Peoria, Illinois. By 1910, "Caterpillar" tractors were being used in the Northwest in combination with the combined harvesters to reap about 65 acres each day instead of the 25 acres using horses and mules. It was hoped that soon it would be possible to cut, thresh and bag the grain and to plow, harrow and seed the land in one operation. Later, in 1913, the Holt concerns were merged as the Holt Mfg. Co. with capital of $3 million. The affected companies included:

Holt Mfg. Co. of Stockton, California
Holt Caterpillar Co. of Peoria Illinois
Best Mfg. Co. of San Leandro, California
Houser & Haines Mfg. Co. of Stockton, California
Canadian Holt Co., Ltd. of Calgary, Alberta, Canada
Aurora Engine Co. of Stockton, California

The Weber Gas & Gasoline Engine Co. of Kansas City, Missouri, entered receivership in the middle of February, 1909.

In September 1909, the General Motors Company, recently incorporated with a capital stock of $12.5 million, reincorporated in New Jersey and increased the value of the stock to $60 million. The General Motors Co. was organized as a holding company for Buick Motor Co., Cadillac Motor Co., Oldsmobile Co., Oakland Motor Co. and

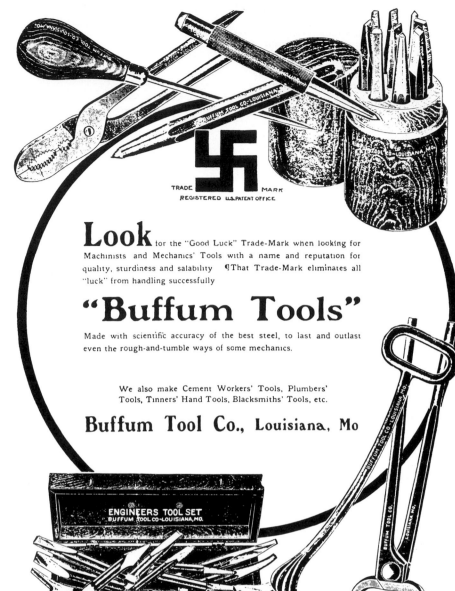

Reliance & Rapid Truck Companies. On November 1, Cartercar was added to the group of General Motors Companies.

In 1909, Bleriot flew across the English Channel and the Wright Brothers Aeroplane sold for $7,500, but prices were expected to drop. "But then, that's the way the automobile started."

The first Paris Aerial Navigation exposition took place on October 18-31, 1909. The Black Crow Manufacturing Co. was incorporated at Babylon, Illinois, to manufacture airships by the owners F.W. Moore, Isaac Hubbell and Maggie Hubbell. People expected flight to be within the reach of everyone by 1925.

Henry H. Timken died on March 15, 1909 of heart failure at age 78. Noble M. Davidson of Ada, Ohio, said to be the inventor of the traction engine in 1877, died on March 29. Davidson searched for and found his original traction engine, purchased it and had it shipped to his home just before his death. Alden C. Millard, publisher of Millard's Directory, died on June 23 at his home in Independence, Missouri. Publication of the Millard's Directory was continued by the Implement Trade Journal. E.M. Murphy died on September 4 of apoplexy.

TRUST THE TRUSS

"The crying need of the farmer today (1910) is a reasonably priced all-purpose tractor which shall not only be capable of plowing, but which in addition shall be able to perform the other offices of the field, farmyard and road." In January 1910, the Minneapolis Steel & Machinery Co. selected the Joy-Willson Co. of Minneapolis to develop a four cylinder tractor and to build five samples. Walter J. McVicker was the engineer of the project and later became the engineer for the Minneapolis Steel & Machinery Co. In 1911, the Minneapolis Threshing Machine Co. contracted with McVicker to design a tractor for that company. McVicker finally quit Minneapolis Steel & Machinery Co. in about 1918 to begin work for the Minneapolis Threshing Machine Co. The first Oil Pull tractor was sold on February 21, 1910, but one hundred were sold by the end of the first year.

EDDIE RICKENBACKER IN THE FIRESTONE-COLUMBUS RACER.

Plowing by Man-Power.

The accompanying illustration shows Ira Nunn, an employe of the Omaha office of the J. I. Case Plow Works, pulling a two bottom Case gang, cutting a 6-inch furrow. This unique method of demonstrating the extremely light draft of their plows was recently adopted by the Case people, and Ira has been kept pretty busy during the last few weeks, visiting county and district fairs. He puts on this stunt in connection with a "plow talk" by J. A. Hamilton, one of the factory experts. When Ira, who is 6 feet in height and weighs 225 pounds, begins to get into his harness, the farmers begin to gather around. By the time he has made a test or two there is usually a pretty good crowd of spectators and standing room is at a premium. Then Mr. Hamilton takes the matter in hand, and in an interesting and convincing manner, tell his hearers all about Case plows. Then Ira gets busy and "shows" those from Missouri. He is capable of pulling a two-bottom gang a distance of 100 yards, although he admits "it gets him a little" to do it. Pulling a plow cutting a 6-inch furrow is no snap at any time, but just think

back a week or so and remember some of those blistering afternoons when it was uncomfortable just to sit around. Then imagine Ira out there in the hot sun hitched to a plow, doing a horse's work. This is one place where that familiar line, "A boy can do it," won't fit.

Ira Nunn is just an ordinary individual and possessed of only ordinary powers. While it is no great task for him to lift the plows out of a two bottom gang and carry them away on his shoulder, there is many another man can do the same thing. He says that pulling a plow in the ground is hard work, but then he is used to hard work, and doing his turn once or twice a day on the fair grounds is better than handling stock in the warehouse from 7 a. m. till 5:30 p. m. The crowd shown in the photograph shows how the stunt "took" at Sac City, Ia., recently.

Firestone Makes Fast Time.

The Firestone-Columbus racing car, driven by E. Reichenbacher for the Racine-Sattley Co., Omaha, carried off a majority of the honors at the two days' meet held in Omaha Saturday and Sunday of last week under the auspices of the Omaha Motor Club. The Firestone was entered in nine events and won eight of them, having fallen down on the five-mile obstacle race, which was won by the Velie. In this event the drivers were required to stop their cars, kill their engines and start again after each mile. The interest shown in the races was extremely gratifying to the promoters. The announcement was made that the professional Eastern drivers would appear as scheduled, and the crowds filled the grand-

stand and the infield and it has been estimated that fully six hundred automobiles were parked inside the inclosure. Summaries of the two days' events are as follows:

SATURDAY.

Ten miles, for fully equipped touring cars, for Western Automobile Supply Co. trophy: Cadillac (Reim) first, Midland (Ashley) second. Time: 11:45. Protested by Chalmers (Frederickson). Judges withhold decision.

Ten miles, for cars selling for $2,000 or less, for Storz Brewing Co. cup: Firestone-Columbus (Reichenbacher) won, Velie (Stickney) second, Cadillac (Reim) third. Time: 11:25 2-5, 11:27, 11:43 2-5.

Ten miles, for cars selling for $3,000 or less: Firestone-Columbus won, Warren-Detroit second, Chalmers third. Time: 11:07 3-5, 11:54 1-5, 12:07 2-5.

Five-mile obstacle race: Velie won, Chalmers second. Protest entered by Chalmers.

Ten miles, for motorcycles: Indian (Bell) won, Indian (Huth) second Indian (Gamble) third. Time: 10:25 1-5, 10:35, 11:03 3-5.

Twenty-five-mile free-for-all: Firestone-Columbus won, Cadillac second, Midland third. Time: 26:57¼, 30:21 1-5, 31:23 1-5.

SUNDAY.

Five miles, for stock cars: Chalmers (H. E. Frederickson) first, Midland (Ashley) second; Stevens-Duryea (Harry Woodruff) third. Time: 6:05, 6:24 3-5, 6:31 2-5.

Five miles, for cars selling at $1,000 or less: Hudson (William Bruner) first. Time: 6:25.

Ten miles, for cars selling at $1,500 or less: Firestone-Columbus (E. Reichenbacher)

first, Hudson (William Bruner) second, Hupmobile (Walter Smith) third. Time: 13:13 4-5, 13:22, 13:32.

Ten miles, for cars selling at $2,000 or less: Firestone-Columbus (E. Reichenbacher) first, Velie (J. Stickney) second, Cadillac (Nygard) third. Time: 11:06, 11:13.

Ten miles, for motorcycles: Indian (George Gamble) first, Excelsior (Otto Ramer) second, Excelsior (Ralph Bates) third. Time: 11:48 1-5, 11:56 3-5, 12:18 3-5.

Flying start mile, for Rome Miller trophy: Firestone-Columbus (Reichenbacher) first, Chalmers (Frederickson) second. Time: 1:07 3-5, 1:08 3-5.

Twenty miles, free-for-all: Firestone-Columbus (Reichenbacher) first, Chalmers (Frederickson) second, Velie (J. Stickney) third. Time: 21:45 2-5, 22:14½, 22:24 2-5.

The Luebben Cylindrical Baler.

It is universally conceded that the right method of taking care of the alfalfa crop has up to date, not been figured out. No crop suffers so much through frequent handling as alfalfa because the leaves, which are the most essential part of the plant, are lost through repeated handling. Then again, too much ground is used in stacking, especially where there are several crops harvested in one year, as is the case in this territory.

After fifteen years of experimenting along the line of a baler for taking care of alfalfa H. H. Luebben of the Luebben Baler Co., Beatrice, Neb., has placed upon the market a machine that succeeds in saving not only the alfalfa but the expense of putting the product into the stack.

The secret of this machine is that it makes a round bale with an open core in the center of the bale. This enables the farmer to take the alfalfa direct from the windrow and put it into bales without any danger of spoiling through heat.

Bailing from the windrow saves the 20 percent loss which is unavoidable in any other method of taking care of the alfalfa more for alfalfa put up with the Luebben baler than for ordinary baled alfalfa. The bale that is made by the Luebben process is

LUEBBEN BALE, SHOWING HOLE IN CENTER.

waterproof and can be stacked up in the field and left to the inclemency of the weather without any bad effects whatever.

It is confidently expected that this machine will revolutionize the process of putting up alfalfa.

To Push Our Foreign Trade.

An association was organized in New York city on the 12th inst., for the purpose of furthering the export business of the United States. The various manufacturing companies interested have a combined capitalization of $250,000,000, and include such concerns as the International Harvester Co., Studebaker Bros. Mfg. Co., the Westinghouse Co., Swift & Co., N. K. Fairbanks Co., American Steam Pump Co., National Cash Register Co., Yale & Towne Mfg. Co., E. I. Dupont de Nemours Co., Welsbach Co., Eastman Kodak Co., Henry Disston & Sons, Oliver Typewriter Co. and Victor Talking Machine Co. The object, as stated, is to bring into close touch the exporting manufacturers of this country and the manufacturers themselves. It seems to be a fact that while a few of our manufacturers have systematically and successfully gone after trade in certain sections of the outside world, in other sections that business has been left almost entirely to the tender mercies of the European manufacturers, who have taken advantage of their opportunity and established their trade firmly and profitably, without competition from our shores. Among other direct objects sought will be the establishment of equitable freight rates which will enable the American manufacturers to compete, the protection of trade marks, and practical assistance in cases of clashes with foreign customs authorities.

There would appear to be a field for practical endeavor for the new organization.

One of the notable instances of improvement in methods of farming, even to the extent of revolutionizing previously accepted ideas, is exemplified in the invention and use of the manure spreader. Not only has its value to the worn-out farms of the older sections of the country become generally recognized, but within the past five years it has been introduced into the West, a campaign of education as to its value and use has been carried on until both farmer and implement dealer have come to regard its sale and use as an established factor in the conduct of every well-regulated farm.

The writer recalls his experience of a few years ago when soliciting advertising contracts from certain manufacturers of manure spreaders. These gentlemen were selling their product in limited amounts east of the Mississippi river, and a very few in Iowa; but they ridiculed the idea of ever being able to introduce this line of machinery into the trans-Missouri country. "Why," they said, scornfully, "those farmers out there don't know what manure is for!" The same manufacturers have since found The Weekly Implement Trade Journal a valuable medium through which to interest the dealers in all sections of the West in an article which the agricultural colleges and experiment stations, through the farm papers, have taught the farmers to appreciate and to buy.

It is in acknowledgment of the debt the implement manufacturers and dealers owe to these agricultural colleges and experiment stations—the "highbrows" of agriculture, to use a term of the period—that we give space in this issue to a symposium on the manure spreader, from the pens of professors in the state agricultural colleges of Nebraska, Kansas, Missouri and Iowa. These articles, written especially for the Implement Trade Journal, show the esteem in which the spreader is held by these men and institutions which are educating the farmers of this country to make a real success of their calling; to elevate their work to the dignity of a business; and to enable them at once to keep pace with the increasing demand for their products and with the necessity of greatly increasing the yield of their crops in a measure proportionate to the great enhancement of land values. Their articles are all well worth reading, and comprise, altogether, probably the most notable symposium on the use and value of the manure spreader ever compiled by any publication for a single issue.

In this connection, also, is an article by Mr. Floyd R. Todd of the Kemp & Burpee Mfg. Co., Syracuse, N. Y., which concern was the first to build a manure spreader for the trade and to perfect the mechanical principles which are the basis of all those which have followed. An illustration is shown in connection with Mr. Todd's article of this first spreader, which adds interest to the story.

A word as to the practical side of the spreader business to the dealer:

There has been some complaint of some manufacturers passing the dealer by and selling direct. Some of these complaints undoubtedly have been well founded; others have been due to the apathy of the dealer in the matter of a proper appreciation of the possibilities of the business to him, and a refusal to assume any responsibility therein,

although he was perfectly willing to claim and receive a commission on sales in which he had no part. But the business is getting on a more satisfactory basis all round, and both dealers and manufacturers are finding out that manure spreaders are becoming as staple as wagons—but there is no reason why they should not carry a much more satisfactory profit. We believe that the manufacturers and the jobbers desire to protect the dealers, if the latter are inclined to meet them half way; and this is as fair a proposition as any reasonable minded man can expect or ask.

Quite a number of manufacturers present their lines of spreaders in this issue, for the consideration of our readers, and we commend their ads and their descriptions of their lines to your thoughtful consideration.

Remember, in this connection, that along with manure spreaders naturally go a number of co-ordinate lines for which there is a steady, or coming demand, such as manure carriers and loaders, sanitary cow stalls and stanchions, cream separators and cow milkers, etc., all of which represent directly a good profit and which indirectly mean better farming and consequently more money for the farmer to spend with you for other lines, while at the same time adding to his own wealth and to that of the nation.

Use of the Manure Spreader in Nebraska.

By C. K. SHEDD, University of Nebraska, Department of Agricultural Engineering.

When Nebraska soil was first broken up, it was said to be inexhaustible. The early settlers thought that there was no possibility of so rich and deep a soil becoming barren. In the early days the application of manure did not seem to be a paying proposition; at least, there was no apparent beneficial result except getting the manure away from the farm buildings. After the land had been farmed a few years, however, the application of manure began to show visible increases in crop yields.

At the present time manure, when properly applied, will give beneficial results in nearly every part of the state. The profit from applying manure is, of course, much larger in the eastern part of the state, where the rainfall is heavier and the land has been farmed longer.

There is no data available to show the number of manure spreaders in use in Nebraska at the present time, or the number

sold every year. It is a safe guess, though, that a majority of the farmers in the eastern half of the state use manure spreaders.

Correspondence by the Department of Agronomy and Farm Management, University of Nebraska, shows that of the farmer who are enough interested in progressive farming to keep in touch with the Experiment Station practically all are applying barnyard manure to their farms. Only a few cases have been reported where beneficial results were not observed. These cases are generally from the western part of the state where the climate is dry and manure must be applied as a light top dressing.

About three-fourths of the farmers mentioned above are applying their manure with the manure spreader. Farmers using spreaders seem to be unanimous in considering it a good investment, not only because it saves disagreeable hand labor, but also because it applies the manure to the land in much better shape.

The Manure Spreader in Kansas.

By C. F. CHASE, Assistant in Farm Mechanics, Kansas State Agricultural College, Manhattan, Kan.

Many Eastern farmers hold that manure applied to the land with the manure spreader is worth 10 to 20 percent more than when applied by hand. The average Kansas farm produces 200 tons of manure each year. One man can haul this to the field and spread in 28 days at a cost of $84, allowing three dollars a day for man and team. By using the spreader the same manure can be put on the land at an expense of $60, allowing besides the expense of labor 15 percent depreciation on machine and 10 percent interest on money invested in the spreader.

When the manure is fibrous and breaks in chunks as it is loaded, it is next to impossible to break up these chunks when spreading by hand. The manure spreader in most instances will tear these chunks into moderately fine particles that are easily worked into the ground. It is certainly true that manure applied by the latter method is worth much more than when applied by the farmer.

For convenience let us take the Eastern farmers' estimates, thus: A ton of manure when properly applied is worth at least $1.50, but by poor application it decreases

THE FIRST SUCCESSFUL MANURE SPREADER.

in value 15 to 30 cents per ton. We will lose then by hand spreading from $30 to $60 worth of manure on the average farm each year. The loss is due to leaching processes, the gases as ammonia escaping into the atmosphere, and to the uneven distribution. The facts are, when manure is to be spread by hand, a greater loss is sustained by allowing the manure to burn, rot, and wash from the barn lots than from any other cause. But, supposing every farmer cleans his barnyards of manure each year, we find he can save from $54 to $84 by using the spreader. Enough to pay for the spreader in two to three years.

These estimates, made with much care, show cause for the growing popularity of the manure spreader. It is only a question of a few years until every up-to-date farmer of our commonwealth will own a manure spreader, or at least a share in one. The farmers of Western Kansas and Nebraska, we are glad to say, are losing that erroneous idea that the application of manure to the land is dangerous practice. They are finding that the danger lies in the condition of the manure and the method of application, but not in the manure itself.

Scientists who have studied micro-organisms hold that the application of manure does more than add humus and plant food to the soil. They say it stimulates the growth of bacteria present in the soil and also adds certain other bacteria which in turn liberate or make available increased quantities of plant food. This will account for the favorable results obtained from the very light applications of manure, and in turn brings up another strong point in favor of the manure spreader; that is, the possibility of spreading the manure thin. To spread manure thin by hand requires an increased amount of labor and patience, but with the spreader it is just as easy to spread it thin as thick, though it will require a little more time, as more ground has to be covered.

During the past year there have been something over forty manure spreaders sold by retailers at Manhattan, Kan. If other towns have sold as many in proportion to population, over 3,000 spreaders were put in use in Kansas last year. There is not any better sign of continued prosperity in Kansas than this. It is not only economical, but by liberal use of the manure spreader we are paying our land for the crops it has furnished us, we are conserving our natural resources, we are saving the fertility of our land for our sons and daughters. No more economical and useful machine has been produced recently by our manufacturers than the manure spreader—the automobile and flying machine not excepted. Now that the machines are durable and operate successfully with moderate care, their use is highly recommended by the agricultural colleges.

In conclusion we may say: First, the manure spreader is entirely practical. Second, on the ordinary farm it will pay for itself in two or three years, and on a stock farm furnishing five hundred tons of manure each year it will pay for itself the first year. Third, with average conditions manure applied with the spreader is much more valuable than when applied by hand. Fourth, the manure spreader is the one great machine which is materially aiding the Western farmer in conserving our natural resources.

The Handling of Manure.

By C. B. HUTCHINSON, Missouri Experiment Station, University of Missouri, Columbia, Mo.

The saving of manure is a matter which is becoming more and more important for the farmers of the Middle West each year. There was a time when manure was of little value, so far as the immediate results were concerned, but on most lands that time has long since past. The decreased productiveness of our soils under grain and timothy hay farming, together with the great increase in the value of these lands in recent years, demands more careful systems of farm management if these soils are to pay interest on their valuation. In these systems the farmer should figure manure worth at least $2 per ton, and he should get that much or more from it by its proper handling and application. As a matter of fact, a ton of barnyard manure will frequently bring $3 or more in increased crop yields during the years following its application when its benefit can be observed.

Many farmers who feed practically all their crops on the farm still get little value from the manure produced. Too frequently they feed in a sheltered grove on a well drained hillside, where the manure is largely washed away and where such accumulations as occur are in the timber so that little return is secured from them. Again, it is customary to feed around the barn with no protection being given to the manure, and it frequently leaches until it loses half its fertilizing value. It is just as necessary, therefore, that the manure be properly handled and gotten back on the fields where it belongs, as it is that the cattle be fed at all, so far as the keeping up of the land is concerned. The time is upon us when every effort should be made to preserve and to return to the fields the fertilizing material derived from the crops fed if we are to maintain the fertility of our lands.

Where animals are fed in stables or sheds the least possible loss in handling manure will occur where it is hauled directly to the fields day by day as it is made. If this is not possible, it should be hauled at least once a week, although many farmers do not find even this practicable under the conditions that exist on most farms. Many farmers think, too, that where manure is scattered months before it is plowed under, much of its fertilizing value is lost; but accurate experiments have shown that this loss is very slight and very much less than where it is allowed to lie around the lots.

The next best system of saving the fertilizing constituents of manure, where animals are lot or stable-fed, is to feed under an open shed where the manure is allowed to accumulate and is kept moist and tramped down compactly by animals so as to exclude the air so far as possible. Such manure loses very little by fermentation and, since it is under cover, it is not reached by rain. It can be left in such a shed throughout the winter, or even until the middle of the summer following, without serious loss. Manure is best applied to corn land, and where crops are systematically rotated and corn follows clover and timothy, the manure can be applied to this sod in late summer, thus working in to excellent advantage in such a plan.

Somewhat more value from the manure can be secured by applying it as a top dressing and working it into the soil by disking thoroughly, but on most farms in the corn belt this is not so practicable as applying it to sod land and turning it under. Where top-dressed it is necessary to do the manure hauling at a time when the work on the farm is pressing. It often comes at a time when the excessive tramping of the ground incidental to the application of the manure is injurious to the land. If the ground is wet, the damage done by puddling the soil is often greater than the beneficial effect produced by the manure. Where the manure is applied to grass land and plowed under, it can be hauled at most any time during late summer, fall, or even during the winter, thus enabling the farmer to get a large part of the manure on the land soon after it is made and so minimize the loss of its fertilizing constituents.

The use of a manure spreader in handling the manure will meet many of the difficulties encountered in getting it back on the fields, and will be found advantageous on most farms. Most farmers believe that the greatest value to be derived from a spreader, however, is in the saving of labor. As a matter of fact, this is only one of the advantages offered by the use of such an implement. An advantage that is even more important in many ways is the fact that manure scattered evenly and rather thinly over a wide area will give very much more return per ton than where scattered irregularly and on the thinner places, as is usually done when distributed from a wagon. This increased return is due to the fact that there is less loss in fermentation when put on thinly, and also to the fact that a considerable share of the value of manure comes from the addition of beneficial bacteria to the soil, so that when this takes place over a wide area it is much more beneficial than when limited to a small area. Where the manure supply is limited, as is the case on most farms, it is a much better practice to spread the manure thinly over a large area than to apply it heavily on the thin places.

Another value of the manure spreader lies in the fact that a man who has his money invested in an implement of this sort will take better care of the manure on the farm. He will not allow it to lie around and leach, but will get it back on to the fields where it is needed. The scattering of manure by hand is a tedious matter and farmers very often neglect to haul it out on this account.

The manure spreader is one of the most economical implements on the farm, and farmers should understand more thoroughly the advantages to be derived from its use. No farm of one hundred acres or more where much stock is kept and fed around the buildings should be without this implement.

Origin and Rise of the Manure Spreader.

By FLOYD R. TODD, Vice-President Kemp & Burpee Mfg. Co., Syracuse, N. Y.

About the middle of the last century efforts were made by various inventors to perfect some machine which would satisfactorily spread manure. Some little experimenting was done along this line extending over a period from 1850 to 1877, but this was of a feeble nature and produced no practical results. In the meantime the necessity for the mechanical distribution of manure, both from the standpoint of labor saving and securing more efficient results, became recognized. Mr. J. S. Kemp, at that time residing in Magog, Can., finally developed a practical machine. It was patented in this country on May 1, 1877, and in the following year the Kemp & Burpee Mfg. Co. of Syracuse, N. Y., began work looking towards its commercial introduction. In 1880 the first successful machine was produced commercially by this company. This machine is now carefully preserved on their sample floor at Syracuse, N. Y., and the illustration thereof is found in these pages.

The beginning was necessarily on a limited scale. The tool was absolutely a new one to the American farmer, and its introduction difficult. The distribution of manure had been made at times when the farmer had nothing else to do. The soils had not become depleted to a point that economy in such distribution was thought necessary. Western lands were so fertile that the agriculturist believed that manure did them more harm than good. The farmers in this section of the country were drawing their manure into sluices and waste places and throwing it away. Therefore, during the first fifteen years in which the Kemp & Burpee Mfg. Co. labored to introduce this machine much discouragement was met with and progress was slow. During this period the machine itself was necessarily in an experimental stage.

It was soon discovered that a manure spreader is used under the most unfavorable conditions, at times when no other implement can be used on the farm. It is used in the fall, spring and winter, when the ground is either rough or frozen or very soft from the spring thaws. It was also found that manure varies much in weight and that frequently a farmer loading his machine high would put two or three tons into an ordinary sized spreader. The first machines constructed were not strong enough to meet these conditions. It took years to strengthen every part of the machine so that it would stand the strain incident to its use. One part was strengthened and mechanically improved only to find weaknesses somewhere else. This is the process of evolution through which every tool must go, and this, together with the apathy on the part of the farmer, made progress slow.

During these first years of endeavor there was no competition to help educate the consumer to the usefulness of the tool and, taking all things together, the progress up to 1895 was not flattering and comparatively few machines had actually been sold, but the seed had been sown. Farmers who had purchased manure spreaders discovered that whereas they were a profitable investment from a labor-saving standpoint alone that this was the least of their merits. They discovered that the available supply of manure was made to cover from two to four times the territory as when spread by the old hand method, and that the manure was so thoroughly pulverized and evenly distributed that this lighter covering produced even better results. Thus the trade was educated and the demand increased.

In the meantime new blood had been gotten into the mechanical end of the business. Mr. Robert Love, in charge of the experimental department of the Kemp & Burpee Mfg. Co., commenced to rapidly improve the mechanical construction of the machine. One of the most difficult problems that he had to contend with was the excessive strain put upon the spreader when the machine was first thrown into gear and the cylinder compelled to revolve with the load against it. This was a problem which the Kemp spreader always had to cope with and was the one thing yet to be overcome to make the machine absolutely practical. A device was then constructed by which when the machine was thrown into gear the cylinder was automatically slid back out of the load, relieving it from strain and permitting it to commence its operation free of the load itself. The adoption of this device in the late '90s marks an epoch in the construction of spreaders, and from that time on the question of mechanical perfection was solved. Then the growth became rapid. Mr. Love devoted his entire attention to other parts of the machine, perfecting and improving them, and in 1903 brought forth the latest, up-to-date product, the "Success" machine, this being constructed with a heavy, steel-pinned chain cylinder drive, the beater-freeing device above described, and the elimination of the clutch in throwing the machine in and out of gear. Since that time, the "Success" spreader has been further improved by the addition of roller bearings, lightening its draft by from 20 to 30 percent, the construction of a worm and gear case permitting these parts to run in oil, and making the positive feed of the bottom, and many refinements in detail which make the present finished product.

About the time that this scheme for relieving the beater was adopted, competitors commenced to come into the field, and for the last ten years the growth in demand for manure spreaders has probably been as large

as that of any new farm tool ever introduced, with the possible exception of the twine binder, and has even approached the early trade of that machine in volume.

Much has been said in current publications recently regarding the serious problem that confronts the farmer in increasing the productiveness of his acres so that the increased population may be provided for in their future needs. Such leading authorities as James J. Hill of the Great Northern and W. C. Brown of the New York Central have been prominently before the public in their advocacy of improved methods of farming, methods which will make the productiveness per acre greater and maintain fertile conditions of the soil. It has been found that the average production of wheat per acre has been decreasing rapidly, that the demands of the increased population were not being supplied by the increased output, and that the farmer must be educated to more carefully conserve those elements which increase and maintain fertility in the soil if this great demand is to be supplied. To accomplish these things the Eastern farmer has resorted to commercial fertilizers; but, unfortunately, commercial fertilizers do not add one pound of humus to the soil, and in addition to this are very expensive. The Western farmer has not yet realized this necessity, but he will in the near future unless he more carefully conserves and better distributes his manure supply.

It is acknowledged by agricultural authorities that manure is the best fertilizer that is known. It not only adds to the soil those needed plant foods, potash, phosphorus and nitrogen, but it also adds abundant quantities of humus, which is necessary to maintain its mechanical condition for conveying this plant food to the roots of the growing plant.

It would be impossible to fertilize the boundless Western acres by hand; first, because such a method would not be practical on account of the problem of securing labor; second, because hand distribution can not be carried on so as to cover the number of acres that must be supplied with fertility. Therefore, the Western farmer must conserve his supply of manure and must spread it in the most economical way. The manure spreader is the only tool that answers the purpose. Soils once depleted are built up slowly. It is much easier to maintain soil fertility than to bring it back after the fertility is once gone. The supply of manure and straw on the ordinary farm is sufficient to maintain this fertility and not only enable the farmer to produce his present crops, but to increase productiveness on every acre in the future if this material is only properly distributed. There is no subject today that requires more careful attention on the part of the farmer than that of properly distributing manure. There is no subject in which the dealer should be as closely interested as in this problem. From a dealer's standpoint it is not only a question of the profit received on the sale of the manure spreader, but it is the proposition that every time he sells such a machine to his customer he puts in the hands of that customer an implement that will permanently increase the produc-

tiveness of the soil, make him a more prosperous and enable him to more largely increase his purchases of necessary farm tools, by which the dealer lives.

It has been clearly demonstrated in the past that the territories where manure spreaders are used generally are the most prosperous. The dealers who sell the largest number of manure spreaders develop this prosperity and share in its returns; and independent of any profit that may be secured from the sale of the spreader itself, every dealer should put greater effort in the introduction of this machine than any other tool that he handles. In so doing he will be building for his own future, and most important of all, will be discharging an obligation which he owes the community by placing in the hands of every farmer the ability to permanently maintain and increase the productiveness of his soils.

A "New Idea" in Spreaders

The New Idea spreader is well named. Its peculiar construction appeals at first glance to even the casual observer, while to the farmer who appreciates the value of manure on his land and who studies the machine, the New Idea spreader reveals its superiority with increasing emphasis.

It is gearless, consequently the possibilities of breakage are reduced to a minimum; and breakage is one of the troubles which in the past have brought grief to users of some other spreaders and thrown discouragement and distrust on them.

The first important difference in the New Idea way of handling manure is that it thoroughly pulverizes the manure before spreading it; and the cylinder then actually spreads it—does not simply throw it out on the ground covering a swath the width of the box. The New Idea is so designed and constructed as to completely pulverize any and all kinds of manure, the pulverizer serving also to take the weight and strain off the cylinder, leaving it free to do its work on the pulverized manure in better shape than is possible under any other condition, while the third appliance, the spreader proper, scatters it uniformly to such an extent that one operation covers two full rows of the corn field. The small sizes of spreaders of other makes cover but the width of one row, while the larger sizes cover a wider swath and still not wide enough for two rows, making the distribution of manure on corn land awkward and difficult.

These features, together with the gearless construction, strength and simplicity and light draft, appeal readily to the intelligent farmer and make it very easy for the New Idea agents to sell New Idea spreaders either in new communities or in competition with other styles.

The spreader is manufactured by the New Idea Spreader Co., Coldwater, O.

THE OLD IDEA WAY

THE NEW IDEA WAY

"Red Devil" Portable Drill Press.

Fifield's "Red Devil" automatic-feed drilling attachment for metal workers has been added to the already extensive line of the Smith & Hemenway Co., 108-10 Duane street, New York City. With this portable drill press the makers assert it is possible to penetrate any metal, in any position, with perfect ease and that it is absolutely automatic in its feed action. No pressure is necessary, as the automatic feed device in connection with the chain draws the drill forward steadily, with positive, even and continuous feed, not allowing the drill to lose its "cut" in the metal, a fault found in some drills. The spindle which holds the drill proper revolves within a cylinder which in turn revolves inside of a yoke. To a projection from the latter is attached the chain which passes around the metal to be bored, and fastens in another projection from the other side of the yoke. By turning the thumbscrew seen in the cut, friction is enforced against the cylinder, without which the spindle and cylinder would turn together, affording no feed. By applying the friction, however, the cylinder is held while the threaded spindle advances under the grip of the chain.

"RED DEVIL" PORTABLE DRILL PRESS.

A Fly Exterminator.

"Swatting the fly" has no doubt resulted in reducing the fly population some millions, although the official census report has not yet been made public.

But swatting flies one at a time is about as slow as trying to irrigate a big ranch with a bucket spray pump.

There is a preparation, however, that will effectually kill and keep away flies from the barn and cow stable.

"Will-Kill-Flies" is handled by the Stowe Implement Supply Co. of Kansas City. This

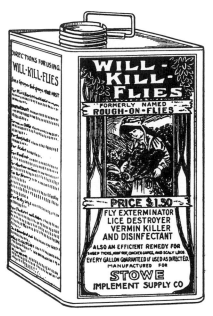

"WILL-KILL-FLIES."

company has handled this product now for three years and reports that it gives excellent satisfaction, proof of which is the fact that dealers who order once usually order again.

The preparation is used with a small tin sprayer, and a very small amount of it will keep flies off cattle and horses. Some almost incredible stories are told by dairymen who have used it, of the increase in milk from their herds after spraying their cows and relieving them of the worry and exertion of fighting flies.

Speaking of this product, a member of the company stated that this preparation should not be confounded with the cheap "dope" advertised as fly killer. The principal ingredient of this preparation (which is imported) is used by the Government about its forts and arsenals; it is one of the best disenfectants and deodorizers known.

It is valuable in the poultry house, as it is also a lice exterminator. There is a nice profit in it for the dealer, and it is a good hot weather specialty.

Trend of Automobile Prices.

An interesting tabulation of the average automobile prices since 1903, prepared by the American Automobile Association, shows that automobile prices rose from 1903 to 1907, but that since that time the price has gradually declined.

This decline, according to the association's experts, has not been brought about by any radical reduction in the price of motor cars, but by the great increase in the manufacture and sale of machines selling for $1,500 or less. In the early days of the industry a car selling for $1,500 was rare, while now the greater number of machines sell for under that figure.

The figures show that the trend in manufacturing has been to give more each year for the same list sum, rather than to make a cut in the selling price.

There has been a tremendous increase in the making of what are termed moderate-priced cars, and a normal and healthy increase in the number of higher priced machines.

From $1,133.37 as the average price for cars in 1903, the average ran up to $2,137.56 in 1907, and since that it has decreased until the first six months of 1910 shows $1,545.93 as the average price.

Dynamite in Wheat Shocks Owners.

Several owners of threshing machines in Dubois county, Indiana, have received anonymous letters informing them that dynamite has been placed in shocks of wheat on certain farms and warning them not to thresh it. The authorities have been asked to investigate.

THE AVERY TRACTION STEAM SHOVEL AT WORK.

In 1910, the J.I. Case Threshing Machine Co. made arrangements to sell the entire output of the Pierce Motor Co. also of Racine, Wisconsin. The automobile previously called the Pierce-Racine was called the Case. The J.I. Case Threshing Machine Co. also built a ten mile long road between Kenosha and Racine, Wisconsin, to prove the value of good roads to the people of the vicinity. This was to be "part of a magnificent lakeshore drive extending far northward from Chicago, including the famous Sheridan Road." The Case company was serious about the speed business as evidenced by the assembled team to build a racing car to assault the record held by Oldfield and his Benz and to build three aeroplanes to compete in the principal aero races. The J.I. Case Threshing Machine Co. also began to handle the road building products (ditchers, graders and drags) of the J.D. Adams & Co. of Indianapolis, Indiana.

The John Deere Plow Co. began to sell tractors made by the New-Way Motor Co. of Lansing, Michigan. The motor was actually two 4-hp opposed, air-cooled, gasoline engines mounted on a steel chassis. The 3,600 pound tractor had two forward speeds, one reverse and it was claimed that it could pull a two-bottom gang with 12-inch plows.

The International Harvester Co. recorded sales of $90 million in 1910. Massey-Harris purchased the Johnson Harvester Co. of Batavia, New York, and the Oliver Motor Car Co. was organized with a capital of $300,000 to manufacture the Oliver delivery wagon in Detroit, Michigan. The Thomas B. Jeffery Co., manufacturer of the Rambler automobile, was incorporated with capital of $3 million in accordance with the will of the late Thomas B. Jeffery. All of the stock is owned by the family.

Count the Times a Horse Rolls.

To see a horse when out at pasture rolling on the ground and endeavoring to turn over on his back is a common sight, but how many people have noticed that in doing this he observes an invariable rule? The rule is that he always rolls over either at the first or third attempt—never at the second—and more than three attempts are never made. In other words, if the horse succeeds in rolling over at the first try, well and good—that satisfies him. But if the first attempt is a failure the second one always is. Then he either rolls quite over at the third or gives it up. He never makes a fourth. If horses are rolling on sloping ground they usually roll uphill. This is more easy of explanation than the strange custom regulating the number of attempts. As to this no adequate reason has ever been offered. Will those ingenious people who tell us why a dog turns around before lying down and why ducks walk behind each other in a string instead of abreast explain why a horse never makes four attempts to roll over and never succeeds at the second?—Team Owner's Review.

122   **1908-1913**

Winnipeg 1910 Motor Competition.

A careful examination of the official report made by the judges of the 1910 Winnipeg motor competition shows conclusively, the International Harvester Co. contends, that the I H C tractors excel in all points that go to make up a simple, practical, and economical machine for farm use.

Durability:—That the I H C tractors are durable is attested by several thousand satisfied customers in all parts of the world, who have been using these machines for several years.

Simplicity:—A comparison of the I H C single-cylinder tractors of the various sizes and the large two-cylinder tractors with any other single or multiple cylinder tractor on the market shows clearly why simplicity can be justly claimed for the I H C line.

The I H C single-cylinder internal combustion engine, the manufacturers claim, is without question the simplest prime mover on the market. The I H C two-cylinder motor is necessarily more complicated, but skillful designers have reduced the number of parts in this machine to the minimum consistent with efficiency and durability. The four-cylinder, high-speed motors have their uses. They are well adapted to automobiles and for other power purposes where the load varies and great flexibility is required, but tractors require strong, durable motors, that will pull to their full capacity every hour in the day and every day in the year.

Fuel Economy:—Reference to the judges' report of the 1910 Winnipeg Motor Competition shows that the I H C tractors delivered more horse power hours per gallon of fuel than any other tractors in their respective classes. Next to durability, fuel economy should count for more than anything else. The I H C tractors not only showed better fuel economy in the brake test, which by the way is the test in which the greatest accuracy is obtained, but in the plowing test the superiority of I H C tractors was further demonstrated by the low fuel consumption per acre as compared with all competitors. The I H C tractors have shown their superiority in this respect, not only in the three annual contests held at Winnipeg, but in all contests held in other parts of the world, and in the hands of hundreds of satisfied purchasers.

Drawbar Horse Power:—A satisfactory traction engine should deliver a large per cent of its brake horse power to the drawbar. A glance at the figures issued by the judges of the Winnipeg Motor Competition, shown herewith, shows clearly and conclusively, say the manufacturers, that I H C tractors are built on scientific lines and excel all others in the percentage of power delivered at the drawbar. They ask that the intending purchaser who has hauling or plowing to do compare the wonderful efficiency in drawbar pull shown by the I H C tractors with the results obtained by other machines tested in this competition. These are official figures, but equally as good results are obtained every day by the many users of I H C engines.

Self-Heating Sad-Irons.

Foremost among gasoline and alcohol self-heating sad-irons are the Improved Modern irons, manufactured by the Modern Specialty Co. of Milwaukee, Wis.

Modern irons are the size of the regulation irons, weighing approximately six pounds. They have none of the disagreeable features of the many kinds of heating irons on the market—no pumps, wicks, wire or other troublesome attachments. They are odorless, never stick, streak or soot, and are adapted to use on the finest fabrics or coarsest materials.

The construction of the burner diverts the heat direct to the bottom and distributes the heat evenly over the base. The heat can be regulated to suit the most exacting requirements. Being of simple construction, they can be taken apart in an instant, easily cleaned and operated, which is greatly appreciated by the housekeepers. The several parts are all interchangeable, a most notable feature, and each part can be supplied when necessary through long usage, etc., direct through the mail.

The tanks are made of seamless drawn copper, brazed under our personal supervision, and subjected to a water pressure test of 300 pounds. Every iron is tested O. K. by competent experts before leaving the factory and guaranteed to work perfectly.

Careful experiments, say the manufacturers, prove them to be the most economical heating irons on the market. They do away with hot stoves, expensive gas and electricity, and represent a portable laundry in themselves, a special delight to travelers, summer sojourners, and hundreds of places where all the metropolitan conveniences are not available. There are thousands now in use, giving perfect satisfaction and they are rapidly taking the place of all other makes and kinds of sad-irons.

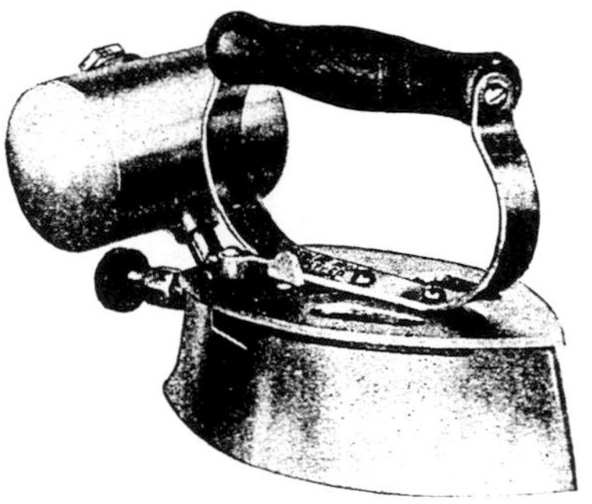

IMPROVED MODERN SAD-IRONS, 1910 MODELS.

CLASS	Entry Number	MAKER'S NAME	Engine Data: Cylinders: Number	Diameter	Stroke	H.P.: Nominal	Specified Brake	Weight, lbs: Total	Brake Test: H.P. Developed	H.P. Hrs. per Imp. Gal. Fuel	Total Running Time	lbs. Water Evaporated per lb. Fuel	Mean Effective Press. in Cylinder	Plowing Test: Plows: Number	Width	Make	Miles Traveled	Acres plowed	Time, minutes	Acres per Hr.	Average Draw Bar Pull, lbs.	Fuel used, lbs.	Fuel used lbs. per Acre	Water Used per Acre	D.B. pull per 1" Width of Furrow	Fuel used Imp. Gals. per Acre	Draw Bar H.P.	D.B.H.P. / Brake H.P.	D.B.H.P. Hrs. per Imp. Gal. Fuel	Total points allowed in judging Engines
A. Internal Combustion 20-B.H.P. & under.	1	International Har. Co.	1	8	14	15	18	10,500	15.28	9.94	2 hrs.	3.63	69.6	2	14"	Oliver	12.49	3.61	299	0.72	1705	81.5	22.58	7.72	60.8	3.23	11.39	75%	4.87	306.7
	5	Avery Co.	4	$4\frac{3}{4}$	5	12	36	6,000	14.16	5.58	2 hrs.	.21	28.08	3	14"	P. & O.	7.14	3.20	255	0.75	1980	75.5	23.59	1.12	47.1	3.37	8.87	62.5	3.49	275.3
B. Internal Combustion 21 to 30-B.H.P.	2	Avery Co.	1	12	18	25	25	12,000	20.69	5.71	1.24	1.27	31.7																	
	3	Goold, Shapley & Muir Co.	2	$7\frac{1}{2}$	10	20	28	11,000	28.13	9.48	1.57m.	0.27	72	6	12"	Verity	8.92	6.65	369	1.08	3250	152.5	22.93	.81	45.1	3.28	12.58	44.7	3.55	292.2
	4	International Har. Co.	1	$8\frac{3}{4}$	15	20	23	14,200	22.11	10.86	2 hrs.	3.51	82.6	4	14"	Oliver	8.92	5.00	222	1.35	2450	76.5	15.3	6.00	43.8	2.19	15.75	71.1	5.33	329.3
C. Internal Combustion Over 30-B.H.P.	6	Birrell Motor Plow Co.	4	$8\frac{1}{2}$	10	22	45	22,000	27.21	5.86	2 –	0.48	32.5		14"	Moline	3.57	2.73	159	1.03	—	93	34.00	3.66	—	4.86				
	7	Gas Tractor Co.	4	6	8	25	45		34.66	6.93	2 "	0	50.8	6	14"	Cockshutt	12.49	10.6	375	1.70	4550	179.5	16.93	0	54.16	2.42	24.24	69.8	5.91	334.0
	8	Goold, Shapley & Muir Co.	2	$9\frac{1}{2}$	13	30	45	17,730	34.86	6.34	"	0.57	49.3	6	14"	Cockshutt	10.70	8.82	390	1.36	4200	240.5	27.26	.60	50.0	3.89	18.44	52.9	3.49	274.6
	9	International Har. Co.	2	9	14	45	55	20,990	46.49	11.78	"	2.71	60.06	10	14"	P. & O.	12.49	17.59	416	2.54	7350	260	14.78	6.76	52.5	2.11	35.30	75.8	6.59	350.1
	10	Kinnard Haines	4	$7\frac{1}{2}$	8	40	60	Approx. 19,000	49.76	10.88	"	0	56.32	6		Jno. Deere					Test not completed									
	18	Gas Tractor Co.	4	$6\frac{1}{2}$	8	20	60	17,500	51.84	10.83	"	0	64.2	7	14"	Jno. Deere	12.49	12.43	408	1.83	5400	199.5	16.05	–	55.1	2.20	26.44	51.1	6.32	354.4
	11	Rumely Co. (Kerosene)	2	10	12	25	50	Approx. 26,700	46.8	6.88	2 hrs.	.80	52.0	8	14"	Rumely	10.70	12.01	356	2.02	5500	329K 36	27.64	2.3	49.1	3.49	26.43	56.5	3.71	not scored

Imp. Gal. of Fuel = 7 lbs. gasoline or 7.9 Lbs. Kerosene

U. S. gal. = 231 cu. in. - Imp. gal = 277.27 cu. in.

TABLE SHOWING RESULTS OF FACTORY COMPETITION AT WINNIPEG.

DANGERS OF THE FUTURE—COLLISION IN MID-AIR.

The Menace of Mid-Air.

Progress has its hideous side. New blessings bring new calamities. Not so long ago a runaway horse was enough to turn the rosy cheek pale, but now it is a mere incident to be laughed about in comparison with the overturned automobile. Horror comes with shipwreck, but there beneath waits the water to save all that may float. The derailment of a train is a thing to inspire terror, yet danger of death is small beside another catastrophe that certain bold and ingenius moderns have made possible.

Man flying and man bound to earth are as far apart as the eagle from the worm. The one is master of boundless space while the other crawls, or at best trundles over the ground in some clumsy vehicle. And yet the worm has advantage over the eagle.

Though the crawler runs some risk, he does not have to hazard the dread disaster in mid-air—where nothing but emptiness lies between the bird-man and the safe earth that he spurned. Aye, in spite of the fact that his hand has beaten back the power that held him to the sod, the instant that hand weakens, gravity, once more mighty, hurls him in silent and terrible anger to the ground, a poor, broken thing.

And such a fate has snatched the two night-flyers as pictured by the artist. Imagine the exhilaration that was theirs as they skimmed through space, Lord and Lady of the Air. And then the crash! Their biplane crumples like a frail moth against the iron weathervane of the upthrust steeple, a thousand times more to be feared than the sharp rock that affrights the mariner. Inevitable death is their portion. And what death is so momentarily terrible as a fall from a great height?

Their forms will be picked up and cared for by the sorrowful crawlers whose kinship they vainly sought to deny.

Curtiss' Remarkable Flight.

Until the past few days, in spite of the brilliant feats performed on both sides of the ocean by aeroplanists, the question of the feasibility of long cross-country flights was unanswered. Glenn H. Curtiss, spurred on by the New York World's offer of $10,000, undertook to make the flight from Albany to New York City and win the money.

Curtiss' new aeroplane is fitted with cylindrical floats and an air-tight canvas bag running the length of the wood strut which connects the front and rear wheels. This machine is very similar to that which which he won the Bennett cup last fall at Rheims. It differs from the latter in having somewhat larger rudders and balancing planes, and also in the extension of the upper plane 30 inches beyond the lower plane at each end of the machine. This differential plane idea was first tried a short time ago by Henry Farman, and was embodied in the machine used by Paulhan in his flight from London to Manchester. It tends to give the aeroplane a certain amount of inherent transverse stability. The total supporting surface in the main planes is 236 square feet, and the weight of the machine complete is 950 pounds, including aviator, fuel, and oil. As a result of this the weight carried per square foot of supporting surface is 4.02 pounds. This means that the machine must travel at a speed of 40 miles an hour along the ground before it will lift. In order to attain this speed, a powerful 8-cylinder motor of 50-h. p. is used. The 7-foot diameter, 6-foot pitch propeller mounted upon the engine crankshaft makes 1,100 r.p.m., while the machine is in flight, and gives a pull, when the machine is held stationary on the ground, of over 300 pounds. This is believed to be somewhat better than the thrust obtained from the Gnome 50-h. p. motor used by Farman and others in their aeroplanes.

THE CURTISS AEROPLANE WITH PONTOONS FOR USE ON WATER.

Gen. Ellsworth D.S. Goodyear died in 1910. He invented a process for making hollow rubber goods and had assisted in the process to make hard rubber. William L. Casaday, inventor of the Casaday plow and nearly 200 other tools, died at his winter home in Ocean Park, California, on December 10. Casaday was the organizer and president of the South Bend Chilled Plow Co.

The automobile industry was only about 15 years old, but had already made a major impact on the nation. It was reported in 1910 that there were an estimated 350,000 automobiles in the United States, compared to 7 million horse drawn vehicles. The automobile industry employed 200,000 people, paid $60 million for rubber, steel, iron and aluminum, accumulated $25 million in freight charges from the railroads each year and consumed about 50 million gallons of gasoline. "The total capitalization of Michigan's 45 automobile factories is nearly $50 million, and these factories gave employment to about 50,000 men in addition to some 20,000 employed in the manufacture of automobile parts and accessories." It was hoped that high number fatalities would be reduced "when the motor car is less conspicuously the travel of the rich and the ruthless instrument of the speed-mad."

Silk top hats were going out of style because automobile travel was difficult without damage to the hats. "Toppies" were giving way to derbys and caps. Mechanical milking machines were also taking over from muscular effort previously required by dairymen.

Experiments in the cultivation of hemp was being carried on near Fayetteville, Arkansas, in 1910 to help relieve the problem of binder twine shortage.

Growth of American Hemp.

This paper has had a good deal to say about hemp generally, but not so much about hemp grown in this country. According to a recent report issued by the Department of Agriculture approximately 20,000 acres of hemp are grown each year in the United States, and this acreage is increasing.

It appears that most of it is produced about Lexington, Ky., in the bluegrass region of that state, says the Washington Herald. About 600 acres are grown in the vicinity of Lincoln, Neb., and an area of about the same acreage in the Sacramento valley of California. Experiments in growing hemp have been made in several states with sufficient success to assure a wide expansion of the industry within a few years.

Hemp is a plant of the mulberry family, cultivated for the production of a soft bast fiber. This fiber, gray if dew-retted, or light yellow if water-retted, is also called hemp. In a strict sense the name "hemp" is correctly applied only to this plant and its fiber.

Hemp is cultivated commercially for fiber production in Russia, Italy, Austria, Hungary, Germany, France, Belgium, Turkey, China, Japan, and the United States. Russia produces more for export than all the other countries.

Hemp requires about 110 days for its growth. It should have a rainfall of at least ten inches during this period. It has not been grown commercially under irrigation. If the level of free water in the soil is within five to ten feet from the surface, as is often the case in alluvial river bottom lands, and the character of the soil is such that there is good capillary action to bring the water up, hemp will not suffer from drouth, even should there be very little rainfall. Hemp is uninjured by light frosts. It may therefore be sown earlier than oats and harvested later than corn.

Practically all of the hemp produced in Kentucky is dew-retted. It is spread on the ground, either from the gavel, shock or stack, in rows with the stalks side by side, and not more than two, or at most three, stalks in thickness, the butts all even in one direction. It is left in this manner from four to twelve weeks, or until the bark, including the fiber, separates readily from the woody portion of the stalk. The stalks are then raked up and set up in shocks to dry. As soon as dried they are ready for breaking.

Much of the hemp produced in Kentucky is still broken by the old-fashioned hand break, but this method is not recommended for introduction into any new locality because it requires a degree of skill that would be difficult to secure in laborers not accustomed to the work. Even in Kentucky the newer generation of laborers do not learn to break hemp, and this is one of the principal reasons that the industry is not carried on there to a greater extent. At least six different kinds of machines for breaking hemp and preparing the fiber have been in use during the past three years, and some of these prepare the fiber very much better than the hand break.

The yield of hemp fiber ranges from 500 to 2,000 pounds to the acre. The general average yield under ordinary conditions is about 1,000 pounds to the acre. Yields are sometimes estimated at 150 pounds of fiber for each foot in height of the stalks, and also at 20 percent of the weight of the dry-retted stalks; but estimates based on these factors alone may be misleading, for slender stalks yield much more fiber than coarse ones.

All of the hemp fiber produced in this country is used in American mills, and increasing quantities are being imported. It is used for making gray twines, "commercial twines," carpet warp, and ropes of small diameter.

The twenty-five mills in the United States using hemp fiber are mostly in or near Boston, New York, Philadelphia, Cincinnati and San Francisco.

The average price paid during the last twenty years by local dealers to the farmers in Kentucky for the rough fiber, tied up in hand-made bales, has been about 5 cents a pound. The prices during the same time for the fiber sorted, pressed in bales, and delivered at the mills as ordered, have ranged from $130 to $175 per long ton.

The market is occasionally overstocked with low-grade hemp or tow, but there is little danger of an oversupply of good, strong, well-cleaned fiber.

The Sensation in an Aeroplane.

What is the sensation when one goes aloft in an aeroplane? Here is how the Wright brothers describe it, according to a writer in the Columbian Magazine:

"It is peculiarly exhilarating, and at the beginning, for most persons, full of suppressed excitement. The machine rises swiftly yet lightly from the monorail along which it is pushed at starting. For a minute the earth seems a blur beneath you, but as you ascend the landscape and terrestrial objects detach themselves more clearly. At an elevation of, say, a hundred feet you would be unconscious of any movement whatever but for the wind that fans your cheeks—and whisks off your hat if it is not held securely. The operator pulls a lever, the aeroplane tilts to one side and makes a sharp turn to right or left, but you are not jerked about in your seat as you would be in an automobile, or even in a railroad car. Now you are facing about, toward the point of departure. The ground, far below, seems suddenly to be rushing along at a terrible speed, although the wind against your face has not perceptibly changed. You are now going with the wind.

"As you approach the earth the conductor stops his motor while the machine is still high in the air. It has been whirling with deafening sound, but in your excitement you did not notice it until it ceased. The aeroplane dives downward, obliquely, and alights after a glide of perhaps a hundred feet. Although it may descend at the speed of a mile a minute, there is such an absence of shock that it is impossible to know the precise moment when it touches the ground."

JOHNNY AIKEN, THE FAMOUS AUTO RACER, BECOMES AN AVIATOR.

Gas and Gasoline Engine Department

The Pierce Motorcycles for 1910.

Views of the Pierce four-cylinder and single-cylinder motors for 1910 are shown herewith. This motor is a replica in miniature of the engine used in the Pierce Great Arrow automobile. The most notable improvement in the four-cylinder model is the addition of the two-speed and free engine contrivances, which gives the machine the same advantages as in the automobile. The aim of the manufacturers was simplicity, and they appear to have achieved it. Combinations have been sought which will take the machine over all roads, whatever their nature, without causing the motor to labor. The ratio is $4\frac{1}{2}$ to 1 on the high speed and $7\frac{1}{2}$ to 1 on the low speed. The gears are shifted by a lever attached to the left side of the frame within convenient reach.

PIERCE 4-CYLINDER. PRICE $350.

The same lever which shifts the gears controls the clutch, which in turn engages or releases the motor. The advantages of the free engine are manifold, for it is often desirable to run the motor independently of the machine. In starting, stopping on account of congested traffic, and in starting on hills, or coasting down, the free engine will prove of great service, as well as add to the delights of riding.

The cylinder dimensions of the new model have been increased from a bore of 2 3-16 inches and a stroke of $2\frac{1}{4}$ inches to a bore of 2 7-16 inches and a stroke of $2\frac{3}{8}$ inches. This is an increase of one-third in the cubical displacement. To comport with the increased dimensions the fly-wheel will have $1\frac{1}{2}$ inches more diameter. The minor changes include a slight alteration in the oil base, so as to eliminate any possibility of the pump not being fed. All bearings will be fitted with removable bushes. The crank shaft will be hardened, and drop forged connecting rods wil be used.

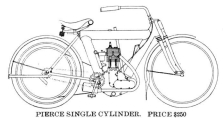

PIERCE SINGLE CYLINDER. PRICE $250

The single cylinder machine embodies many of the scientific principles and constructions contained in the four-cylinder; for instance, the large-size tubing combining frame and tanks will be utilized, and no clevice will be used in the front forks. The Bowden wire control will be used on both machines.

Hints as to Kerosene.

"Referring to the influence of kerosene on lubricating oil," says an expert of a prominent oil company, "it is enough to say that oil is ruined for the purpose of lubricating if it is thus diluted. Proof of this is given in the plan by which kerosene is used in clutch oil, if the clutch will not hold. In a word, lubricating oil in the clutch is too slippery, and kerosene is added to diminish the unctuousness on the one hand, and a still more important property on the other, namely, the ability of lubricating oil to persist in remaining between two surfaces under great pressure. In the case of the clutch, it is desired to have the oil squeeze out from the discs with ease. This ease would be ruinous to a crankshaft bearing. But this is not even half told; the flash point of the lubricating oil is of the greatest importance, and this point will be much altered if the oil is tinkered with. The flash point of kerosene is low in comparison with the same point of lubricating oil. Constantly feeding kerosene to the cylinders of a motor, even if the same is 'sneaked' in with the lubricating oil, is bound to result in the unbalancing of the explosive mixture, and it is plain that the question of 'carbon deposit' will be rendered the more prominent."

Test for Lubricating Oil.

An easy and simple test for lubricating oil, that any novice can make, is to place the oil to be tested in a small porcelain or glass cup filled with sand and set it in a metal dish, so that the surface of the oil in the cup is about level with the surface of the sand in the dish. The receptacle containing the oil should be covered, the cover having two perforations $\frac{1}{4}$-inch in diameter. One of these holes is for the introduction of the thermometer and the other to provide an exit of the vapors given off. The bulb of the thermometer must be clear to the bottom of the cup by at least $\frac{3}{8}$-inch. Heat the metal dish by means of a Bunsen or gas flame, noting the rise in temperature of the oil by means of the thermometer. After the oil reaches a somewhat critical temperature, which can be determined by a preliminary test, pass a lighted taper over the open hole in the cover, carefully noting the thermometer reading at the instant at which a flash is observed. The temperature so noted, checked for accuracy by one or more repetitions of the run, is the flash point of the oil.

Water in Gasoline.

"The symptoms of water in the fuel tank are not too generally known, and there is nothing about them to identify the cause—they are simply misfiring, more or less chronic and more or less pronounced, according to the degree of the dilution," says a writer in Motor Print. "Most motorists afflicted by this trouble are probably duped just as I have been and have run through the whole gamut of tests. First, I test the cylinder individually, to find all are equally affected; then I spend a solid hour on the magneto, carbureter and valves, all without finding anything amiss. Finally, in desperation, I empty the tank of some six gallons of gasoline and pour in the spare can, when the engine immediately responded to the next pull of the handle, and ran like a clock. In my case there is no doubt one of the cans just bought had contained a very large percentage of water, possibly pure water. This trouble may afflict any motorist at any time, and since the only safe test for it is to empty the tank and refill with fresh gasoline, if any is available, it is obvious-

ly a trouble to avoid. There is only one certain method of avoiding it, and that is to carry your own funnel perpetually on the car, and to test the gauze of the funnel to make sure it is fine enough to retain water. At first sight one would say that if there is water in the tank there will be sufficient token of its presence in the carbureter to assist diagnosis, but I have not found it so, and if the carbureter is periodically cleaned out, few motorists will look at it when misfiring occurs until a lot of time has already run to waste. Another precaution of a similar type is always to sniff gasoline when it is poured in.

It Sings Easy.

The latest popular waltz song is one that has just come to hand with the compliments of the Marshall Oil Co. of Marshalltown, Ia., sole distributor of French Auto Oil. It is unique in that the title of the song is "French Auto Cylinder Oil," and it gets right down to business from the start. We had one of our stenographers try it on the typewriter, and actually the machine needed no lubrication from the instant she touched the keyboard. The words are by A. A. Holthaus and the music by Charles L. Johnson, and the song is copyrighted by the Marshall Oil Co. The title page carries a view in colors of a pretty race between a roadster car and a Wright aeroplane, and as both are presumably using the Marshall Oil Co.'s product, honors appear to be even. A copy will be sent free on application.

Grades of Gasoline.

Representatives of the Standard Oil Co. have recently been notifying manufacturers that, commencing about April 15th, it will be impossible for the company to supply gasoline of a better grade than 64° test, except in the largest cities, says a Brooklyn daily. This means that the tourist who is going across country cannot figure on filling his tanks with anything better than this low grade of gasoline. This is a condition which will cause no end of inconvenience to many motorists, and designers of motor cars will have to take cognizance of the new conditions.

The clipping might have mentioned, also, manufacturers of internal combustion farm engines and tractors as being affected by this change. As a matter of fact, this announcement simply makes public the result of a gradual change which has been in progress for a decade. Ten years ago the gasoline in popular use ranged from 62 degrees to 76 degrees Baume, the greater quantity used being from 70 degrees to 74 degrees. The test has gradually declined at the rate of about 1 degree a year during the decade, and it is a safe prediction that the greater part of fuel gasoline sold in the next decade will run close to the lowest limit at which oil is still rated as gasoline, namely, 62 degrees Baume.

In the past, the distillates just heavier than 62 degrees, have been sold partly as such, partly in a mixture with lighter gasolines to make a heavier and poorer product, and partly in a mixture with heavier kerosenes to make an oil with a lower flash point, hence less desirable for illuminating purposes. Until the development of oil engines capable of using both the heavier and intermediate oils efficiently, there was no established market for the oils, which were too volatile for safe use in illumination, and too heavy for successful carburetion in the existing types of gasoline engines. For fuel purposes, the dividing line between kerosene and gasoline is rapidly disappearing. Owing to the impossibility of supplying enough high grade gasoline to meet the demand, the grade is being lowered to include a larger and larger proportion of the heavy oils, which occur in greater abundance.

Even in the face of this expedient, the proportion of oils refined as kerosene and distillate is not only outrunning gasoline eight or ten to one, but outrunning the demand for heavier oils in much greater degree. Eventually kerosene also will run

heavier, but oil as low as 35 degrees Baume, has been used successfully in a traction engine designed especially for the purpose of handling the heavy oils. The heavier the oil, the greater its heat value per gallon, and the problem which automobile and engine makers face is that of utilizing this heat.

For the information of those who may not be familiar with the terms used to designate the various grades of oil, it may be said that the gravity test involves the use of an arbitrary Baume scale, graduated in reverse order from the specific gravity of liquids. The following table shows the quality in degrees Baume at 60° Fahrenheit, the specific gravity as compared with water, and the weight in pounds per U. S. gallon.

Fuel	Baume Test degrees	Specific Gravity	Weight, lbs. Per Gallon.
Gasoline	76	.679	5.66
"	70	.702	5.85
"	64	.722	6.02
Kerosene, 120° "Water White"	49	.784	6.53
Kerosene, 150° "Water White"	47.5	.789	6.58
Fuel Oil	35	.850	7.08

Many automobiles now on the market will handle the lower grades of distillates without difficulty, except, perhaps, in starting. With the present types of carburetors there will be, of course, more carbonization. To some extent this may be helped by feeding an ounce of wood alcohol into each cylinder at the end of a run, and allowing it to exert its solvent action over night. Much of the carbon will then be blown out at the next start.

All signs point to the general necessity for vaporization of the heavier oils, and manufacturers are on the alert for anything promising results in this direction.

Horseless Funerals.

The other day Detroit had a funeral, the like of which the world had never before seen. To be sure, the man died in the old-fashioned way by simply breathing his last, but after that incident, the obsequies proceeded along perfectly modern lines, with one exception. Following the services at the house, the procession to the cemetery began —and such a procession! Not a horse's hoof was heard to fall and not a single be-tiled Jehu graced the seats of the carriages, for the carriages were all automobiles, as was the hearse itself. Of course, every exhaust was muffled, and thus silently and swiftly the cortege rolled out to the brink of the grave. There things reverted to the old style again, for the body was buried 'neath the sod as in days agone, and not cremated, as would have been the case had those in charge adhered consistently to modernity. Herewith is illustrated a motor hearse built by the Crane & Breed Mfg. Co., of Cincinnati, Ohio.

Kansas City has followed Detroit with the second motor funeral on record. W. J. Osborn of the Fidelity Oil Co., and a member of the Automobile Club and the Chauffeurs' Club, who was killed in a fall from a Cutting racer on its return down-town from the Elm Ridge races Saturday night, was buried Tuesday afternoon in Forest Hill cemetery, and not a horse was seen in the procession. The Rapid motor ambulance that took him to the South Side hospital after the accident was converted into a hearse and the body was followed to the grave by his many friends in local motordom, riding in automobiles.

FORECAST OF 1911 MODELS

Influence of Torpedo-Body Type Will Be Felt, Although No Other Radical Changes Are in Sight.

In the first place, there will be no radical changes in the 1911 automobile models as compared with those of this year. In spite of the foregoing statement, however, the average observer will see some striking alterations in body design that may appear to be a distinct departure, yet in reality will be but the development of tendencies first manifested several years ago and non-essential as far as the mechanical construction of the car is concerned. Some makers, indeed, have stated that their cars for next season will differ in no vital detail from those placed on the 1910 market. At the same time there will continue to be a wide diversity of construction among the different makers.

Perhaps this departure in outward design is best exemplified in the 1911 model of the open-body touring car announced by the H. H. Franklin Mfg. Co., Syracuse, N. Y. The predominating feature of this model is the influence of the torpedo type of body, first introduced two years ago, and the unbroken fore-and-aft lines of the body itself. This latter feature is permissible in this case on account of the fact that the Franklin is an air-cooled machine, allowing the hood to slope down gradually to the front and eliminating the visible radiator. Doors are furnished for both front seats and tonneau, which will be an added comfort to the driver in the winter. More room is provided in this model between the front seat and dash and a sharper pitch gives a piratical rake to the steering column. Indeed, the effect of the whole is that of a long, low, rakish, speedy and withal handsome machine. With its shiny, black body, one misses the Jolly Roger which it seems should fly at the wind-shield or whatever happens to be the automobilist's equivalent for "peak," as a sailor knows the word.

In general, wheel bases are unchanged, and this applies also to the arrangement of the springs, says the Scientific American. The weight of an automobile makes the problem of its suspension a difficult one, and is responsible for the absence of the easy riding qualities of a horse-drawn vehicle. It is inevitable that improvements in this part of the construction will be made, but there are few indications of serious work being done in this direction.

In regard to the engine, the four-cylinder type retains the foremost position, although a number of makers have abandoned it, and are producing six-cylinder cars exclusively. These are of sizes that develop 30-h. p. and over, but inasmuch as the European designers are turning out six-cylinder cars as small

THE FRANKLIN 1911 MODEL OPEN-BODY TOURING CAR.

as 12-h. p., it is not surprising to learn that some of the American makers have similar designs in view.

The horizontal double-opposed engine, which is a distinctly American type, is losing in popularity, and a number of makers whose reputations have been built on it have announced four-cylinder models for 1911. As an illustration of the diversity of design a car may be mentioned that will shortly make its appearance, of which the engine is of the eight-cylinder V-type, and also of a car equipped with a single cylinder engine with a stroke that is twice the bore; the engine is reported to develop 40-h. p. at 2,400 revolutions per minute. Cars having engines of this character were the racing sensation of Europe in 1909, and developed extraordinary speed.

The "block" type of engine, in which three or four cylinders are cast and handled as a unit, is decidedly in favor, for the construction results in economy of weight, size and cost. In four-cylinder engines of this type the crank shaft is supported on but two bearings, and because of the strains to which this subjects the parts, a high grade of material and workmanship is necessary for success.

The year 1910 saw the abandoning of the tained circulation system, and this will be practically universal on the 1911 models. The oil is contained in a compartment that is usually integral with the crank case. An interesting feature is a float resting on the oil, and operating an indicator located in plain sight on the engine. This is a great improvement over the former designs that provided a gage glass located inaccessibly on the lower side of the crank case.

The great advances that have been made by the manufacturers of carbureters have induced many automobile makers to abandon their own designs in favor of commercial types. It may be added that an exten-

sion of this practice would result in an improvement in many otherwise good cars.

Jump spark ignition has become practically universal, with a high tension magneto as the source of current. In many cases a battery and coil system will be provided in addition, operating either on the same plugs as the magneto, or on a separate set, but in any case, this is only to render easier the starting of the engine. The battery coil is usually so arranged that the pressing of a button will produce a spark in the cylinder that is on the firing stroke, and if mixture is present the engine thus becomes self starting. The use of high tension magneto on small engines permits the spark to be fixed; that is, no arrangement is necessary for varying the relation between the magneto and the crank shaft when the engine speed changes. This greatly simplifies the control, and the system has been widely adopted.

The simplicity of the thermo-syphon or gravity system of water circulation is an attractive feature, and for 1911 its use will be much more general than was the case in 1910. With properly designed passages and an efficient radiator, it is in no respect inferior to the force system, and its first cost is, of course, considerably less. Those makers who still use the force system have adopted pumps of the centrifugal type, which may now be considered as the standard.

The multiple disk clutch has not lost in popularity, but it is noticeable that many of the designs have been simplified. As far as the change speed gear and transmission are concerned, they differ but slightly from the 1910 models, and such changes as have been made are for an increase in accessibility. That there is room for improvement in this direction is indicated by the number of instances in which it is still necessary to remove twelve nuts or even more in order to gain access to the gear case.

This matter of accessibility is one that some makers have studied, while others have neglected it to an astonishing degree. For example, the draining of the crank case at regular intervals is of such importance to the life of an engine that a manufacturer would be supposed to render the operation an easy one, but in some cases it is necessary to remove the sod pan and to use a wrench to unscrew the drain plugs in order to accomplish it. An instance of this sort may be extreme, but it is a fact that when drain cocks are provided they are usually so located that it is necessary to get under the car in order to reach them. In a French car that was exhibited at the Automobile Show in January, 1909, the drain cocks could be operated by a handle located on the upper side of the crank case. It is strange that so simple and inexpensive a device has not been generally adopted.

A notable advance over the designs of previous years is the providing of the oil tank with an opening of such size that oil may be poured in without the aid of a funnel. The radiator manufacturers might well adopt a similar design, and make it the rule to provide openings large enough to permit water to be poured in from a bucket. A few cars are so arranged, but it is far from being standard practice.

The new cars will in many cases show an increase in the wheel sizes, but this is an advantage, as it improves the easy riding qualities and reduces the tire maintenance costs. Some concerns will supply demountable rims, while a far larger number will make some form of quick detachable rim the standard equipment of their cars. In some of the higher priced cars, air pumps driven by the engine are provided for tire inflation, and this is a feature that is deeply appreciated. The pneumatic tire is the *bete noir* of the automobile, and any device that renders its handling less laborious is sure of speedy adoption.

An Auto Lawn Mower.

An inventor in Pennsylvania has come to the rescue of the citizen who is doomed to urge a lawn mower around his yard a certain number of hours every few days throughout the summer. The Pennsylvanian has played the role of rescuer by devising a motor attachment which may be used on any lawn mower. This motor will not only operate the blades, but will propel the mower, and all the operator has to do is to tag along behind and guide it aright.

A MOTOR-DRIVEN LAWN MOWER.

The New Science of Fertilization.

After much experimentation Dr. G. Earp-Thomas of Bloomfield, N. J., has carried the art of cultivating and preserving bacteria to a point where every farmer in the country can make use of pure bottled bacteria, alive and healthy and ready for use, which he feeds to his seeds before planting, instead of working expensive nitrogen fertilizer into the ground. The bacteria on a good crop will thoroughly fertilize the ground for at least three years without reinoculation. Dr. Earp-Thomas collects healthy bacteria wherever he can find them already flourishing, takes them to his laboratory, puts them into glass jars with a gelatinous plant food and legume seeds, and tests their power under scientific conditions. There he can watch the formation of the nodules on the roots and select only the healthiest bacteria for distributing. In preparing the bacteria for farm and garden use, a needle is thrust into this pure breed of bacteria, and comes out laden with thousands of them. These are quickly transferred to another bottle containing a bed of jelly, which preserves the bacteria for years. The neck of the bottle and the needle are, during the process, passed through flame, to destroy any foreign substance. The bottle is corked with a patent rubber cork through which a glass tube runs, so that air can reach the bacteria. The tube is stopped with cotton, which prevents the entrance of foreign germs. In this bottle millions of bacteria breed, exerting themselves to absorb nitrogen from the air which filters in through the cotton. The jelly contains no nitrogen—the bacteria work to get it from the air and so keep healthful and active.

The jelly soon becomes alive with bacteria, and the farmer can get his nitrogen fertilizer for all the clovers, all the beans, all the peas, alfalfa, all the vetches and peanuts—a different kind of bacteria for each, which can be purchased, like medicine, by the bottle. Pure cultures of active, vigorous nitrogen-gathering bacteria, which need simply to be mixed with sugar and a little water, to be shaken well and poured on the seed before planting, cost less than $2 per acre. This process entails no waste of valuable time, no expensive nitrogen fertilizer; but instead a maximum of benefit to the present crop and improvement to the soil for years to come, with a minimum of expense and labor.

Dynamite the I. H. Co.'s New Building.

At midnight of the 7th inst. several tons of steel window frames at the new building of the International Harvester Co., on West 31st and Rockwell streets, Chicago, were blown to scrap with dynamite, setting back the construction work several weeks on the building, which is to be used for the manufacture of farm tractors, etc. Business agents of the sheet metal workers' union are believed to be responsible.

Auto Delivery

The field for the sale of the International Auto Wagon is unlimited.

Here is a partial list showing the wide use of the International Auto Wagons.

Automobiles	Grocers
Awning Makers	Garage
Bicycles	Hospitals
Blacksmiths and Wheelwrights	Hotel
	Hardware
Butchers	Hackmen
Bakers	Ice Cream
Brewers	Implements and Fertilizers
Bottlers	
Busmen	Ice
Coal Dealers	Livery
Contractors	Mineral Waters
Cotton Mills	Millers
Confectionery	Nurserymen
Carpenters	Oil Dealers
Carpets	Produce
Carriages	Painting and Paper-Hanging
China	
Delicatessens	Plumbers
Dry Goods	Roofers
Department Stores	Real Estate
Dairymen	Sea Foods
Dumb Waiters	Seeds
Electric Supplies	Sporting Goods
Electric Companies	Stoves
Flour and Feed	Teas and Coffees
Furniture	Tinners
Farmers	Telephone Companies
Florists	Traveling Salesmen
Gas Fixtures	Upholsterers
General Stores	Vacuum Cleaners

Every business man or farmer in your vicinity is a prospective customer.

The International Auto Wagon gives as satisfactory service as the highest-priced commercial car and costs much less, both in first cost and up-keep. You can sell it at a profit.

This car runs every month in the year. Snow and mud do not keep it out of service.

Bodies to suit the different requirements of purchasers are often put on.
The International can be used as a pleasure vehicle. This makes it a business and pleasure vehicle combined.

INTERNATIONAL HARVESTER COMPANY OF AMERICA
CHICAGO - - USA
(INCORPORATED)

In 1911 the United States Supreme Court forced the breakup of Standard Oil and the American Tobacco companies, a fire on March 25 at the Triangle Shirtwaist Factory in New York City killed 145 and the first running of the Indianpolis 500 mile race was won by a Marmon car. Great Britain began building a battleship; the first powered by an internal combustion (Diesel) engine. The Avery Planter Co. began building tractors and the Bates Tractor Co. was begun in Lansing, Michigan, and the J.I. Case Threshing Machine Co. began to build gasoline tractors. Bates tractors were designed by Madison F. Bates who was a pioneer engine builder of Bates & Edmonds Motor Works. The Cleveland Farm Implement Co. was formed in Cleveland, Ohio, the Foos Mfg. Co. of Springfield, Ohio, changed their name to the Bauer Brothers Co. and Dowagiac Mfg. Co. changed their name to Dowagiac Drill Co.

THE AVERY CO.'S NEW GASOLINE TRACTOR.

Wonders of the Gnome Motor.

By WILLIAM B. STOUT.

Success in aviation really began as a practicable proposition with the advent of the Gnome rotary-type motor, which since its first great successes has added unto itself glory, holding nearly all records at the present time and possessing advantages over any other type.

Naturally a motor so radically different from anything that engineers are used to has met with opposition on the part of many designers who have been more prone to follow accepted lines of design than go to the time, expense and delay of developing something new and different.

The designer of the Gnome—M. Sequin by name—rather than sacrifice strength for lightness in his design, sought to develop a machine which by its compactness and fewness of parts would attain a minimum of weight with a maximum of strength, and in the use of the strongest materials rather than the lightest he has succeeded in producing a motor which develops almost 100-horse power in a weight of less than 300 pounds.

In general arrangement the motor is similar to the Adams-Farwell, formerly built and originated in Chicago, but it has been developed and refined to a point far beyond the

Cylinders Rotate About Shaft.

These cylinders rotate about the shaft at the center of the crank case, supported on annular ball bearings, this shaft having on it, inside the case, a crank. The shaft is held still and the crank with it.

Attached to the crank are the connecting rods, R, one of these, shown in K, being the master rod and the other six bearing on it so that all really fasten to the same crank.

At the other end of the connecting rods are the postons, P, and as the motor turns you can see that these pistons turn, too, with the crank pin as a center and in the circle I have shown dotted at A.

Thus these piston weights revolve in a true circle and will give no vibration, since all are of equal weight.

The cylinders revolve about the shaft C, so that these, too, in rotating revolve in a true circle, B, and hence form no vibrations. Thus, no matter how fast the motor may turn, the masses are always balanced and the motor is free from vibration, a thing of no small moment in a flying machine, where vibration means distortion of propeller and planes, and thus loss of efficiency.

Having seven cylinders, there are three and a half explosions every revolution, so that the power is fairly steady, and "propeller flutter," the bane of the wooden screw, is well done away with compared to other types of motor.

Lubrication is the chief problem in the design, the machine throwing expensive oil like a sprinkling cart, and only a part of this can be caught again. Thus the motor uses almost as much oil as gasoline, an expensive thing for the driver.

Steel, especially of the chrome nickel variety, is not an ideal lubricating material, not having the pores to retain the oil that cast iron has, nor any free graphite in its makeup, so that wear is bound to be comparatively rapid in spite of oil. Hence the Gnome of today is guaranteed for only 300 hours' flying.

It may be that this motor, when developed, will be the flying machine motor of the future, but undoubtedly it is the most interesting today, especially interesting to its makers from the fact that it has made them something like a half million of dollars in less than three years.

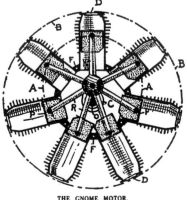

THE GNOME MOTOR.

American original. In appearance the motor is simplicity itself.

First there are seven cylinders sticking out of a central narrow crank case like points of a seven-pointed star.

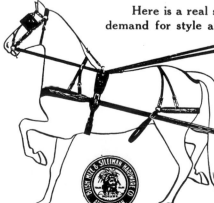

John Deere merged their branch houses, factories and the parent company, forming Deere & Company with a capitalization of $50 million at Moline, Illinois. Included in the merger were:

Deere & Co. of Moline, Illinois
Deere and Mansur of Moline, Illinois
Deere & Webber Co. of Minneapolis, Minnesota
John Deere Plow Co. of Kansas City, Missouri; St. Louis, Missouri; Omaha, Nebraska; Dallas, Texas; Denver Colorado; Portland, Oregon; Spokane, Washington; Oklahoma City, Oklahoma; San Francisco, California; New Orleans, Louisiana; Indianapolis, Indiana; Baltimore, Maryland; Syracuse, New York; Winnipeg, Manitoba, Canada.
Dain Mfg. Co. of Ottumwa, Iowa
Fort Smith Wagon Co. of Fort Smith, Arkansas
Kemp and Burpee of Syracuse, New York
Marseilles Co. of East Moline, Illinois
Moline Wagon Co. of Moline, Illionis
Richardson Mfg. Co. of Worsester, Massachusetts
Syracuse Chilled Plow Co. of Syracuse, New York
Union Malleable Iron Co. of East Moline, Illinois
Van Brunt Co. of Horicon, Wisconsin

Later, Deere & Co. purchased the Davenport Roller-Bearing Wagon Co. of Davenport, Iowa. Also in 1911, Deere and Co. installed a refrigeration plant at their Moline facility to cool the water from their new artesian well for drinking and tempering. The cost of the cooling equipment to lower the water temperature to about 40 degrees was about $25,000. The Moline Plow Co. pur-

chased the McDonald Bros. Scale Co. of Newcastle, Indiana. The Morton-Heer Tractor Co. of Freemont, Ohio, introduced a Four Wheel Drive tractor

model powered by a Heer engine with two opposed cylinders. Chris Heer applied for 4WD patent, but didn't receive patent number 1,104,537 until 1915. The name was changed to Heer Engine Co. in Portsmouth, Ohio. The M. Rumely Co. of La Porte, Indiana, merged with Gaar Scott of Richmond, Indiana, and the Advance Thresher Co. of Battle Creek, Michigan. The Minneapolis Threshing Machine Co. had entered into a joint venture with the Advance Co. in the American-Abell Co., but managed to slide out and leave Rumely the loser. The Western Malleable & Gray Iron Mfg. Co. of Port Washington, Wisconsin, changed their name to the Turner Manufacturing Co. and continued to manufacture Simplicity engines until the end of WW1. The Fish Bros. Mfg. Co., long time manufacturer of wagons, suffered a disasterous fire about two years before, but financial problems in 1911 were reported to be caused by lack of operating capital. The Fish Brothers Wagon Co. continued, mostly selling the already built wagons, until 1913 when it officially closed its doors. By 1911 many automobiles were

1908-1913

THE FIFTY-BOTTOM OLIVER GANG OPERATED BY THREE RUMELY "OIL-PULL" ENGINES.

the product of companies who were attracting the attention of people with money, not old fashioned wagon builders. The Studebaker Corporation, who sold E-M-F "30" and Flanders "20" automobiles, inaugurated credit buying of their automobiles by "farmers and other responsible buyers." Up to this time, the automobile industry was unique in that it was strictly cash business. Two years later, Studebaker would expand the line to include three new Studebaker models, the "25", "35" and "Six." It was reported in September 1911 that the Waterloo Gas Engine Co. had decided to manufacture tractors and had increased capitalization to $500,000 to begin expanding.

LIVE DEALERS SHOULD INVESTIGATE THE PIONEER "30," FIRST IN GAS TRACTION

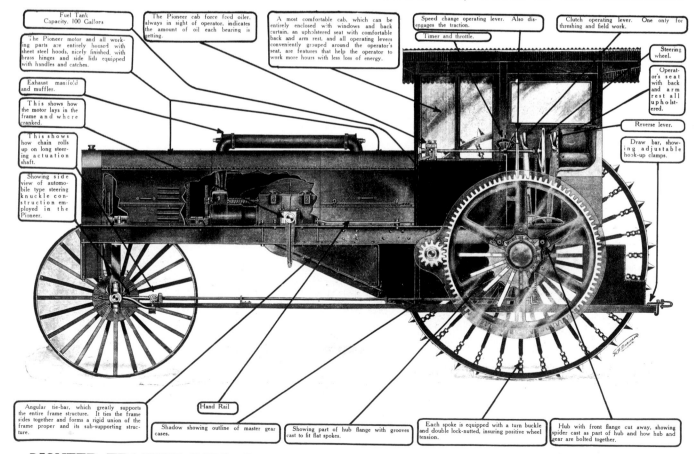

Fuel Tank Capacity, 100 Gallons

The Pioneer motor and all working parts are entirely housed with sheet steel hoods, nicely finished, with brass hinges and side lids equipped with handles and catches.

Exhaust manifold and muffler.

This shows how the motor lays in the frame and where cranked.

This shows how chain rolls up on long steering actuation shaft.

Showing side view of automobile type steering knuckle construction employed in the Pioneer.

The Pioneer cab force feed oiler, always in sight of operator, indicates the amount of oil each bearing is getting.

A most comfortable cab, which can be entirely enclosed with windows and back curtain, an upholstered seat with comfortable back and arm rest, and all operating levers conveniently grouped around the operator's seat, are features that help the operator to work more hours with less loss of energy.

Speed change operating lever. Also disengages the traction.

Timer and throttle.

Clutch operating lever. One only for threshing and field work.

Steering wheel.

Operator's seat with back and arm rest all upholstered.

Reverse lever.

Draw bar, showing adjustable hook-up clamps.

Angular tie-bar, which greatly supports the entire frame structure. It ties the frame sides together and forms a rigid union of the frame proper and its sub-supporting structure.

Hand Rail

Shadow showing outline of master gear cases.

Showing part of hub flange with grooves cast to fit flat spokes.

Each spoke is equipped with a turn buckle and double lock-nutted, insuring positive wheel tension.

Hub with front flange cut away, showing spider cast as part of hub and how hub and gear are bolted together.

PIONEER TRACTOR MFG. CO., Winona, Minn., Corner Front and Carmona Streets.

SMALL AUTOMATIC SYSTEMS OF REFRIGERATION

Motor Driven and Using Ammonia as a Refrigerant.

By WILLIAM E. STONE.

Medical authorities agree that in the freezing of natural ice the germs of disease which may be in the water are imprisoned and stored up, instead of being exterminated by the low temperatures which produce the ice, as claimed by some writers who have never given the matter the thought and investigation it merits. Natural ice is put before the public as "pure washed" ice, but the fact usually remains that it has never been analyzed to show that there are disease germs in it, and the real meaning of the term "pure washed" is that the ice has been thrown on same floor and a stream of water turned on it to remove the sawdust or straw from the outside and make it look clean. The ice is delivered from door to door, washed again, dropped into a tank of spring water and the result is "ice-water," with all the disease germs and impurities still contained in it. It looks good, tastes good, and therefore the public concludes that it must be good. It is put in refrigerator boxes, the germs are set free by the melting of the ice, lodge on the food products or impregnate the walls of the refrigerator, and are eventually taken into the systems of the users.

Artificial ice and refrigeration, on the other hand, are entirely free from these dangerous qualities. Ice may be made and storage rooms or small refrigerators cooled in the creamery, market, dairy and in the home by the same machine, and without any attention whatever other than taking the blocks of ice out of the tank and refilling the cans. These systems, as the name implies, are completely automatic. Any desired temperature can be maintained in any number of insulated cold storage compartments; ice can be made in any desired quantities, all from one machine, and without any attention whatever. The system is complete in itself; automatic mechanical devices, positive in operation and always on the alert, control the system from beginning to end. When the compressor is in operation, automatic oiling devices carry lubricant where it is needed at every stroke of the piston. An automatic water regulating valve supplies cooling water to the condenser, always limiting it, however, to the actual requirements, and stopping the flow entirely when the compressor stops, thus

A HOME AUTOMATIC REFRIGERATING PLANT.

preventing waste and insuring minimum water bills. If, through any accident due to the closing of a valve or stoppage of water supply, the pressure in the condenser becomes abnormal, an automatic high-pressure cut-off rings an alarm gong and stops the compressor, thus removing all danger and making it absolutely safe to operate the system indefinitely without attendance. As the temperature within the cooled compartment falls to the desired point, an automatic starting and stopping device, actuated by a sensitive automatic thermostat, breaks the motor circuit and stops the compressor. When the temperature rises a fraction of a degree the thermostat again actuates the starting and stopping device and starts the compressor. As the pressure in the expansion coils is reduced on starting, an automatic ammonia expansion valve begins to feed ammonia to the expansion coils, and, maintaining a uniformly high back pressure at all times, allows the system to operate at its greatest efficiency, producing the greatest amount of refrigeration for the smallest possible expenditure of power and water. These operations are gone through many times during

the twenty-four hours, keeping the rooms at the same temperature, and running only when the cooling work is needed. When the compressor stops, the cost, both for electric power and for water, stops also. Ice and refrigeration can actually be obtained by this system for about one-half the cost of natural ice required to do the same work in a less satisfactory manner.

Melting ice will cool a storage box or refrigerator to only about 45° F., unless the melting is artificially accelerated. At this temperature meats and perishable products cannot be kept during the warm weather for more than a few days. The moisture caused by melting ice assists in the rapid decomposition of food products, and causes the induced on starting, an automatic ammonia expansion valve begins to feed ammonia to the expansion coils, and, maintaining a uniformly high back pressure at all times, allows the system to operate at its greatest efficiency, producing the greatest amount of refrigeration for the smallest possible expenditure of power and water. These operations are gone through many times during the twenty-four hours, keeping the rooms

at the same temperature, and running only when the cooling work is needed. When the compressor stops, the cost, both for electric power and for water, stops also. Ice and refrigeration can actually be obtained by this system for about one-half the cost of natural ice required to do the same work in a less satisfactory manner.

Melting ice will cool a storage box or refrigerator to only about 45° F., unless the melting is artificially accelerated. At this temperature meats and perishable products cannot be kept during the warm weather for more than a few days. The moisture caused by melting ice assists in the rapid decomposition of food products, and causes the insulation to deteriorate rapidly. Automatic mechanical refrigeration will reduce and maintain a uniform temperature in the refrigerating box at any desired limit, even at zero degrees F. The cold air being absolutely dry, all articles of food can be preserved for long intervals, depending upon the degree of cold in the box. All labor, moisture and the annoyances connected with the handling and storing of ice are done away with, and the ice storage space becomes available for food-stuffs to be cooled. The refrigerator can be located wherever desired, without regard to accessibility for charging or filling with ice, and several compartments can be cooled, even though they may be far apart. A coil of pipe placed in the drinking water tank produces a better cooling effect than ice, and does away with all danger of disease from this source.

These plants have been installed in hotels, private residences, markets, packing houses, ice cream and candy factories, with the best of success, and are especially suited to the needs of those who use from 500 to 20,000 pounds of ice in twenty-four hours.

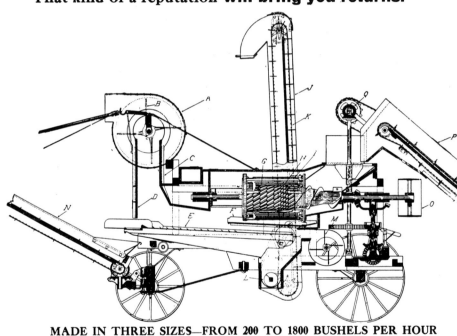

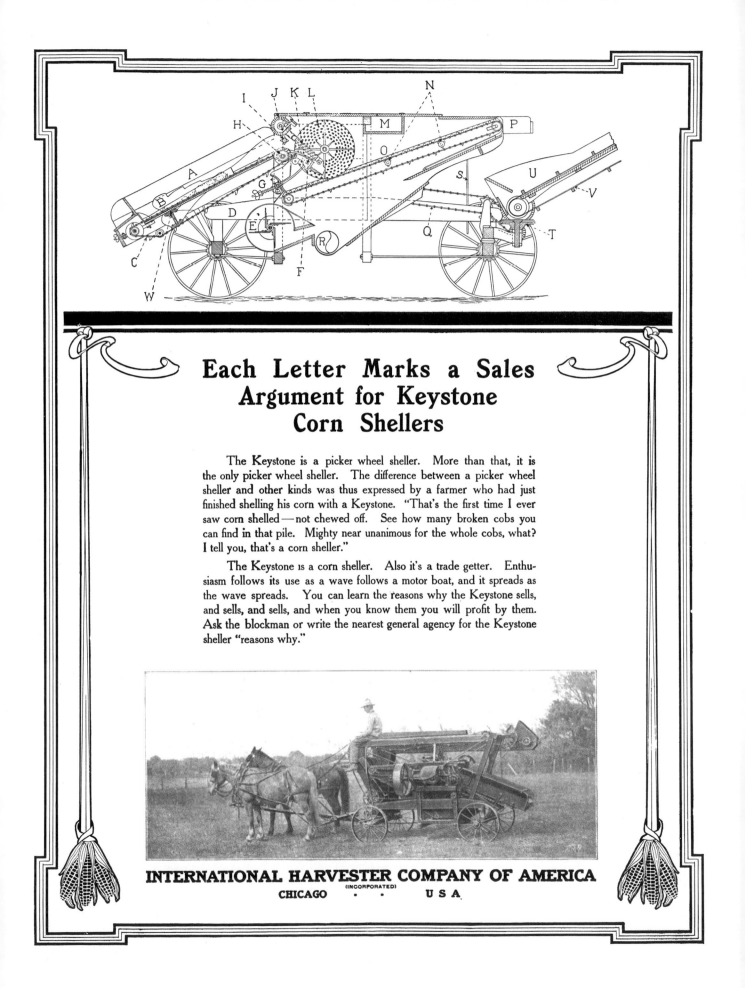

Each Letter Marks a Sales Argument for Keystone Corn Shellers

The Keystone is a picker wheel sheller. More than that, it is the only picker wheel sheller. The difference between a picker wheel sheller and other kinds was thus expressed by a farmer who had just finished shelling his corn with a Keystone. "That's the first time I ever saw corn shelled—not chewed off. See how many broken cobs you can find in that pile. Mighty near unanimous for the whole cobs, what? I tell you, that's a corn sheller."

The Keystone is a corn sheller. Also it's a trade getter. Enthusiasm follows its use as a wave follows a motor boat, and it spreads as the wave spreads. You can learn the reasons why the Keystone sells, and sells, and sells, and when you know them you will profit by them. Ask the blockman or write the nearest general agency for the Keystone sheller "reasons why."

INTERNATIONAL HARVESTER COMPANY OF AMERICA
(INCORPORATED)
CHICAGO · · USA

LUEBBEN VACUUM CURING MACHINE.

The Luebben Vacuum Curing Machine.

The Luebben Vacuum Curing machine is the last addition to the manufactured lines of the Luebben Baler & Vacuum Curing Co. This concern is located at Beatrice, Neb., where for a number of years it has produced the well-known Luebben hay and alfalfa cylindrical baling machine. The new vacuum curing machine is a radical departure from anything heretofore seen in this market and is most interesting to all who are engaged in the raising and the sale of hay and alfalfa.

In the preparation of hay, grass and alfalfa for market the farmer has long been obliged to "cure" his stock in the open field, where, subject to wind, rain and heat, a considerable loss is almost certain to occur. In order to bale the hay and prevent "sweating" after the process is completed, the hay must remain exposed to the air until sufficient amount of moisture has been drawn from it to render it thoroughly dry. This takes time and is often unsatisfactory when rainy seasons set in shortly after cutting. With the Luebben vacuum curing machine the process of nature is carried out with identically the same results, but the necessity for long air exposure is entirely eliminated. By means of artificial vacuum and heat a specified amount of moisture is drawn from the bales in a very short time.

The outfit is mounted upon a large truck and consists of a small gasoline engine, two large "curing" tanks of galvanized steel, a vacuum pump and over-head cable conveyer. As seen in the accompanying illustration, the tanks are so arranged that one may be in use while the other is being unloaded and replenished with the bales. The machine is for use in connection with the Luebben cylindrical baler, only, and will not cure solid, pressed bales. As soon as the hay is placed upon the ground, the cable conveyor is let down, the bales attached and hoisted up over the first tank, then lowered down into the same and the cable removed. The top is closed and the tank made airtight. On the inside of each tank is a weighing scale and on the outside a vacuum dial indicator. The engine is started and the vacuum pump put into operation. As the moisture is drawn off in this manner, the weighing scale shows the lightened weight of the hay. The process is continued until the desired amount of moisture has been withdrawn. The tank is then opened and the hay removed. While this is in action, the vacuum pump is connected to the other tank, rendering the operation of curing a continuous one. Two men can easily cure three tons an hour with the machine.

When the Luebben baler and curing machine are used together, the cutting, baling and curing of hay may be done in a single day. Not only is the resulting bale more fresh and consequently of a higher market value, but the leaves of the alfalfa, usually crumbled or lost entirely in the ordinary manner of curing, are retained perfectly with the stalks. A difference of from $4 to $5 per ton is to be gained by the use of the vacuum curing machine. The hay, once cured, will positively remain impervious to attacks of heating or sweating. Critical and laborous tests at the state agricultural schools have demonstrated the absolute practicability of the Luebben vacuum curing machine.

Richmond
MADE STRONGER
LASTS LONGER

Price $1,200

Why use adjectives when facts are more impressive? The pictures show that the Richmond is attractive and adapted to many needs. The Richmond is made entire in its own factory. The parts that should be strong are easily made strong. The shafts are heat treated. The cast parts are 25 percent steel. The body cannot sag. The doors will always fit. Breakage is reduced to a minimum. The car is light in weight, but the construction is strong. It will last.

★★★★

You don't like to pump up a tire. This pump is attached to the Richmond engine. It will fill any tire in two minutes.

Pioneer Implement Co.
COUNCIL BLUFFS, IOWA
GENERAL WESTERN AGENTS

Industry pioneer Peter Kells Dederick died January 18, 1911. Dederick was issued about 300 patents for hay presses and was considered the dean of the hay press industry.

New Mexico and Arizona become states and Wilber Wright died in 1912. The Titanic struck an iceberg and sank in the North Atlantic on April 15, killing 1502. The Department of Justice charged International Harvester Corporation with being a monopoly in restraint of trade and demanded its dissolution. International met government demands in 1918 after a long legal battle. Later in 1912, International purchased corn planter patents and patterns from Chambers, Bering & Quinlan Co. of Decatur, Illinois. In 1912, almost every company and industry was being watched very carefully. The leather industry was suspected of being a trust and would soon be investigated. E.I. DuPont de Nemours Powder Co. of Wilmington, Delaware, was forced by government action to divide into the Hercules Powder Co. and the Atlas Powder Co.

Allis-Chalmers Co. was having financial trouble and was appointed receivers to help correct the problems. As a result of an insolvency suit brought by its creditors, the United States Motor Co. was also assigned a receiver who reorganized the company. Later in the year, when Walter E. Flanders was appointed president, the United States Motor Co. absorbed the Flanders Motor Co. The following year, the Maxwell Motor Co. bought the United States Motor Co.; however, in 1912 the United States Motors Co. was the $42.5 million holding company for:

Alden Sampson Mfg. Co. of Detroit, Michigan
Brush Runabout Co. of Detroit, Michigan
Columbia Motor Co. of Hartford, Connecticut
Courier Car Co. of Dayton, Ohio
Dayton Motor Co. of Dayton, Ohio
Gray Motor Co. of Detroit, Michigan
Maxwell-Briscoe Co. of Tarrytown, New York
Providence Engineering Works of Providence, Rhode Island

The New Idea Spreader Co. purchased the entire business, including patents, patterns, stock and equipment, of the Rollman Machine Co. of West Manchester, Ohio. Manufacturing operations for the Rollman transplanter were moved to the Coldwater, Ohio, plant of the New Idea Spreader Co. The Agricultural Credit Co. was formed in New York City with an initial capitalization of $4.4 million to purchase and hold for investment promissory notes, obligations, book accounts and claims from

"LI." THE MIZZOURI HOUN' DAWG MOUNTED ON A VELIE AUTOMOBILE.

manufacturers. In separate purchases during 1912, the Emerson-Brantingham Co. acquired:

American Drill Co. of Marion, Indiana
Gas Traction Co. of Minneapolis, Minnesota (Big 4 Tractors)
Geiser Manufacturing Co. of Waynesboro, Pennsylvania
LaCrosse Hay Tool Co. of Chicago Heights, Illinois
Reeves & Co. of Columbus, Indiana
Rockford Engine Works of Rockford, Illinois

SPEED

On January 13 of this year a tiny monoplane whirled rapidly from the earth up into to atmosphere, straightened her course and then, with the ease of the swallow, sped gracefully and quickly along the path of air current around four high towers. Again and again this mechanical bird flew around the aerial course, banking high on the turns, then settling to a level on the straight-away. At last the bird-machine was pointed earthward and light as a feather it descended gently to earth, the engine ceased its battery of pops and a small, odd-looking Frenchman jumped

out of the chassis, swung his arms rapidly to restore the circulation in his limbs, numb with cold, and was carried from the field upon the shoulders of the enthusiastic crowd. Such was the making of the latest world's record, that of speed in the newest type of transportation—the aeroplane. Verdrines, at Paris, on that eventful day had driven his mount 88.33 miles within the hour, smashing the former figures set by the late E. Nieuport, at Mourmelon, France, on June 21, 1911, of 82.73 miles per hour.

In the make-up of mankind generally there seems to be an innate desire or craving for speed—to go fast and then faster. From the earliest history of our country there is recorded many an attempt by various individuals to go faster than our sedate forefathers were wont to go. From the primitive race of men on foot to the straining wheel of a monster racing car, is a long step forward. In fact, the advance of civilization in the world is well marked in the records set by mankind for speed. China, probably the last of the great empires to adopt the modern customs recognized in this day and generation, is credited with not a single speed record, while France, England and America, representing collectively the highest degree of civilization at this time, stand holders of practically every known record. The slow-going Chinese has probably never troubled his head as to whether he could go fast or not; but with the introduction of the steam locomotive, electric railways and possibly the automobile, he has awakened to the realization

Recognized World's Record to Date.

Event.	Name.	Place.	Date.	Miles Per Hour.	Time Per Mile.
Automobile	Bob Burnam	Daytona	Apr. 23, 1911	142	:25 2-5
Motorcycle	Glenn Curtiss	Daytona		136.4	:26 2-5
Locomotive	Plant System	Jacksonville	Mch. 23, 1901	120	:30
Aeroplane	Verdrines	Paris	Jan. 13, 1912	88.33	:39 4-5
Bicycle	George Cramer	Los Angeles	June 33, 1910	62	:58 1-5
Motor Boat	Dixie IV	Marblehead Bay	1911	45.32	1:19 1-5
Horse	Salvador	New York	Aug. 28, 1890	37.8	1:35 1-2
Roller Skating	Kearney	Cleveland	Jan. 26, 1910	24.5	2:27 1-5
Ice Skating	C. Hamilton	Chicago	1908	23.53	2:54 4-5
Running	J. Paul Jones	Cornell	May 27, 1911	14.46	4:15
Rowing	Edwin Henley	Newark, N. J.	July 11, 1892	14.02	4:28
Walking	International	London, Eng.		9.20	6:52
Swimming	C. M. Daniels	New York		2:56	23:40 2-5

that speed is possible and profitable. While it may be many years before China will threaten the world for speed honors, yet the modern vehicles now being installed in the new republic are doubtless the beginning of a new standard of life in the land of the yellow men. America, with the great mixture of humanity from which to pick real champions, today holds the lion's share of honors. France and England run a very close tie for second position, while Spain, Norway, Sweden and Italy comprise nearly the balance of the list.

The gasoline automobile is supreme in the speed world. Last April the redoubtable Robert ("Bob") Burnam in his 200-h. p. Benz sped like greased lightning across the sands of Florida's beaches for two miles, and a Warner electrical timing device indicated that he had attained and held for the brief period of less than a minute the remarkable speed of 142 miles per hour, or exactly 25 2-5 seconds per mile—a grand new world's record.

Long yeares before Burnam was even attempting to make a mile-a-minute, a slim young man mounted an odd-looking contrivance and clinging low over the handle bars rode for several miles at the rate of 136.4 miles per hour, or 26 2-5 seconds per mile. That man, now renowned in the aviation world, was Glenn H. Curtiss and the machine he rode was an eight-cylinder motorcycle of his own making. Though virtually a two-wheeled automobile, to this motorcycle probably belongs the credit for first having traveled two miles in less than a single minute.

Third in line, the steam locomotive is an important factor in the records of 1912. For many years its supremacy was unchallenged, and as recently as ten years ago a Plant system engine, with stripped tender and light ballast, ran over a specially constructed track near Jacksonville, Fla., at two miles per minute, or actually 30 seconds per mile. To the ordinary traveling public, this speed is at least thrice too swift, but were an announcement of the maintenance of such speed on all roads in the United States made it is hardly to be expected the public highways would be thronged with parties in transit on foot, horseback and automobile, while the trains whizzed by with empty coaches and Pullman cars.

The aeroplane, which stands fourth at this writing, is the machine which will probably shatter speed records most rapidly. Little more than out of the experimental stage, its real ability is practically unknown, and with the development of efficient aviators Verdrine's record seems doomed to fall sooner than that of any other record now recognized.

In Los Angeles on June 2, 1910, George Cramer, behind a motor pacing machine, rode

A NEW CALIFORNIA AUTO TRACTOR.

one mile on a bicycle in :58 1-5, an average of 62 miles per hour. Even with the tremendous advantage of "broken wind" and motor pacing, the enormous physical energy to perform such a feat is incomprehensible to many and Cramer's famous ride will stand out as the greatest achievement in the annals of bicycle history.

An American power boat, the swiftest hydroplane ever launched, is responsible for the record of 1:19 1-5 for a mile. In nautical terms, this practically equals sixty knots an hour. This record was made when the Dixie IV had beaten all her foreign competitors for international honors, in a special race.

The equine follows the motor boat. Salvator, of the early '90s, ran a mile in 1:35½, a record that has never been lowered in the twenty years since it was established, and which will probably remain as such for years to come.

The various forms of roller skating, ice skating, running, rowing, walking and swimming, in their respective order, complete the list. Of course, in the figures compiled by this paper, there are no records of many forms of travel, such as sail boating, skiing, sliding, ostrich riding and log shooting, for the simple reason that no definite figures on these are available. But it is doubtful if any one of them would seriously menace the published records.

The world cries for more speed—more speed. And records are shattered, only to fall again before the onslaught of some new champion. Faster and faster action takes place, greater and greater are the speeds attained, until we in this strenuous age pause to wonder when the limit of human and mechanical endurance will be reached. With great steam engines pulling trains of passenger coaches across the continent at better than a mile a minute; with automobiles

whisking along at a pace unconceived before; and with the great American people themselves living such a rapid pace, it is apparent that the limit has not yet been reached. When will the time arrive? Not until, at least, such men as Burnam, Curtiss, Verdrines and their kind have lost the cunning of their hands and the nerve of their souls will the rush towards annihilation of time and distance cease.

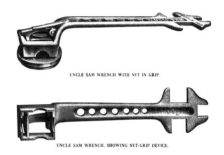

UNCLE SAM WRENCH WITH NUT IN GRIP.

UNCLE SAM WRENCH, SHOWING NUT-GRIP DEVICE.

No. 169 Uncle Sam Buggy Wrenches.

The Richards-Wilcox Mfg. Co., Aurora, Ill., is placing on the market a nut-grip vehicle wrench known as the No. 169 Uncle Sam. The accompanying illustrations show the wrench, the nut in the grip, and also show the patented grip device. This is a very valuable feature. It is to be readily seen that the Uncle Sam will save a lot of time in greasing buggies. The wrench is made of malleable iron finished in black baked Japan and made in three sizes for ⅞-inch, 1-inch and 1⅛-inch nuts. As these Uncle Sam wrenches can be sold at such low prices it would be well for dealers to write to the Richards-Wilcox Mfg. Co. for further information.

STUDENTS DOING PRACTICE WORK ON THE ERECTING FLOOR.

Charles W. Nash was elected president of the General Motors Co. Nash had been manager of the Buick Motor Car Co. The Imperial Automobile Co. bought the Buick truck manufacturing plant at Jackson, Michigan, and the Lauson Frost King engines for 1912 were equipped with a rotary magneto, "eliminating the need for batteries, spark coils, switches, etc." The Minneapolis Steel & Machinery Co. contracted to build 500 Model "30-60" tractors for the J.I. Case Co. in 1912. The Moline Plow Co. purchased the Monitor Drill Co. of Minneapolis, Minnesota, and the Henney Buggy Co. of Freeport, Illinois. Rumley in La Porte, Indiana, purchased the Northwest Thresher Co. of Stillwater, Minnesota, manufacturers of the Universal tractor. The name of the tractor was changed to "Gas Pull." Several other companies were also purchased by Rumley in 1912 and the company name was soon changed to Advance-Rumley Co. The Waterloo Gasoline Engine Co. bought the cream separator plant from Lisle Mfg. Co. of Clarinda, Iowa. The first Waterloo Boy Model "R" tractor was sold in 1912, but a total of 118 Model "R" tractors were sold by the year's end.

Epidemics of spinal meningitis, black leg and other diseases destroyed large numbers of horses, cattle and other livestock. Loss of the large numbers of draft animals just before the fall harvest lead to the increased sale of tractors. The American Express Co. was one of the companies that exchanged their small horse-drawn wagons for commercial automobiles and motor trucks.

Potash was discovered in the ancient "Borax Lake" located in the Mojave Desert of southern California. The Italian experimenter Francesco de Bernocchi, succeeded in sending pictures by wireless using the Marconi process. The Great Yuma Tunnel (or syphon) was undertaken by this country's Reclamation Service to irrigate 100,000 acres of arid land and the French invented roller skates powered by a pair of two cylinder gasoline engines.

The harvest of 1912 closed the year as one of the most successful for the farmer, equipment dealer and the country. Woodrow Wilson was elected president.

Dr. Gustaf DeLaval, inventor of the continuous cream separator, died at his home in Stockholm, Sweden, on February 3, 1913. His death was a result of an operation for cancer from which he had been suffering for many years. Stephen Bull, one of the founders of the J.I. Case Threshing Machine Co., died on November 15, A. Montgomery Ward died on December 7 and William Deering died on December 14.

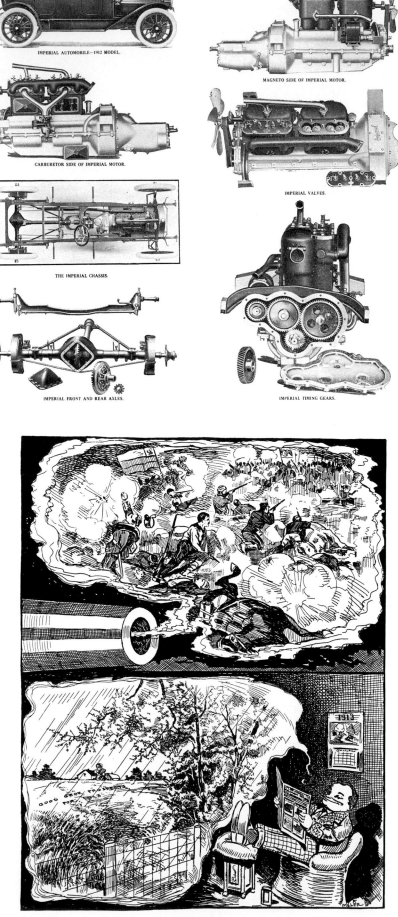

IMPERIAL AUTOMOBILE—1912 MODEL.

CARBURETOR SIDE OF IMPERIAL MOTOR.

THE IMPERIAL CHASSIS.

IMPERIAL FRONT AND REAR AXLES.

MAGNETO SIDE OF IMPERIAL MOTOR.

IMPERIAL VALVES.

IMPERIAL TIMING GEARS.

MORE CLOUDS—EUROPEAN AND AMERICAN.

The New <u>Chevrolet</u> Is Here

THE PRODUCT OF EXPERIENCE

Type H-4 "Baby Grand" Touring—$875 f. o. b. Flint, Mich.

CONDENSED SPECIFICATIONS:

Wheel Base—104 inches.

Motor—Powerful, quiet and simple. Four-cylinder, valve in the head. 3¹¹⁄₁₆x4 inches.

Cylinders—Detachable head, cast en bloc.

Valves—1½ inches, enclosed.

Connecting Rod Bearings—2¼x1½ inches.

Crankshaft Bearings—Front, 2¾x1½ inches; Center, 2x1 31-32 inches; Rear, 3⅛x2 inches.

Cam Shaft Bearings—Front, 2⁹⁄₆x1⁷⁄₆ inches; Center, 2x1 11-64 inches; Rear, 1⅞x1¼ inches.

Oiling System—Splash with positive pump; Sight-feed on dash.

Carburetor—1-inch, improved double-jet; pressure feed through automatic pump.

Ignition—Simms high-tension magneto.

Clutch—Cone, leather faced, with adjustable compensating springs.

Transmission—Selective type, three speeds forward and reverse.

Cooling—Thermo-syphon, with fan.

Axles—Front, I-beam; rear, semi-floating.

Brakes—Service, external contracting. Emergency, internal expanding.

Front Wheels—Ball bearing.

Tires—32x3½ inches, Q. D., straight side.

Drive—Left side; center control.

Steering Gear—Worm and worm.

Body—Touring type, gasoline tank hung in rear under body.

Gasoline Tank—16 gallons capacity.

Finish—Chevrolet gray body, with black and nickel trimmings.

Weight—Completely equipped, with gasoline and oil, 2,100 lbs.

Equipment—Mohair tailored top and cover, windshield, five lamps, Prest-O-Lite tank, horn, speedometer, complete tool equipment, including jack and pump. Electric lights and generator equipment, with electric starter, $125.00 extra. When car is equipped with electric light and starting system, a coil and distributor will be used for ignition, instead of magneto.

Large Six $2500—Little Six $1285—Little Four Roadster $690

Type H-2 "Royal Mail" Roadster—$750 f. o. b. Flint, Mich.

One of the most important features is the motor, which is of the valve-in-the-head-type, with valves enclosed, making for extreme silence and cleanliness (cover removable by two nuts).

The removable cylinder head is another characteristic for a valve-in-the-head motor. The spark plugs are set in the head casting at an angle very accessible.

The exhaust manifold is integral with the head casting, the outlet being through a single pipe, eliminating back pressure; also the intake manifold on the opposite side is bolted to the head casting.

The piston and connecting rods may be removed through the top of the cylinders.

The drive is through a leather-faced cone clutch and selective type 3-speed transmission.

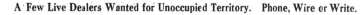

A Few Live Dealers Wanted for Unoccupied Territory.   Phone, Wire or Write.

SOUTHWEST MOTOR CO.

DISTRIBUTORS

Sixteen-Sixteen Grand Ave.   Bell, 1888 Grand.   Home, 3755 Main.   Kansas City, Mo.

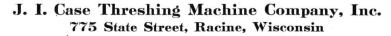

The WEEKLY IMPLEMENT TRADE JOURNAL

PROSPERITY NUMBER

GILSON

Johnny·on the·Spot

Retails for $32.50   with good Profit to the Dealer

1½ h.p. HOPPER COOLED.   1¼ h.p. AIR COOLED.   Can Furnish Truck.

This Finest of All Small Engines

was designed especially to help Gilson dealers to meet the demand for light duty engines. You can sell one with every washing machine, cream separator, churn, grindstone, or other light-power machine that leaves your store. Get your orders in quickly. Deliveries begin June 1, 1913.

Gilson Mfg. Co., Port Washington, Wis.

This Is I H C Tractor Year

JUDGING from the number of prospects now figuring and the percentage of sales made, this is I H C tractor year.

A little judicious work on your part, a close study of the local situation, the planting of a few helpful suggestions, a thorough study of your tractors to enable you to meet arguments, and tractor business is sure.

The beauty of tractor business is that, once started, it usually comes fast. I H C tractors demonstrate quickly their labor-saving and money-making qualities. When a wide-awake farmer sees his neighbor getting ahead of him, he develops an active interest at once.

If you haven't already started your tractor campaign, now is the time. If you want the line that insures the greatest return for your work, see the blockman or write the general agent.

International Harvester Company of America
(Incorporated)

CHICAGO   U S A

Parcel Post began January 1, 1913, federal income tax was authorized by the 16th Amendment to the United States Constitution and the first tractor demonstration near Fremont, Nebraska, was held on September 8-12 and drew 4,000 farmers to watch the performance of 42 tractors. The demonstration was called the National Power Farming Demonstration the following two years. In 1916, a series of four day demonstrations were planned by a committee of the National Tractor and Thresher Manufacturers Association to be held in Dallas, Texas; Hutchinson, Kansas; St. Louis, Missouri; Cedar Rapids, Iowa; Bloomingdale, Illinois; Madison, Wisconsin; Fargo, North Dakota; Aberdeen, South Dakota. After 1916 the demonstrations began loosing effectivenesss as a sales tool and slowly died off, except in conjunction with fairs.

In 1913 the tax savings for companies with less than $50,000 was great enough to cause some companies to divide into several small companies which were really divisions. The Advance-Rumley Co. formed Rumley Products Co. on April 4, as a sales arm for tax purposes. In January, the International Harvester Corp. was formed to take over the business and properties of the International Harvester Co. in foreign countries. During 1913, several senior

management people left the International Harvester Co. to work for M. Rumley Co., including H.A. Waterman, who became manager of design and manufacture, J.M. Robinson who became sales manager, M.R.D. Owings who became Vice President and L.W. Ellis who became advertising manager. Alexander Legge became the general manager of the International Harvester Co. in May. The Minneapolis Steel & Machinery Co. signed a contract in December to build 4600 Bull tractors for the Bull Tractor Co. The Parrett Tractor Co. was incorporated in Ottawa, Illinois, and the Ingersoll Electric Vaporizer Co. was incorporated in Chicago, Illinois, to manufacture and sell tools and other devices. The Stewart-Warner Speedometer Corporation was orgainized by the merger of the Warner Instrument Co. of Beloit, Wisconsin, and the Stewart & Clark Mfg. Co. of Chicago, Illinois. In April, the Massey-Harris Co. purchased the Deyo-Macey Engine Co. of Binghampton, New York. Massey-Harris then moved the Deyo-Macey plant to Weston, Ontario. The Moline Plow Co. purchased the Adriance, Platt & Co., manufacturers of haying and harvesting machinery. The Moline Plow Co. became the 5th largest farm equipment company in the world when it reorganized with capitalization of $30 million. The Monarch Tractor Co. began in Water-

town, Wisconsin, the Russell Grader Mfg. Co. began in Minneapolis, Minnesota, and the Nilson Agricultural Machine Co. began in Minneapolis, Minnesota. The name was changed within months to Nilson Farm Machinery Co. The design of the Wallis "Cub" introduced in 1913 would set engineering style of tractors for many years. John N. Willys, president of Willys-Overland Co., purchased the plant, stock and all rights of the Edwards-Knight Motor Car Co. of Hartford, Connecticut. Tractors were given credit for defeating the boll weevil by allowing economic drainage of land thereby permitting crop rotation.

The Easy Washing Machine Co. became insolvent and entered receivership in October, 1913. Many automobile and truck models introduced early in 1913 were not equipped with electric starters, but the demand for them increased to such and extent that they were added. Other changes in motor car "fashions": Wider and longer springs to absorb the jolts of the road, eliminating exhaust cut-outs and larger engine castings to reduce noise. Foreign markets were encouraged for used automobiles. The *Hardware Trade Journal* was incorporated as part of the *Weekly Implement Trade Journal* on a fortnightly basis, to serve the growing needs of the implement dealer of rural America.

An End to the Conspiracy?

The conviction and sentencing of thirty-six men, mostly high union labor officials, for complicity in the wide-spread dynamiting outrages which have terrorized the country for several years past, ought to have a wholesome effect in opening the eyes of the really honest workingmen to the fact that they have been made dupes and fools of long enough by a set of blatherskites and criminals, and to turn their attention to the perfectly legitimate work of their unions. We believe that the labor union has a legitimate place and an important mission, notwithstanding the fact that the repeated acts of violence and duplicity by the leaders who have foisted themselves on the main body of workingmen who really work, have caused many of their employers to look with contempt, and even hatred, on the whole scheme of labor organization. Surely the men who work with their hands have as much right to organize to secure justice and to better their condition as have the men who happen to be their employers. There is no "divine right" of a "ruling class" any more than kings rule by divine right; and every man has as good a right as every other man to shape the course of his life as he pleases, so long as it does not interfere with the rights of anyone

else. He has a right to get what is coming to him. The main trouble with unionism has been that it has not stopped there, but has presumed to insist that other men, who do not belong to the union—sometimes, even, those who are unionists, but who do not happen to acknowledge allegiance to their particular brand or branch—shall not have their rights. It is this spirit which has been promulgated, not by the rank and file, the "really honest workingmen" above referred to, but by the men who have forced themselves to the front and corraled the offices. In these positions they have found ample opportunities for graft, and, being men of low morals, or no morals, they have been quick to seize these opportunities, and to exploit them to their own advantage, and to retain their offices for years through a practical conspiracy of graft, where, posing as the representatives of real labor, they have exacted tribute from employers too timid to fight them. Failing in this, they have grown so bold that they have resorted, first to the boycott and then to that anarchist's weapon, dynamite, to gain their point. These unspeakable scoundrels are well rid of, though their punishment is extremely light.

The pathos of it all is that the mass of workingmen have been but little benefited

for all their loyalty and the sharing of their wages with these grafting anarchists. There is no question, in our mind, that the contributions levied upon the wages of the union privates have usually been paid with the notion that they were contributing to a legitimate and righteous cause, although the more intelligent of them should have realized that they were being duped. But there is also little human possibility that the higher union officials, all the way up—or is that the proper designation?—to Gompers, have been cognizant of the facts, and it is to be hoped that the probe will not be withheld, but will be pushed to the quick. If the laboring men of this country are made of the right sort of stuff they will insist that this be done, no matter who cries "ouch."

Incidentally, perhaps the most remarkable feature of the prolonged trial at Indianapolis is the wonderful memory of Ortie McManigal for the mass of details by means of which he enabled Detective Burns and the Government prosecutor to wind the threads of evidence around the defendants until they were simply unable to move hand or foot. If the suggestion that he be released and employed as a Government detective to ferret out evidence against the rest of the gang is followed, it can but strike terror to their dastardly hearts.

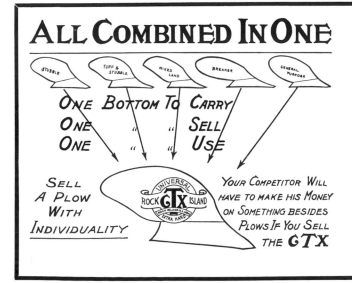
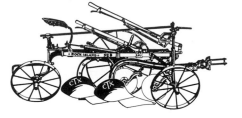

MECHANIZATION OF THE FARM

THE ABOVE IS AN EXACT REPRODUCTION

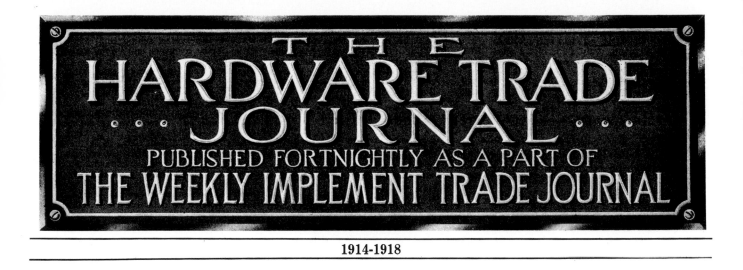

THE HARDWARE TRADE JOURNAL

PUBLISHED FORTNIGHTLY AS A PART OF
THE WEEKLY IMPLEMENT TRADE JOURNAL

1914-1918

The "Great War" served the purpose of showing the nation about the usefulness of cars, trucks, trailers and track laying equipment. The shortage of men and draft animals and the need to increase farm production forced the nation's farms to mechanize during the war. The country's farms were ready to accept all of the new farm equipment that was manufactured during the war and the returning "Doughboys" demanded even more gasoline powered equipment after the war. Much more obvious to the nation was how the women of America were able to fill jobs previously reserved for men.

Urges Farmers to Sell Horses.

The Avery Co., Peoria, Ill., sees in the present European war an excellent opportunity for the farmers to change to power farming. The company recently through its "Dealers' House Organ" advised its dealers to urge their customers to sell surplus horses and mules now, while the war has boosted the market to its present high prices. The following is from the "Dealers' House Organ":

The State Department at Washington has just issued this announcement:

"American citizens are at liberty to ship all articles whatsoever to the nations engaged in war."

As a result, tens of thousands of horses and mules are now being bought and rushed to Europe. Newspaper reports say that the French government has just contracted for 46,000 horses; that over 18,000 horses have already been shipped to England; and that shipments are being made daily from the larger cities all over the Central and Western states.

If a farmer keeps his surplus horses through the winter, he will have to feed up his high priced grain instead of selling it.

Urge your customers to sell their surplus horses now at high war-time prices. Then they can also sell their grain at high war-time prices instead of having to feed it up into idle horses all winter.

Then sell them tractors to be delivered early in the spring to replace their surplus horses. Right now is the American farmer's golden opportunity to make the change from horse or mule farming to tractor farming under the most favorable conditions.

He can make money now by selling his horses and grain at high prices, and next year he will also make more money, for a tractor will reduce his farming expenses and enable him to raise larger crops.

EUROPEAN WAR AND AMERICAN BUSINESS

FIVE great powers and three lesser ones are at war in Europe. Before this reaches the reader, other nations may be offering their men—and their women—for the sacrifice. The vast and bloody operations of Alexander, Caesar, Napoleon, and of our own Grant and Lee, that made such epochal chapters in world history, are probably due to be dwarfed. The conflict of the titans is on, yet the people of earth are slow to appreciate its dread import. Like a bad dream it seems; like some imaginative melodrama of the impossible future.

An emotion that began but a few days ago with the killing of a man and woman in the obscure capital of an obscure dependency, has swelled into a wave of international madness that demands the death of myriads of Europe's best. Nation after nation leaps into the red vortex. Tens of millions are—thinking?—no! —*feeling* in terms of slaughter. "Beat them back!" is the cry. "Havoc!" is the answer. And the god of battle grins. It is good to see brothers at brothers' throats.

For forty years and more Europe has been compared to a group of armed camps, each coveting more arms, more men and more room; each jealous of all the others; each unable to conceive that excess of armament is both a fool's burden and a criminal menace. Now that it has come, the explosion appears to have been inevitable, albeit those who said so several weeks back were known as alarmists. Throw a lighted match into a powder magazine and there can be but one result. A pistol cracked at Serajevo and gigantic guns belch forth their destruction over all Europe. Over-armament is teaching its hideous lesson.

What a contrast the relations between the United States and Canada present! Cross the line at any point along that three-thousand-mile boundary and you will scarcely encounter a uniform. But for the flags that fly over the custom-houses there is little to remind one of the passage from one sovereignty to another. Great inland seas lie between the Provinces and the States, yet never a ship of war is to be seen. From Maine to British Columbia no fortification threatens. No soldiers strut. No

evidence of hostility intrudes. The sanity of peace prevails.

Yet war in Europe is a fact, and as a fact it must be faced; for its effect will be wide. Such a catastrophic condition that has set a whole continent aflame cannot but be felt throughout the world. It is simply another terrible proof that all peoples are interdependent. Business is bound to be disturbed. Wealth that has been built up through normal years will be dissipated in a flash; will be spent as powder, projectiles—and men.

Of course credits will be shaken, even in this country. But the recuperative powers of America are so wonderful that there need be but little cause for apprehension. Our Government has taken every precaution possible. The new system of currency, a tremendous influence for stability in itself, is going into effect. Europe must suffer, but only a fraction of her suffering can spread to this side of the Atlantic.

Solely from the business viewpoint, the war across the sea is but an inconvenience for American business, great or small in proportion to the degree of interest any one business may have in the European market. As nearly as a nation can be, we are self-sufficient, and there is nothing of timidity in the sentiment of thanksgiving which Americans must feel over that solid fact.

At the head of our country stands a great and good President, who has already offered the good offices of this Government toward the restoration of peace abroad. This is what he says:

"The situation in Europe is probably the gravest in its possibilities that has arisen in modern times, but it need not affect the United States unfavorably in the long run. Not that the United States has anything to take advantage of, but her own position is sound and she owes it to mankind to remain in such a condition and in such a state of mind that she can help the rest of the world.

"I want to have the pride of feeling that America, if nobody else, has her self-possession and stands ready with calmness of thought and steadiness of purpose to help the rest of the world. And we can do it and reap a great permanent glory out of doing it, provided we all cooperate to see that nobody loses his head."

This Cruel War's Abuse of Our Old Friend "Bob Wire"

When Joseph F. Glidden, a farmer of De Kalb, Ill., back in 1872, got the idea of making wire fences with barbs on them, he had no more harmful design than to teach horses, cattle and hogs, by the pricks they might receive, that wire fences were meant to keep them in or out, says the New York Times. When Uncle Sam, on Dec. 24, 1874, gave Farmer Glidden the Christmas gift of a patent on his new device, his idea was heralded to the world. The Western prairies, with their lack of fencing materials, had tried single strands of wire, but they availed little and the whole consumption of wire for fencing in 1874 was only fifty tons. Glidden's barbs made the cattle think, and the farmers soon saw their worth. In ten years the wire fences had increased 10,000-fold, and in ten years more its growth had been the foundation of the wire trust. But Glidden

reaped small reward from his invention until Feb. 29, 1892, when the United States supreme court upheld his claims and he was able to collect royalty on all the fences that had been strung before. He lived fourteen years to enjoy it, and died in his home town in 1906 at the age of 93.

Quite naturally some animals inclosed by Glidden's fencing gashed themselves on the barbs. Just as naturally men and boys tried to climb over or under these fences and had their clothes and their flesh torn. These wounds upon man and beast and the suddenness with which Glidden's barbs halted all living things came to the attention of military men, and the barbed wire entanglement of which we now read almost every day in the war news was born. And it may be said right here that soldiers who have been halted by wire entanglements while making a charge or maneuvering for a new position, say the devil never invented anything nastier.

Possibilities seen by American military students in barbed wire were soon carried to the armies of Europe and engineers in every country in the world were put to work devising means for using this new device. Natural forerunners of the barbed wire entanglement had been in use from the earliest times. Roman soldiers had defended their positions with abatis. They had held off their barbarian enemies by felling trees, sharpening the ends of the branches, and massing them with their points turned away from the Eternal city. Fraises—sharp-pointed piles—had been planted in the earth in front of armies for their enemies to wound themselves against or to halt the onrush of a charge till the piles could be removed or scaled.

Nobody outside of the European armies now at war knows how they are using barbed wire entanglements or in what form they are building them, for the engineers of each army are constantly devising new methods, and these new ideas are not divulged even in times of peace. But the dispatches tell of cavalry and infantry running headlong into meshes of unyielding steel thorns that rouse the imagination to the horror of the wounds they inflict.

One use for barbed wire that seems to be new is reported from Belgium. There certain roads that it was desirable to have made passable to the people of the country were made impassable to an army by building zigzag fences from side to side. The peasant, going to market, might pass by traveling slowly and double distance, but an army could not thread such a maze and must halt to destroy it.

DOWN GOES THE OLIVE AND UP GOES THE WHEAT.

MEXICO, THE BATTLE-TORN

Despite the Cares of Business the Eyes of All Implementarians Are Turned South of the Rio Grande

THE eyes of the world are now on battle-torn Mexico—particularly American eyes. "Huerta, the usurper," as the Constitutionalists call him, stands at bay, threatened from the north by the victorious rebel army of Pancho Villa, from the South by Zapata and his bandits, and from the east by the well-trained, well-armed and eager American sailors, soldiers and marines who hold Vera Cruz. Unless the diplomatic "good offices" of Brazil, Argentina and Chile bring about peace, an American army of invasion will soon be in possession of the principal Mexican cities. But it won't be in possession of the swamps and mountain fastnesses. If expert opinion is to be regarded, it will take months and perhaps years of hard, guerilla fighting to come into complete possession of the country and perfect the "clean-up."

What sort of a country is this Mexico? Well, it's a very rich country and a very various country. Would the United States annex it? As the Mexicans themselves are fond of saying, *quien sabe?* Who knows?

There is this much that is certain: If we annex Mexico, we shall have to annex the Mexicans. If this were not true, annexation might be a desirable thing. But there are fifteen million of them. All but a very few are in the depths of ignorance, superstition and poverty. And the small but powerful upper class is very careful to see that they are kept there. Does such a people offer promising material for good American citizenship?

A Little Speculation.

Then there is the moral side of the question, which in these queer days always has to be considered. Formerly, if a country wanted to do some annexing, all it had to do was to go out and annex, providing it had the power, and no questions were asked. But things are different nowadays

Should it be necessary for an American army to enter Mexico as invaders, its first task, naturally, would be to conquer the country. The next thing on the program would be the setting up of a Mexican government "deriving its powers from the consent of the governed." As soon as such a government were firmly on its feet, the American troops would probably be withdrawn and Mexico and the world in general would be that much better off. Perhaps before they left the Yankee occupants of the country would insist on establishing a school system, as they have done in the Philippines.

Of course, this is all mere speculation, but it is a kind of speculation that implement dealers as well as all other interested Americans are exercising. After Mexico were once occupied and pacified, it is altogether probable that a certain jingo element in this country would insist that "the old flag must never be hauled down." Several powerful daily papers are already advocating that we go "on to Panama." Some of the more incorrigible spread-eaglists, indeed, would not stop short of the South Pole. Notwithstanding such perfervid sentiment, the decent thing to do would be to help the Mexicans develop the best that is in them and when the time comes aid them to express it governmentally.

The farmer in the field is a builder; the soldier in the field is a destroyer.

Belgium Sees a U. S. "Atrocity."

Belgium, through Henry Cartin de Wiart, minister of justice, points out to us one of our own "atrocities," says the Chicago Post. "The keynote of American life is waste," said this visiting Belgian. "On our ride from Montreal to Chicago we saw hundreds of miles of fertile lands lying fallow. We saw orchards and fields with ungathered products rotting on the ground. We saw miles of young trees being destroyed by fires started by engine sparks and left to burn unnoticed. Everywhere the farms and residences were divided with wooden fences that contained enough lumber to build the homes of an empire. The waste of America is not confined to materials. Never can a European believe the magnitude of wasted labor in America unless he visits the cities. Hundreds of thousands of men, whose energy might be applied to production, remain in enforced idleness. In the country, wasted lands; in the cities, wasted men. On the trees and plants, ungathered food; in the centers of population, hungry people. Those are the economic elements of American life. Why do not the statesmen here address themselves to bringing about an adjustment that will cure these evils?"

Photographic Supplies Take the Price Elevator

European War Has Curtailed the Importation of Picture-Taking Chemicals.

Some hardware dealers, with a tendency toward expansion, have branched out into the sale of photographic supplies. These will be interested in the startling increases in the price of many chemicals essential to the photographer in the development of pictures that have been recorded since the war stopped the importation of these chemicals and compounds which come from abroad. With prices mounting higher daily, some of the photograph galleries making a specialty of cheap pictures are facing the prospect of going out of business because of the impossibility of getting supplies to carry on their work.

Camera supply houses are unable to sell any quantities of these chemicals, having on hand only a limited supply, which they are keeping for their own developing and printing work, and are only selling in small amounts to their regular customers. Customers who formerly bought these chemicals in lots of one pound and more are now fortunate if they can buy an ounce or two, and very few sales of as much as a pound are being consummated by the dealers.

Hydroquinone, a chemical compound used in developing, has advanced from the normal price of $1 a pound to $5, and, in some shops, as high as $10 a pound. Further advances are anticipated, in view of the news received by one of the dealers from Chicago that the material was selling there for $1 an ounce. The same dealer also received word from Cleveland that one house in that city had sold their entire stock of hydroquinone at a price of $18 a pound. Hydroquinone is a coal tar product, imported from Germany. One dealer in photographic supplies believes that the American manufacturers would no doubt attempt to fill the breach with compounds of their own manufacture, if the stress continues for long.

Another coal tar product which is imported from Germany for use in the development of pictures is noted. This chemical has more than doubled in price during the past week, and is now selling at $12.50 a pound, with still further increases in prospect. Bromide of potassium, silver bromide and other bromides used in the development and printing process have increased in like proportion.

The paper stock used in making the higher grades of papers for printing pictures is mostly imported from France and Germany, and the stopping of these imports will shortly mean that photographers will have to fall back on the lower grade papers manufactured in this country.

HOLT CATERPILLAR HAULING A 30-TON SIEGE GUN IN GERMANY.

The Holt Caterpillar in War Service.

The Holt Caterpillar Co. of Stockton, Cal., and Peoria, Ill., has furnished the Implement Trade Journal with an interesting photograph of which the accompanying plate is a reproduction.

The plate shows a Caterpillar tractor hauling a German siege gun. These powerful guns were the surprise of the European war. The secret of their existence was closely guarded by Germany, and this picture was secured by the Holt Mfg. Co.'s representative at the risk of being shot for a spy. The gun weighs more than thirty tons. Practically every Caterpillar in Europe has been confiscated for war use.

Putting the Auto to Work on the Farm

AN AUTO TRACTOR PULLING A 6-BOTTOM GANG PLOW.

Losses of animals during the war were unbelievable. One source indicated that 7 million horses were killed in the first seven months of the war. The inability to breed, break and transport new animals as fast as they were being killed resulted in a shortage of animals both at home and on the battlefield that created a need to use gasoline, kerosene and steam engines for power. It was noted that about 5 years were necessary to produce a horse that could do farm work and that the horse continued to eat and required care during the seasons when he wasn't working. The tractor drank fuel only when it worked. By 1915, steam traction engines were on the way out and didn't show up at the larger tractor demonstrations.

At home, automobiles were expanding the borders of rural America. The period between 1910 and 1918 was marked by unprecedented increases in truck sales (from 6,000 to 227,000) and a national campaign for "good roads" was in full swing. Increased production was needed to supply our fighting men as well as those of our allies and, because of the war, there were fewer to grow the produce. Modernization was the way to produce more. Manufacturers developed new equipment, improved old designs and vigorously marketed older product lines. Those who saw the chance for success and who had enough money rushed to become dealers. Opportunities were nearly boundless, but the fate of both the manufacturer and the dealer was controlled by the purchasing trends and success of the farmer.

Following the war, shortages were filled by the willing farm equipment manufacturers. Some designs, such as the track laying tractors and heavy duty trucks, were not considered practical before the war, but returning "doughboys" knew the practicality, because of their experiences during the war.

An interesting comment from an article in the Implement Trade Journal, 1916, says that trucks will be important as soon as they were "countryfied." The role of trucks before the war had been limited to city quality streets. These city trucks were used for hauling farm equipment from the railway station to the dealerships. The military showed manufacturers and soldiers how a truck could be built and used in the country. Track laying equipment was being developed as a possible answer to the problems of support and traction. Crawler type tractors would, of course, be used later to improve the roads throughout America.

Latest reports from Belgrade indicate that the cartridge crop in the valley of the Danube will be greater than was expected earlier in the season.

During the past few years automobiles have been utilized for all sorts of purposes other than for the general use for which they were originally intended. Coincident with the automobile and commercial vehicle development, the gasoline engine has been greatly improved and has been found so useful and satisfactory that it is now the modern, and also what might be termed the standard means of producing power. Gasoline engines have replaced steam to such an extent that the modern farm tractor is now rapidly replacing the steam traction engine as a source of power on the farm.

During the past three years about 90 percent of the different makes and styles of farm tractors have been brought out. It has been found that on most farms a tractor of almost any size or weight can be used to some extent, but on the average sized farm it has been found doubtful as to whether or not such a tractor can be used with a resultant net saving which will warrant the farmer in making the necessary investment. It is evident that with a farm of a given size the usefulness and consequent profit derived from the use of any tractor investment can be readily determined.

If, for example, an investment of, say, $2,000 is made in a farm tractor, and the farmer finds that this tractor enables him to make a net saving of $500 per year as compared with the use of additional horses and labor, then it is evident that this $500 saving represents only 25 percent of the original investment of $2,000; after deducting, say, 6 percent for interest and 15 percent for depreciation, we have left a net profit of only 4 percent on $2,000, or $80 per year as actual profit. If the same tractor work could have been done with a tractor investment of only $800, then a charge of 6 percent for interest and 15 percent for depreciation would amount to only $168 per year, thus leaving a net profit of $332, or a little more than 41 percent on the tractor investment.

In order to secure for the farmer all the advantages and profits which possibly can be derived from the use of a farm tractor the Auto Tractor has been produced. The Auto Tractor is claimed to be so simple and inexpensive that the tractor investment has been reduced to a minimum, and at the same time its simplicity makes it practical for any farmer or any member of his family to operate it.

The Auto Tractor is a machine or attachment for operation in connection with an automobile of any size or make, so designed that it can be easily attached to the automobile in about five minutes, and detached from it in about two minutes, leaving the automobile ready for its usual road service.

When the automobile is attached to the Auto Tractor the rear wheels are raised from the ground so that they can rotate as flywheels. The power is transmitted to the gearing of the tractor from pinions mounted on each of the rear automobile hubs. The gear ratio is such that when the rear auto-mobile wheels are running at the speed of twenty-five miles per hour, the tractor will be geared down to about two miles per hour. By thus reducing the tractor speed, and operating the automobile engine and transmission at its normal speed, it is found that a 20-h.p. automobile can do the work of from four to six horses; a 30-h.p. automobile can do the work of from six to nine horses; a 40-h.p. machine, from ten to fourteen horses, and a 60-h.p. machine, from eighteen to twenty-five horses.

Every possibility of developing strains on the automobile parts has been carefully taken care of. An auxiliary cooling system is provided, and is necessary on account of the automobile moving at a speed of only two miles per hour with a consequent lower air velocity through its radiator. A belt attachment is provided at the rear end of the tractor to furnish power to any kind of stationary machinery.

The advantage of the Auto Tractor, in addition to its low first cost as compared with ordinary complete tractors, is in its light weight, which permits operation on soft or plowed ground, or in hilly fields. Ample traction is secured by means of conical shaped spurs which are furnished to suit conditions in various parts of the country.

The Auto Tractor is now furnished in only one size, suitable for operation in connection with all automobiles, ranging from 20 to 90-h.p. However, this season the Auto Tractor Co., Niles, Mich., is bringing out a smaller, lighter, and cheaper Auto Tractor for use on Ford cars and all other automobiles, including those in the 30-h.p. class or smaller.

If automobiles can be used as a substitute for horses in doing the heavier farm work, then farmers certainly need look for no further excuse for spending their money for automobiles.

Glass That Won't Fly When Broken.

A very ingenious process for producing glass which will not splinter and fly about when broken has been devised, according to the Scientific American, by a French inventor. One side of each of two glass plates is covered with a thin coating of gelatine, a thin clean plate of celluloid is laid between these two surfaces, and the glass and celluloid are then united by strong pressure into a single pane.

The "triplex pane" thus produced is said to lose little in transparency while it gains the valuable quality of not splintering when struck or pressed against. While it does not become unbreakable the fragments do not fly, being held by the gelatine layer, and hence no damage is done to bystanders from flying pieces. Tests showed that the most violent blows with a hammer merely caused radiating cracks and concentric rings of cleavage.

Buy and Sell Hardware "Made in America"

War Gives Dealer Opportunity to Remove Prejudices Against Home-Made Goods—A Chance to Help in a Movement That Will Become Nation-Wide.

HOWEVER much we may deplore the havoc and loss attendant upon the European war, we must feel grateful for the benefit to the American mind in freeing it from the foolish prejudices regarding American-made goods. For it has long been evident that American industry has been hobbled by the popular antipathy for American-made goods. We have come to look for the foreign trademark before we allow ourselves to place the stamp of approval upon any article. All this despite the fact that in accepting foreign goods for our daily use, we are sacrificing the superior merit of goods made in America.

Now that we are dependent upon our own manufacturers we will be forced to recognize merit, where heretofore there has been a careless lack of appreciation. The pro-European prejudice has had its influence on nearly every branch of American manufacture and domestic trade. Milady's toilet articles must bear a French name, our silks and linens, and even cotton, must come from foreign mills before it is the "genuine" article. Unless our glassware or china comes from foreign lands its merit is questioned.

Even the hardware trade is visibly affected. Where is there a dealer who has successfully refuted in the minds of his customers the popular belief that cutlery—knives, razors, scissors and shears—has the quality unless it bears the name of a foreign factory? And yet American knives are not surpassed by any product from Sheffield and are vastly superior to the German goods in quality, temper and finish.

Some Facts About Cutlery.

The failure a few weeks ago of Hermann Boker & Co., New York, one of the largest and oldest hardware importing houses in America, serves to call attention to the conditions surrounding the demand for European goods. This failure was a surprisingly swift result of our severed industrial relations with Europe. For seventy-five years this firm had been in business, representing exclusively many foreign manufacturers, as well as operating a factory of its own in Germany where the well-known "Tree Brand" of cutlery was made. The I. Wilson Co. of England, manufacturers of butcher knives and steels, were also represented through the Boker company.

Every well-posted hardware man understands, most likely by reason of an unpleasant experience in trying to weaken it, the strong hold the I. Wilson & Co. goods have had on the American consumer, particularly the expert butchers in the large packing houses and city butcher shops. That their prejudice is largely the result of habit and custom, rather than fair-minded judgment, cannot be doubted when a certain skinning knife made in America is given preference by those same experts who demand a knife of foreign make for every other phase of their work.

But despite this fact, American manufacturers up until the present time have found it useless and expensive to attempt to wrench the control in these commodities from their foreign competitors, although hardware men generally acknowledge the American knife to be every bit the equal of the I. Wilson product.

And so in many lines we have erected an artificial barrier against goods made in our own land. We have been protecting foreign manufacturers. We have been paying Europe thirty-five million dollars annually for manufactures of iron and steel, and yet our own country is underlaid with coal and iron. We have been sending a few thousand dollars worth of dye stuffs to Europe yearly and paying a few millions for their return in the finished product. We have been paying Europe millions of dollars for the products of the cotton we send to their mills.

Why We Have No Cotton Market.

It is a curious fact that while ninety percent of the world's cotton supply is raised in America, sixty percent of it is made into cloth in European mills. The reason for this economic inconsistency is easy to explain when we recognize the prejudice in favor of European-made products. Had this false demand for foreign goods been removed a generation ago America today would have a market for the South's cotton.

Further elaboration is unnecessary. We know the condition we are now facing and we know our duty as patriotic Americans. It would seem now that the European upheaval by shutting out foreign competition will accomplish what years of thoughtful effort has been unable to do in convincing Americans of the

---

BUY GOODS

MADE IN AMERICA

KEEP AMERICAN WORKMEN BUSY

KEEP THE MONEY AT HOME

---

merit of the goods made in our own country, and changing the prevailing fallacies regarding American manufacture.

But such a change in public sentiment is not to be wrought of its own efforts. American manufacturers, distributors and merchants must cooperate in realizing the greatest good from the present opportunity. When this is done the hobbles of prejudice will be removed from American manufacture; the outflow of millions to Europe will cease; more money will be made for ourselves as a people; more work will be afforded those who want it; and our own prosperity and prestige will be increased enormously.

So why not a Made-in-America movement, since we are all interested in its benefits? Why not a nation-wide slogan that will be sounded from ocean to ocean, from the Lakes to the Gulf; a slogan that will be sung by Eastern manufacturer and by the dealer in every city and hamlet of the nation? It is a patriotic move, and it will sell American-made goods.

How the Hardware Dealer Can Cooperate.

New York business men already have a "Made-in-America" movement, a powerful organization in fact, representative of every line of trade, and maintaining a permanent headquarters. Among the first to get behind the movement and give it momentum was the American Hardware Manufacturers' Association. The association has already designed a window card for distribution among its members, urging the American people to buy goods made in America. A reproduction of this design appears in connection with this article. One manufacturer already has ordered 25,000 of these cards to distribute among his customers—jobbers and retailers. These are furnished members at 38 cents a thousand, and may be obtained from the secretary.

Here is an opportunity for the retail hardware dealer, no matter where he may be located within the boundaries of the United States.

Preach a "Made-in-America" sermon. Appeal to the patriotism of your customers. The American people are patriotic politically, and they can be made patriotic commercially. They have been buying foreign goods through a misapprehension, which you can dispel. They have not known that American genius has devised machinery by which American goods is manufactured more uniform in quality than can possibly be done by the hand labor methods so prevalent in Europe.

You may be handling but few lines made in foreign countries. Foreign goods may be but a small portion of your stock. But talk America anyhow. Every time you make a display in your window, have a neat window card painted appraising the public that every article in that window is made in an American factory. Put similar cards on your counters, or in your show cases. Enlist the aid of your local editor, getting him to call attention to the superiority of goods "Made in America." Speak a patriotic message through your display space.

It is unnecessary to arouse any prejudice against foreign goods. The "Made-in-America" movement will get the public to consider the merits of home-made articles. There are and always will be some goods in which foreign factories will excel, certain products which are unique, which are manufactured in Europe because of the conditions peculiarly adapted to their manufacture. There will be goods in this line which we must have.

But the hardware dealer must help in the movement which will cease to cause the public to prefer an imported label to a superior product.

Here is an excellent idea borrowed from a popular national weekly, which could be embodied in a hardware dealer's window card:

Give American Goods a Trial. American Manufacturers Want No Favors. All They Ask Is Fairness. The Next Time You Buy Anything—No Matter What It Is—See That It Is Made In America.

Ford Motor Co. May Make Tires.

The Ford Motor Co., Detroit, Mich., recently purchased a tract of forty acres of land south of Akron, O. This is taken in many quarters to mean that the company will engage in the manufacture of tires for its own cars, instead of buying by contract as has been done in the past. It is a rumor of long standing in rubber circles that Mr. Ford expected at some time to make his own tires and the recent land purchase is believed to confirm this rumor.

SAD PROSPERITY

War is rumbling again in the Balkans. Every great power of the Old Country may furnish a battlefield. Lives, homes, cities, crops, all that men build and fondly foster, will be sacrificed to the Red God. Devastation will drink deep. Masters of manslaughter will move their pawns and thousands will topple bleeding. And to the end that one set of human beings may crush another set of human beings!

They tell us that these things will make business better on this side of the Atlantic. The price is too high. We would not pay the peace of Europe for the prosperity of America. It will be sad prosperity, indeed, if it comes.

Many of the products previously available only to city dwellers were beginning to be sold in smaller towns through farm equipment and hardware dealers. The farm equipment dealer was the center of much rural buying that included much more than farm implements. Farm implement dealers showed their interest in selling automobiles and automobile manufacturers were quick to respond by explaining the advantages of selling their brand of cars. Missouri had about 4,000 registered cars in 1914, but the number increased to 68,815 in 1915. It was reported that 32 cities had automobile "Jitney" service in 1915. The word "jitney" had been in common usage for a generation or so around circuses and indicated a 5 cent piece, which was the cost for riding the "busses." In Los Angeles, California, 1050 jitneys were operating.

Mohawk Wick Oil Cook Stoves

As Delightful as a Gas Range   No Trouble to Operate

Concentrates Heat Under Cooking Vessels and Does Not Heat Room.

Consumes Four Hundred Gallons of Air to Only One Gallon of Kerosene.

Furnished with or without Shelf.

**Made by
Skilled
Workmen,
and the
Result
of
Careful
Experiments.**

**Every
Stove Is
Thoroughly
Tested
Before
Shipping
and Is
Bound to
Give
Satisfaction.**

The Mohawk Four-Burner Oil Stove.

The Burner is the Most Important Part of any Oil Stove. The Mohawk Burner is Far Superior to Any on the Market. It Automatically Generates Gas from Kerosene, Mixing it with Air. This Insures Perfect Combustion. It Burns Like Gas, Giving an Intense Hot Flame. It is Odorless and Smokeless. The Construction is Durable and Efficient.

**Every Burner Is Equipped with Chimney Raiser.
Send Us Your Orders and You Will Receive Prompt Shipment.**

Can You Guess These Slogans?

Many implement manufacturers and jobbers employ the use of slogans which find conspicuous place on all their advertising literature. Some word or phrase characterizes the line almost as forcibly as the copyrighted trade-mark and is to be seen in every trade paper advertisement, circular, company letter-head and even on the article itself. The constant use of the slogan through many years of national advertising has made the lines of many manufacturers well known and the slogans are household words. Whenever the line is mentioned the slogan is thought of and, conversely, the appearance of the slogan suggests also the line. Many hardware manufacturers religiously adhere to the trade phrase and in course of time have made their products standard the world over. "Hammer the Hammer" is the well known phrase to be seen on all advertising of the Iver-Johnson Arms Co. "The recollection of quality remains long after the price is forgotton" is at once associated with the Simmons Hardware Co. Practically all automobile manufacturers resort to slogans, which are almost as well known as the cars themselves. "No hill too steep, no sand too deep," and "Ask the man who owns one," classifies the Jackson and Packard automobiles at once. To list the slogans of all the automobile manufacturers would fill a page.

To Whom do the Following Slogans Refer?

Look over the list of a few of the slogans given below and see how many of them you can connect with the implement manufacturers or jobbers. You have read advertisements of many of the lines represented and doubtless most of the slogans are familiar to you. Try your luck on these:

1. All ——— goods are good goods.
2. Take off your hat to ———.
3. Have you an ——— contract?
4. The one best line for the one best dealer.
5. The name tells a true story.
6. Suckers that work.
7. The boy that gets the business is the
8. Don't be afraid, it's an ———.
9. The baler for business.
10. If it's "Star" brand it's the best.
11. Hardware shelving with brains.
12. The engine of national supremacy.
13. The house that honest dealing built.
14. The foot-lift line.
15. World's oldest and largest makers of potato machinery.
16. The engine that breathes.
17. The farm workhorse.
18. Has made its way by the way it's made.
19. Standard the world over.
20. First huller builders in the world.
21. Power farming machinery.
22. 33 years in the pump business.
23. The knot with a grip that will not slip.
24. ——— sprayers, big payers.
25. The latest word on farm equipment.

Tennis and Golf Balls Cheaper.

The tariff reduction on rubber goods has resulted in a decided drop in the price of tennis and golf balls as compared with last year. Purchasers of larger quantities of these rubber balls have found the retail price reduced $1 a dozen on all grades of balls reported. Tennis balls which cost $5 a dozen last year cost $4 a dozen now. The golf balls of the regular $9 variety are selling for $7.50 and $8. Tennis players paid $1.20 for three balls of the best quality last year. The same balls this year cost $1 for a set of three.

Your Government May Become Your Competitor

Department of Agriculture Makes a Cooperative Buying Experiment in Pennsylvania That Ignores the Mercantile Middleman

If a report from Washington is to be given credence, the Department. of Agriculture through its new Rural Organization Service proposes to supplant the implement dealer as purchasing agent for the farmer. An experiment in cooperative buying of farm equipment has been made at Schellburg, Pa., a hamlet about five miles off the railroad, under the direction of an expert attached to the Rural Organization Service, A. B. Ross by name. L. H. Goddard, an official of that bureau, appears to have supervisory authority over the movement and is said to have stated that farm implements, seeds, etc., offer a peculiarly attractive field for this latest piece of paternalism on the part of the Department.

Heretofore the Rural Organization Service has been principally concerned in helping the farmer market his products. Now it seems to have directed its attention to the other side of his ledger and offers to do his buying for him.

Planning to Widen the Scheme.

There appears to be much of menace against the legitimate implement trade in the movement, although it has not been extended far enough to offer real cause for alarm. At the same time, the bureau is said to be planning to introduce the scheme throughout the country.

Inasmuch as the Department has hundreds of farm advisers stationed in many states upon whom it would probably fall to extend the enterprise, it assumes proportions that will no doubt give rise to thorough investigation on the part of the organized dealers and manufacturers in the farm equipment trade.

A ridiculously impracticable feature of the scheme contemplates inducing the manufacturers to send out exhibits of their goods to be shown at every railroad station long enough before the shipping season to give the farmers a chance to make their choice. Such displays, it is contemplated, would be made once or twice a year in buildings owned or rented by the farmers' exchanges.

The slaughter across the water is enough to make a self-respecting hardware merchant go out of the gun business in disgust.

Here Are the Keys to Those Slogans.

1. All P. & O. goods are good goods.
2. Take off your hat to Myers.
3. Have you an Oliver contract?
4. The John Deere Plow Co. of Omaha.
5. Superior grain drills.
6. Hayes Pumps.
7. Waterloo Boy (gasoline engines).
8. Don't be afraid, it's an Anchor (buggy).
9. Ann Arbor hay press.
10. Hooven & Allison (binder twine).
11. W. C. Heller Co. (hardware shelving).
12. Field engines.
13. Kretchmer Mfg. Co., Council Bluffs, Ia.
14. Emerson-Brantingham Co.
15. Aspinwall Mfg. Co., Jackson, Mich.
16. Gade engine.
17. The Heider tractor.
18. International Harvester Co. (auto trucks).
19. Peter Schuttler wagons.
20. Birdsell Mfg. Co., South Bend, Ind.
21. Rumely Products Co.
22. The Deming Co., Salem, O.
23. Keystone Steel & Wire Co. (fencing).
24. Bean sprayers, big payers.
25. Implement Trade Journal Co., Kansas City and Omaha.

Bureau Backs Farmers' Exchange.

It is the farmers' cooperative association that the bureau proposes to make the buying unit for the territory around each shipping station. Such associations, of course, would be controlled by the farmers, but the buying part of it, according to the plan, would be under the auspices of the Rural Organization Service.

It is said ten thousand letters are being sent out to farmers, bankers and manufacturers inviting suggestions for the plan. Of course, the mail-order manufacturers and distributors have been included in that canvass No doubt they will commend it.

There may have been a very good reason for selecting Schellburg as the place for carrying on this experiment. Judging by the hazy reports, the department has taken good care to see that as little publicity as possible has been given the scheme. Schellburg in all its isolation would naturally lend itself to this policy. Officials of the bureau seem to be highly pleased with their experiment in Pennsylvania and anxious to extend it into other fields.

Only Cash Transactions Made.

Cash represents the central idea of the scheme. This or its equivalent must be deposited in bank before any goods can be moved. The negotiable bill of lading also enters prominently into the plan. This must be properly endorsed and surrendered to the railroad before the goods will be delivered.

Farmers wishing to buy in carload lots must have their orders and money in a specified bank at a specified time. The bank certifies that the money is awaiting the draft of the shipper. Then the orders and certificates are sent to the manager of the farmers' exchange, who in turn forwards the facts to the shipper, asking him to ship on his own order, attaching a draft to the endorsed bill of lading and together with the invoice.

No money is handled by the exchange. The cash transaction takes place between the farmer and the bank and the bank and the shipper. No books are kept by the exchange other than the necessary simple records.

Comparative Cost of the Tractor and the Horse

In selling tractors to a farmer, the dealer must show the tractor's commercial advantages. The farmer must be shown that a tractor will furnish power at a lower cost than horses. The following tables show the daily operating cost of a 15-30-h. p. oil tractor as compared with the cost of a day's work with a horse.

Cost of 15-30-h. p. Tractor.

Depreciation, figured on 1,000 days' work	$ 2.00
Interest at 6%, charging off $200 each year and figuring 100 days' work	.66
Fuel, 35 gallons, at 9c	3.15
Lubricating oil and grease	1.00
Repairs	1.00
Wages and board of tractioneer	3.50
Hauling water and fuel	.49
Total daily cost	**$11.80**

The Cost of Horsepower.

Total cost of feed	$ 80.21
Interest, figuring 6% on $180.00 and allowing $14.40 yearly depreciation	5.84
Yearly depreciation	14.40
Care and shelter	17.50
Interest and depreciation on harness	2.00
Shoeing	1.00
Veterinary and medicines	1.00
Total yearly expense	**$121.95**
Daily expense (100 days per year)	$1.22
Wages of driver (one man to three horses)	.66
Daily cost of operation	**$1.88**

Comparative Costs.

The table below shows the cost of tractor power as compared to horse power and the percentage of the cost of horsepower saved on belt work and tractive work in general and on tractive power as applied to three common farm jobs.

	Tractor	Horse	Percentage of horse costs saved
Cost of belt work per horsepower hour	.039	.188	79%
Cost of tractive horsepower hour	.079	.188	58%
Cost per bu. in hauling wheat 10 miles on average roads	.023	.075	69%
Cost of plowing per acre under ordinary conditions	.984	2.336	56%
Cost of harvesting per acre	.273	.353	23%

How These Figures Were Obtained.

The price of 15-30 tractor is figured at $2,000. This price is only approximate, but it is sufficiently high to thoroughly serve our purpose.

Few tractors have been used long enough to wear out, but those that have been in the field for four or five years and that are still in good condition bear out the conclusions at which most power-farming experts have arrived— that there are at least 1,000 ten-hour days' work under the iron hide of every gas tractor.

The question of fuel and the question of lubricating oil are local ones. The quantities of these required will vary just about as much as the prices. The quantity will depend upon the skill with which the tractor is operated and the conditions under which it works. This figure, 35 gallons of fuel for a 15-30 tractor, is right in line with the estimates made by most of those who have figured the fuel question. Fuel oil is placed at 9c per gallon. This is undoubtedly high where distillate or some of the other baser fuels are used, but it is an easy matter for anyone to get prices on fuel and lubricating oils and figure these items for themselves. An allowance of $3.50 a day is made to cover wages and board of the tractioneer, and 49 cents a day to cover the cost of a team to supply the tractor with fuel and water. No allowance is made for shelter as in very few instances is it necessary to provide a special building for the tractor.

Figuring Horse Costs.

The first estimate in computing the cost of horsepower is the cost during the hundred days the horse works. This is generally agreed to be on average year's work. A bushel of oats will make five average feeds and a bushel of corn will make 12 average feeds. Using 60c as the price per bushel of corn and 40c as the price per bushel of oats on the farm, we find that it will cost 24c per day to feed a horse on oats or 15c per day to feed him on corn. A mixed ration is always better, so we will average the cost of corn and oats, which gives us 19½c per day as the cost of grain for a working horse. A horse will eat about 1½ tons of good timothy hay during 100 working days. Pricing this at $12.00 per ton, which is a low price, it will cost $15.00 for hay to supply a horse for 100 days. A horse will therefore consume $34.50 worth of hay and grain during 100 days per year he works. A horse will not require so much grain nor quite so much hay when not working, even though he be kept in the barn all the time. It is also likely that horses can be turned out to pasture part of the time they are not busy, which reduces the cost of keeping them, so it can be figured that it costs only one-half as much to keep a horse when he is idle as it does to maintain him when he is working. Figuring it this way, it will take $45.71 to supply a horse with food during the 265 days per year he is idle. The total yearly cost of feed is found by adding $34.50 for the working days and $45.71 for the 265 idle days, which gives a total of $80.21.

No estimate is made for bedding, but horses return something in the way of fertilizer which will probably offset this item, so it is reasonable to leave it out of our calculations.

The figures for care and shelter, veterinary, shoeing and depreciations are taken from many sources—mostly from the Minnesota Bulletin on horse costs. The price of $180.00 is based on market reports covering the first part of the year. Prices at the present time average a little higher than $180.00.

For yearly depreciation, $14.40 is extremely low. The use of this figure means that every horse can do 12½ years' work, or in other words, that he must live and be able to do a full day's work until he is 15½ years old. Investigation will show that 15-year-old horses are as scarce as the proverbial hen's teeth, in any locality. The allowance for feed is higher than the figures shown by some government feed tests.

The Comparative Results.

In estimating hauling costs, it is figured that a tractor will haul 500 bushels of grain at one trip and will make one ten-mile trip per day. A three-horse team will haul 75 bushels. Twelve acres has been allowed as a day's work for the tractor at plowing, and 2½ acres as a day's work for a three-horse team. In harvesting, it is 16 acres per day for three horses hitched to a seven-foot binder and 80 acres as a day's work for the tractor, pulling five seven-foot binders, are considered conservative figures. In estimating tractor costs in harvesting, $10.00 is allowed for the wages and board of five men, one on each of the binders.

Ten hours' work has been considered a day for both the tractor and horses, but it is well to bear in mind that a tractor does not tire and will work 12 or 14 hours or even 24 hours per day when this is desirable.

His First Lesson

The farmer of today proudly teaches his son what his own father taught him— to use a John Deere Plow.

Only the 15-30 tractor has been considered. A 30-60 tractor will show an even greater saving when compared to the cost of horse work and is undoubtedly a better size where the farm is large enough or a sufficient amount of custom work can be found to keep it employed.

Taking it all in all, the horse gets a little the best of it in the figures we have tabulated. It is preferable to have it this way because these figures are meant to show the advantages of a tractor over horse power. It is an easy matter for anyone to figure costs. If farmers will do so and will count all of the items, they will, in most cases, arrive at a higher total cost of keeping a horse than the one used here.

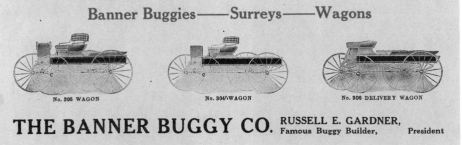
Traditional farm implements were the backbone for expansion into some new-fangled product lines. The implements sold during this period could often be drawn either by animals or tractors and belt driven equipment could be powered by a stationary engine, a steam traction engine or one of the "new" gasoline tractors. It was during this time that it was noticed that implements such as plows that were to be powered or drawn by animals should be lighter to save the animal and powered equipment had to be built heavier to reduce breakage.

The farm implement dealer was recognized as the person who was familiar with the most modern equipment available, so it was logical that his patrons would ask his opinion about other new products. An article published early in 1916 said that it was believed that tractor dealerships would someday be important. A later quote said that "The season of 1915-1916 will go down in farm equipment history as one of the most momentous of all. One of the chapter headings covering the developments taking place might well be 'Curtailing Terms; or the Emancipation of the Implement Man.' And the other important heading might as well be 'Marketing the Tractor; or How the Dealer Adopted a New Machine.' The emerging farm tractor manufacturers used durability runs, contests and conventions to gain attention. Practicality was stressed as well as value. The horse drawn buggy and wagon companies were busy during the war years supplying the armies of the various warring nations.

Minnesota lumberman, Frederick Weyerhauser, died at his winter home near Pasadena, California, on April 4, 1914. His company's lumber land was said to be equal to the state of New York. Col. Max Mosler, founder of the Mosler Safe & Lock Co. died later in April and Richard W. Sears, co-founder of Sears, Roebuck & Co., died on September 28. Robert H. Foos died on October 11. Foos, with his father and brother, acquired controlling interest in the Evans Mfg. Co. of Springfield, Illinois, which was later known as the Foos Mfg. Co. The Bauer Bros. Co. later bought the Foos Mfg. Co.

An Electric Incubator.

Charles Cuykendall, manager of the Atlantic Canning Co., Fremont, has invented an electric incubator which poultry men believe will revolutionize the incubator business. The machine is operated automatically with the use of a thermostat arrangement and keeps the temperature uniform. Electric globes are used to heat the interior and when the temperature reaches a certain point the current is turned off automatically. The cooling of the atmosphere in the interior causes the current to be turned on and the temperature on the inside rises.

HOW A WOMAN MADE GOOD AS A DEALER

Plucky Oklahoma Widow Gives a Modest Account of Her Successful Fight Against Odds in the Implement and Hardware Business

By MRS. H. O. DUNCAN, Oktaha, Okla.

MRS. H. O. DUNCAN.

I DO not hold that woman was designed especially for business pursuits, nor that business should be remodeled for her convenience, but I do think that girls should be given a training for life's work equal to that given to boys—and then they are fitted for whatever comes to their lot, whether it is housekeeping, storekeeping or something else.

I had the advantage of a business college training and when Mr. Duncan and I were married we became real partners. He was foreign traveling representative for a Chicago corporation, and I accompanied him to various European cities, after which we re-crossed the Atlantic to Pernambuco, Brazil, then visited several other Brazilian and one Uruguayan city, finally making headquarters at Buenos Ayres, Argentina, whence we went several hundred miles up the Parana River to the interior of the country. I could do little to assist him at this time, however, except to reduce the various moneys he handled into their relative values in English gold—in which money there was always a basis of exchange—and then into American gold, in order that the accounts could be properly presented to the office at home.

When we came to Oklahoma in 1903 and started in the hardware business here in Oktaha, I was given the care of the books and most of the correspondence, but the strenuous life of years of travel over land and sea and the sudden changes from one extreme of climate to another, had already told on Mr. Duncan's health. It was not long until I found myself getting more into the work; and, as much of my married life had been not unlike that of a petted, indulged child, I was glad to feel myself capable of relieving him of responsibility and work.

But One Thing to Do and She Did It.

I did not know by sight a six-penny nail from one of forty-penny size, nor a hame strap from a check line, but I did not hesitate to ask questions of any one to whom I did not object to showing my ignorance, and I shall never forget the kindness and patience of the traveling men who taught and helped me.

Business was not all a dream, however. On an occasion when Mr. Duncan had gone to a health resort for six weeks and had been away about a month, I went out to try to buy a farm, met with an accident and broke my arm. I assure you it was quite awkward and trying—but there was only one thing to do and that was to "stay on the job," and I did it. But Mr. Duncan was not in a condition to be advised of what had happened to me, so there was the plan to be worked out to keep him away a month longer than at first intended. Yes, I got the farm. The owners of it heard of my accident and came to me.

Farm Girl Explains Implements.

The first season I undertook to handle implements it seemed to me that I could never fix in my mind the difference in the uses of breaking plows, shovel plows, middle-breaker plows, etc., and my husband found me so inexcusably stupid that he thought it useless to talk much about them. Finally a girl from Arkansas who had worked in the field explained to me that the ground was first worked with a turning plow, and how the other plows were used in their turn. She did this with great care and enthusiasm as though it were a privilege, and I shall always remember her gratefully. Her language was so free from grammar or form that it was funny to listen

to—but she could teach me! And why was this not a lesson in humility?

In the winter of 1907, owing to Mr. Duncan's declining health, I was compelled to take entire charge of everything, and we moved into one half of our new building, renting the other half as a general store, as we had until this time occupied one rather small room. Not long afterward I traded for the building we had been occupying, to use as a wareroom.

In 1909 our tenant moved and I put in a stock of furniture. Afterward I added another warehouse, 25 by 85 feet, and bought a stock of buggies and wagons, making our stock, with what we already had, quite complete.

I, however, do not take to myself any great credit for having been successful. I did what any other woman would have done had she been able, and have not become wealthy as I understand many Oklahoma hardware men have. I have followed some certain lines of action that have kept me going forward instead of backward, among which may be noted:

What Keeps Her Going Forward.

I buy only such merchandise as I am reasonably sure of selling, and do not over-buy.

I discount my bills strictly, which I think creates confidence in me.

I never sell anything without a profit unless it is damaged or a close-out.

I sell for cash or insist on a settlement by note *at the time* of making the sale.

I consider my customers as guests of honor when in the store and show them every courtesy and attention I know how to command.

I try to keep stock complete and advertise in the newspaper and by personal letter copied on a duplicator.

I have tried to learn cost accounting, so as to avoid the mistakes that are so often made by merchants who do not know the cost of operating a business.

I believe in vacations as an investment and have not missed a year in going back to the scenes and friends of my childhood—and the cool lake breezes.

And last, but not least, I am always as good as my word, and never allow a customer to be dissatisfied with any purchase from the house.

I see no reason why the implement and hardware business is not entirely suited to women nor any reason why they should not succeed as well as men who use the same

effort and ability. It is a very interesting work and quite scientific. I always visit other hardware stores when I have a chance and make a point, also, of visiting factories that are making the goods we sell. This is very helpful in learning to sell the goods.

The Annual Christmas Tree.

One feature of my work which has proven very satisfactory is the annual Christmas tree which is given to the public. They expect it, and it is advertised in advance, together with the fact that every one is invited to come and make use of it in distributing presents to their friends, and that a present will be on the tree for every child that is under a certain age. This always includes a large number of children who will have no other Christmas present (though no mention is made of such a thing), and the event is very popular generally.

As I said before, I do not lay claim to having made a great success, because I have not become wealthy, but what success has come to me has been due to work and close application to business. One can hardly fail in an undertaking if one keeps everlastingly at it and uses average judgment.

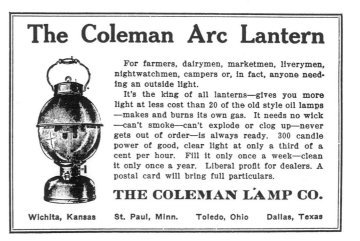

Among Motorcycle Riders and Their Mounts

Every Enthusiast Swears by His Own Machine and None Can Break His Loyalty.

Once in a while when things seem to be going wrong the man who manufactures or handles a certain brand of machine sits down and wonders gloomily whether the name of that brand has any appreciable weight, whether the customer cares anything for it one way or the other. Of course, he only indulges in such doubt when things seem to be going wrong. In his less bilious frame of mind he knows that the make of a machine counts heavily.

Apropos of this, an interesting insight into the relations of motorcyclists with each other and with the motorcycle dealer is to be had by reading the following extract from Freeman Jargo's account of the organization of a local motorcycle club in his town, which appeared in a recent number of Motorcycle Illustrated:

I believe we had fourteen riders at our organization meeting, and six of these mounted the same make of machine. It was the latter condition that brought on our first hitch. The six riders referred to felt that they were bound together in a sort of vague brotherhood by the fact that their machines bore the same name, and they began to act as a unit. The vote of one was the vote of all, and in little or no time the riders of other machines were thrown upon the defensive. Two riders of the same make on this side voted together, three owners of another make over yonder held out stubbornly as one man, two others mapped out a little policy of their own and the last member simply rampaged to and fro, voting against every proposal, and accusing the riders of all the other machines of snubbing him because he was the sole representative in the club of his make of machine.

Each Make Had Its Own Squad.

It happened that our club was organized late in the spring, and after it had been in existence about a month a cool-headed officer managed to launch plans for a Sunday run to a small amusement place. What's more, he succeeded in having all the boys in line when the trip was started.

He paired them off carefully, being certain to have each rider teamed with the owner of another make of machine, but before we had been on the trail half an hour each make of machine had its own little squad, and the rider of the lone mount was burning up the road ahead in an effort to prove that he did not need the companionship of the riders of any other make.

There was conversation of a sort as we rode along, of course, but nevertheless the clan lines were always evident. There seemed to be a woeful lack of enthusiasm.

THE UNDESIRABLE MEMBER OF THE MOTORCYCLE CLUB.

Where Diplomacy Triumphed.

In this fashion we reached the turning point of our run, and a good dinner helped to relieve the tension. After the cigars were going, the cool-headed officer very humbly offered to exchange mounts for a short spin with a rider of a rival make, hinting that his own machine was not up to its best work and that in new hands its failings might be discovered. They made the exchange, rode off smiling and, strange as it may seem, they returned smiling.

They had established a bond of sympathy, and from that moment they had plenty to chat about. The cool-headed officer then started a tactful juggling of machines, until every rider present had enjoyed a short jog on another member's mount. He called it a new system of "trouble-hunting," but he really intended it for trouble-chasing, and that's just the way it worked.

Mechanical suggestions were soon passed about freely, makes of machines were forgotten for the moment and conversation turned to the motorcycle in the abstract. At this juncture I detected a covert smile about the lips of the cool-headed officer, and I knew that his first battle had been won.

The Influence of Agents.

One of the knottiest problems we had to solve, and it undoubtedly presents itself to every club that is organized, concerned our dealer members. We had three dealers on our list, each handling several makes of machines, and as might be expected there was considerable rivalry between them. Each was on the lookout for any slight advantage that might be gained by one of the machines he represented, and the other two were keen to see that no such advantage was allowed. Indirectly, the club was a business proposition with the dealers and they could not be blamed for seeing the situation reduced to dollars and cents.

As is usually the case, we held our first meeting at the establishment of one of these dealers, and we felt called upon to mention his place of business when we gave a story of the meeting to the reporters on the following day. This went against the grain with the two other agents, and not until we had held meetings in their stores and mentioned their names in the local papers were they fully appeased. They looked upon the newspaper story as a form of advertising, and naturally each wanted his share of the publicity.

Do not infer from this that all dealers are unreasonably mercenary; they are not, and they usually make the most practical and energetic club workers, but in view of the fact that the whole enterprise affects their livelihood it is natural that they should want all the consideration obtainable. What's more, they are entitled to it, but always with the understanding that they should not seek to play club factions—if the organization is unfortunate enough to have any—against one another with a view of working up trade.

MANY FEMININE FANS RIDE THE MOTORCYCLE.

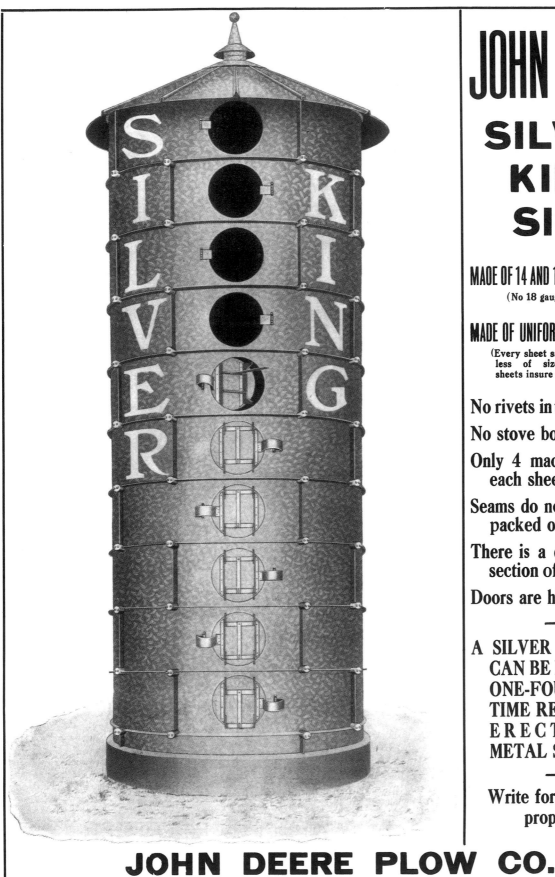

JOHN DEERE
SILVER KING SILO

MADE OF 14 AND 16 GAUGE PURE IRON
(No 18 gauge metal used)

MADE OF UNIFORM SHEETS, 6¼x2 FT.
(Every sheet same size regardless of size of silo—small sheets insure rigidity)

No rivets in the Silver King

No stove bolts

Only 4 machine bolts to each sheet

Seams do not have to be packed or cemented

There is a door in every section of a Silver King

Doors are hinged

—

A SILVER KING SILO CAN BE ERECTED IN ONE-FOURTH THE TIME REQUIRED TO ERECT OTHER METAL SILOS

—

Write for our agency proposition

JOHN DEERE PLOW CO.

DENVER   KANSAS CITY   OKLAHOMA CITY

Features Worth Having

THE FAMOUS CATERPILLAR TRACK,
STEADY FOUR-CYLINDER MOTOR,
EASIEST POSSIBLE STEERING,
TURNING AT RIGHT ANGLE,
NO SLIPPAGE,
ECONOMICAL TRANSMISSION OF POWER,
TWO SPEEDS,
SPRING MOUNTING,
STRADDLING OF ROWS IN CULTIVATING,
LESS GROUND PRESSURE THAN HORSE OR MAN

The Baby Caterpillar

30-Brake—20-Drawbar Horsepower

The no-front-wheel Baby Caterpillar has the light ground-pressure, the easy-steering and short-turning habit, and the long-wearing build that farmers on small acreages have been waiting for. It is the one tractor that brings mechanical power down at last to everyday work on the small farm. They can plow, harrow, seed, cultivate, harvest, haul with it—and find it handy for any job, and powerful enough for profitable work.

Its long, wide tracks hold it up—it cannot pack the most finely cultivated field. With its widest track it has only 3.4 pounds of weight per square inch on the ground. A 1,200-lb. horse has 21.4 lbs. to the inch while pulling. A man walking presses harder than the Caterpillar. Farmers can use this tractor when all other power is idle.

Send for specifications—they show the same splendid construction that has made the bigger Caterpillar models a success for the last ten years. The Baby Caterpillar has a record of success in the field—all the advantages of mechanical power without the drawbacks of the ordinary small traction engine.

It will pay you to handle this real general-purpose tractor, and we are glad to establish new agencies in open territory. Our terms are liberal to reliable implement dealers. No cooperative associations need apply for agents' terms.

This year can be just as good a tractor year for you as 1912 was, if you become a Holt agent. Caterpillar owners have not been disappointed in what they got out of their tractors. They are ready to buy more, and to recommend the Caterpillar to their neighbors. Nearly two thousand of them are using Caterpillars now—one size or another—and getting in more days of work each week than round-wheel engine owners. The Caterpillar output has grown steadily for ten years without one set-back. We want you to join with us in pushing 1914 business.

Let us get acquainted. Our folder EA38 describes our machines and gives our selling policy.

The Holt Manufacturing Company built the first successful track-laying tractor. The endless track, with its sure grip and big bearing surface, first made the traction engine fit for **universal** use as a farm power to replace horses.

The creeping motion of the track suggested the name "Caterpillar." We protected it as our trade-mark. It is ours, and ours alone. There is no other Caterpillar but the Holt.

Caterpillar outfits for plowing, cultivating, hauling, threshing, and a hundred other uses, are found at the furthermost points of civilization, from Nome, Alaska, to Cape Horn, South America, and from St. Petersburg on the east to Manila on the west.

Every continent knows the Caterpillar. Its owners are the highest type of farmers everywhere. It has its imitations, all known as "the Caterpillar type," but it stands alone, by name, by quality, and by results. **There is just one Caterpillar, and Holt builds it.**

THE HOLT MANUFACTURING COMPANY,
(Incorporated)
PEORIA, ILLINOIS.

New York Office: 50 Church St.

SALES AGENTS.
Spalding Deep Tilling Machine Co., Denver, Colo.
Lininger Implement Co., Omaha, Neb.

The Baby Caterpillar Turns Easily in Snow, Sand or Mud.

On June 28, 1914, the Archduke Francis Ferdinand was murdered in Sarajevo, marking the beginning of acts leading the world into the Great War or World War I. The British used the Holt "Caterpillar" and called it the "Tank." By 1915 more than 2000 Holt "Caterpillar" tractors were being used. Some of the "Tanks" were highly modified, but many were only slightly different than those sold to road builders following the War. The United States placed orders for armor plate for the battleships California, Mississippi and Idaho. Near the end of 1914, Mme. Thebes, a well known seeress, predicted that peace would come to Europe in 1915. Everyone hoped for an early end to the war.

The Smith-Lever Act was enacted on May 8, 1914, creating Cooperative Extension Service and the system of Federal reserve banks began in November. On December 1, congressional bill H.R. 18891, better known as the "War Tax," went into effect. The Implement Trade Journal Company purchased Millard's Implement Directory.

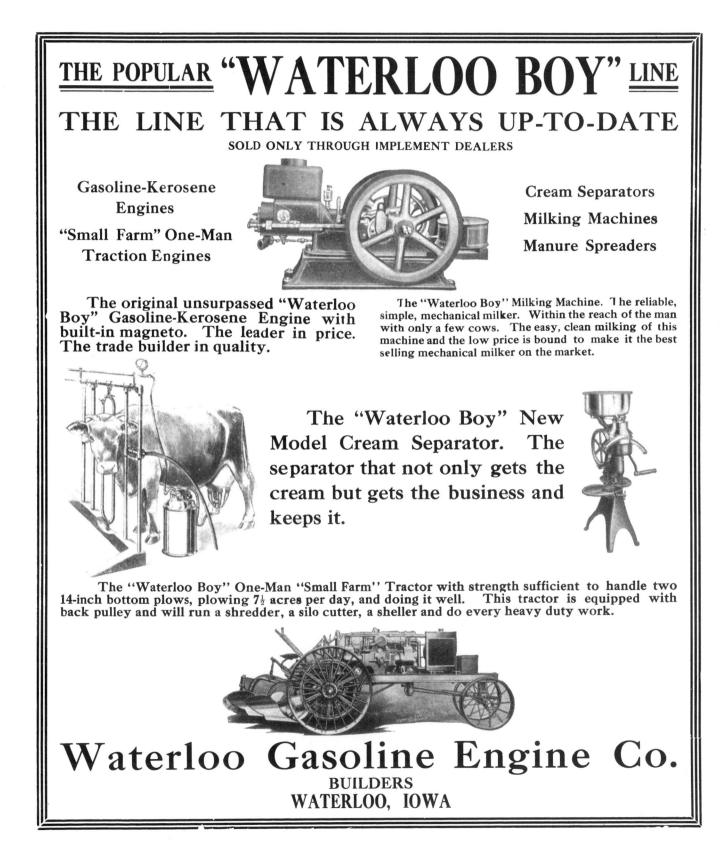

THE POPULAR "WATERLOO BOY" LINE

THE LINE THAT IS ALWAYS UP-TO-DATE

SOLD ONLY THROUGH IMPLEMENT DEALERS

Gasoline-Kerosene Engines

"Small Farm" One-Man Traction Engines

Cream Separators

Milking Machines

Manure Spreaders

The original unsurpassed "Waterloo Boy" Gasoline-Kerosene Engine with built-in magneto. The leader in price. The trade builder in quality.

The "Waterloo Boy" Milking Machine. The reliable, simple, mechanical milker. Within the reach of the man with only a few cows. The easy, clean milking of this machine and the low price is bound to make it the best selling mechanical milker on the market.

The "Waterloo Boy" New Model Cream Separator. The separator that not only gets the cream but gets the business and keeps it.

The "Waterloo Boy" One-Man "Small Farm" Tractor with strength sufficient to handle two 14-inch bottom plows, plowing 7½ acres per day, and doing it well. This tractor is equipped with back pulley and will run a shredder, a silo cutter, a sheller and do every heavy duty work.

Waterloo Gasoline Engine Co.
BUILDERS
WATERLOO, IOWA

During 1914, Advance-Rumely sold many Oil Pull tractors in Canada on credit and crop failures caused the owners to default on their payments. Bankers demanded their money from Rumely, which forced the company into receivership on January 21, 1915. The company which was reformed as Advance-Rumely Thresher Co. by September of 1915 and the Rumely family was forced out. Equipment manufacturers were directly dependent on the success of the farmer to pay the manufacturer. The manufacturers were also at the mercy of the steel, twine and railroad industries, especially during the early days of the war. More money could be made by these industries by providing services and war goods to the combating countries.

The Allis Chalmers Mfg. Co. of Milwaukee, Wisconsin, began building a tractor at its West Allis, Wisconsin, plant in 1914. The Allis Chalmers 10-18 tractor continued in production until 1918. The Bull Tractor Co. of Minneapolis, Minnesota, began in January, 1914, with the biggest sales promotion yet. The Bull tractor was manufactured by the Minnesota Steel & Machinery Co. until 1917 when they canceled the contract. The Bull Tractor Co. finally stopped trying to sell the tractors in the 1920's. The Minneapolis Steel & Machinery Co. also contracted to build the "T-C" (Twentieth-Century) tractor for the Grain Growers Co-op in Canada. The "Twin City" was their own tractor line. The Mayer Bros. Co. introduced the "Little Giant" tractor, then later changed the company name to Little Giant Co. The Dodge Brothers were incorporated in Detroit, Michigan, for $5 million to build automobiles. About 1914, the Sampson Iron Works began production of the Sampson tractor in Stockton, California. Oil companies were scouring the world for new sources of petroleum. Of particular interest was the Island of Timor in the Pacific that seemed to have an inexhaustible supply of oil and it was noted that natives were paid 10 cents a day.

Henry W. Putnam, an inventor of barbed wire and horseshoe nails, died on January 30, 1915. Putnam was one of the wealthiest men in San Diego, California, where he had retired.

A Tool for Reboring Is Invented.

A tool for reboring cylinders has been invented by E. F. Heiser, Kingston, Mo., which if successful will fill a long felt want in machine and repair shops. Reboring now requires expensive machinery and is attended with quite long delays.

When anybody tells you to go to war, it's time to feel insulted.

OMAHA'S NEW TRIPLE-RADIUS AUTO RACING TRACK

TOMORROW'S CAR TODAY

The Logic of the Case Car from the Dealer's Standpoint

It is an acknowledged fact among dealers that the public is more determined than ever to buy automobiles from the basis of inbuilt merit.

Likewise, there is a feeling among many makers that they would like to raise the prices of their cars to meet this new attitude of the buying public. Some have already raised their prices. Others would like to.

With these two principles established in the motor-world—new recognition comes to Case.

First: Because of our consistent adherence to a policy of producing a car of conscientious merit. Through all these years we have built well—making price secondary. From year to year, as we have progressed, we have steadily decreased the price of Case Cars—never at any startling declines, but just as much as could be afforded without risking their merit.

Second: Because we are prepared again to lower the price of our 40—maintaining that same standard of excellence which has made our name. The new Case 40 is a better car than we ever built before. It could have been made much lower in price if we had been willing to gamble with our reputation.

The buying public, we know, will be satisfied more in the long run by a continuance of Case policy. Hence the great value to dealers.

ANNOUNCING THE NEW CASE 40 $1090

Reputation is nowadays selling cars. People first want their confidence well founded, then they seek a car of dignified grace. If anyone wants to go deeper—you can go through the Case 40 with a microscope, revealing at every point intelligent design and good, honest construction. This is the car, we believe, that will last and make friends and money for dealers.

Case dealers, under our new agreement, can now profit to the fullest extent by the increased Case demand. In short, the Case proposition offers you the following:

1. A meritorious car.
2. A reasonable price.
3. A reliable company.
4. An International Sales Organization.
5. An attractive dealer's agreement.

Case dealers now have a great opportunity to please the public and make friends. The announcement of our new 40 and the subsequent advertising will tell people of Tomorrow's Car Today. More and more people are coming to appreciate the value of Case reputation.

We shall be glad to explain fully to you our new proposition. We invite correspondence in relation to the establishment of new Case dealers. We still have a number of splendid opportunities, but, since we have an exceptional proposition, we are looking for exceptional representatives.

J. I. Case T. M. Company, Inc. 564 Liberty Street, RACINE, WISCONSIN
Founded 1842 (361)

A Few Points for Your Guidance

Wheel Base: 120 inch.
Motor: Four cylinder, bore 3⅞ inch, stroke 6 inch, cylinders cast en bloc integral with crank case, L Head, 40-45 B. H. P.
Westinghouse ignition, starting, lighting.
Lubrication—Force feed to crank shaft and cam shaft bearings; splash to piston pins and cylinder walls.
Carburetor of special design, with feed by gravity from cowl tank, dash adjustment.
Radiator—Cellular type, with thermo-syphon circulating system.
Clutch: Cone.
Transmission: Selective, three speeds forward and one reverse; three point suspension, in unit with power plant, left hand drive, center control, Timken bearings; Spicer universal joint.
Axles: Rear—Weston-Mott; ¾-floating, with spiral bevel gears; torque and drive thrust taken by torque tube to rear end of transmission through a ball and socket joint; pinion shaft provided with two Bock roller type, bear-

ings. Front—I beam, designed and built by Case; Timken bearings; I-beam section, steering arms, steering knuckles and king pins all of special chrome nickel steel—forged, heat treated and machined in our shops.
Frame: Designed with exceptionally deep section, greatest depth at center where front hanger of cantilever spring is suspended.
Springs: Rear—Cantilever, 50 inches long, 2½ inches wide; attached to rear axle by means of universal joints, which take all side play, allowing springs to do full spring duty—an exclusive feature in construction.
Wheels: 34x4 inch, Artillery type, with Goodyear detachable, demountable rims.
Body: All steel, with removable upholstery of genuine grain leather. Front seats divided, and are adjustable forward and backward, as are the clutch and brake-pedals. Finish—Brewster green, with ivory stripe.
Equipment: One-man top, with dust hood and quickly adjustable side curtains. Stewart-Warner Speedometer. Windshield—Rain vision, ventilating. Tires—Goodyear 34 inch, non-skid on rear. Motor-driven horn. Regular tools, tire repair kit, etc., etc. **Price:** $1090.00, f. o. b. Racine.

The Sign of Mechanical Excellence — The World Over — CASE

Buick

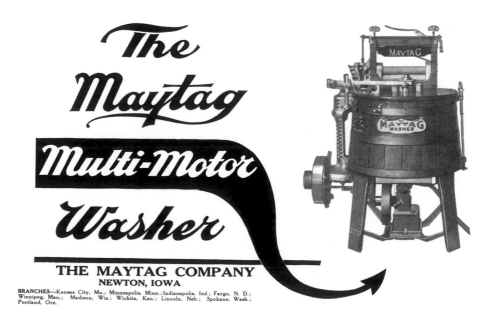
1914-1918

First Move Toward Tractor Standardization

Nearly thirty men connected with the tractor business gathered at the Sherman hotel in Chicago, March 25 to endeavor to locate matters of agreement in tractor design that would simplify the building of tractors. One point that was brought out, and which illustrates very nicely what the standards committee of the National Gas Engine Association is attempting to do, was that it often requires eight or ten sizes of wrenches to fit all of the nuts, bolts and cap screws of a single make of tractor. If the sizes of these were reduced to something like two or three standard designs, one or two wrenches would do the work. All through the manufacture of farm tractors this same diversity of design is apparent. Such looseness of construction complicates the machines, increases the cost of manufacture, adds to the work of caring for the tractor and adds quite materially to the problem of renewals and renewal costs.

Spark Plugs First Discussed.

The object of the meeting was stated by Mr. Kratsch, who said that the committee was desirous of consulting the various manufacturers preparatory to presenting to the coming convention in June recommendations for adoption along various lines. The first matter presented was that of spark plugs. Some manufacturers have used 1½-inch plugs, but the majority seem to use the ⅞-inch, 18 U. S. thread plug. Discussion developed the fact that for foreign trade metric threads were needed, so that any spark plug at hand could be used. It was proposed to adopt as N. G. E. A. standard the ⅞-inch 18 thread plug, with the recommendation that for foreign business the tractor manufacturer use a bushing, tapped for metric threaded plugs. There seemed to be quite a diversity of opinion as to the advantage of using a bushing. One manufacturer stated he believed all engines should be equipped with the bushing anyway, then when constant taking out of the plug had caused the threads to strip a new bushing could be inserted. Without the bushing, it is sometimes difficult to retap the hole and get a new thread without breaking into the water jacket space. Another manufacturer opposed this with the objection that a bushing takes a plug further away from the water jacket anyway and lets the plug become heated easier. On motion, the ⅞-inch, 18-thread, plug was recommended for adoption for all gas engine use. The committee will report this recommendation to the annual convention in June, when it will be acted upon. The matter of using a bushing was not determined upon.

The next matter to be disposed of was the adoption of standard magneto practice, as recommended by the Society of Automobile Engineers, this relating to various dimensions of magnetos which make it possible to take any standard S. A. E. magneto and place it in use on a tractor without altering the bracket. During the discussion of this question, there arose the matter of impulse starters and couplings between the magneto and engine. F. B. Williams made plain the many difficulties in attaching a magneto to an engine so that there will be no lost motion, so that there will be no undue wear upon the magneto bearings, and mentioned as the requirements for a magneto coupling strength, alignment of the shafts, flexibility, adjustability, ability of the members to slip so that magneto may be adjusted readily, and reasonable cost. A number of makes of magneto couplings on the market were shown. Mr. Menges proposed that a recommendation be made to provide over-all dimensions of the couplings, so that individual designers might solve the problem as they found best for themselves. After considerable discussion, the mat-

ter was referred back to the committee for further investigation, to offer a standard of this sort at the next convention.

An Address on Screws, Threads and Taps.

The meeting then adjourned for lunch and after convening again E. H. Ehrman of the Chicago Screw Co., addressed the meeting as the representative of the Society of Automobile Engineers. Mr. Ehrman explained his connection with various standards committees of the S. A. E. and the A. S. M. E., and told of some of the difficulties in the adoption of a standard for screw threads, taps and die tolerances, etc. A bolt may be designed for certain work in steel and a different design will be better for cast iron or aluminum. It was pointed out by Mr. Menges that the automobile manufacturer is working under somewhat different conditions than the tractor manufacturer. A farmer using a tractor will take a large, heavy monkey wrench to a ¼-inch cap screw and twist it off, so that it is better to use perhaps nothing less than a ½-inch screw on a tractor. The automobile engine manufacturer is working perhaps where space is more cramped, and the smaller screw will be better for him. Automobile motor design, in other words, must meet a different set of requirements from the tractor motor design. Mr. Schwer agreed with this, at the same time pointing out that there are many places in tractor design, particularly with four-cylinder motors, where the requirements are as close for tractor as for automobile work.

The S. A. E. standard for carburetor flanges, etc., was adopted.

Mr. S. M. Walker brought up the subject of standard belt speeds, mentioning what had been done by the N. G. E. A. in the line of adopting a standard belt speed. Since 1913, however, the manufacturers of power-driven agricultural machinery have increased their speeds so that today 1,800 feet per minute of belt speed seems to be about the average. Mr. Walker proposed, however, that the committee adopt 2,000 feet as their recommendation. The object of a standard belt speed is to permit the engine manufacturers to adopt certain standard sizes of pulleys that will give this speed. Then the power-driven machinery manufacturers place upon their machine pulleys that at this speed of belt will run the machinery at the correct speed. As a result, the purchaser of an engine buys an engine and it has on it just the right size pulley to run all of his machinery. Otherwise he may have to change some of the pulleys on some machinery, or run the risk of operating it at incorrect speeds. Many cases of engines reported to be without sufficient power have been remedied by changing the pulleys to secure a proper relation. Mr. Schwer stated that in his opinion a higher belt speed was required when machines to be run consisted of such as corn huskers, grain separators, etc. Mr. Menges stated that for running cream separators, pump jacks, etc., a lower speed was required. He believed that the only standard that could be effectually adopted would be, not a standard belt speed, but a standard pulley diameter for engines operating at certain speeds. For instance, an engine at 300 r. p. m. would have a certain pulley diameter, and an engine at 550 a different pulley diameter. Messrs. Buffington and Glover proposed to limit the discussion to tractor conditions, as that was the purpose of this particular meeting, and all agreed that a speed of about 2,250 would be right for these conditions, to include team portables. It is expected that the sub-committee which was formed later will take up this matter with the various tractor manufacturers, and have this particular

subject thoroughly gone over, to be taken up at the annual convention.

Standard dimensions for clutch pulley shafts and spiders was a matter on the program, and it was easily to be seen that this, too, constituted a subject to be divided into two groups; one for tractor action and one for farm engines. In fact, so far as tractor design is concerned, there is apparently no possibility of a standard being formed for clutches owing to the multitude of designs and the different needs. Some manufacturers of tractors are using cone clutches, some band clutches, some clutch at motor speed, some reduce speed, some use their traction clutch for thresher clutch, etc. A sub-committee to consider the matter as relates to farm engines was appointed by the chairman, consisting of A. L. Herkenhoff, F. J. Lemley, T. Hanson, Charles Druschell, and T. T. Morgan.

Chairman Kratsch reported that a number of agricultural colleges had asked him to bring up the question of a standard plow hitch, but discussion of this proved that at the present stage of tractor practice no uniformity can exist. It was moved and adopted that a committee of one be appointed to gather information upon current practice, to be submitted to the members of the association in a data sheet.

"Made in U. S. A."

It is impossible to please everybody. Even branding our exported merchandise with "Made in U. S. A." is being criticized because it is alleged that we would be guilty of plagiarism of a German idea, says the New York Commercial. "Made in Germany" has become a valuable brand; but the Germans did not originate the idea. English manufacturers have used distinctive trade marks and have labeled their merchandise "Made in England" or "Made in Great Britain" for two centuries, and manufacturers of fine goods have always coveted the right to describe themselves as purveyors to or makers for royalty. The best known cutlery maker in Sheffield stamped his razors with "Real Old English Razor" a century ago. Americans can set up a hereditary right to follow such trade practices.

Such faultfinding should not be allowed to influence our exporters. To brand a package or an article with "Made in America" is a sign of honesty. Makers of reliable merchandise take pride in their work and their reputations. Even uncivilized tribesmen have learned to look for certain trade marks on guns and cutlery because a trade mark is a guarantee of quality. "Made in U. S. A." or some similar designation should become a valuable trade asset for American manufacturers just as the trade marks of certain English and German houses are worth more in many cases than the plants in which the goods are made.

We have already moved far along these lines and some of our leading manufacturers are reaping rich rewards from the combination of good trade mark and high and uniform quality. The trade name of a certain line of biscuits made in this country is estimated by experts to be worth millions of dollars. It is rated higher in commercial circles than the name and trade mark of any of the old and well known English makers of similar goods. This proves that we can do what the Germans and English have done and we can do it even better, as this biscuit company has done. The field is open and it is wide enough for it embraces the greater part of the world today.

The Common Sense Gas Tractor Co. began on February 25, 1915, in Minneapolis, Minnesota. Mr. H.W. Adams, designer of the tractor, also taught operation of tractors, emphasizing Common Sense tractors, of course. The Moline Plow Co. purchased Universal Tractor Mfg. Co. of Columbus, Ohio, on November 13, 1915. The John Lauson Mfg. Co. began to build tractors, the Parrett Tractor Co. moved to Chicago Heights, Illinois, the Boss Tractor Mfg. Co. of Detroit, Michigan, changed its name to the Chief Tractor Mfg. Co. and Sandy McManus Incorporated changed its name to Interstate Engine and Tractor Co. It was rumored that Henry Ford was going to build a cheap tractor.

JOHN LAUSON AND THE ACTIVE CLASS IN HIS SCHOOL OF TURNERS

Ford Tractor Prices Not Ready.

Detroit, Mich.—To the Implement Trade Journal: We have your recent favor seeking information in regard to the Ford farm tractor.

We have no information to give out concerning this product, as it is not yet ready for the market. We have the tractor all right. It has been thoroughly tested for several years, and it will do the work for which it is intended. We expect to sell it at a very low price, but the question of manufacturing them in the quantities which will be called for has not yet been decided upon. And until we are able to place this tractor on the market, so that deliveries can be made promptly, we do not wish to say anything further about it.

The Ford farm tractor will, in all probability, be placed on the market through the regular selling organization of the Ford Motor Co. The agents of Ford cars will likely be the agents for the tractor. But, we repeat, we cannot at this time even anticipate just when we will be ready to announce the tractor prices and the time of delivery.

Ford Motor Co.

FORD MAY PRICE IT AT $200

New Farm Tractor Won't Cost More Than $250—Big Motor Company Admits Its Intent.

Small farm tractors are going to be manufactured and marketed for use next year by the Ford Motor Co., Detroit, Mich. This much-denied fact, first announced to the implement trade April 3 by the Implement Trade Journal, was officially admitted by the Ford company in a letter to this publication received last week. The communication, signed by C. A. Brownell of the Ford advertising department, was in response to a query sent by the Implement Trade Journal, which enclosed a clipping from the Indianapolis Star containing an account of a lecture delivered April 27 by John R. Lee, sociological expert of the Ford company, who declared that the company was about to bring out a tractor selling for $250. Mr. Brownell's letter says:

"This farm tractor has occupied the attention of Mr. Ford for the past three years, during which time all sorts of tests have been made, under all sorts of conditions—on all kinds of hills, ravines, level roads, etc. It has been used in autumn and winter and spring

and summer and has been brought to such a point that Mr. Ford is now willing to hazard his reputation by placing it on the market.

"Mr. Ford's hope is to be able to sell the tractor for $200; so Mr. Lee was conservative when he fixed the price at $250. It may be $250, or $225, or even $215, but Mr. Ford hopes to produce it in such large quantities that the price can be brought down to $200. Mr. Lee was also conservative in limiting the power of the tractor to do the work of 6 horses. It will do the work of 12 horses."

In rather amusing contrast to this official affirmation of the Ford company's intentions was the earlier emphatic denial made to the Implement Trade Journal by the company that it intended to manufacture a tractor. March 29 the Implement Trade Journal queried the Ford company as follows: "Persistent rumors are afloat in this part of the country, indicating that your company is about to bring out a new farm tractor. If there is any good ground behind these, may we ask that you give us the facts in detail in order that we may make an announcement to the trade, which, of course, would be very much interested in such news?"

This was the Ford denial, written March 31: "Your letter under date of March 29 is before us and we note therefrom your statement that persistent rumors are afloat in your part of the country to the effect that the Ford Motor Co. is about to bring out a new farm tractor. Nothing could be more ridiculous. This is on par with the report that we anticipated making a motorcycle. At the present time the Ford Motor Co. is approximately 40,000 orders behind demand, and for this reason we have found it necessary to suspend all advertising for the time being, believing that under existing conditions publicity is not our necessity. Would thank you for anything you may be able to do in the future to refute this rumor in your vicinity."

But the Implement Trade Journal had already gathered reliable information elsewhere that justified publishing the definite announcement that the Ford company had entered the farm tractor field of manufacture.

1915 was the Nation's first billion bushel wheat crop. The Arrowrock Dam 22 miles from Boise, Idaho, was completed at a cost of $2 million less than estimated and one year earlier than expected. Dedication of the dam was on October 4. It was noted that if the English pound sterling, so-called con-

tinued to go down in value and the American dollar continued to go up abroad, that "the dollar would soon be well known from one end of the world to the other."

Uncle Sam Wants Farmers for His Land.

Uncle Sam is looking for several hundred practical farmers to take up homes on the irrigation projects he has been building in the West. The land is free, but the law requires settlers to pay their share of building the irrigation system, and for this reason a moderate capital is necessary. A practical farmer with from $1,500 to $3,000 should have no trouble in acquiring one of these farms and putting it in successful cultivation.

Under the new Extension Act the settlers are allowed twenty years in which to pay for their water right, and no interest is required on deferred payments. Details concerning opportunities and terms will be furnished upon request by the statistician of the reclamation service, Washington, D. C.

The farms are located in Idaho, Montana, South Dakota, Nebraska, Wyoming and Nevada, and offer opportunities for citizens to establish homes in a growing country. Adjacent farms are under cultivation, railroads have been built, schools and churches established, telephone and rural free delivery are available, and most of the hardships of pioneering already have been overcome.

Alfalfa is the big crop, although grain and sugar beets are profitable and in some sections truck farming pays well. Livestock and dairying are the principal industries.

The Chinese, Hindoos, Hottentots and other heathen should club together and send a few missionaries over to Europe to teach the natives the rudiments of civilization.

Real Barbs on "War" Fence.

A manufacturer of Kokomo, Ind., has been awarded a contract for furnishing one of the European armies with barb wire to be used in fortifications. This wire will have barbs more than an inch long wound on a one-eighth inch steel rod. The barbs are to be placed in sets of four, one inch apart along the steel core, the points extending out at right angles. Barb wire entanglements have come into general use in the European war for holding the enemy at bay.

Automobile Supplies a Logical Hardware Line

Manufacturers and Jobbers Prefer the Dealer as the Sales Channel for This Highly Profitable and Growing Business.

ONE profitable line of trade toward which the hardware dealer is giving more attention all the time is automobile supplies and accessories. There are many reasons why this line of goods should be handled through the hardware trade and no bonafide excuse can be found for dodging it.

In the first place this business already has assumed large proportions. The automobile supply and accessory distributors in Kansas City estimate their business volume last year at $5,000,000. The demand already is here: it isn't necessary to work it up.

Within the territory reached by the Missouri river distributors there are approximately 275,000 automobiles. Missouri in 1914 had registered 54,467 cars; Kansas, 52,408; Iowa, 106,-087; Nebraska, approximately 40,000, and Oklahoma, approximately 30,000. Every car owner in these states is in the market for automobile supplies.

People don't have to be rich nowadays to own an automobile. The family of only ordinary means can afford a car of some sort. The tendency of the past two years has been in two directions: To develop higher power cars, but principally to bring automobiles within the reach of everyone. Automobile prices have been continually revised downward.

Farmers Buy the Most Cars.

In this section of the United States the great majority of farmers are car owners. Take Kansas as an example. There is not a town in that state of 100,000 people. But three cities have more than 20,000. It is primarily a rural state. In 1914 there were 20,000 more cars than in the previous year, and the farmers are not through buying yet. The state produced 181,000,000 bushels of wheat, which exceeded the total yield of Canada. Another big yield soon will be harvested. The money comes in a lump and they invariably become motor enthusiasts. The automobile houses are expecting the trade in Kansas this year to assume higher proportions than ever before. And the conditions that obtain in Kansas hold good in other states which buy their supplies from Missouri river points.

AN ATTRACTIVE AUTO HORN   AUTO MIRROR

Another encouraging feature of the trade is the fact that nearly one-third of the cars used in this section are Fords. They tell all kinds of stories about this car. They even spell it with a small "f." But it comes a long ways nearer than any other on the market to being the "universal car." Ford cars come with less equipment than any others. And while Henry Ford has become a multi-millionaire through the tremendous sales of this car, others have become millionaires by merely furnishing the accessories. The Ford is the popular car with the farmer, and for that reason the accessory trade will be unusually heavy in those states where agriculture predominates as it does in the Missouri Valley.

The Sales Field Is Growing Constantly.

There is certainly a glowing outlook for the far-sighted hardware merchant who is handling motor accessories. With the growing use of motor cars there is every reason why the sales possibilities of accessories should have the earnest consideration of the trade. The automobile industry is one of the nation's business giants, though only an infant. Every addi-

tional car means a new customer. More than half a million cars, made by 450 factories, were sold in 1914.

The motor accessory business has developed fast. So fast in fact that few dealers have realized the tremendous strides it has been making. It is a line that has grown of necessity. It is but a few years since the automobile itself was a rival of the proverbial mother-in-law on the pages of the comic weekly. The manufacture of cars keeps automobile factories busy. The demand is so great they have no time to turn out equipment. So within a very few years has come the accessory industry, second only to the automobile itself in importance.

The Hardware Dealer Is Preferred.

Ask the manufacturers or distributors of accessories, whom they prefer to sell their goods to the consumer. Their reply in ninety-nine cases out of each hundred will be the hardware man. The accessories are closely related to other hardware articles. Why should a hardware dealer sell a wagon wrench, and permit some competitor to sell all automobile wrenches? Why should he sell bicycles, motorcycles, gasoline engines and the accessories that accompany these lines, and then refuse to handle automobile tires, batteries, magnetos or spark plugs? It is just as natural that he should handle motor accessories as that he should handle cutlery.

The hardware dealer already is established. He has the building, the shelving, the fixtures, the organization—in fact he fills every requirement. He is responsible, can obtain the necessary credit and can compete with any other person who would handle motor accessories. He can get the goods on open account. He is selling other lines and does not need to make heavy profits at one season to carry him through the year.

So important has the automobile accessory business become that no hardware jobber would

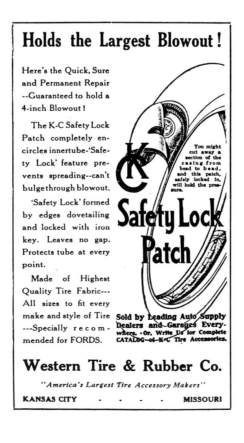
think of conducting his business without a large accessory department. In fact within the past five years this department has taken rank with the most important, and what is better yet, it is still growing.

The trade in tires alone is assuming enormous proportions. No matter how substantial a car may be, or how completely equipped it comes, the demand for tires will always be in evidence. Punctures and blow-outs will come in the best of tires. They generally come when least expected, and unless the driver carries an extra one he will drive on the rim to the nearest source of supply and expect his demand to be instantly supplied.

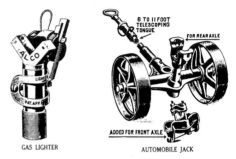

6 TO 11 FOOT TELESCOPING TONGUE   FOR REAR AXLE

ADDED FOR FRONT AXLE

GAS LIGHTER   AUTOMOBILE JACK

A Line That Carries Profits.

Both tires and the other accessory lines carry substantial profits. In fact it is one of hardware's most profitable lines. Profits on different items range from 25 to 50 percent and on some lines even higher. This is on the cost price with the cost of business thrown in for good measure.

Accessories for which the most calls will come are found in this list: Tires, tire gauges, tire covers, tire holders, tire protectors, tire chains, tire patches, tire lugs, tire boots, spark plugs, electric tail lamps, electric attachments for oil and gas burning lamps, electric dash side lamps, jacks, tire tools, shock absorbers, oil cans, cleaning brushes, bulb boxes, tire inner-liners, tool boxes and kits, ignition cable, vulcanizers, spark plug terminal clips, demountable lug wrenches, acetylene, burners, horn bulbs and horn reeds, gas tank keys, tire pumps, brake lining, license brackets, head lamps, automobile clocks, lamp brackets, valve lifters, goggles, bumpers, electric and bulb horns, battery connectors, goggles, gas connections, exhaust horns and whistles, volt ammeters, etc.

It is unnecessary to buy monkey or other kinds of wrenches as these can be supplied out of the regular hardware stock. Other small items which will be included but which are small and can be bought at a very nominal price are cementless patches, taper pins, cotter pins, hexagon head cap screws and hexagon nuts, semifinished, lock washer, set screws, castellated nuts, etc. The dealer should also provide himself with a high quality lubricating oil, grinding compounds, graphite mixtures, greases and waste.

Many dealers have found it advantageous to add to their service to car owners by providing free air for their tires. Others have an accessible gasoline supply. A self-measuring gasoline supply tank extending above the sidewalk will enable car owners to drive up and have tanks filled without leaving their seats, a service that is sure to be appreciated. If the driver can see a nice assorted stock of accessories displayed in the window, while the dealer fills his car he is receiving an impression and a bit of information that will be remembered in time of need. That hardware store then becomes his depot for automobile supplies.

No Longer a Fad.

The automobile is no longer a fad. For ten years past a lot of people have been vainly trying to argue the motor car into the bicycle class of unstability, but all the while by leaps and bounds the automobile industry has been growing bigger and bigger. It has passed through some of the most crucial moments in the nation's financial history and has flourished under even the most adverse of these circumstances. So again, we say, as a fad the automobile is a failure.

Fifty percent or more of the retail implement dealers in this territory have had a vision of what might come and are entrenching themselves more strongly every day in the automobile game by lining up with the popular medium-priced cars to meet the growing demand. For some years this number has been finding in the automobile an efficient profit raiser.

When an Automobile Is Worn Out.

From time to time someone starts a discussion as to the life of an automobile and the usual conclusion that is reached is that the car, with proper care, will run from 50,000 to 80,000 miles. Five or six years ago it was generally understood that a machine that had 10,000 miles to its credit was ready for the junk heap.

The 443-mile, day and night road race was run July 4 and 5 from Los Angeles to Sacramento, Calif. The best time made in this race over rough mountain roads was a little less than 11½ hours. The car that won second place was not a racing machine. It was said that the owner and driver, Ed. Waterman, bought the car after it had seen considerable livery service, for $50, put it in trim himself and drove it in competition with new cars costing as much as $5,000, and won $2,500 with it.

In the Reliability Run from Minneapolis, Minn., to Glacier National Park, Mont., the winning machine was a 1909 chain drive model that had 100,000 miles to its credit beore it started on the run. The run was under the supervision of the American Automobile association and was, as the name implies, a test of the reliability of the cars; some of the best known and highest priced cars on the market were in the contest. If a car is properly cared for and kept in good order the fact that it has run 100,000 miles or is an old model is not proof that it is worn out. Many of the cars that were new in 1908-09 are giving as good service as some of the 1913 models. It isn't the miles that a car has run that determines whether it is worn out or not; it all depends on the care that the machine has had.

A Lighting Outfit for Ford Cars.

A lighting outfit for Fords or other small priced cars is being put on the market for 1915 trade by the Henricks Novelty Co., Indianapolis, Ind., which is known as the Ford Lighting Outfit. In making this outfit the company has taken into consideration the objections voiced against previous efforts in this line, such as high cost and difficulty of mounting. The cost of the Ford outfit is well within the reach of the average owner of Fords or other low-priced cars, and the installation on the Ford especially is so simple that the man of average ability can install the system complete within half a day.

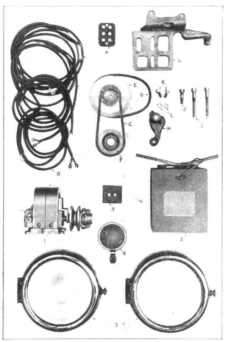

THESE PARTS MAKE UP THE OUTFIT.

The Ford outfit comes complete, even to the wires, which are cut to length with terminals attached. A terminal block is provided, which makes is exceedingly simple, and which does away with the necessity of making soldered joints or splicing the wire. The bracket is of malleable iron, and while light, it is amply strong for the purpose intended.

A "V" belt drive is provided throughout, including the fan belt which does away with the annoyance of the fan belt coming off and also gives a positive drive and one that will not slip. All of the bolts, screws, lock washers and everything needed for installation are furnished. Two 9½-inch head lamps constructed with bulbs, and extra good tail lamps, a Cutler-Hammer dimmer switch and a special 6-60 lighting battery. The generator is complete with automatic cut-out, has ball bearings throughout, including the governor. The generator charges a battery at 6 to 7 amperes, therefore has ample current for 35 to 40 candle power light. The current from the battery may be used for ignition if desired.

Full instructions for installation are given with each outfit. The Henricks company will be glad to furnish further information, together with special terms to dealers, upon request. An outfit of this character with the prestige of the manufacturer's name is sure to command attention.

To Make Over All Old Fords.

Every Ford car now in service is to be "remanufactured" provided it needs such treatment, according to the latest announcement of Henry Ford. A uniform charge of $65 will be made which will cover every needful repair disclosed by a close inspection. It matters not how much repairing may be necessary, the charge will remain the same. The remanufacturing will cover bent fenders, scraped paint or enameling, cracked isinglass or even a new motor block, or any other needed repair. Ford branch managers are instructed to go after owners who do not seem to be desirous of this opportunity. Mr. Ford says he expects this plan to cause him a heavy net loss, but he expects to profit in the added repute his product will gain.

Blitz Ignition System.

The Blitz ignition system for Ford automobiles is the product of the Electrical Specialties Mfg. Co. of Omaha. The accompanying illustration shows the manner of its installation. It is designed especially to increase the power of the Ford car and at the same time make it start easier and run smoother. The manufacturers state that the Blitz ignition system prevents the sticking of platinum points, burnt out timers, missing and back firing. On the other hand, it assures a uniform spark and perfect timing.

The Blitz will handle the alternating current of the Ford magneto as well as battery current. It delivers at each spark plug a spark of the same intensity and the same duration and in each cylinder at exactly the same angularity of the crank shaft; or, in other words, absolute synchronism.

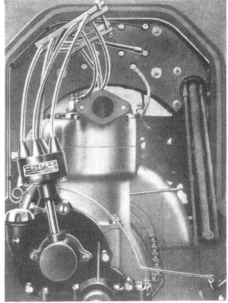

SIMPLE INSTALLATION OF BLITZ IGNITION SYSTEM

Hudson's Aluminum Running Boards.

One of the greatest sources of annoyance to owners of Ford cars is the running board which frequently gets and stays rusty. Ford running boards are painted. When the paint wears off the iron is exposed to the elements and rust is a natural result. The Ford Specialty Co., Philadelphia, Pa., has placed on

HUDSON'S ALUMINUM RUNNING BOARD

the market Hudson's aluminum slip-on running boards, which do away with all troubles of this kind. They are made of aluminum, which does not rust and which always presents a pleasing appearance. The car owner merely slips on these running boards. They fit perfectly and no bolts or clamps are required. When once on they are secure and do not rattle. A set of two retails for $3.50 netting the dealer a substantial profit.

The war, bloody as it is, does not necessarily puncture the Peace Movement. It's simply another powerful proof that those back of that movement are right.

It all goes to prove that if a country totes a lot of guns long enough some of them are going to go off.

THE FORD LIGHTING OUTFIT INSTALLED

Coupler That Converts Ford Engine.

Ford automobiles for stationary power purposes is an achievement attained by means of the "Omaha coupler" which has just been placed upon the market by the Omaha Coupler Co. The device is readily attachable to any model Ford car and at a comparatively slight cost the full 22-h. p. automobile engine may be converted to operating power machinery of almost any character within capacity of the motor.

By driving out the pin which holds the ordinary coupler in place, the crank is removed. The coupler, consisting of a short nickel-chrome shaft with sectional, quick-detachable, steel-pressed pulley at the "outside" end, is then ready for mounting. It is inserted through the same journal and the patented coupler connecting the pulley shaft with the crank shaft is held in place by pin and peg. Taps on both sides of the radiator are removed and on these same bolts the brackets are connected with two other bolts clamping onto the fender braces. On the front end of these brackets the bumper bar is bolted. Special adjustment for lining up the shaft is furnished with the coupler.

The pulley and shaft do not run continuously but are so arranged that by adjusting a thumbscrew they may be meshed only when pulley power is desired. The crank handle fits into the pulley and operates exactly the same as on an unequipped machine, whether the engine is running "free" or is furnishing stationary driving power. It is claimed that the pulley is sufficiently strong to operate an ensilage cutter, a grain elevator, a hay press, an irrigating pump, a small thresher, a feed mill and many other farm machines. The attachment, packed for shipment, weighs but 60 pounds. The bumper is included in the outfit which sells f. o. b. Omaha for $30. The Omaha Coupler Co. is located at 403 Ware block. The coupler will be sold through retail implement dealers.

A Motor Wheel for Bicycles.

An invention which it is predicted will come into wide use next spring and summer is the motor wheel, a simple little device which converts any bicycle in a few minutes into a very fair sort of a motorcycle. It consists of a small wheel upon which is mounted an exceedingly compact power plant consisting of a one-cylinder, four-cycle gasoline engine complete with carburetor, magneto, driving gear and gasoline tank. This wheel can be attached in a few minutes to any bicycle by means of a pivot bracket which is clamped to the frame at three points. The motor wheel carries a heavy, double tube, clincher motorcycle tire. The power from the motor drives this wheel and its own weight on the ground affords sufficient traction to push the bicycle. Upwards of one hundred miles have been covered on one gallon of gasoline.

The motor which furnishes the power for this motor wheel is interesting on account of its light weight and compactness—the entire device weighing less than fifty pounds. The cylinder is 2¼-inch bore with 2¼-inch stroke. A high tension magneto is used and a simple automatic float feed carburetor of special design.

The bicycle is under as perfect control as though the device were not attached. Starting, stopping and regulating the speed are regulated by a lever attached to the handle bars.

The motor wheel is mounted on a pivot so that when the bicycle leans over it cants the same way, the planes of the two wheels being always parallel. While the attachment does not interfere in any way with the rider's perfect control of the bicycle, it helps materially in preventing side slipping on greasy pavements.

The device is manufactured by the A. O. Smith Co. of Milwaukee, Wis., the world's largest manufacturers of automobile parts.

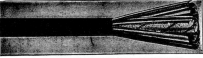
Turkeys and the Whip Industry.

When whalebone first was used in the manufacture of whips about thirty years ago it sold for $2 a pound. Since then the price has advanced to $15 and $18. No one knows how long the supply will last. Whales are becoming scarcer every year. It takes about 100 years for a whale to attain a size where it is of commercial value. The balkiest horse will die long before that time. The prohibitive price was responsible for the more general use of rawhide centers. Rawhide whips are strong and durable yet lack the fine elasticity of whalebone. French horn, gutta percha and spring steel wire failed to put the elasticity into the rawhide product. Nothing was found to fill that want until Charles H. Clark of the Featherbone Whip Co., Westfield, Mass., discovered a by-product in the quills of the turkey. Straight rawhide whips droop in damp weather when they become almost useless. And turkey quills have surmounted this obstacle. These turkey quills, split and wrapped in continuous layers around a center of hard rawhide, make a whip Mr. Clark calls "Anti-Whalebone." The quill covering he calls "Quillbone." The whip he produces is waterproof, and has the feel, swing and elasticity of the whips that were so popular when whalebone was sold at prices that permitted its use in the manufacture. Clark has made possible a whip that defies the elements, and it is on the market at a price that would put the old whale out of the running if he still cavorted in the waters of competition at $2 a pound.

The cut open section of an "Anti-Whalebone" whip illustrates the heavy, hard twisted rawhide centers used in these whips, surrounded with the strips of "Quillbone" (heavy enameled quills) which entirely cover the rawhide and extend the entire length of the whip, making these whips absolutely waterproof, and giving strength and great elasticity to them, and keeping them straight, so that

ILLUSTRATING THE STRONG CONSTRUCTION OF "ANTI-WHALEBONE" WHIPS.

they swing, feel and wear like whalebone, and are just about as near an approach to genuine whalebone as can be produced.

A stronger and better wearing material than common cotton cloth to be used in drop or lash top whips was next looked for by Clark, who produced a heavy, flexible "Quillbone" cord, made from quills, which is used in all drop top "Feathertone" whips, giving them strength and elasticity and providing a strong re-enforced loop for the snap.

THE MOTOR WHEEL ATTACHMENT FOR BICYCLES IS BECOMING POPULAR.

QUESNELL? Yes, C. Quesnell, that's his name. But there isn't anything else queer about him, for the wheat-growing section of the Pacific Northwest takes him and his inventions mighty seriously. Mr. Quesnell is the inventor of the one-man combined harvester. He has invented other harvesters, too, but this latest one is bidding for the attention of the small-farm wheat-grower of the "Inland Empire" just now.

Some years ago Mr. Quesnell brought out a combined harvester which required eight horses and cut nine feet. Now he has taken a long step forward by developing his one-man outfit, which requires but four horses and cuts 7½ feet.

Farmers of the Middle West know very little about the combined harvester. This remarkable machine cuts and threshes small grain at the same operation. Of course, the straw is left in the field. Some outfits are pulled by horses and others by tractors.

In his one-man outfit Mr. Quesnell has endeavored to emancipate the wheat farmers of the Pacific Northwest from the help problem, and from the delays and expense of the threshing season. The builders of the new Quesnell combined harvester assert that it makes it possible for the farmer to reap what he sows with the same horse power that put in his crop.

The machine operates much like a header. The horses are hitched behind it. As it is cut the grain is carried by a draper directly back from the cutter-bar to the cylinder, which is six feet wide. There it is threshed and deposited on a roller sieve, which separates and cleans the grain. The straw goes on over and out the back of the machine on the ground. The tailings are elevated back over the draper and through the machine.

Approximately twenty bushels are held in the hopper. The pause for emptying the threshed grain into sacks or other receptacles affords the horses time to rest, that is, if it isn't pushed by a tractor. Unless a gas engine is used as auxiliary power to operate the cutting mechanism, six horses are required.

One man, who drives the horses and works the stand as with a header, cares for all the machinery. There are but three levers to manipulate. One raises and lowers the platform. One levels the cutter-bar independent of the position of the platform. The third adjusts the sieves.

Combined harvesters are very much the rule in the "Inland Empire" among the big wheat ranchers. The smaller grain farmers use binders and have their wheat threshed from the stack or shock just as is done in the Middle West. It is the hope of Mr. Quesnell that with his new machine the smaller farmers will take advantage of their opportunity to save labor.

Thousands of binders, of course, are sold to the Pacific-Northwestern farmers every year by the harvester companies. To these machines the grain-growers owe much of their success. And they will probably be sold in large numbers for years to come.

But just now the agricultural observers of that part of the country, so far as they are concerned with the harvesting of small grain, are watching the development of this type of small combined harvester which has met with so much encouragement and which is new to the country. Also they are observing with keen interest the application of more and more mechanical horsepower to the operation of the larger combined harvesters. With the former the farmer is enabled to reduce the number of horses on his farm as well as the amount of his help.

With the latter, where he has the capital and the acreage, he is enabled to harvest an enormous volume of grain at a comparatively small expense. For the Easterner the sight of one of these huge machines cruising across a wheat ranch, cutting and threshing the grain as it goes, is one never to be forgotten. But this latest Quesnell machine, of course, is built on a smaller scale.

May End Dishwashing Drudgery.

At last mere man has invented a better kind of dishpan. How very long we have used either a high, round pan or a low, oval one, neither of which was the shape of size of the sink. But we have now a vast improvement—namely; a square dishpan made of very heavy retinned ware, according to the Philadelphia Public Ledger. It is much more like the shape of the sink and, in addition, has the peculiar feature of a little plug at the bottom with a piece of wire netting so that the plug need only be drawn out when the dish water needs to be changed. In this way, the water is strained and food particles prevented from going down the sink.

Another improvement is a dishmop fitted with rubber tubing to the faucets of the kitchen sink, so that hot water actually goes through the mop, thus constantly supplying fresh water. A special rack comes with this device into which each dish is set, then washed, or, more properly, sprayed with the mop, which connects either with the hot or cold faucet, so that the temperature of the water can be changed at will. This is a sanitary method which reduces considerably the labor of dishwashing.

Those who have room for a large, permanent dishwashing machine will be interested to know that there has been a modification of this device and that there is now on the market a kind of wall dishwasher. The best one is operated by current and consists of a tank set next the wall and connected with the regular plumbing. The dishes are packed in a wire basket within the tank, and by turning a button a force of water is sprayed continuously over them. This new model can be operated by hand, but it is not as satisfactory as when current is used.

At last the drudgery of dishwashing is to be reformed, so the woman to whom it is a bugbear need only look around and adapt some modern improvements to reduce her labor.

Bicycle Wood Rim Industry.

The manufacture of bicycle wood rims was started twenty-one years ago, and in less than three years this entirely displaced the all-steel rim by its merit. The wood rim is resilient and springy and enhances the life of the tire, and, when made properly of fine selected straight-grain maple, has proven a very serviceable and lasting construction for bicycles. The percentage of straight-grain maple suitable for bicycle wood rim strips from each 1,000 feet of lumber cut does not exceed from 12 percent to 15 percent, therefore it was necessary to install a large wood-working manufacturing plant to work up the balance of the lumber after selecting the wood rim stock from it. The price of the raw material for bicycle wood rims has advanced 40 percent during the past five years.

A Steel That Won't Rust.

One would think that since steel had been manufactured so many centuries, that any great discovery in regard to its manufacture would be impossible, but Commerce Reports, issued by the Government, states that a firm in Sheffield, England, has put upon the market a new kind of steel that is non-rusting, unstainable and untarnishable. This steel is especially adapted for table cutlery, as the original polish is maintained after use, even when brought in contact with the most acid foods, and it only requires ordinary washing to cleanse. It is claimed that it retains a keen edge much like that of the best double-sheer steel, and, as the properties claimed are inherent in the steel and are not due to any treatment, knives can readily be sharpened on a "steel" or by using the ordinary cleaning machine or knifeboard. It is expected it will prove a great boon, especially to large users of cutlery, such as hotels, steamships and restaurants, and to housewives, who have largely abandoned steel knives and substituted silver plated ones because of the work it requires to polish them. This kind of steel costs twice as much as the common sort.

A Safety Cracker for the Fourth.

Agitation for a safe and sane Fourth of July has turned the attention of the inventive world to noise producers that give all the old time thrills with none of the old time dangers. One of the very best devices of this kind on the market is the Barler safety cracker, which is distributed by the Richards & Conover Hardware Co., Kansas City. This cracker has a magazine handle ten inches long, made of steel, which will not bend or break with use. This handle is enameled in bright red, the enamel being baked on. This handle has a slide for feeding the ammunition, the ammunition consisting of slips of paper. Attached to the end of the handle is a specially designed rubber ball fastened with a reinforced rim. The air ball fits into a steel cap and can be renewed. The ammunition is fired by the compression of the air in the ball when it is struck suddenly against any solid object, a loud explosive noise resulting. The Barler cracker is 10½ inches long and is durable. It retails for 25 cents, together with sufficient paper ammunition for 500 shots. The dealer makes a liberal profit. After the furnished ammunition is used the youth can prepare his own ammunition. It can be "shot" at the rate of one shot a second.

THE BARLER SAFETY CRACKER

"Women and Children First."

The National Farmers' Union has registered a protest against the working of women and children in the field and is waging an effective campaign against this form of agricultural slavery. "The census enumerators tell us that there are 1,514,000 women who work in the fields in the nation," states a recent communication issued by the Union, "and of this number 400,000 are sixteen years of age and under.

"What is the final destiny of a nation whose future mothers spend their girlhood days behind the plow, pitching hay and hauling manure, and what is to become of womanly culture and refinement that grace the home, charm society and enthuse man to leap to glory in noble achievements, if our daughters are raised in the society of the ox and the companionship of the plow? In rescuing our citizens from the forces of civilization, can we not apply to our fair nation the rule of the sea—women and children first?"

Effect of War on American Industries

THE blow in the face received by American industries through conditions brought about by the European war has acted as a tonic, has forced the nation to create new branches and enlarge the scope of existing phases of manufacture, opened the way to utilize, on a vast scale, great natural resources of the United States, and induced manufacturers and merchants to expand their markets into foreign fields with prospects of permanent results, says the Bureau of Foreign and Domestic Commerce, of the Department of Commerce, in a forecast of the effect of the war on the industrial future of the country.

American ingenuity has been applied with success to the making of articles previously imported, and among those who have shown conspicuous ability in meeting the situation, an important place is given to Thomas A. Edison, "America's scientific wizard," who has had a great part in the enterprise and initiative required to build, at a moment's notice, some of the new American manufactures required by the emergency.

A review of the chief industries ministering particularly to the temporary needs of the belligerents across the Atlantic shows that the final outcome will be a very material addition to the manufacturing plant of the United States. Part of this plant will be simply anticipatory of the normal growth of the country's mechanical equipment; part must lie idle in time of peace, but is a distinct asset in the national preparation for an adequate defense against attack; the remainder furnishes at once products needed in the healthy expansion of the chemical industry of the country.

Uncomfortable, But Good for Us.

Less conspicuous and spectacular, but of far greater permanent value, is the impulse given to the manufacture on American soil, with American raw materials, of a variety of articles for which we have hitherto been dependent upon foreign skill and enterprise. In a more or less uncomfortable way, we have suddenly been brought to recognize the unwisdom, the folly, of shipping vast amounts of the crude material of our farms, forests, and mines 3,000 miles across the ocean, and buying it back in a manufactured form at a vastly enhanced price. We have likewise come to recognize the absurdity of allowing many natural products of the tropics, of South America, of the Far East, to find their way to Europe, and of paying foreign intelligence and skill to transform them into articles of daily need in our lives.

American ingenuity, adaptation, inventive talent, scientific attainments, and general enterprise have promptly rallied to meet widespread demands, and establish on our own soil the permanent manufacture of a number of wares, some of minor, others of major importance. The return of peace will see them well rooted and able to withstand foreign competition.

The Bureau of Foreign and Domestic Commerce points to the course of events that followed the cutting off by war of the aniline imports from Germany and the supply of potash from the same source, with the resulting tremendous impulse given to the expansion of domestic manufacture.

Can Furnish Our Own Potash.

Of the domestic potash supply it is stated that large amounts of the compounds of this element are present in the vase beds of kelp floating on the waves of the Pacific, close to the western littoral of the country, that each year the waters of the Pacific coast are producing a crop in which potash salts possessing a normal value of more than $90,000,000 are readily available for use in agriculture and the arts. Now a dozen companies are engaged in the campaign. Not only the inexhaustible supplies in the waters of the Pacific, but also the remarkable deposits in the arid waste about Searles Lake in California, and the valuable alunite of Utah are being rapidly transformed into standard, commercial grades. A year or two hence we may be able to fertilize our broad acres with American potash exclusively, while another year or two may see us free from dependence upon dyes of foreign make.

The Bureau advises the business men of the United States that the present time is opportune for them to study the Latin American markets, to get in touch with the people of the countries, and thus to open the way for extensive business operations. In other countries also there are unprecedented opportunities for the extension of foreign trade, and with the indications that we are entering upon a period as a creditor nation, we are in a position, as never before, to invest our capital in industries and developments in foreign countries.

It does not believe that the cost of production in the warring countries of Europe will be lowered as a result of the war, or that there will be danger from that source to the holding of new markets already gained. Experience has shown that it is apt to be higher instead of lower after the close of the war, with higher interest rates, higher wages, and higher prices in the warring countries. Surveying the whole field, it may justly be said that the world's conflict has been of unmeasured value to American industry as a whole.

New Record Wheat Prices.

The highest price ever attained by Nebraska wheat on the local market was reached Tuesday when a car of ordinary 60-pound hard winter was sold at $1.56, and later several cars of No. 2 were sold at $1.55, while a car of the dark turkey variety sold at $1.54.

These prices represent an advance of from 4 to 6 cents over those paid Saturday, and were much the highest ever quoted for hard winter wheat in this territory. The winter grain sold Monday as high as Durum has ever sold, $1.51. Rye advanced 3 cents, corn from 1 to 2 cents and oats from a ½ to 1 cent.

Total receipts of grain Monday were very heavy, 594 cars being reported in, of which 493 were corn, much of which came from Iowa.

$1,000,000,000 in War Goods.

According to a long article in the New York World of March 1 $1,000,000,000 is a conservative estimate of the value of the trade in arms, ammunition and war supplies between the allies of Europe and the manufacturers in the United States during the first year of the war.

It has not been possible for American manufacturers to furnish war supplies to Germany, because of the interruption of trade relations with that country. During the past six months the allies have bought from the United States $400,000,000 worth of war supplies. Many of the manufacturers in the United States hold contracts with the British, French and Russian governments which have at least nine months, and, in some instances, two years longer to run.

Virtually all war material destined for the use of the allies is now shipped by the American manufacturer or producer to Canada, where it is transshipped in British bottoms to England. Even war supplies for France and Russia are delivered in Canada, shipped to England and distributed from there.

Japs Imitate American Tires.

That the great imitative powers of the Japanese have been to a large degree responsible for the rapid modernization of the Orient, since they place duplicates of the latest mechanical devices on the Oriental market at exceptionally low prices, is the opinion of William J. Gorham, formerly president of the Gorham-Revere Rubber Co., which recently consolidated with another large concern under the name of the United States Rubber Company. Gorham is now connected with the new concern as one of its managers, but he has disposed of his stock in all rubber manufacturing factories. Gorham claims to have been the first man to introduce manufactured rubber into the Orient, and he tells interesting stories of his experiences at the time when no rubber was manufactured in the Far East.

"For many years the United States usurped the rubber trade of China and Japan," he says, "but finally the Japanese began manufacturing exact replicas of the imported product, and because of cheap labor, a near source of raw rubber and a duty imposed upon foreign importations, the Japanese were able to underbid their American competitors and soon monopolized the trade of the entire country. When the Japanese enter a business it is with maximum energy and a determination to sweep their competitors before them.

"Undaunted by the acts of the Japanese manufacturers, which practically cut off the rubber trade between the two countries for a time, American manufacturers saw an opportunity to introduce solid rubber tires for the native jinrikshas with patent tire preparations, and on presenting the new tire into the countries of Asia, American manufacturers became popular once more. Japanese manufacturers, not to be outdone, eventually began the manufacture of these tires and slowly but surely the foreign goods were forced from the market. The introduction of the pneumatic jinriksha tire by Americans came next, and after brief popularity in the Orient, it, too, was forced off by Japanese manufacturers.

"So it is with every piece of modern machinery which makes its appearance in their country. The Japanese study the object with the minutest care and after having thoroughly learned the machine they duplicate it and place one equally as good on the market at a much smaller cost. The big cry of the Japanese at present is for gas engines, and while there are a number of American engines in operation there, the parts of the machine are too complicated to be easily understood, and it is taking a long time for the Japanese manufacturers to duplicate it."

Gorham asserts that although the Japanese are running American manufacturers from their country, they are nevertheless doing a great work in producing modern appliances so cheaply and making them popular in the Far East. The country is progressing rapidly in civilization, he declares, and will be soon ranked among the leading industrial nations of the world.

Ford Cars for European War.

President Henry Ford of the Ford Motor Co., Detroit, Mich., has announced an order from one of the European belligerent nations for 40,000 cars for immediate shipment. These will be used in army service.

Remington Doubles Its Plant.

The work of erecting an addition to the plant of the Remington Arms & Ammunition Co., which, with machinery, will cost more than $1,000,000, is being rushed in Ilion, N. Y. The demand for rifles for use in the war makes the early completion of the new plant, which will double the company's output, a matter of urgent importance. It is reported that the Remington company has closed a contract for supplying a great quantity of rifles to one of the European powers, but the closest secrecy is maintained concerning the entire matter.

Denatured Alcohol
May Soon Supplant Gasoline

It was only a few years back when we considered any gasoline that did not show a gravity test of at least 72 degrees Baume as entirely worthless for gas engine fuel. Now they tell us that the gasoline this year will be lower in gravity and have a higher boiling point than ever before. Just now the gravity test of gasoline is around 58-60 degrees, but as the summer advances it will drop to 55-56 degrees, and then perhaps to 53-54 or lower. At the same time, the boiling point will climb toward the 500 mark.

This gasoline, for that is what they call it, will bear little resemblance to the gasoline of eight or ten years ago. It will in fact appear more like kerosene, for, as the refiners put it, they are going to "take the very heart out of the kerosene" to make this gasoline.

However, despite all this, high-grade kerosene, or low-grade gasoline, whichever you choose to call it, will work very well. It will undoubtedly take a little more spinning of the motor when the engine is cold, but it seems to be a case of use it or nothing. We have become, perhaps unconsciously, accustomed to using this low-test gasoline, for last year most of our gasoline did not test over 58-64 degrees.

While the price of gasoline is considerable higher in price now than it has been at any other time during the past few years, it will be recalled that it reached a higher level four years ago and then dropped back. The dealer should not become alarmed to any great extent about this talk of fifty to sixty cents a gallon for gasoline, nor allow it to influence his customers in holding off on purchasing engines. Even should the price of gasoline reach these prices, the farm engine would still be the cheapest prime mover, for the amount of fuel a farm engine consumes in ten hours is very small and a gallon of gasoline will operate a 5-h. p. engine quite a number of hours.

To be sure, the price of gasoline, like that of any other commodity, is governed by the law of demand and supply. On account of the European conflict the amount of gasoline demanded abroad has materially increased, and at the same time the production of crude oil has fallen off to a considerable amount. Then again, the number of tractors, farm engines and automobiles has greatly increased during the last year and they all require fuel for operation.

However, with plants using the Rittman process of distillation now in operation, which will increase the production of gasoline from 15 to 45 percent; with two dozen or more bills before Congress inquiring into the gasoline situation, to say nothing of the Federal Trade Commission on the job, the dealer has little to fear in the way of gasoline going a great deal higher in price. At the same time, there is little hope that it will reach the low selling price that formerly obtained.

In many ways, denatured alcohol is an ideal internal-combustion engine fuel, and if Congress would only wake up to the fact that the farmer is anxious to know something more about the practical distillation of alcohol from his waste products, and would authorize the Agricultural Department to set up a few demonstrating stills, a solution would be found which would greatly relieve the situation as far as the farm engine and tractor are concerned. It has been demonstrated that this fuel can be successfully used in farm engines and the only reason that further development has not been made in alcohol engine construction is the fact that the price of gasoline has been sufficiently low to make it the cheaper fuel. However, that time seems to be passing.

Tractors to Replace Men.

The enlistment of so large a proportion of the young men of Canada in the British army is said to be causing a serious shortage of farm hands in Canada, and as a result there is going to be a very large increase in the demand for tractors to replace the men.

War May Affect Baseball.

If trade with Russia is not continued, there will be no more baseball. Curtis Guild, former United States ambassador to Russia, made this announcement at the recent luncheon of the American Manufacturers' Export Association, and then went on to prove his statement. The only leather which will not stretch under sudden impact comes from the hides of Siberian ponies, and cannot be gotten elsewhere. These hides are imported by the United States mainly for the covering of the regulation league ball. Without them there would be no baseballs or baseball.

The Gramm Motor Truck Co., Lima, O., has received an order for 2,000 three and four-ton Willys-Overland trucks from the Russian government.

Tungsten Now Going Up.

Closing of the Panama Canal is said by manufacturers to be responsible for the marked scarcity of high speed and tool steels. Tungsten, which enters largely into these grades of steel, has been very scarce since the outbreak of the European war, but recently it was announced that a cargo was on its way from Japan. Advices received later, however, changed the situation, as the ship, because the canal was closed, had been diverted around Cape Horn, and was not expected to arrive until after the first of the year. Meantime, it was said, tungsten had advanced from 75 cents a pound to $8.50.

Aerial Boat Goes to England.

The aerial boat America, the one in which Lieut. Porte had expected to fly across the Atlantic, and two others of like type were shipped to England recently on board the Mauretania, to be used by the British government in war service.

The America was originally built for Rodman Wanamaker at Hammondsport, N. Y., for the purpose of making a trans-Atlantic flight, but he declined to finance the undertaking and refused to accept the machine after it became necessary to postpone the race.

The accompanying illustration shows the new third gasoline motor on flying boat America. After trying many schemes in order to make the twin motors and attached propellers lift the desired five thousand pounds to the water surface, so that the boat might plane swiftly on her bottom and rise into the air, it was decided to add a third 90-h. p. motor like the others, with a direct connected propeller, and to add beside the bow the early planing fins which had proved so efficient shortly after the America's launching.

It is stated that with these fins and two engines nearly forty-four hundred pounds total weight had been lifted, and carried in successful flight and it was thought that the third engine would add fifteen hundred pounds lift, one thousand pounds of which could be available for carrying gasoline. The third motor and propeller was mounted on top of the upper plane, near its center of lift, the propeller jutting forward of the leading edge and sending its stream back parallel to those of the twin propellers without damage to the original arrangement of propellers and control surfaces. After installing the third motor, the America rose from the water without difficulty and flew satisfactorily, but no flight was attempted, as the object of this latest trial was to get her to lift from the water with her full load, which has been the result sought for in the long series of experiments made with the craft since she was launched.

In this trial the total load lift was six thousand two hundred and three pounds, which is considerably in excess of what she will be expected to carry on her long flight, and provides a margin for additional fuel for the third motor. It is pointed out that including its third engine and hydroplaning surfaces the America weighs thirty-five hundred pounds and on her trans-Atlantic flight her cargo would have had to include Lieut. Porte, weighing one hundred and sixty-eight pounds, the mechanician, Hallet, weight one hundred and forty-five pounds, and gasoline fuel of eighteen hundred pounds, also engine oil one hundred and seventy-six pounds and food and instruments weighing one hundred pounds, making a total of two thousand three hundred and eighty-nine pounds.

From Petroleum 200 Percent More Power

Discoveries of Dr. Walter F. Rittman Are of the Utmost Importance to All Who Burn Gasoline, and Many Others.

DR. WALTER F. RITTMAN'S discoveries, which are expected to revolutionize the manufacture of gasoline, toluol and benzol, have been pronounced by Government scientists and chemists for oil and powder companies as absolutely practicable and of great economic saving.

This young Government scientist gave a technical description of his processes the other day. He said:

"The oil is passed in the form of vapor into a hot tube which is under a pressure varying from ninety to five hundred pounds a square inch and at a temperature of about 450 degrees Centigrade. This 'breaks' the kerosene molecules up into gasoline molecules. The process is similar to the popping of corn and very simple. The old way was to distill off the gasoline from the liquid and when that was done no more gasoline could be obtained.

"By my process we break up the residue, the big molecules that were not used heretofore to get gasoline. It virtually begins where the old process ends. Double the the amount of gasoline is obtainable by the breaking-up process. The residue solids, which sell for about three cents a gallon, are converted into gasoline, which sells for about twelve cents a gallon."

The scientist explained that the same process is used largely in obtaining toluol and benzol.

"If this Nation is put in a national stress," said Doctor Rittman, "we can supply by my process sufficient toluol and benzol for the high explosives from which we are now cut off by the war abroad.

"I do not claim to cheapen the process of producing benzol or toluol nor, for that matter, do I claim to have lessened the cost of deriving gasoline from petroleum. My process makes it possible, though, to get gasoline from cheaper oils and residues and, therefore, the cost is lessened in this way."

Process Will Be Used at Once.

Two concerns, one an independent oil company of Pennsylvania and the other a large powder manufacturing corporation, who sent their chemists here, announced at the conclusion of the demonstration that the processes would be used at once. Secretary Lane agreed for the sake of actual commercial perfection that Doctor Rittman should supervise he installation of his patented process and that the Government should get the benefits of the results for its high explosives and the commercial world the discovery in other respects.

Following is the announcement given to the public by the Department of Interior:

"These processes," said Secretary Lane, "are fraught with the utmost importance to the people of this country. For some time the Standard Oil Co., through the great amount of money at its command, through its employment of expert chemists and through its extensive organization, has had a big advantage over the independents in the protuction of gasoline, this company having a patented process that obtains for it as much as three times the amount of gasoline from a given quantity of petroleum as the independents now obtain. There are two or three other large corporations that have an efficient process for the manufacture of gasoline, but the independents as a whole have never been able even to approach the results obtained by the Standard Oil Co.

For the Use of All the People.

"Now the Federal Government, through the efforts of Dr. Rittman, proposes to make free for the use of all of the people of this country who wish it, a process that is confidently expected to increase their yields of gasoline from crude petroleum fully 200 percent and perhaps more, such results having repeatedly been obtained in the laboratory. It is claimed by Dr. Rittman that his process is safer, simpler and is more economical in time than processes now in use and these are economic factors of great importance. With a steadily increasing demand for gasoline for automobiles, motor boats and engines, this fortunate discovery comes at the proper time.

"It is but two years ago that the automobile industry, fearful that the supply of gasoline might not be adequate for its rapidly expanding business, offered through the International Association of Recognized Automobile Clubs, a prize of $100,000 for a substitute for gasoline that would cost less than gasoline. Happily the urgency of this situation has passed and at the present time there is a plentiful supply of motor fuel to meet immediate demand. This new process adds to the hope, that in spite of the wonderful growth in the use of gasoline, there may not be any shortage in the future.

"It indicates an increased production of gasoline from the present production of petroleum—an output of 50,000,000 barrels instead of 25,000,000, as under the present methods. It will render free for use to all, the results of that efficient and intelligent research which has heretofore been only at the command of the wealthy. I am led to believe that it will not only be of inestimable value to the refiners commanding but limited capital as well as those of wealth, but also to the hundreds of users of gasoline. When it is realized that the gasoline industry each year in this country yields products amounting in value to between $100,000,000 and $150,000,000, the importance of this discovery is seen.

"The second process discovered by Dr. Rittman may prove of much more value to the country than the first, in that it suggests the establishment of an industry in which Germany has heretofore been pre-eminent—the dye industry, and also promises indirectly a measure of national safety of incalculable import. Among the necessary ingredients of high explosives used in modern warfare toluol and benzol are in the first rank.

Also Furnishes High Explosives.

"Heretofore these products have mainly been obtained in Germany and England from coal tar, and the explosives manufacturers have had to depend largely on the supply from these sources in the making of explosives. I understand that some toluol and benzol have been obtained from American coal and water-gas tars, but this supply does not begin to satisfy the present demands.

"The Federal Government now proposes to obtain the toluol and benzol from crude petroleum also. I am further informed that these products can be produced from practically any American petroleum and that the supply can be made sufficient not only for the entire American trade but also for other purposes. This process has gone far enough to indicate that the two products can be produced at a reasonable cost.

"The real comforting thing, however, is that we have the knowledge that this new source of supply is at the command of our people, and that in time of great national stress, if the nation is ever called upon to defend itself, we will be able to manufacture the most efficient and most powerful explosives known in warfare.

Good for Our Dye-Stuffs Industry.

"Were it not for this discovery, it is possible that in such an emergency, we might be compelled to rely largely on the greatly inferior explosives that were used in the time of our Civil War and this would spell national disaster.

"Dr. Rittman concludes from his experiments that this process may become more economical than the German method of obtaining these products from coal tar, as this process not only makes toluol and benzol, but also gasoline in considerable quantities. He intimated to me the possibility of the value of the gasoline being an important factor in paying the costs of the process. If this should prove to be true, it may result in eventually giving the United States a supremacy in the dye-suffs industry that has for some time belonged to Germany, since toluol and benzol are the source of many of these important dye stuffs that are used in the silk, cotton and woolen industries. It would also tend to prevent disturbance of the great industries engaged in the manufacture of silks, cottons and woolens in such extraordinary times as we are now experiencing, for we would be able to supply them with the necessary dyes."

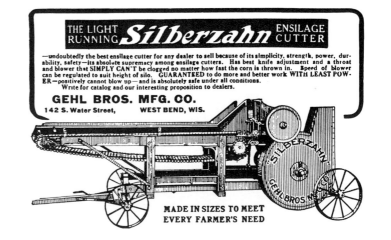

Implement and Automobile Dealers
─Big, New Field─

THE successful motor cultivator has come. Light, strong, flexible, ready for a hundred uses. But most of all to lift the big load of cultivation in corn, cotton, tobacco, orchards and truck. This motor cultivator works in all kinds of soils and under all kinds of conditions. It is manufactured by a big company in a big factory and guaranteed.

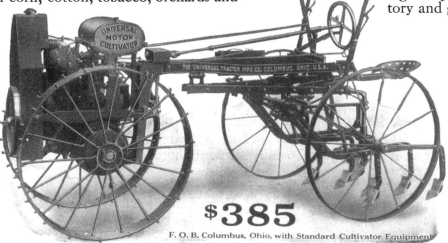

$385
F. O. B. Columbus, Ohio, with Standard Cultivator Equipment

Universal Motor Cultivator

THE dealer's big opportunity. To show it is to make sales. Unlike the automobile it will make money for your customer every day and is the biggest step yet taken to solve the labor problem.

A Hundred Uses

The Universal Motor Cultivator cultivates corn, cotton, tobacco, fruit, sugar cane, potatoes and truck (between and astride the rows).

Plants corn, pulls a mowing machine, a hay rake, a drag, a set of disks, a roller and any of a dozen light draft farm tools. Sprays trees and plants. Pumps water, cuts wood, runs a cream separator, feed grinder, corn sheller, fanning mill, washing machine, churn or any light draft farm appliance.

The Ideal Machine to Get the Quick Dust Mulch after Rain That Means Crop Insurance

The Universal Motor Cultivator is not an experiment put out by some enthusiastic inventor. It is being manufactured by big business men in a big factory. There is plenty of money back of it.

From January 1st the Universal will be heavily advertised in the big national and state farm publications, producing thousands of live inquiries for our distributors and dealers.

Already inquiries from interested farmers are coming in by hundreds from the little advertising we have done. As our advertising increases in volume these inquiries will grow in number.

As the inquiries come in they are being forwarded to the dealers to close the sales.

It is already selling big. Get your territory arrangements made now.

No New Mechanical Principles

The Universal is no experiment. Few new mechanical principles are involved in its construction. It is a direct application of principles that have been tried and tested in the automobile and motor truck field.

Developed from the Modern Light Automobile – Not from the Heavy, Crushing Traction Engine

Built to Last—It has a 3½ in. bore, 5 in. stroke, twin cylinder, long stroke motor, automatic lubrication by plunger pump, automatic thermosiphon radiator, Holley carburetor, automatic governor to take care of the changes in the load, Atwater-Kent ignition, worm gear transmission running in oil bath, expanding ring clutch, one speed—forward and reverse. Develops 10 to 12 h. p. on brake (equal to pulling power in the field of an ordinary team). Weight 1,000 pounds.

Costs the Customer Less than a Good Team–Works Faster–Works Constantly

Think what this means. The Universal will do most things the average team will do. It will work 16 to 24 hours a day when work presses. Your customer can put on a headlight and plow corn all night by shifting drivers. And when the Universal doesn't work it doesn't eat. In the winter it will do all of the small power work of the farm.

That means that you are not selling your customer a luxury but a modern necessity that will make money for him. Get into this field. It will be a winner for you.

The Demand Is Universal

Any man who can afford to own a team will be interested in this machine. You won't have to drag them in to make them buy. Jump on the seat and go down the first good road and you'll get orders. It answers the demand of any kind of farming. Your canvassers will make it hum.

Write for the Proposition NOW

Here is a bigger proposition than the automobile ever was—with more money in it for you.
Send for Book F.

The Universal Tractor Mfg. Co., 2044 South High Street, Columbus, Ohio

Pulling Spike Tooth Harrow

Grinding Feed

Mowing

Shaded portion shows Staude
Mak-a-Tractor and Special
Staude Radiator

**This Wonderful
Opportunity for Dealers**

$195 and Your Ford Car
makes a
Guaranteed Powerful Tractor

THE most sensational money saving application of low cost power ever developed for the farm. A real tractor—powerful—efficient—at *less than one-third the cost* of any other that will do anywhere near an equal amount of work—that's what every up-to-date farmer can get in the Staude Mak-a-Tractor.

22 rated horse power—*the greatest draw bar pull for the weight ever built* and greater strength in every unit of construction, in proportion to weight and work required than in any other tractor.

The Ford power plant and drive system is time tried and tested. It is practically indestructible. The Staude Mak-a-Tractor cannot injure these wonderful mechanical units. It puts no greater load on them than when they are used as touring cars. It merely reduces the speed and increases the pulling power in proportion, giving 11 to 1 reduction in speed and a corresponding increase in power utilizing it to reduce labor, save money and increase efficiency in all heavy farm work.

Staude
Mak-a-Tractor

Plowing	Spreading Manure	Mowing	Packing	House moving
Listing	Pulling trees	Hoisting Hay	Drilling	Loading logs
Cultivating	Grubbing	Hay Loading	Irrigating	Stretching wire
Disking	Pulling stones	Pulling Binders	Grading	Ditch digging
Crushing clods	Smoothing	Pulling diggers	Dragging	Spraying
Pulling Stumps	Rolling	Hauling crops	Leveling	Trucking

The Master Design of a Mechanical Genius

The Staude Mak-A-Tractor is the master stroke of a mechanical genius. For *years* the name Staude has stood for the best that mechanical skill could develop. Now E. G. Staude has turned his mechanical ability towards designing a practical working unit which utilizes the highly developed power plants of automobiles for tractor purposes. The result is an exceedingly simple, yet wonderful machine which is rapidly revolutionizing the application of power for farm work.

All over the country in every line of work on the farm, Staude Mak-A-Tractor has demonstrated in a full season's operation that it is the most efficient—lowest cost—power in the world. *Everywhere farmers* have used it and found that it would *easily* do the work of four horses and *and it costs no more than one.* They have driven it out of their fields and in *twenty minutes* changed their Fords back to touring cars and driven away to town. So our own factory tests have been proven by *actual service tests — results,* not claims, are the real protection and guarantee of satisfaction for owners.

Install It in Twenty Minutes
(Patent Applied for)

Simply remove the rear wheels and rear fenders on Ford—clamp the Staude Mak-a-Tractor on to the car with four bolts and you get the tractor ready for use.

Leave the Mak-a-Tractor channel irons on if you want to when you change back to the touring car—they will not interfere with riding or appearance—or you can remove them in five minutes.

No holes to bore, nothing to change. All you need for tools is a wrench, a jack and a wheel puller. We furnish the puller—the work takes 20 minutes.

All the Power—All the Time
(Patent Applied for)

The Staude roller pinions, integral with the brake drum, replace the rear wheels of the car. The pinions mesh into the semi-steel gears in the tractor wheels. Power is applied near the outside rim of the tractor wheel—there is no torsional strain on hub or spokes. And the Staude Mak-a-Tractor Axle is *back of the car* axle—an exclusive construction. The driving pinions *push* the tractor wheels *down*—no power is wasted—you get all the power all the time.

So wonderfully is the power applied that the Ford used with the Staude Mak-a-Tractor not only runs in high gear in the hardest work, but it actually starts on high.

The Henry Ford Tractor.

The latest report concerning the much touted Henry Ford tractor is that the dove candidate for President has purchased 80 acres of land between Newark and New York City and will build four million-dollar-unit plants for the manufacture of his new tractor. But the question arises—granting the truth of the report— Why the seaboard location? Has he given up the idea of competing with the 200 or more tractors at home, and is he intending to go after the foreign trade as soon as he succeeds in inducing the soldiers to quit the trenches and go home to their Christmas dinners?

The Newton Wagon.

Happy Farmer Tractor $495

Blue Ribbon Auto Trailer

—— *Manufactured By* ——
Durant-Dort Carriage Co., Flint, Mich.

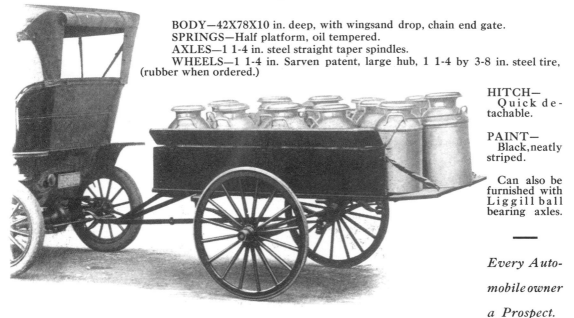

BODY—42X78X10 in. deep, with wingsand drop, chain end gate.
SPRINGS—Half platform, oil tempered.
AXLES—1 1-4 in. steel straight taper spindles.
WHEELS—1 1-4 in. Sarven patent, large hub, 1 1-4 by 3-8 in. steel tire,
(rubber when ordered.)

HITCH—
Quick detachable.

PAINT—
Black, neatly striped.

Can also be furnished with Liggill ball bearing axles.

———

Every Automobile owner a Prospect.

In 1916, there were more than 175 companies making tractors, but only about 50 could supply tractors to their dealers in any quantity, the remainder were really in the experimental stage of design. The Rock Island Plow Co. purchased the tractor operations from Heider Manufacturing Co. at Carroll, Iowa, on January 1. The Rock Island Plow Co. had been selling Heider tractors since 1914. The Cleveland Motor Plow Co. was begun in Cleveland, Ohio, by Roland H. White and others. Mr. White was the head of the White Motor Co. The name was soon changed to the Cleveland Tractor Co.

HOW THE 1916 MODELS CLASSIFY AS TO CYLINDERS

Cars Made Only with 4 Cylinders	Cars Made Only with 6 Cylinders	Cars Made Only with 8 or 12 Cylinders	Cars Made with Various Numbers of Cylinders
Allen	Buick	Cadillac	Abbott
Arbenz	Cameron	Cole	Apperson
Argo	Chalmers	Daniels	Auburn
Bell	Chandler	Enger	Briscoe
Biddle	Davis	Hollier	Empire
Brewster	Dorris	King	Fiat
Case	Franklin	Packard	Herff-Brooks
Chevrolet	Glide	Peerless	Jackson
Crow	Grant	Ross	Jeffery
Detroiter	Halladay		Kissel
Dispatch	Haynes		Lenox
Dodge	Hudson		Lexington
Dort	Kline		Lozier
Elkhart	Locomobile		Mitchell
Farmack	Luverne		Monitor
F. R. P.	Madison		National
Ford	Marion		Oakland
Harvard	Marmon		Oldsmobile
Hupmobile	McFarlan		Overland
Interstate	Moon		Partin-Palmer
Maxwell	Owen		Pathfinder
Mecca	Paige		Pilot
Mercer	Paterson		Pullman
Metz	Pierce-Arrow		Regal
Moline	Premier		Reo
Monroe	Republic		Saxon
Morse	Singer		Scripps-Booth
S. J. R.	Stewart		Simplex
Spaulding	Sun		Standard
Sterling	Velie		Stearns
Stutz	Westcott		Studebaker
Vixen	Winton		
White			
Willys-Knight			
Woods			

The Campmobile.

"The Call of the Field, the Stream, the Starlit Sky and Pleasure Unalloyed" is the title of an interesting pamphlet just published by the Cozy Camp & Auto Trailer Co., Indianapolis, Ind., which is descriptive of the company's campmobile. The campmobile is an automobile trailer which can be utilized for regular trailer work or can be used as a camp for outings, etc. It is particularly useful for long overland trips, though handy for an outing of an afternoon or night. When used for outing purposes it is fully equipped with beds, etc., and is as elaborate as the most up-to-date camp could be.

The name of the *Weekly Implement Trade Journal* was changed to *Implement & Tractor Trade Journal*. The first *Cooperative Tractor Catalog* was published as the April 8, 1916, issue of the *Implement & Tractor Trade Journal* to illustrate and describe the tractor models available to dealers. The listing of tractors was so popular that a separate "Red Book" was produced from the information contained in the April issue.

In 1916 it was reported that the average expense to run a farm was $1500 or about $7.50 per acre. This expense included depreciation on buildings and machinery and labor value of the farmer and hired men, but did not include interest on the land. Interest was said to be 5% and "on $150 land, $7.50 would be as much as all other expenses."

Poor Gasoline Causes Trouble.

Poor grades of gasoline are causing a great deal of grief to motorists this season. Gasoline of a quality that creates no end of trouble in cars two and three years old that have not the most modern appliances for heating the fuel is now being sold.

A misconception is current regarding so-called "low test" and "high test" gasoline. High test gasoline is not always the best, neither is any gasoline of very low test successful as fuel. High and low test pertain only to specific gravity. Low test is heavier and consequently doesn't lift so uniformly nor as lightly, nor break up so readily as high test. But it is quite possible for high test gasoline to be adulterated and of a poor grade.

Fuel trouble is, of course, more noticeable in cold weather than in hot. Gasoline vaporizes better at a great heat and the newer type carbureters are equipped with larger hot air intakes from the heated exhaust manifold. These great-

ly aid carburetion. But even with these there is still much dissatisfaction with poor grades of fuel.

A good test for gasoline is to place a few drops of it in the palm of the hand. If it evaporates cleanly and leaves the hand absolutely dry it will prove good fuel; if it leaves the hand oily or wet, refuse to accept it. Low test gasoline will evaporate readily, but it sometimes leaves oil deposits.

A FIELD FOR INVENTION

One of the most attractive opportunities awaiting the inventor just now is an all-purpose carburetor that will successfully handle kerosene and the heavier distillates. The alarmingly increasing cost of gasoline is causing a good many persons to think seriously of the future; for, while no doubt the price will fluctuate according to circumstances, it seems improbable that former low prices will ever again prevail.

Of course, it is true that even at present prices of fuel the tractor and the gas engine are effecting economies in the work done; and that there are certain motors that appear to be getting along very well on kerosene as a fuel—some, indeed, on heavy distillates and even on crude oil. Yet with these facts before us, the inventor who succeeds in developing an appliance that shall extend the satisfactory use of the lower grades of fuel as suggested so as to make them generally available to users of internal combustion engines, has a fortune awaiting him—if he is a good collector.

Ran on Water and Green Liquid.

Lewis Enricht, a German chemist of Farmingdale, L. I., it is reported, recently ran an automobile six miles with no fuel except two quarts of water and two ounces of a greenish liquid of his invention. There was a five-gallon gasoline tank back of the dashboard. Attached to the cap of the tank was an electrode connected with a two-cell dry battery. There was no other electrical equipment and the tank

was empty. Enricht declares that by his discovery he can produce automobile fuel for approximately one and a half cents a gallon. He says his invention utilizes the extraction of hydrogen from water.

Water Versus Gasoline.

Just how much, or how little, there may be in the stories emanating from New York concerning an alleged invention of Louis Enricht by which by the addition of a little powder —some reports say a "green liquid"—water of the common or hydrant variety may be transmogrified into motor fuel is problematical. Reports further have it that Henry Ford may corral the output for his "tinlizzies," also that it is to be backed by B. F. Yoakum, formerly of the Frisco railway company. Nevertheless, skeptical though they may be, every car or tractor or gas engine owner who is paying 20 cents or more for gasoline will stretch his credulity to the cracking point hoping it may prove true.

The Creeping Grip (Bullock) was among the early builders of track laying tractors. The Dunning Motor Implement Co. was renamed Denning Tractor Co. of Cedar Rapids, Iowa, on March 7, 1916. The Ford Tractor Co. of Minneapolis, Minnesota, was appointed a receiver to dispose of the company's assets. This company was not connected to the Ford Motor Co. or Henry Ford & Son of Detroit, Michigan, in any way. The Heer Engine Co. was reorganized as Reliable Tractor Co. at Portsmouth, Ohio, in 1916 and the Nilson Farm Machinery Co. moved to Waukesha, Minnesota and changed its name to

ANNOUNCING

THE DIXIE FLYER

The Season's Best Offering
At the Louisville Auto Show

THE Dixie Flyer represents the best thought of engineers prominently identified with the automobile industry since its inception. In its design and structure every good feature which the past has developed, every new process, every new metal that offered promise of greater efficiency has been considered.

Our designers have been striving for an ideal and when you see the car you will agree that they have attained it.

The Dixie Flyer is not an untried structure. It has been nearly two years in development. Every conceivable metal-trying test which the ingenuity of engineer and mechanician could invent and thousands of miles of road service have proven its **absolute dependability**.

DISTINCTIVE features of the Dixie Flyer are: Mechanical simplicity reduced to its lowest terms, in this respect rivaling the advantages of the "electric," a control so simple in its mechanism and operation that a child can readily understand it. Distinctly a car which your wife will find pleasure in driving.

Convenient and roomy. Every detail of the carefully appointed interior harmonizing with the grace of line and beauty of the exterior finish imparts that satisfying sensation which only quality and quiet elegance can impart.

Power a plenty—and some to spare. Smooth, continuous, positive power, instantly responsive to your merest touch. Riding qualities that are not surpassed in any. "Distinctly the season's best offering."

A Car That Will Thrill You with the Pride of Ownership—Anywhere—in Any Company

Dixie Motor Car Co., Louisville, Ky.
(INCORPORATED)

Completely Equipped $775 F. O. B., Louisville, Ky.
DYNETO ELECTRIC SYSTEM OF STARTING AND LIGHTING
A Money Making Car—Some Open Territory Left for Aggressive Dealers

If you already have **satisfactory** selling arrangements on a car of competitive price—do not write.

If not—wire or write us and we will send you the details of a profitable connection.

Four-cylinder, high-speed motor; Dyneto electric starter; 112-inch wheelbase; demountable rims; 32x3½-inch tires. One-man top, with quick detachable side curtains and hood and complete equipment.

Sell This Tried and Proved Tractor

Don't run chances or invite trouble by experimenting—take hold of the tractor that's an out-and-out **proved success**. The "INGECO" has demonstrated its reliability by consistent performance in actual farm work—under all sorts of conditions. It makes good —and gives your customers the kind of service they are looking for. Built in a big modern factory by a responsible company specializing in the manufacture of power equipment.

The "Ingeco" Improved Tractor

A simple, strong, practical one-man machine built for both traction and belt work. 2-cylinder, horizontally opposed engine, magneto equipped, with automobile type radiation. Reliable transmission. Handles from two to four plows according to speed and soil conditions.

Write at once for our exclusive proposition — it means big, profitable business for you.

International Gas Engine Co.

108 Holthoff Place,
Cudahy, Wis.

(Suburb of Milwaukee)
Central West Branch:
1007 Farnam St., Omaha, Neb.

Replaces 8-10 horses

No side draft

No danger of skidding

20-10 H. P.

The Allis-Chalmers 10-18 H.P. Farm Tractor

Built by the Largest Engine Works in the United States, Gives Dealers a Commanding Advantage.

Here is a tractor that is built right and backed right—a machine that represents the finished work of the best designers and largest engine builders in America. It has behind it the reliability that inspires confidence in your trade and insures you against disappointments. The Allis-Chalmers 10-18-H. P. Farm Tractor meets all the power needs of the average sized farm, both for tractor and belt work. Efficiency methods in its manufacture make possible its sale at an extremely attractive price. The service offered by our sales organization is far-reaching and contributes to the complete satisfaction of both dealer and purchaser.

See Our Exhibit at the Convention

The Allis-Chalmers 10-18-H. P. Farm Tractor merits the careful consideration of every implement dealer who wishes to enter the profitable field of tractor selling under the most favorable conditions. See this great tractor at Kansas City during the Implement Dealers' Convention. Satisfy yourself of its practical mechanical advantages. Find out how cooperation with the big Allis-Chalmers plant in its sale will benefit your business.

Exclusive agencies for this great Farm Tractor are now being closed up. Write for details. There may be an opening in your territory.

Allis-Chalmers Manufacturing Co.
Milwaukee, Wisconsin

SPECIAL FEATURES.

DRIVE WHEELS—Two 56-inch diameter, 12-inch face.
DIFFERENTIAL GEAR—Enables you to turn easily in any direction.
TRANSMISSION—Enclosed gears running in oil insure long life.
CLEARANCE—25 inches.
MOTOR—2-Cylinder, long stroke; opposed type; 5¼-inch bore; 7-inch stroke; powerful, well-balanced; economical.
FRAME—Cast steel, rigid and strong.
GUIDING DEVICE—Automatic.

The LAUSON TRACTOR

A REAL TRACTOR IN TWO SIZES 15-25 AND 20-35

"Built Up to a Standard Not Down to a Price"

In putting on the market a line of Tractors we have endeavored to build a Tractor that would give the farmer real service, Back Up the Lauson Reputation for Quality Machines and be a Profitable Selling Proposition for the Dealer.

A glance over the following specifications will convince you that the Lauson Tractor is right.

ALL ENCLOSED and EASILY ACCESSIBLE. **Medium Weight with Maximum Power.** Selective Type Sliding Gear Transmission with Hyatt Roller Bearings. TWO SPEEDS forward and reverse, giving a wide range of service on the large or small farm.

SPECIFICATIONS OF THE LAUSON TRACTOR

15-25	20-35
Rated Tractive H. P., 15. Rated H. P. on Belt, 25.	Rated Tractive H. P., 20. Rated Belt H. P., 35.
Motor, Four Cylinder, Valve in Head.	Motor, Four Cylinder, Valve in Head.
Transmission, Sliding Gear. Bearing, Hyatt Roller.	Transmission, Sliding Gear. Bearing, Hyatt Roller.
Speeds, Miles per Hour, 2½ High, 1¾ Low.	Speeds, Miles per Hour, 2½ High, 1¾ Low.
Pulley, 550 R.P.M.; Diameter, 14 inches; Face, 8 inches.	Pulley, 440 R.P.M.; Diameter, 20 inches; Face, 8 inches.
Front Wheel Diameter, 32 inches; Face, 6 inches.	Front Wheel Diameter, 40 inches; Face, 8 inches.
Drive Wheel Diameter, 54 inches; Face, 12 inches.	Drive Wheel Diameter, 66 inches; Face, 16 inches.
Wheel Base, 80 inches; Length Over All, 133 inches.	Wheel Base, 108 inches; Length Over All, 161 inches.
Width, 74 inches. Height, 7 feet 7 inches.	Width, 84 inches. Height, 8 feet 7 inches.
Approximate Road Weight, 5,500 pounds Tanks Filled.	Approximate Road Weight, 7,500 pounds Tanks Filled.
Approximate Shipping Weight, 5,300 pounds.	Approximate Shipping Weight, 7,300 pounds.
Capacity Gasoline Tank, 20 gallons.	Capacity Gasoline Tank, 20 gallons.

Owing to the limited Tractor Capacity this year, we urge all dealers who want Lauson Tractors to place their orders early.

Write for Tractor Bulletin and Special Dealer's Sales Proposition

Frost King Portable, 3½ to 28 H. P.

52 Monroe St.

If you do not have our Engine Catalog No. 18 ask for it. We have a complete line—1½ H. P. Special Pumping engine, 3½ H. P. to 28 H. P. Portable, 2½ H. P. to 50 H. P. Horizontal Stationary, 35 to 100 H. P. four cylinder verticals—an engine for every purpose.

The John Lauson Manufacturing Co.

1½ H. P. Frost King Junior.

New Holstein Wis.

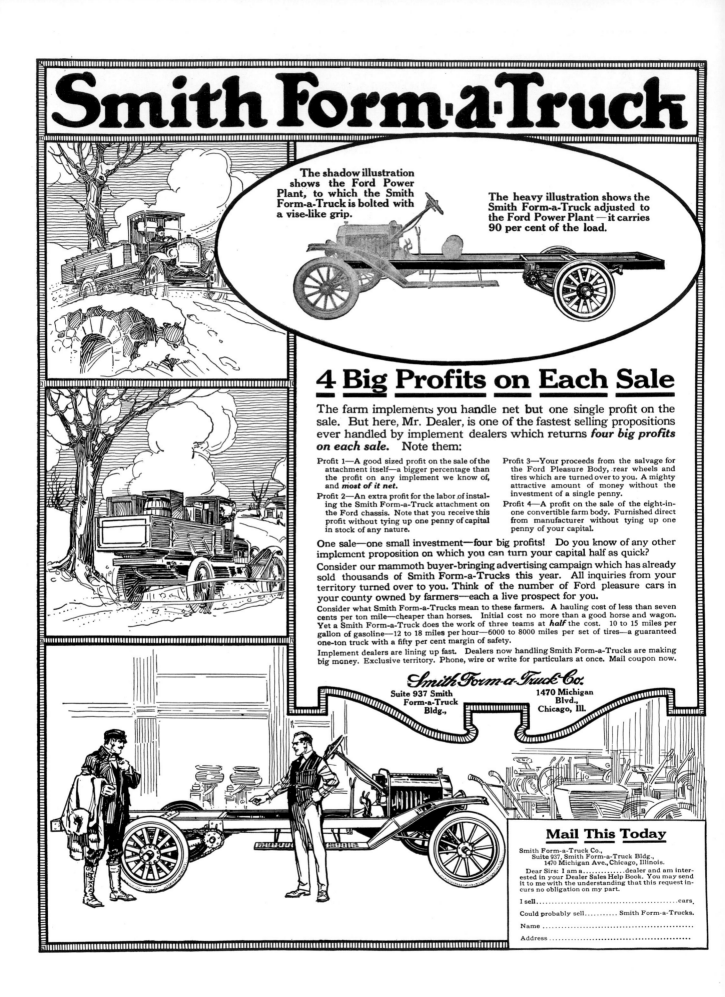

Smith Form-a-Truck

The shadow illustration shows the Ford Power Plant, to which the Smith Form-a-Truck is bolted with a vise-like grip.

The heavy illustration shows the Smith Form-a-Truck adjusted to the Ford Power Plant — it carries 90 per cent of the load.

4 Big Profits on Each Sale

The farm implements you handle net but one single profit on the sale. But here, Mr. Dealer, is one of the fastest selling propositions ever handled by implement dealers which returns *four big profits on each sale.* Note them:

Profit 1—A good sized profit on the sale of the attachment itself—a bigger percentage than the profit on any implement we know of, and *most of it net.*

Profit 2—An extra profit for the labor of instaling the Smith Form-a-Truck attachment on the Ford chassis. Note that you receive this profit without tying up one penny of capital in stock of any nature.

Profit 3—Your proceeds from the salvage for the Ford Pleasure Body, rear wheels and tires which are turned over to you. A mighty attractive amount of money without the investment of a single penny.

Profit 4—A profit on the sale of the eight-in-one convertible farm body. Furnished direct from manufacturer without tying up one penny of your capital.

One sale—one small investment—four big profits! Do you know of any other implement proposition on which you can turn your capital half as quick?

Consider our mammoth buyer-bringing advertising campaign which has already sold thousands of Smith Form-a-Trucks this year. All inquiries from your territory turned over to you. Think of the number of Ford pleasure cars in your county owned by farmers—each a live prospect for you.

Consider what Smith Form-a-Trucks mean to these farmers. A hauling cost of less than seven cents per ton mile—cheaper than horses. Initial cost no more than a good horse and wagon. Yet a Smith Form-a-Truck does the work of three teams at *half* the cost. 10 to 15 miles per gallon of gasoline—12 to 18 miles per hour—6000 to 8000 miles per set of tires—a guaranteed one-ton truck with a fifty per cent margin of safety.

Implement dealers are lining up fast. Dealers now handling Smith Form-a-Trucks are making big money. Exclusive territory. Phone, wire or write for particulars at once. Mail coupon now.

Smith Form-a-Truck Co.

Suite 937 Smith Form-a-Truck Bldg.,

1470 Michigan Blvd., Chicago, Ill.

Mail This Today

Smith Form-a-Truck Co.,
Suite 937, Smith Form-a-Truck Bldg.,
1470 Michigan Ave., Chicago, Illinois.

Dear Sirs: I am a..............dealer and am interested in your Dealer Sales Help Book. You may send it to me with the understanding that this request incurs no obligation on my part.

I sell...cars.

Could probably sell.......... Smith Form-a-Trucks.

Name ....

Address ....

The Tractor a War Necessity.

England led the way for the adoption of the farm tractor as a war machine. Threatened by isolation from the food-producing countries of the western hemisphere by the submarine, she took heroic measures to protect herself from famine.

The minister of agriculture organized an army of farm tractors, placed headlights on them, and plowed day and night for weeks and weeks in the spring of the year. When the harvest was gathered, and England found she had enough food to withstand any blockade of submarines for another year, credit was given the tractor as the most effective weapon of warfare yet discovered.

France had the same experience. The government subsized the manufacture of tractors and organized schools for their operation.

Woman Repairs Farm Machinery.

Burlington, Ia., is the first city in the United States to boast of a woman blacksmith. This unique distinction falls to the lot of Miss Turka Hawke, and her specialty is mending farm machinery. Miss Hawke recently received a diploma as a graduate blacksmith from the Iowa State College. She is opening a shop in Burlington expressly for farm machinery, in which she specialized in college.

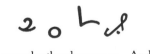

"The Name Tells a True Story"

SUPERIOR

Superior 10x8 Plain Single Disc Drill with Wood Wheels

Superior 11x7 Fertilizer Double Disc Drill with Wood Wheels

Superior 10x8 Fertilizer Single Disc Drill with Wood Wheels

Superior 12x7 Plain Single Disc Drill with Gang Press Wheel Attachment

Superior Low Down Disc Press Drill with Tongue Truck

Superior 18 x 7 Single Disc Plain Drill with Steel Wheels and Seat

Superior Drills stand up to the work, year in and year out, satisfying particular farmers everywhere on earth.

Catalogues and Prices are yours for the asking.

Every Style   Every Size

Superior 12 x 7 Single Disc Plain Drill with Steel Wheels and Seat

Superior Double Run Force Feed

Superior Open Delivery Single Disc with Adjustable Wing Shield

Superior 9x8 Fertilizer Spring Hoe Drill

Superior 9x7 Plain Pin Hoe Drill

Superior Plain Five-Disc Drill

Superior Fertilizer Five-Hoe Drill

Superior "Front Delivery" Double Disc

The American Seeding-Machine Co. Inc.
SPRINGFIELD, OHIO,
PARLIN & ORENDORFF PLOW CO., General Agents. *Write nearest branch.*

SUPERIOR DEALERS—Tear this Page out and Tack it up.

THUS defines the Mexico (Mo.) Ledger: "A pacifist is a person who believes in fighting his friends instead of his enemies."

CAN you hate the perpetrators of the German atrocities without rancor? We, too, acknowledge the same difficulty.

WE care not how broken his English may be —an American's an American wherever he may have been born.

War Boosts Motor Body Trade.

Considerable increase in the demand for special bodies for Fords and other small cars has been noted since the beginning of the war. With wagons and trucks becoming harder to get, there has been a gradual trend in favor of the special body. Thousands of old Fords are scattered throughout the country, which have outlived their usefulness as pleasure cars, and whose motors are still in working order, for which the special body affords the means of equipping a serviceable small truck with but little expense.

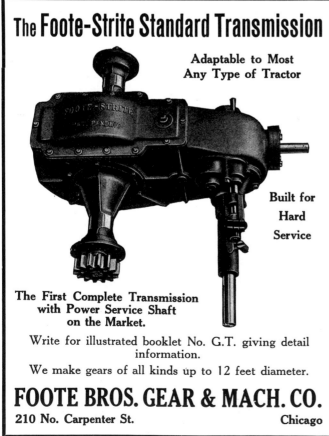

WITH characteristic efficiency Germany has proven itself the biggest boob country in the world, in addition to a number of far more disgraceful kinds of superiority.

ANOTHER patriot suggests that the kaiser be taken to Los Angeles to act as a target for the pie-spattering movie commedians.   But wouldn't that be a criminal waste of pastry?

ANOTHER patriot suggests that the kaiser be taken to Los Angeles to act as a target for the pie-spattering movie commedians.   But wouldn't that be a criminal waste of pastry?

Standardizing Engine Sizes.

In this day of standardization it is high time that the gas engine should be brought under reasonable rules of standardization.   The current National Gas Engine Association Bulletin calls attention to existing conditions in the following cogent editorial:

It is confusing to the purchaser—these fifty-seven varieties of engine sizes.   Did you ever read the list over? $1\frac{1}{2}$, $\frac{3}{4}$, $1\frac{1}{4}$, $1\frac{3}{4}$, 2, $2\frac{1}{2}$, $2\frac{3}{4}$, 3, $3\frac{1}{2}$, 4, 5, $5\frac{1}{2}$, 6, $6\frac{1}{2}$, 7, 8, 9, 10, 12, 14, 15, 16 and 18.

Every time a manufacturer brings out a new engine, it looks as though he thought it would be an excellent idea to make it an odd size.   Probably thought it would attract attention.   Why they didn't include 11 and 13 horse power we don't know.   They should be to make the list complete and odd.

Each size requires a different set of patterns; special jigs; machinery and tools.   It requires extra repair parts.   In other words some manufacturers have so much money tied up in patterns, etc., of special sizes of engines that they cannot make money.

We are making an excellent start on standards.   Let us standardize on sizes.   Five sizes should be enough to meet every requirement.   Dealers can carry a larger stock and you can make better deliveries.   And best of all you will make money.

There will be some "kicks" at first, but it is largely a matter of education.   There will also be manufacturers who will "kick" but they will soon fall in line.   You are building engines to make money as well as to please your customers.   This takes care of both.

Self-steering devices make it possible for tractors to accomplish "almost human" performances in the plowing field. At a recent Kansas demonstration several machines attracted much attention by dispensing with their operators for several minutes at a time. A man would start the tractor down the furrow, then dismount and let it go on alone. At the other end another man would climb into the seat, turn the machine around on the headland and send it back solus. Thus the two operators acted in about the same capacities as the pitcher and catcher in a baseball battery. It is related that a Nebraska tractor owner takes his shotgun with him when he plows. Occasionally he sees signs of quail or rabbits. At such times he allows the tractor and gang of plows to go on their way unattended while he strolls along the hedge with dog and gun stalking for birds and bunnies.

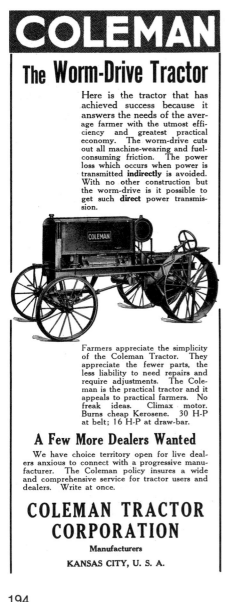
The German submarine, that "sailed into Newport, Rhode Island and afterwards torpedoed several European merchant vessels," interrupted American commerce especially in the steel industry.

The United States severed relations with Germany in 1917 and signed the declarations of war on April 6. Wheat prices in the dust bowl area jumped to $2.06/bushel from an earlier price of $.91 and prices remained above $2.00 until 1920. Some farm equipment was advertised during the war as just "what's needed to win the war."

More track laying tractors entered the market, Gehl began to build tractors and Ford started to manufacture the first mass-produced Fordson tractors. More than 7000 Fordsons were sold to the British in 1917, before introducing them to the American farmer the following year. The "Flying Dutchman" trade mark of Moline Plow Co. was removed from advertising because of possible pro-German connotations during WWI.

Deere and Company purchased the Waterloo Boy Tractor Co. of Waterloo, Iowa, on March 18, 1918. Ford introduced Fordson tractors to the U.S. market

When Distress Calls
the Red Cross Answers "HERE!"

NOW the Red Cross calls! The annual Christmas Roll Call of members will echo throughout the land the week of December 16th to 23rd.

Membership in the Red Cross now is more than duty—it is an honored privilege, and an evidence of loyalty. When that Roll is called, your conscience, your sense of right and justice, your love of country and your devotion to the highest ideals of unselfish service all suggest that you answer "HERE!"

All you need is a heart and a dollar

These entitle you to membership for one year.

When you wear your button, signifying that you are a member, you will not be asked to join again this year—it means that you have answered the Roll Call.

Join—be a Christmas member—but just join once.

Our soldiers and sailors look to the Red Cross for comforts. They have never been disappointed.

The Red Cross looks to you for the moral support of your membership. Answer "HERE!" when the Roll is called.

Join the Red Cross

Contributed through Division of Advertising

United States Gov't Comm. on Public Information

This space contributed for the Winning of the War by
IMPLEMENT & TRACTOR TRADE JOURNAL

Wear Your Button   Fly Your Flag

and was an instant success, toppling International Harvester from number one position. Within two years, Fordson tractors dominated, capturing two thirds of the American market. It is interesting to note that the International Harvester Corporation introduced power takeoff (pto) this year, but couldn't match the low Ford price. The International Harvester Co. agreed to divest themselves of the Osborne, Champion and Milwaukee lines as a settlement for an antitrust suit. Emerson-Brantingham Co. purchased the Osborne line.

The General Motors Corporation purchased a plant in Janesville, Wisconsin, to produce the Sampson tractor. The tractor works was incorporated as Janesville Machine Works. The success of the Fordson was noticed by everyone. especially in the midst of the financial difficulty experienced by some. The Nilson Tractor Co. office in Minneapolis, Minnesota, moved into the building formerly occupied by Bull Tractor Co. The Nilson Tractor Co. entered receivership in late 1918 and the assets were sold in 1919. The Turner Manufacturing Co. also entered receivership in 1918. In August, the Little Giant Co. began in Mankato, Minnesota. This was really a name change from Mayer Bros. Co. and Little Giant continued to build tractors until about 1927.

On November 11, 1918, an armistice ending war with Germany was signed at 5:00 A.M. in Foch's railway car in the forest of Copiegne.

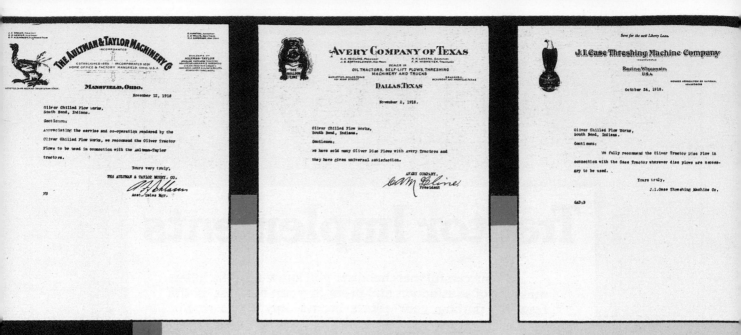

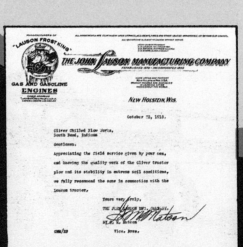
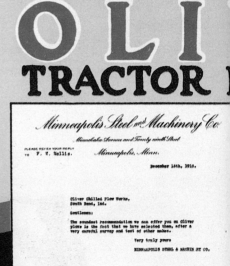

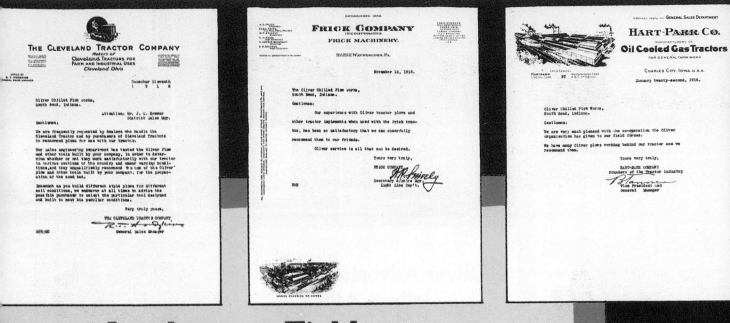

...ractor Implement Field

...nts of These Leading Tractor Makers

What does all this mean to you as dealer?

Simply this—in the Oliver name on the tractor implements you sell you have a business-building asset, you cannot afford to overlook. It means that these implements measure up 100% in material quality, utility, and prestige.

You want to tie up with a winner—a line that is worthy of your ambitions and your greatest energy, with an organization that offers you unlimited cooperation in the selling of its line.

You have it in Oliver—not only the prestige of a name with a long and honorable history, but a name that in this newer age of

power farming signifies all the enthusiasm, vision and energy of youth.

Best of all, the Oliver organization can make good in action this same enthusiasm, vision and energy. This organization includes 18 branches, 53 transfer stations, 350 field men and a nation-wide dealer organization, all imbued with the Oliver spirit and all sharing Oliver success.

We invite you to write us in detail regarding your plans. We will be pleased to answer any questions you may wish to ask, to give you full information regarding the Oliver line, Oliver advertising campaigns, Oliver selling helps, or the Oliver organization.

for the World" South Bend, Indiana

...YER

...PLEMENTS

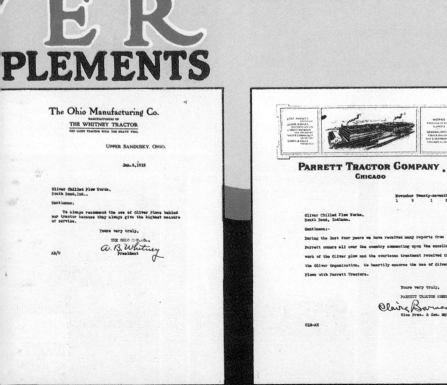

1919-1923

PRODUCTION INCREASED
FASTER THAN DEMAND

PEACE AND THE IMPLEMENT INDUSTRY

Manufacturers of Farm Operating Equipment Believe that the Re-
adjustments Will Be Gradual—Entire World Must Now Be Fed—
Reconstruction Period Will Create a Great Demand for Machinery.

SAWYER-MASSEY THRESHER

100 PER CENT
SEPARATION

FOR TRACTOR
POWER

Number 1
22 × 36

Number 2
28 × 44

MANUFACTURED BY

SAWYER MASSEY COMPANY, LTD.

Manufacturers continued to produce equipment at the fast pace necessary to meet the country's earlier needs, necessary during the war, only to find the challenges of overproduction. There had never been a problem of producing too much, only the problems of enlarging plants and starting new facilities to increase production. Neither the American farmers nor the equipment manufacturers thought that production could ever catch up with demand. Men, who had returned from the war with money, paid high prices for land and the newest equipment. Some even borrowed heavily against overvalued land and artificially high crop prices. Prices increased during 1919, but demand slacked after the shortages caused by the war were filled. The resulting surplus led to lower prices and later a postwar recession. Reductions in exports caused prices of farm produce to drop even more. The downward spiral led to financial disaster for many. From 1920 to 1921 farm commodity prices fell 53 percent and between 1929 and 1932 prices fell 56 percent more.

The 18th Amendent to the U.S. Consitution (Prohibition) was ratified in 1919 and on July 15, the Tractor Test Bill was passed into law by the Nebraska legislature. Testing began in the fall on a Twin City 12-20 tractor, but was not completed until March 21, 1920, because of the weather. A Waterloo Boy Model "N" tractor was given the distinction of being listed as Test No. 1, the Twin City was listed as Test No. 19. Early tests on the Waterloo Boy and Case tractors had to be revised because of objections by the Nebraska State Railway Commission, making tests numbers 8, 9, 10 and 11 (for Oil Pull 30-60, 16-30, 12-20 and 20-40 models) the first tests officially released. In 1919, most combines cut a swath 12 feet wide and track laying (crawler) tractors were the rage. Many iron, steel and manufacturing plants were crippled by strikes and an iceless icebox called a "Frigidaire" was demonstrated by the General Motors Corp.

The Bates Machine & Tractor Co. was formed by the merger of the Joliet Oil Tractor Co. of Joliet, Illinois, and the Bates Tractor Co. of Lansing, Michigan.

On May 20, 1919, the Denning Tractor Co. was sold to the General Ordance Co. of New York, New York. The tractor was renamed "National" then "G O". The General Motors Corp. purchased rights to the "Jim Dandy" motor cultivator and began production of it as the "Iron Horse" model of the Sampson tractor. Walter P. Chrysler was promoted from president and general manager of the Buick Motor Co. to vice-president in charge of all General Motors Corp. operations and assistant to the president.

There is Nothing Experimental About the G O Tractor

MOST dealers prefer to handle a tractor which has proven its usefulness in actual experience, rather than one which has only theory to recommend it.

For eight years the G O Tractor has been meeting the requirements of farmers all over the country. It has long passed beyond the experimental stage.

Hard, merciless work on all kinds of land, in all kinds of weather, has demonstrated that the friction drive of the G O Tractor is best adapted to the varying conditions met with in all-round farming.

It permits of six speeds, forward and reverse—enough to give the right plowing speed in any sort of soil, the right speed at the belt pulley for any machine it may be called upon to run.

Its gears and bearings run in oil, enclosed in dust-proof casings.

Where many tractors mire themselves, the G O Tractor pulls itself out by applying all its power to the free wheel. For this same reason, it turns more readily than other tractors.

These are facts which any farmer will understand. They convince him quicker than unproved generalities.

To the man who is capable of handling it, the G O Tractor agency is a valuable business proposition.

The General Ordnance Company—*Tractor Division*

All inquiries pertaining to ordnance and other products, except TRACTORS, should be addressed to Dept. 33.

Western Sales Office and Factory, Cedar Rapids, Ia.

Executive and Eastern Sales Office, Two West 43rd Street, New York.

Eastern Factory: Derby, Connecticut.

Designed by L. A. LaFOND, one of the Greatest Tractor Engineering Experts in America.

The Farmer's "War Tank" Will Help Win the War!

Just as the manufacturer must speed up the output of his plant, so must the farmer speed up the output of his farm. Scarcity of good farm labor makes it necessary that he substitute motor-power for man-power.

In the Pan Tank-Tread Tractor we are confident that we have perfected the greatest farm motive power machine in the world. It is the Tractor that will play one of the most important parts in the winning of this great war—the farmer's "War Tank"! Like the Pan Car, the Pan Tractor represents the work of some of the greatest brains in the engineering world. There is no tractor just like it.

Manufactured by the
PAN MOTOR COMPANY
Saint Cloud, Minnesota, U. S. A.
Pan-Town-On-The-Mississippi

Motor, Buda; farm tractor type, cylinders en bloc.
Stroke, 5½ inches.
Bore, 4¼ inches.
Carburetor, Kingston.
Ignition, high tension k. w. magneto.
Governor, centrifugal ball type.
Tractor, H. P., Twelve on Drawbar.
Belt, H. P., 24.
Speeds, Three Forward and One Reverse.
Weight, approximately 3,500 pounds.
Traction Surface, 888 square inches.
Diameter Traction Drive Wheels, 37 inches and 12 inches.
Width of drive wheels, 12 inches.
Turning Radius, 12 feet.
Type of Clutch, Borg & Beck dry plate disc.
Bearings, All Hyatt.
Cooling, Honey Comb type Radiator, 20-inch fan, centrifugal water pump, radiation 15,000 square inches.
Steering Gear, clutch and brake levers telescope.
Fuel capacity, kerosene, 14 gallons, gasoline 1½ gallons.
Diameter of belt pulley, 14 inches.
Width of belt pulley, 8 inches.
Speed of belt pulley, 900.
Draw bar, Adjustable.
Total over all length, 90 inches.
Total over all, width, 58 inches.
Total over all, height, 66 inches.
Minimum ground clearance, 18 inches.

TANK-TREAD PAN TRACTOR

Automatic Safety Switch

THE Apco Mfg. Co., Providence, R. I., well known maunfacturers of Ford attachments, has placed on the market an attachment for the Fordson tractor which is designed to do away with the trouble caused by the tendency of the tractor to tip over backward under certain conditions.

The attachment is known as the Apco automatic safety switch, and it is designed absolutely to prevent any accident of this kind by shutting off the ignition current when the front wheels rise to a predetermined height. It is constructed on the pendulum principle, which it is declared will not fail to operate and at the same time it will not swing and break the circuit when the tractor is going over uneven ground.

The switch is located near the magneto plug at the side of the engine and held by two of the crank case bolts.

The installation takes about five minutes. The material is malleable iron, brass and the best bone fiber for insulating, and the finish is Fordson gray, so that when the switch is installed it looks like the regular equipment. The weight is five pounds.

THE SWITCH WITH CONNECTIONS

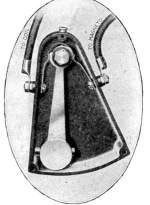

WITH COVER REMOVED, SHOWING PENDULUM

In 1919, it was decided to manufacture wagons with wheel tread width of 56 inches which was the standard for both the automobile and motor truck industries. It was said "it is no use to fight the inevitable. It is easier to follow the 'beaten track' than to cut a new one."

The International Harvester Co. purchased both Parlin & Orendorff of Canton, Illinois, and the Chattanooga Plow Co. of Chattanooga, Tennessee, effective July 1. The J.I. Case Threshing Machine Co. purchased the Grand Detour Plow Co. of Dixon, Illinois, also effective on July 1. The Grand Detour Plow Co. was originally begun by John Deere. In 1919, the Wallis Tractor and J.I. Case Plow Works merged using the J.I. Plow Works name. H.M. Wallis was a relative of Jerome Increase Case, but these two companies were completely different than the J.I. Case Threshing Machine Co.

The Hession Tiller & Tractor Corporation of Buffalo, New York, changed the name of their farm and road tractor to "Wheat" and adopted the slogan "World's Standard." The Moline Plow Co. changed their name to the New Moline Plow Co., then purchased the Independence Harvester Co. of Plano, Illinois. The Monarch Tractor Co. was reorganized as Monarch Tractor Inc. Late in 1919, the Common Sense Tractor Co. changed the name to Farm Power Sales Co., then finally stopped operation in 1920.

The Schofield Auto Tractor of Kansas City, Missouri, was but one of many companies offering an attachment to convert the Model "T" Ford to field work in 1919. The International Harvester Titan 10-20 sold for $1000, the Fordson sold for $885, the Sampson Model M sold for $650, the four wheel drive Nelson 15-24 sold for $1765 and the price of the LaCrosse 12-24 tractor was cut from $1250 at the start of 1919 to $895 in November. The Nelson Corporation of South Boston, Massachusetts, entered receivership later in the year.

Have you cornered your share
of the Wheat Tractor market?

The Wheat Tractor dealer in your territory is going to make money—

1. Because the **Wheat** is the standard tractor for the farm—and is practically 100% in performance.

2. Because the road wheel feature is exclusive, and is a tremendous selling feature.

3. Because the price is right.

4. Because our contract is such that you can make money.

If you haven't already acted—if you haven't written, or wired for information about the Wheat Tractor in your territory—do it **now**—today—before someone else beats you to it.

Rubber tired wheels easily substituted for the heavy cleated wheels, make a truck that will pull a string of loaded farm wagons at a speed of 10 miles an hour.

You must act now — if you want the territory for this real tractor. Write, wire or telephone us today

HESSION TILLER & TRACTOR CORPORATION, BUFFALO, N. Y.

The Twin City 12-20 tractor featured four valves for each cylinder, overhead valves and twin cams. The Midwest Engine Co. of Indianapolis, Indiana, manufactured a heavy duty four cylinder truck and tractor engine with a bore of 4.5 inches and stroke of 6 inches. Full pressure lubrication through the hollow crankshaft was considered innovative, but other features included roller cam followers and rocker arms as well as a vacuum controlled oil pressure relief valve. Oil pressure was automatically controlled by the manifold vacuum, thus increasing during high load conditions. The International Harvester Co. demonstrated the company's harvester-thresher, which was operated by a large truck motor. The Aspinwall Mfg. Co. of Jackson, Michigan, introduced a strawberry digger that was very similar to the company's potato digger. The machine was used to dig plants for transplanting. Kicker fingers similar to those used for potatoes reduced damage to the plant's root system.

This New 12-20 Kerosene Tractor

equipped with 4 cylinder 16 valve in the head engine

In design and equipment it represents the most modern engineering skill, and is the first tractor on the market with this new-type engine, which gives complete scavenging of combustion chambers with freedom from over-heating, pre-ignition and other troubles.

This new type engine also means lower fuel cost, and greater power, for though rated at "12-20," it delivers 35 H. P. on kerosene and 40 H. P. on gasoline.

In construction, it is based upon years of broad experience, in the tractor industry and actual knowledge of the work a farmer requires of his tractor.

It is manufactured upon the same quality basis as the larger members of the famous TWIN CITY family, and it has proven its performance ability in the farmers' fields.

See It at the Show

Get detailed specifications, and examine the many special features of this new "12-20." This is a 'production' proposition for national distribution from the greatest tractor shops in the country. Dealer connections are being made now.

Specifications:

ENGINE: 12-20 H. P., 4-cylinder vertical, sixteen valve engine, valve in head type; cylinders en bloc; 4¼x6; removable cylinder sleeves.

REAR AXLE DRIVE: Drop forged front axle, automobile type.

TRANSMISSION: Selective spur gear type; two forward speeds, one reverse; direct drive on both forward speeds. All gears enclosed and run in oil.

CLUTCH: Borg & Beck.

BEARINGS: Hyatt high duty bearings. Ball Thrust Bearings.

Spirex Radiator. Oakes enclosed Ball Bearing Fan. 16-in. Belt Pulley, regular equipment.

Bosch High Tension Magneto. Air Cleaner. Holley Carburetor.

Minneapolis Steel & Machinery Company

Manufacturers of the famous Twin City Tractors. Builders: Culverts, Tanks, Water Towers, etc.

BRANCHES: Fargo, Des Moines, Great Falls, Wichita, Denver, Spokane, Salt Lake City, Winnipeg.

Between 1910 and 1920 the number of tractor manufacturers in the United States increased from 10 to 190. Manufacturers made fewer and fewer steam traction engines following WWI until 1925 when virtually all production had stopped. The best sales year for many years to come was 1920, for farm equipment as well as other products.

The 19th Amendment to the Constitution was ratified in 1920, assuring the country's women the right to vote. It was noted that following prohibition, "much revenue has been lost and to get it back, almost all the towns and cities have imposed or are preparing to impose extra taxes on business interests." Robert M. Green, originator of the ice cream soda, died in Philadelphia at the age of 78.

The International Harvester Co. purchased the American Seeding Machine Co. of Richmond, Indiana in 1920 and the Franklin Tractor Co. began making track laying equipment. About two years later, the Franklin Tractor Co. had gone out of business and all assets were sold.

Fordson
TRADE MARK
TRACTOR

Manufactured by HENRY FORD & SON, Inc.

100,000 IN 1919

THE FORDSON TRACTOR is the perfected result of a number of years spent in tests and trial under all sorts of conditions. 40,000 Fordson Tractors were manufactured last year and quickly sold to American farmers who had confidence in Mr. Ford and his organization. Today the demand for immediate needs necessitates production at the rate of 100,000 a year.

The simplicity of construction and ease of operation, coupled with fuel economy and low upkeep, make the Fordson the ideal tractor for use on the average farm.

See the Fordson Tractor at the Fourth Annual National Tractor Show, Kansas City, February 24 - March 1, Booths 209, 210, 211

G. T. O'Maley Tractor Company
DISTRIBUTORS   KANSAS CITY, MO.

When Heat Destroys the Lubricating Oils

By F. M. Buente
Advertising Department, Tide Water Oil Co.

S HE doesn't use the oil, she spoils it!" That was the exclamation of one of the engineers who was testing an old time, small gasoline and kerosene tractor. He was using a cheap grade of oil and his machine was literally destroying the oil in the crankcase.

Since that time tractor design has changed to an astonishing degree. From the lubrication point of view the change is just as great—tractor engineers and most operators know that their power plants cannot operate on low grade lubricating oil.

What happens to oil in the crankcase and on the cylinder walls of a tractor engine is a complex phenomenon of physics and chemistry. The fundamental facts, however, are simple. Oil is destroyed or worn out by heat, not by friction. The tractor engine generates much higher temperatures than any automobile engine, and therein lies the secret of the need of the highest grade oil for tractors at all times.

It is on the cylinder walls that the heat is greatest and for that reason it is there that oil is destroyed or breaks down most rapidly. Almost all the oil on the cylinder which is exposed to the heat of the explosions—2000° F to 3000° F—is of course burned completely away. But the oil on the walls below must not burn, even though the temperature, as with the tractor, is continuously at the 400° F point.

The Danger of Sediment.

If oil is not made so it will withstand heat, it breaks down on the cylinder walls, forming sediment; and a chain of troubles is started. First the tractor overheats; in fact, it may even seize and stop simply because the lubrication on the walls has not been maintained. The "piston seal," as the thin film is known, is broken. This film holds in the gases and helps to deliver the power. Oil that does not resist heat cannot maintain the piston seal.

When the piston seal is broken power escapes, the engine overheats and unburned gasoline or kerosene is permitted to seep through and contaminate all the oil in the cranckcase.

The sediment which is formed when the oil breaks down under heat, is finely divided carbonaceous material which has no lubricating power. It displaces good oil on the bearings, and other metal to metal surfaces, and does untold damage, causing large repair bills, delays and layups.

It has been shown that friction and wear is largely caused by sediment in lubricating oil, that sediment accounts for too great consumption of fuel and oil, rapid carbonization, scored cylinders, loose bearings, and in fact 90 percent of the engine troubles of the tractor.

Direct Cause Is Heat.

Heat is the direct cause of sediment, and only an oil that resists heat can prevent the formation of large quantities of sediment.

Motor-oil testing laboratories have brought out these vital points of tractor lubrication. In the laboratory, at the refinery of the Tide Water Oil Co. on New York Harbor, motor oils and gas engines have been studied exhaustively in an effort to produce an oil that would reduce the formation of sediment to a minimum.

The result of long investigation in this laboratory, road tests, chemical tests and actual use in tractors, is an oil that reduces by 86 percent the amount of sediment formed in a twelve hour continuous run.

The astonishing performance of Veedol, as this lubricating oil was named, is illustrated by the photographs of the two bottles. The left hand bottle contains ordinary oil after a test run in an engine. The right hand bottle contains Veedol after the same test in the same engine. The difference in the amount of sediment produced is plain.

Keep the Crankcase Clean.

Almost the same grade of oil as that used for tractors is now replacing castor oil in all but rotary airplane engines, rotary machines cannot use mineral oil.

The similarity between airplane and tractors is that both keep their engines at high speeds and develop great heats. This not only forms sediment but also increases evaporation in ordinary oils. With Veedol evaporation is reduced from 25 percent to 50 percent for the same reason that sediment is reduced 86 percent—because the oil resists heat.

In handling a tractor it is important to clean out the crankcase every night or morning and put in fresh oil. It is much better to put in additional oil at noon than to put in too much in the morning and hope to run until night without adding to the supply. If you do this, and use an oil that resists heat, you will reduce greatly your cost of upkeep on the tractor.

Killed by Gasoline Fumes.

Casper Wilson, a member of the experimental department of the Heider tractor factory of the Rock Island Plow Co., Rock Island, Ill., was found dead in his garage recently, having been overcome by gasoline fumes.

Mr. Wilson was known to a great many men of the implement trade. both dealers and manufacturers, as he spent a good deal of time visiting among dealers and has also made all the tractor demonstrations in the last four years in the capacity of expert. He is known to his friends as "Cap" or "Shorty." His great number of acquaintances will regret to hear of his untimely death.

"Cap" Wilson has been connected with the manufacture of Heider tractors for eight years. He was with the Heider Mfg. Co., at Carroll, Ia.. for four years before the Heider tractor was purchased by the Rock Island Plow Co., and moved to that city. He was thirty years old and unmarried.

SADLY do we recall the good old days when a man could be trusted alone in a restaurant with a sugar-bowl.

ARMY men, on the whole, haven't contributed much to the mass of information as to why our war program is not being carried out as efficiently as we now want it carried out. But Major General William M. Wright, commandant at Fort Doniphan, is an exception. He blames it on the recent indifference of the people to matters of national defense. And that's where the blame, primarily, belongs. That pill will make us make an awful face, but we won't be good sports until we swallow it.

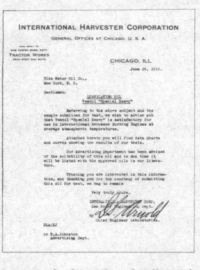

Ordinary oil
after use

Veedol after
use

Showing sediment formed after
500 miles of running

THE INTERNATIONAL HAR-
VESTER CORPORATION TRAC-
TOR 10-20 TITAN

Why tractors require special oil

Tractor manufacturers have learned almost universally that ordinary lubricating oil is certain to make trouble for tractor users.

Engineers of the International Harvester Company were among the first to make exhaustive tests of tractor lubrication and they soon realized that lubrication of farm tractors is more important even than that of trucks and automobiles.

Inferior oil is the cause of 90% of tractor engine trouble. Excessive dilution of the oil supply by fuel, loose bearings, overheating, excessive carbon deposits, knocking—all are directly traceable to poor oil.

The problem of tractor lubrication

The special problem of tractor lubrication arises from the fact that a tractor runs at full engine speed for hours at a time. Tremendous heat is developed. The entire supply of oil may attain a heat of 180° F. Very much higher temperatures than this are reached in the cylinders.

Under this intense heat ordinary oil breaks down very rapidly, forming large quantities of sediment which has no lubricating value.

How Veedol, the scientific lubricant

prevents the formation of sediment is shown by the two bottles illustrated above.

Inferior oil not only produces sediment but permits the fuel to come through and contaminate the supply in the crankcase.

Veedol reduces evaporation loss and for this reason is much more economical per acre ploughed.

After severe tests of Veedol Special Heavy by the International Harvester Company, the engineering department sent the letter reproduced on this

page, recommending Veedol Special Heavy.

Most Fordson dealers recommend Veedol Special Heavy. At the Salina demonstration July 23-24, practically every tractor operated on Veedol.

Send for this book on scientific lubrication

The 100-page Veedol book contains important information which every tractor owner and dealer should have. It describes the functions of Internal Combustion Engines; Transmissions; Differentials; Oils and their Characteristics; Oil Refining. It also contains many illustrations and charts. It describes the special lubrication problems of tractors as well as those of passenger cars and trucks. This is probably the most complete and authoritative work ever written on the subject of scientific lubrication.

When you write for the Veedol proposition send 10c for a copy of this book. It will help you to keep your tractors running at minimum cost.

TIDE WATER OIL CO.

Veedol Department

458 Bowling Green Bldg., New York

Branches or distributors in all principal cities in the United States and Canada.

Raybestos Molded Facing

THE Raybestos Co., Bridgeport, Conn., has recently announced a new type of facing for multiple disk facings, known as the Raybestos molded disk clutch facing.

This improved facing is composed of pure asbestos with a suitable binder, molded under tremendous pressure, insuring absolute homogeneity and accuracy, declared by the company to be impossible with the woven type of facing, which was composed of asbestos spun around a composition wire (the wire to give the required tensile strength), this in turn woven into tape, subsequently formed into a ring.

One claim to advantage of the Raybestos molded facing is that it can be made endless.

From an engineering standpoint, this new facing is declared to be a distinct advancement in clutch facing design, inasmuch as the exacting demands of modern automotive engineering design makes necessary a clutch having greater capacity of power transmission, still keeping the size of the clutch as small as possible.

The Raybestos molded clutch facing, it is claimed, presents approxi-

A RAYBESTOS DISPLAY RACK

mately 20 percent greater frictional surface to the steel disk than the old woven type, resulting in a greater carrying capacity for a given size of clutch. The load carried or power transmitted by any faced clutch is directly dependent on the frictional properties or co-efficient of the facing used in the clutch. The co-efficient of friction of the facing must be such as to carry the required load and at the same time giving ease in pick-up or acceleration without excessive wear. It is therefore vitally important then that the co-efficient of friction be alike throughout the facing as it wears to insure action of the clutch.

The Raybestos facing has a co-efficient of friction of approximately .35 to .40.

IN 99 years Mexico has had 73 governments. And yet they say the Mexican't form a government.

CARRANZA has been called a suicide. Of course! Anybody who accepts the Mexican presidency is.

ON his rather remarkable head Villa has a price of 100,00 pesos, which must amount to all of a dollar and a quarter.

DUNKARDS, meeting near Logansport, Ind., recently decided to bar phonographs from their homes. We have not heard whether this action was taken in the interest of religion or music.

Announcing the
MIDWEST
TRUCK *and* TRACTOR ENGINE

An engine that stands the gaff of tractor service will meet the hardest test to which ANY engine can be subjected.

In designing a truck or tractor engine, good, sound engineering dictates that you start out with that distinct type of engine EXCLUSIVELY in mind.

DEPENDABLE POWER

AT LAST here is a strictly heavy duty internal combustion engine designed strictly *as such*—not a remodeled, built over, or rehashed modification of passenger car engineering practice. In conjunction with the best consulting engineering talent available, we have successfully developed a heavy duty engine really fitted to "deliver" heavy duty service — and heavy duty service alone. To the Midwest Engine Company, with its half century of successful engine building experience, must go credit for effectively divorcing the tractor and truck engine from its two lighter-duty prototypes—the passenger car engine and the hybrid, semi-heavy-duty truck engine which has often failed to make good in tractor service.

A successful tractor engine, with superficial modifications, makes the best possible truck engine. Experience has shown that the converse is not true.

Dependable Power

Proving That the Farm Home May Be a "Sunnyhome"

Exhibition at Milwaukee Shows That Every Luxury Is Available

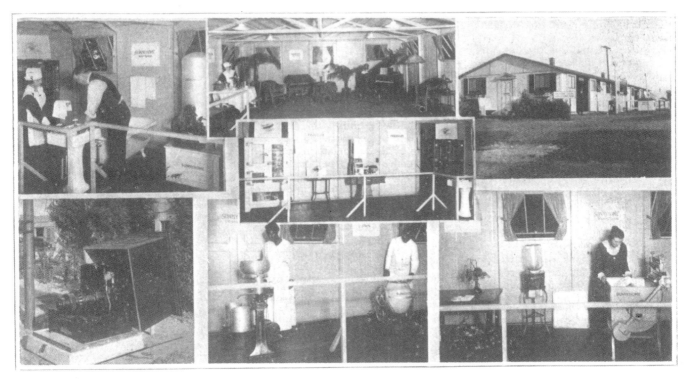

BY MEANS OF THE AUTOMATIC POWER PLANT, LABOR IS MADE LIGHT FOR BOTH THE MAN AND WOMAN OF THE COUNTRY HOUSEHOLD
THAT NEARLY EVERY TASK IN THE HOUSE MAY BE DONE BY ELECTRICITY WAS SHOWN IN THIS UNUSUAL DISPLAY

Proving That the Farm Home May Be a "Sunnyhome"
Exhibition at Milwaukee Shows That Every Luxury Is Available

SUNNYHOME for the farm became a reality at the Wisconsin State Fair in Milwaukee. Sunnyhome is the most recent development of the General Motors Corporation and as exhibited at the fair, demonstrated to the many thousands of visitors the various comforts and economies that electricity brings to the home.

Sunnyhome is the name of the new electric light and power plant of the Sunnyhome Electric Co., of Detroit—a division of the General Motors Corporation. This outfit is a complete power plant in itself, which sits in its own little power house out in the yard. It operates automatically, requiring no attention except filling with gasoline. Once in a year it is charged with lubricating oil. Sunnyhome starts and stops itself; when it is out of fuel it shows a red light, usually located somewhere in the farmer's home.

Plant Runs Continuously.

Because Sunnyhome is automatic, it will run, when necessary, heavy power devices continuously, such as electric irons, cooking utensils, heaters and refrigerators, without any attention from the owner. The battery cannot run down or become overcharged. The power house warms itself automatically in cold weather.

At the fair a substantial building called Sunnyhome was a part of the exhibit. This building was supplied with current from the Sunnyhome power house outside. In the Sunnyhome building all the electric utilities that appeal to the farmer and his wife were shown in operation.

Iceless Icebox Was Shown.

Folks entered one door of Sunnyhome and were greeted with music from an electric piano and in turn were shown the table utilities such as the coffee percolator, the toaster and casserole, then the electric iron, the churn, the vacuum cleaner, the washing machine, the electric fan, the cream separator, the water system which included bath room equipment.

One of the features was the iceless icebox called Frigidaire. This new kind of an icebox does away entirely with the old fashioned methods of refrigeration. It makes its own cold and automatically keeps the icebox at the temperature necessary for the proper preservation of food. In addition to this it makes a sufficient supply of ice for table use. Frigidaire is operated by electricity.

Included in the Sunnyhome exhibit were also utilities for the farm machine shop—the Sunnyhome drill press, the lathe and the grinder and buffer.

WHY GEORGE H. BRETT'S SALES VOLUME INCREASED

Here we have a good interior view of the offices of the George H. Brett Implement Co., Ponca City, Okla. Mr. Brett himself is seen seated. The store was built last year. Since January the business has increased fifty percent over that for the same period in 1918, which was the best season in the history of the house. Mr. Brett blames most of the increase on the new building. He isn't losing much time in regret, however.

Locklear at $1,000 Per.

The thousands who saw Lieutenant Locklear change airplanes in midair at the Nebraska State Fair have indulged in a lot of speculation as to what salary the man got for performing the thrilling feats.

Secretary Danielson of the state fair board has answered the question to satisfy the curious public. For standing on his head on the wings of the plane, hanging by his toes on the axle, running back and forth on the plane, and finally climbing up a rope ladder to another airplane soaring above him, Locklear and his entourage got just $1,000 per day for the four days or a total of $4,000.

This covered everything, Secretary Danielson says. Locklear had to take care of all his own expenses on this amount. He had to furnish his two highly trained pilots to operate both planes while the principal did his stunts. He had to pay the cost of transporting the planes half way across the continent, and all the hotel expenses, drayage and incidentals while at Lincoln.

Changing "Horses" in Mid-Air.

The first time you see a United States cavalryman changing horses at a full gallop, you think it is some daring stunt. That's nothing today. Lieutenant Ormer Locklear, daredevil aviator, is coming to the Nebraska State Fair to show the public how to change airplanes while both planes are gliding around in the zenith.

"Safety Second" is his motto. He is said to be the only man who ever changed airplanes in flight. He is scheduled to do this stunt on the Nebraska fair grounds four days, Sept. 2, 3, 4, 5. Locklear scorns all safety appliances during his hazardous work.

"I am annoyed almost every day," declared Locklear, "with fellows who want me to use their safety packs and parachutes. They seem surprised that I do not use anything to 'break the fall.' The truth of the matter is there is going to be no fall.

"In the first place, a parachute strapped on me would make my fast work impossible, and I am a fast worker while going through my 'aerobatics' and plane-changing. In the second place, I would have a lot of nerve to ask the big salary I get to do my act and then have a feather bed to land on. How would a swimming champion look with a pair of those inflated water wings strapped around his neck?

"Aviators and manufacturers must think less about parachutes for life saving after something has happened to the aviator, and give more attention to preventing that something from happening. Flying is considered dangerous because mistakes are made with such frequence. How much longer will it take to convince the world that if flying is not as safe as any other mode of travel I could not keep going on with my kind of work."

Locklear will present his "aerobatic" act of scampering from the seat of his plane to all other parts of the craft.

An aerial combat between Locklear's assistant pilots, Lieutenants Shirley Short and Milton Elliott, will also be given.

The New Golf Champion.

No more will Colonel Bogey be hindered and tripped up by the imprints of horses' hoofs on the golf links. No more will his long drives be found nestling snugly in the little hollow where stepped some heavy weight horse while pulling the mower. Gone forever is that deep imprint which is a miniature bunker. For the new champion has done away with all that. Although the new champion cannot make the course in par, it fixes it so the budding amateur champ can drive 'em a mile.

The long and short of this is that it is now the fashion for golf clubs to mow their links not with horses but with motor cultivators. Recently, owing to the wet weather, it was found impossible at the Kickapoo golf links at Peoria, Ill., for the golfers to play any kind of a game as long as horses were used to mow the links. There were too many big hoof imprints into which the ball could fall. One of the members of the golf club suggested that they look about for some better method of mowing the course. The method seized upon and tried out was that of pulling the mower with an Avery motor cultivator. This cultivator was one of the standard machines turned out by the Avery Co., Peoria, Ill. It was built for regular farm work such as planting and cultivating row crops and for doing light field and belt work, but when they hitched the mower to the motor cultivator and started across the course, it was seen at once that they had hit upon a method of mowing that would far surpass the horse mower method, both in the saving of time and in the quality of the job.

The cultivator can be used with only slight changes in the attachments, for raking the golf course and doing various odd jobs about the park.

Three of the new harvester-threshers are being used in Sumner County, Kan., this year. W. A. Boys, county agent, has arranged with the operators to keep accurate records of the costs to be compared with harvest expense under the old methods. Because of interest in the "combines" the farm bureau hopes to have accurate records to present to its members.

Who can claim patriotism unless he is a cheerful taxpayer?

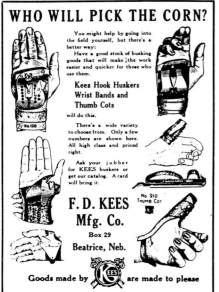

A Tractor Plow With Distinct Features

By W. R. Heilman
Vulcan Plow Co., Evansville, Ind.

THERE are new and striking features on the Vulcan tractor plow. They appeal to tractor users and to dealers, and they make good in field work. This plow may be set for cutting 12, 14 or 16-inch furrows. No added parts are needed. The same 14-inch bottoms are used.

Thus the width of the cut made by three Vulcan bottoms may be 36, 42 or 48 inches—a range equal in width up to that of a 4-bottom, 12-inch plow.

It means, further, that the load which the Vulcan plow puts upon the tractor may be adjusted to match the power of the tractor, according to the condition of the field. In hard, dry ground, the Vulcan may be set for three 12-inch furrows; in usual field conditions, three 14-inch; in light soils, three 16-inch.

This width adjustment is made by simply spacing the beams on 12, 14 or 16-inch centers. All of the cross members of the plow frame are drilled with the extra holes necessary for the adjustment. These cross members are built of strong, special steel, braced for ample rigidity.

So often, we hear the tractor salesman say, "This is a 4-bottom tractor in light soils and it will pull three anywhere." The Vulcan 3-bottom plow with its frame feature fills this demand for adjustment of width of cut. The 3-bottom Vulcan may be converted into the two. The 2-bottom is also adjustable for 12, 14 or 16-inch furrows.

An interesting question which arises is "How does the regular 14-inch bottom handle the 12 or 16-inch width?" At 12 inches, the wing of the 14-inch share sticks out into the open furrow, behind the next bottom ahead. This two inches of share does no work nor any harm.

At 16 inches, the 14-inch share does not cut the two inches beyond the wing end of its cutting edge. However, this two inches of soil breaks loose and is readily turned as the plow raises and turns the furrow slice. The shape of the Vulcan bottom is right for this work. However, if desired, a 16-inch share may be had which is interchangeable with the 14-inch. The moldboard handles the different widths perfectly. Its long sweep and its width will handle the 16-inch cut, and the narrower cuts are of course easier to turn.

Release Hitch.

The Vulcan is equipped with a patented spring release hitch. It protects the plow and the tractor gears from damage when the plow strikes a stump or a rock. By hooking the spring farther out on the trigger arm, the strength of the hitch is increased. The proper setting is at the strength which will pull the plow through the hardest ground in the field. Then, when anything solid is hit, the release trips loose.

The links of the chain come out of the hitch body with a ringing sound, as the plow is left behind. To recouple, the chain is first slipped back into place in the hitch and then the tractor is backed up and the pin dropped through the shackle and drawbar. This Vulcan release is simple, sturdy and satisfactory. It is regularly supplied on the Vulcan tractor plow.

Notched Coulters.

The notched rolling coulter is supplied on special order. A study of its work is interesting and its principle in use has an element of common sense which is generally appreciated by users

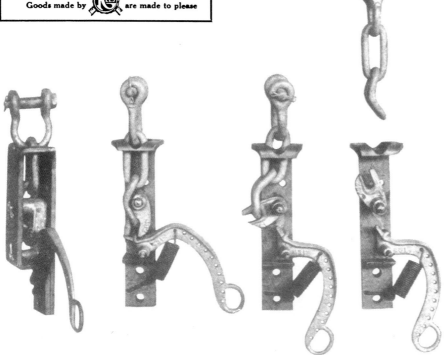

THE PATENTED SPRING RELEASE HITCH IS A VULCAN FEATURE

of gang plows. The purpose of the notches is to take hold of the trash as the coulter rolls and pull it down against the ground so that the coulter cuts it. When gang plows clog full of trash, the source of that trouble is generally the coulter, which pushes loose trash ahead of its cutting edge. This is particularly true of wet tough corn stalks.

Once started to dragging, the trash builds up quickly and packs tight. The notched coulter is of decided advantage in plowing trashy fields. It will cut through where the plain coulter will fail to clear. This statement is based on the writer's field experiences with the Vulcan, studying the action of one notched coulter on a three-bottom plow with two plain ones, and then all notched, as compared with all plain. The notched coulter wins every time.

The notched coulter is a new feature with a very practical usefulness. It is typical of the progressive spirit in plow design, which tests thoroughly first, then swiftly accepts and markets new features which are needed in tractor plowing.

Hitch Shifter.

The Vulcan hitch shifter is regularly supplied as a part of the plow. Its greatest usefulness is for even plowing on hillsides or sloping ground. When turning soil down the slope, we have all noticed how the plow crowds sidewise downward so that the first bottom cuts narrow. The result is that each round is marked in the plowed ground with a gully, because the first bottom has not thrown enough soil to fill evenly. By the lever of the hitch shifter, the plow may be set immediately to cut the full width with the first bottom and to do good, even plowing.

For straightening a crooked furrow, a little more or less may be cut at a desired place until, in a few rounds, the furrow has been brought into line. In hard spots or on up-grades, where the tractor is overloaded, the pull may be eased off by setting the plow to cut narrower, maintaining the depth. The shifter may be operated either when the tractor is standing still or plowing.

The Vulcan rear wheel control permits backing either in the furrow or on the headlands. The power lift consists of a compact clutch on the land axle, with all parts enclosed. It is carried here high above the ground so as not to gather trash. Its action is quick, causing the plows to cut out of the ground at a sharp angle. For entry, the plow drops on the noses of the shares, with bottoms tilted down so that they go to full depth right away. The Vulcan holds its depth in hard ground.

In many a field, it is not necessary to back the plow; but when a rock or stump or old fence post is hit and backing is necessary, then the Vulcan rear wheel holds the wheel in line with the landside so that the backing operation may be accomplished. The point to be emphasized is that when the demand comes upon this plow for the unusual tractor operation, the Vulcan is ready for it.

In plowing out the dead furrow, when both front wheels of the plow run in the furrow, there is ample adjustment by the levers to plow as shallow as desired. When the very trashy spot in the field is reached, the clearance is ready between the bottoms and under the beams to let the stuff through.

THE FRAME ADJUSTMENT FOR 12, 14 OR 16-INCH FURROWS

SPEAKING OF "BUGS"

THERE are still people on the outskirts of the trade who insist that the tractor "has a long way to go," that its development is too much in the nature of experiment. There is just enough of truth in this, although the degree is little enough, to injure the growth of the industry.

As a matter of fact, the progress of tractor development has been so marked that it has gone ahead of what even the incorrigible optimist ever had any right to expect. No, the tractor man has no reason to be diffident about the headway that his machine has made in terms of practical agricultural mechanics.

But there is a phase of tractor development that has been slow enough in all conscience. This is not the tractor itself, but the man who sells it—the tractor salesman. It is in his make-up that much development has to take place before the tractor industry, no matter how meritorious the machine itself may be, can be rid of the grief that arises fundamentally from the selling of it.

Of course it is true that there are, in the aggregate, many fine men who know the business of selling tractors to dealers and farmers and have raised it to the dignity of a profession. That aggregate, however, is altogether too small. Where it is numbered in tens now it must be numbered in hundreds tomorrow.

Manufacturers recognize this need of high-grade tractor salesmanship. The type of man who will agree to "throw in the factory" if the prospect will but "sign up" has no legitimate place in the industry. His work is destructive. The "easy promiser" and the "smooth liar" must be weeded out of the ranks of tractor salesmen before the market can develop in good health. There are still some "bugs" in some tractors, to be sure, but the sooner the "bugs" can be cast out of some of their salesmen the better some tractor manufacturing and selling organizations will progress.

THE HITCH SHIFTER IS FREQUENTLY NEEDED ON HILLY GROUND

THE NOTCHED COULTER IS OF ADVANTAGE IN TRASHY FIELDS

Avery introduced a six cylinder tractor model and the Packard Motor Co. began building the "single six automobile", which was a small version of the popular "twin six", twelve cylinder engine. The Reliable Tractor & Engine Co. went out of business during 1920 and on February 16, the Nilson Tractor Co. was reorganized as Minnesota-Nilson Corp. The name of the Toro Motor Co. was formally changed to Toro Mfg. Co. to properly define the company's work. Toro had been making complete power farm machinery as well as engines since the end of the war. The Dump wagon business of the Studebaker Company was sold to the Western Wheeled Scraper Co. of Aurora, Illinois. Studebaker continued to make farm wagons for a while longer, but production of these stopped before the end of 1920.

The Kentucky Wagon Mgf. Co. of Louisville, Kentucky, bought the complete farm wagon business from Studebaker early in 1921. The Foote Bros. Gear & Machine Co. introduced a new tractor transmission and axle assembly designed for 30 hp models. The new "F.U." transmission incorporated planetary gear reduction. Foote transmissions were standard equipment on many tractors, including Gilson, Illinois, Standard, Turner and Wisconsin. The Hare's Motor Co. was formed as the operating company to control Locomobile, Mercer and Simplex automobile companies. E.S. Hare, founder, was formerly a vice president of the Packard Motor Co. International Harvester Co. announced plans to build "the largest motor truck plant in the world" at Fort Wayne, Indiana. The truck research bureau of the Goodyear Tire & Rubber Co. estimated that the farm market was immediately ready for 800,000 motor trucks. Pneumatic tires were tested on trucks and were soon preferred over the earlier hard rubber type. Strikes affecting rail traffic provided great opportunities for trucks and the new trucking industry. Deere & Co. began a $500,000 improvement to the Waterloo, Iowa, Waterloo Boy plant. The Ford Motor Co. was reorganized as a huge corporation with $100 million capital. The capital stock was held by Henry Ford, Edsel F. Ford and Mrs. Henry Ford. Reorganization was not to open the stock to public sale, but to consolidate the various interests. The Ford Motor Co. was incorporated under Delaware laws because Michigan statutes did not permit organization of companies with more than $50 million in capital.

Pan Chief Found Guilty.

In Federal court in Chicago last Saturday Samuel Pandolfo, president of the Pan Motor Co., St. Cloud, Minn., was found guilty of using the United States mails to defraud in the financial promotion of his company. The charge against him of conspiracy was not sustained. Twelve other defendants were found not guilty.

The head of the Pan company, with other officers of the organization, was indicted early in the year. It was charged that too great a share of the receipts reverted to the president of the company. An endeavor was made to prove that this method of doing business was legitimate and that the firm was on a bona fide basis.

A number of dealers throughout the country had been sold stock in the company.

DISPLAY BOARD FOR BONNEY WRENCHES

MOVING AN AMARILLO, TEX., POOL HALL WAS EASY FOR THIS HART-PARR TRACTOR

Pool Hall Takes a Journey.

Charles City, Ia., Sept. 22.—To the Implement & Tractor Trade Journal: From the enclosed picture you might judge that a Hart-Parr 30 tractor had invaded Mexico and rescued a pool hall from Villa and his bandits. Our distributor, the Southern Border Motor Co., at Amarillo, Tex., did not give us particulars.

At any rate, the building being pulled by the little Hart-Parr 30 weighs 10,000 pounds, billiard balls, soda pop and all. Perhaps the operator of the tractor helped himself to the cold soda pop—he seems to be "taking it cool."

This is another indication of a tractor's adaptability to "hard jobs" of every kind.

A. W. SAWYER,
Hart-Parr Co.

Much Older

They had been up to town to see the latest musical comedy, and were discussing its merits as they traveled homeward in the train.

"I think I liked the bad man best of all," declared the girl. "He was so very natural in everything—and oh, what lovely hair—so black and curly!"

The young man beside her cursed inwardly. He was very fair.

"What did you think of the big chorus of twenty-two?" he ventured to change the subject.

"They were more than that," declared the girl decisively. "There wasn't one under 30, in my opinion."—Answers, London.

GRAIN HEADS ENTER LOW

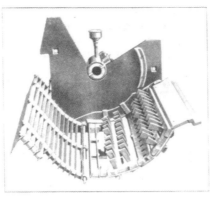

THE CONCAVE AND GRATE

LOOKING THROUGH THE THRESHER

THE CYLINDER HAS 66 TEETH

SHOWING THE TRIANGULAR BEATER

Now What Is Wrong With America?

IN laying the blame for the ills of the country, in the ordinary procedure, it all depends upon who you are. If you are a capitalist, the shiftlessness and the restlessness of employes in general are likely to receive the blame. If you are a laborer, the ding-donged capitalist is apt to have the benefit of your vituperation. If you belong to the great in-between class, the abounding bourgeoisie, you will of course cry to Capital and Labor, "A plague o' both your houses!"

As a matter of fact, everybody's to blame. George F. Whitsett, editor of the Harvester World, published by the International Harvester Co., comes right out and says so in a recent issue of his paper. His editorial entitled "What Is Wrong With America?" is worth the attention of every reader. It follows in full:

"There is no use saying nothing is wrong with America, because something is wrong. There is too much spending and too little earning; too much extravagance and too little economy; too much restless and vague desire and too little real enjoyment. Something is wrong.

"We believe this is what is the matter. America has been builded by a race of people, who knew nothing about the soft life. All they knew was work, denial, hardship, and good straight fun when it was time for it. They knew nothing about taking something which someone else had already made and sitting back to watch the money roll in. The citizens of foreign countries who came in those days knew there were great values here, provided they would pitch in and earn them. They did not come expecting something easy or soft. They came

expecting something good and real and honest, and they got it.

"The present generation of America has forgotten what it has all cost. They see this wonderful business and commercial structure, and they think it all just happened, that it has always been this way. They enjoy these comforts and reap these easy dividends, and think this has always been the order of the day. The people from other countries who have come recently have been told that it is soft over here, that there are loads of money and lots of easy jobs, and that everything is fine.

"What the present generation of Americans and the present class of

residents from foreign countries need is a clear understanding of the kind of thing that has produced America, and a clear understanding that the same kind of thing is going to produce a still greater America. If we could have some of the old-fashioned hard work, and saving and self-denial, we might also have some of the old-fashioned contentment and joy of living.

"The present generation has more than any other has ever had, and it is bored to death. It thinks because everything isn't still softer and still more perfect, that it is all wrong. It is time for a revival of American common sense and American simple tastes and work and saving."

ONE of our traveling friends, a veteran of the "washbowl and pitcher circuit." says that the love between a flock of chickens and an old binder at roosting time passeth his understanding.

How the Farm Loan Helps the Dealer
By C. M. Gruenther
Secretary, Federal Land Bank of Omaha

NEXT to the Homestead Law, the Federal Farm Loan Act is the greatest and most helpful piece of legislation ever passed by our National Congress for the farmers of the United Sates. Over $271,000,000 have already been loaned to the needy and worthy borrowers of the United States through the twelve Federal Land Banks since the law went into effect a little over two years ago.

Funds for these loans are obtained by the sale of bonds issued by the United States Treasury Department and secured by first mortgages on farms. This is the best security in the world and the bonds sell readily to prudent investors, thus insuring a constant and ample supply of funds which is loaned out by the Federal Land Banks to farmers at a low rate of interest.

The Federal Land Bank of Omaha is one of the twelve land banks in this great system. It operates in Iowa, Nebraska, South Dakota and Wyoming, these four states comprising the Eighth District. Over $37,000,000 have been loaned to the farmers in these four states in the last two years. Under this system, all profits belong to the borrower and at this time the cost of the loan to the farmer borrower is a trifle over five percent per annum.

Loans are made for 34½ years, but may be repaid sooner if desired. This long time loan with option to pay sooner, if desired, is really the best feature of a Federal farm loan.

It enables the farmer to work out his destiny successfully and progressively. The threat of foreclosure, or higher interest rates and excessive renewal commissions do not stare him in the face every five years. He pays his low interest dues and about one percent of the principal on his mortgage, semi-annually, and in that way

Terms of the Loan.

the mortgage is gradually killed off. He can then work in peace with his family and devote his energies and earnings to make two blades of grass grow where one grew before.

He is insured against future financial perplexities and as a result he is in much better position to improve his farm with new and better buildings. He will be in the market for better and more farm implements, motor cars, tractors, farm trucks, silos, dairy equipment, farm lighting plants and the numerous other useful and convenient necessities of modern farm life. The home builder is always a better asset to a community and to the business world than a landlord or a tenant. The Federal Land Bank of Omaha stretches out its helpful hand to the tenant and helps him to become the owner of a farm.

Practical and Profitable.

The system is truly cooperative; it is profitable to the borrower, it is scientific; it is practical.

The prospective borrower should consult the nearest secretary-treasurer in his county for complete information. If a secretary-treasurer is not conveniently located for him or if he does not know the name of the secretary-treasurer in his locality, he should write direct to the Federal Land Bank of Omaha, Nebraska, and information will be cheerfully furnished

1. ON THE LOUIS BERTRAND FARM, OAKLEY, KAN.   2. SCHLYER & ARNOLD STORE, HAYS, KAN.   3. A "COMBINE" READY FOR DELIVERY

The power to convince stands unrivaled when life and action are put to work through the filmed demonstration of your products.

Men in Australia
Can Watch This Tractor Plow A Wisconsin Farm.

Whether your prospective purchaser be in Australia or here in the United States—whether you are selling a farm tractor, an article of clothing, an office appliance, a service, or promoting a western project, the power of your sales appeal depends upon a graphically clear explanation. That alone inspires confidence—and men *do not buy without confidence*.

The printed word is plaintive, the spoken word inadequate, the ordinary photograph incomplete, when compared to the filmed demonstration brimming with life, action, and human interest appeal.

When words have failed, the salesman resorts to demonstration. He knows its power, but the usual contingencies bar it from common use. Motion Pictures make your most successful demonstration the *basis* of the sale. They start where ordinary sales methods are forced to stop.

The Rothacker Film Mfg. Co. is your "stepping stone" to bigger sales at less cost. We have successfully filmed the most difficult subjects; have invariably served profitably our customers. Using Rothacker Motion Pictures in Business and Industry is as simple and practical as any other method of selling.

For your convenience and reference we have prepared a booklet which presents Motion Pictures In A Simple and Understandable Way. Booklet will be sent free at your request.

Consultation Without Obligation.

Rothacker
FILM MFG. CO. CHICAGO U.S.A.

1339-51
Diversey Parkway
Chicago, Illinois

MOVING PICTURES

Rothacker

The Senior Specialists In Motion Picture Advertising.

Case P. W. Brings Out a Thresher

THE J. I. Case Plow Works Co., Racine, Wis., has entered the thresher field with a new machine called the Wallis. It is built to order for the Racine firm by the Sawyer-Massey Co., Hamilton, Ont.

The frame of the Wallis thresher is so constructed as to make it as nearly a unit as is possible. It is mortised and tenoned and held or secured by the use of joint bolts and tie rods, these being of sufficient size to withstand any shock that the frame may receive. The lumber used in the frame consists of hard maple or oak. Before assembling the frame it receives a coat of priming, thus covering all tenons and mortises. The remainder of the painting is done after the machine is assembled. The material used in the side of the machine is soft white pine, tongued and grooved and glued together, thus making it a unit, then put in the frame and securely nailed thereto.

The principal bearings on the machines are babbitted bearings, a high quality babbitt being used, thus securing good running and long wear. The rock shaft bearings are gray iron machined bearings. The bearings are of ample dimensions.

Width of the cylinder is 32⅞ diameter over the teeth 21⅜. Number of bars 12, size of bars 1x1½x23⅞. Number of teeth 66, r.p.m. 1,140. Size of standard drive pulley, 8-inch diameter, 9-inch face.

Length of concave 23¾. Number of teeth in one wide concave 21, number of teeth in one narrow concave 11 and in other narrow concave 10. Lock washers are used on both cylinder and concave teeth. The cylinder and concave, that is the spacing of the teeth and so forth, is so designed to do proper heading. A special concave for turkey or marquis wheat is furnished, thus enabling the machine to head properly the most difficult wheat. The full set of concave comprises two wide and two narrow bars. In ordinary threshing it is only necessary to use two wide concaves, making four rows of teeth with a barred grate between the two concaves.

Diameter of beater is 12 inches, r.p.m. 888. The type is three-winged. The construction of it is designed so that it would be impossible to make this beater wind in the toughest straw. Another feature is that instead of shooting straw half way through the separator, it is delivered from the grate down on the straw deck, so that a full separation of the deck is obtained.

The Straw Deck on the Wallis thresher is driven from a rock shaft and connected on each side with a pitman. The Pitman is so constructed at the end that any wear or lost motion can be easily taken up by simply driving in the taper wedge and tightening up the bolt. The deck is equipped with four agitating forks. The motion of these forks can be regulated by either raising or lowering the pitman in the outside rock arm. By raising it up toward the shaft it increases the motion of the forks and by dropping it down, lessens it. The tossing motion of the four agitating forks pitches the straw with increasing speed toward the rear, with the result that the straw is spread out thin and a maximum of separation is secured. Width is 39¼ inches, length 10 feet 6½ inches, separating surface and square feet 34.5.

The grain deck is driven by the same rock shaft as the straw deck but opposite motions, thus making perfect counter balance. The shoe is driven by two eccentrics. It is fitted with an adjustable sieve. The number of vibrations is 252 per minute. Size of the sieve is 37¾ inches by 46 inches; sieve surface and square feet 12. The housing of the grain conveyer is made of sheet metal. In this is fitted the grain auger constructed of steel flighting 7 inches in diameter, r.p.m. 374.

The cleaning fan is four blade type, 22 inches in diameter, driven from the threshing cylinder; r.p.m. 835. This fan is of the overshot type and is provided with adjustable wind boards so that any kind of grain can be taken care of. The outside blast doors of fan are fitted with spring and ratchet so that any desired opening can be maintained.

Trucks are fitted with steel wheels. Size of front wheels is 5x34, size of rear wheels 5x38. We would call particular attention to the steel axle used in this machine. It is constructed of 5 inch, 6½-lb. per foot steel channel, which makes a very strong durable axle. The tongue is of the slip point type and can be either used for team or tractor.

The feeder is built of steel and, while it is light, is sufficiently strong to withstand all shocks and strains. It is fitted with a governor which controls the raddle. If the proper speed of the threshing cylinder is not maintained, the governor automatically stops and starts the raddle, thus preventing feeding being done while the separator is not at normal speed. Each feeder is equipped with an extra sprocket so that it can be made to feed either medium or fast. The feeding mechanism sets on feeding forks operated by crank shaft. This can be adjusted to suit all kinds and conditions of grain. The band knives are of the rotary type, serrated edge. The knife holder is mounted on knife shaft and the knife is securely held in position by the holder. It can easily be removed without disturbing the shaft or the holder.

The windstacker is a side fan type, steel housing, four vane fan with the blade 7 inches wide. Diameter of fan, 33 inches, r.p.m. 967. Diameter of shute 12 inches. It is provided with a worm gear and cable for raising and lowering. The ball bearings on turntable make it very easily operated. It is fitted with a telescopic pipe and revolving hood. This stacker is also equipped with the grain saving device recommened by the United States Government during the war.

The capacity of the Wallis thresher depends upon the grain to be threshed and the condition of the grain. A fair average for wheat is 90 bushels per hour and for oats 135 bushels per hour. The length of the machine, with feeder carriers folded is 22 feet 4 inches. Weight of machine with attachments is 5,438 pounds. Height of machine, to top of tailings elevator, is 4 feet 10 inches. Total width is 6 feet 9 inches, with 10 feet 4-inch wheelbase.

THE WALLIS THRESHER CAN BE OPERATED BY SMALL AND MEDIUM-SIZED TRACTORS

Revised Report on the Waterloo Boy Tractor Has Been Approved and Released by the State Railway Commission

ONE of the first reports sent out by the Nebraska test board was on the Waterloo Boy tractor. The report was withdrawn for revision. Publication in the Implement & Tractor Trade Journal has been withheld until the revised report was approved. All of the reports appearing in the Implement & Tractor Trade Journal have been complete in detail.

Test No. 1, 12-25 Waterloo Boy.

The Waterloo Boy 12-25 hp. Model N tractor, with a rated speed of 750 r.p.m., manufactured by the Waterloo Gasoline Engine Co., Waterloo, Ia., and sold by the John Deere Plow Co., was tested from March 31 to April 9. The first report issued on the Waterloo Boy was withdrawn and this revised report issued. The equipment included a Dixie No. 246 magneto, Schebler Model D carburetor and angle lugs 2¼x2¼x5-16 inches and 16 inches long.

In the rated load test, with load constant, the speed increased from 768 at the beginning of the test to 794 at the end of one hour and ten minutes. A slight change in governor adjustment was then made after which the slowest speed was 748 and the fastest speed was 774 r.p.m. It was necessary to shut off the water feed to fuel mixture for O, ¼, ½ and ¾ loads in the varying load tests. In the maximum load test the governor was set the same as in the rated load test, which gave nearly, but not quite, full opening of the governor valve at rated speed. The water for radiator and fuel mixture could not be measured separately.

In computing slippage, the circumference of the drive wheels was taken at points of the lugs. Kerosene used for the test weighed 6.71 pounds per gallon. The drawbar tests were made with the tractor in low gear. During the complete test of about 44 hours' running, 3⅜ gallons of Mobiloil A were used for the engine and none was added for the transmission. The official report states further:

At the beginning of the limbering up run the oil was drained from the crank case and 4½ quarts of fresh oil put in.

At the end of the limbering up run 5 quarts of oil was drained from the crank case, indicating that some fuel was passing the pistons unburned during this test.

The valve tappet rods were adjusted on April 2, 1920, at the beginning of the rated load brake test.

The tractor was in good condition at the close of this test. There was no evidence of undue wear in any part nor of any weakness which might call for early repairs.

Brief specifications Waterloo Boy 12-25—Engine: Twin cylinder, opposed cranks, horizontal, valve-in-head. Bore 6½-inch, stroke 7-inch, rated speed 750 r.p.m. Chassis: Four wheel. Rated speeds: Low gear, 2.34 miles per hour; high gear, 3.02 miles per hour. Total weight, 6,183 pounds.

The governor on this tractor did not give close regulation of the speed even with the load constant and on varying load the speed regulation was erratic. We do not consider this to be so serious a defect as to disqualify the tractor.

In the advertising literature submitted with the application for test of this tractor we find the following statement regarding the horsepower capacity: "It has ample reserve power for prompt utility when needed." We do not approve this statement for the reason that it is indefinite and, therefore, likely to be misleading.

We also find in this advertising literature some claims and statements which cannot be directly compared with the results of this test as reported above. It is our opinion that none of these statements or claims are unreasonable or excessive except the following:

Page 2. "Drive internal gear, most efficient type—."

"Air takes from high level, insuring no dust."

Page 5. "Air stack brings air to the carburetor from a high level—no dust."

				Fuel Consumption			Water Consumption Gallons per Hour						
BRAKE HORSE POWER TESTS									Test No. 1				
Horse Power Developed	Crank Shaft Speed R P M	Length of Test Min.	Kind of Fuel	Amount Used per Hour Gallons	Horse Power Hours per Gallon	In Radiator	In Fuel Mixture	Total	Temperature Cooling Fluid Deg F	Temperature of Atmosphere Deg F	Humidity %	Barometric Pressure Inches Mercury
RATED LOAD TEST												
25.51	771	120	Kero	3.28	7.78	x	x	0.16	177.2	61.9	36	28.6
Belt slippage 1.54%			Radiator partly covered									
VARYING LOAD TEST												
25.06	750	10	Kero									
25.33	744	10	"									
0.99	713	10	"									
7.31	874	10	"									
12.31	739	10	"									
18.25	732	10	"									
14.88	758	60	Kero	2.35	6.31	x	x	0.21	176.0	61.7	68	28.5
MAXIMUM LOAD TEST												
25.97	724	60	Kero	3.80	6.83	x	x	0.83	179.4	60.7	50	28.4
Belt slippage 1.30%												
HALF LOAD TEST												
15.02	903	60	Kero	2.40	6.25				175.7	64.4	50	28.4
Belt slippage 0.84%												

					Fuel Consumption							
DRAWBAR HORSE POWER TESTS								Test No. 1				
Horse Power Developed	Draw Bar Pull Pounds	Speed Miles per Hour	Crank Shaft Speed R P M	Slippage of Drive Wheels %	Kind of Fuel Used	Amount Used per Hour Gallons	Horse Power Hours per Gallon	Water Used per Hour Gallons	Temperature of Cooling Fluid Deg F	Temperature of Atmosphere Deg F	Average Humidity %	Barometric Pressure Inches Mercury
RATED LOAD TEST TEN HOURS					10 Hours, 0 Minutes.							
12.10	1982	2.29	778.5	11.84	kero	2.858	4.23	0.19	160.6	49	60	28.7
Radiator partly covered 8:20 to 1:30 and again 6:00 to 7:42.												
MAXIMUM LOAD TEST				(98.2ft)								
15.98	2900	2.07	746	17.0	kero	No record	No record		154	50	61	28.8

Hay Rides, Truck Fashion, Are Now Coming Back Into Favor

REMEMBER those old-time hay rides? Hard to beat, were they not? Most of us respond to those questions with a choral, "We'll say we do and we'll say they were!" But don't get the notion that such social functions are no longer held. On the contrary. The young folks are still young. They may have acquired bulbous ideas about spending money for entertainment, but our friend H. C. L. is facing some of them back toward the simpler life. That's why the hay ride is returning to favor. Yet this internal-combustion age has modified it somewhat. Instead of an old hay rack mounted on a rickety farm wagon, the vehicle in greatest demand these days is the motor truck. It affords plenty of room and gets along

in a comfortable hurry, at the same time suggesting none of the effete luxury of the automobile, which would never, never come up to the specifications for a real hay ride. Otherwise, this time-honored function is about the same. It has not yet been found necessary to use imitation hay. There is the inevitable young man who dresses a bit eccentrically and endeavors to live up to his reputation as "the life of the party." In this pictorial instance he is to be seen reclining on the hood of the truck. Several of his social competitors are near at hand. Back of the cab sits the lenient chaperon. The spooning couple also seem to be present, as of yore. Nothing new at all, but the truck

The whole question is one of truck economy. The tire makers derive just as much profit from solids as they do from pneumatics. All the data thus far gathered has strongly favored the air-cushion tire.

From Boston to Frisco it is 7,763 miles by road. Some time ago a fleet of three-ton motor trucks made the back-and-forth trip, under load both ways, and with pneumatic tire equipment. For the whole journey an average speed of 13 miles was maintained; on good roads it was easy for the truck train to keep up to a speed of 25 miles an hour. Each machine was equipped with 44x10 rear and 38x7 front pneumatic tires.

Two trucks of the same make and size, but equipped with pneumatics and solids respectively, were recently tested in a long run. A total saving of $546.61 was effected by the pneumatically equipped truck. In the same running time this machine made a mileage of 6,000 as against 4,800 miles for the solid-tire truck. Other savings effected by using pneumatic tires were: On repairs, $66.46; on gasoline, $197.47; on lubricants, $55.18; on tire depreciations, $84; on driver's time, $161.50.

Motor truck development in the rural market is one of the things on which the farm equipment dealer must keep thoroughly informed. These figures on tire economy are worth filing away for reference.

PNEUMATICS AND SOLIDS

FOR some months an advertising campaign has been going on in behalf of certain rubber companies wherein the gospel of pneumatic tires for motor trucks has been preached insistently. At the start of this campaign, perhaps, there was a feeling in many quarters that the pneumatic tire "propaganda" should be accepted with reservations.

Gradually, however, it has been borne in upon the consciousness of

those closest to the motor truck trade that this "propaganda" is sound; that pneumatic tires are the best possible cushions to put between the wheels of a motor truck and the road. With almost negligible exceptions the principle is being O. K.'d by the motor truck industry. It is an interesting development with which dealers should keep abreast. Only on the very light trucks is it conceded that solid tires make acceptable shoes for motor trucks.

Sentence Is Pronounced on Pandolfo

Another Chapter Is Written in the Pan Case, One of the Most Sensational in the History of Financial Frauds.

Federal Judge Kenesaw Mountain Landis last week sentenced Samuel C. Pandolfo, organizer and head of the Pan Motor Co., St. Cloud, Minn., to serve ten years in a Federal penitentiary and fined him $4,000. Pandolfo was convicted of using the mails to defraud. His attorneys gave notice that his case would be appealed to the United States Circuit Court of Appeals.

The jury found Pandolfo guilty on four counts on an indictment charging the use of the United States mails in a scheme to defraud in mailing letters to prospective stockholders misrepresenting the company's progress and development by stock salesmen.

Along with Pandolfo, other officers and directors of the Pan company were indicted, charged with conspiracy to use the mails to defraud, but Pandolfo alone was found guilty. However, a charge of perjury, growing out of the trial, still hangs over H. S. Wigle, one of the acquitted men in the case.

A Remarkable Case.

In many respects the Pan case is the most remarkable in the history of great financial frauds. Pandolfo and his aids sold approximately nine and one-half million dollars' worth of stock, and during trial of the case in Chicago, there were hard-headed business men and hard fisted farmers—one of whom had put $24,000 into the company—who still believed in him and his scheme.

He had made many assertions as to his belief in the future of the company. For example, he said he believed it would be a greater money-maker than the Ford company of Detroit. He said that $100 invested in the Ford business, early in its history, had grown to a value of one-fourth million dollars. At this rate, the ten million dollars being invested in the stock of the Pan company would have to attain a value of 25 billions.

The Pan motor case attained national notoriety some months ago, following an expose of the stock selling methods of the company through a report issued by the national vigilance committee of the Associated Advertising Clubs of the World. In this report, it was shown that out of each $2 paid for stock in the company, $1 went to Pandolfo for his share as fiscal agent and out of which he was to pay stock selling and certain "promotional" expenses.

Government Was Aroused.

The federal capital issues committee, in recommending to Congress that legislation be enacted for the protection of the public against fraudulent stock selling schemes, cited the Pan promotion as an example of the kind of schemes which should be curbed, quoting extensively from the history of the promotion's operations.

The report of the national vigilance committee of the advertising clubs, which resulted from an extensive investigation by the better business bureau of the advertising club at Minneapolis, concerned itself chiefly with the advertising matter of the Pan company, much of which went through the United States mails.

The Pan Motor Co. is a $5,000,000 affair, and its stock was being sold at "double par" and higher. It advertised that it would take Liberty bonds for stock, and asked the readers of its advertising matter, "Are you interested in making your return 25 percent or better?" The fact that such an appeal did convince Liberty bond holders that they had made a comparatively poor investment is indicated by the fact, brought out in the trial at Chicago, that approximately $1,000,000 of Liberty bonds had been taken in exchange for the stock of the company.

Fraudulent Indorsements.

It was shown that one widely quoted "indorsement" which the company used was from a "write-up" financial journal published in Chicago, and that letters of indorsement were used after they had been repudiated by the writers. It was also shown that there were untruthful statements in the advertising matter pertaining to the skill and former connections of men who were employed by the company as engineers, etc.

In the various statements of the company the item of "intangible assets" has covered large sums. At a time when approximately $2,500,000 had been paid in, about $1,500,000 was carried as "intangible assets," the company contending that these represented good will, though the money was largely paid to the fiscal agent, Pandolfo.

The Pan company has in the neighborhood of seventy thousand stockholders, residing in nearly every state of the United States. Several statements from the Associated Advertising Clubs have indicated that the clubs had no desire to do injury to the company and its stockholders, but believed the advertising matter fraudulent and unfair both to the investing public and to every legitimate business institution.

Introducing Thornlea Knut.

The Weidely Motors Co., Indianapolis, Ind., builders of motors for trucks and tractors, has adopted a new trademark, which is now appearing in all of the company's advertising.

When it was first decided to adopt the "bulldog" trademark, officials of the company cast about for a worthy specimen of dogdom to exemplify the things the word "bulldog" represents, which, according to Webster, are "compact, of muscular build," "remarkable for courage and tenacity of grip," "stubborn," etc.

Through the courtesy of the Warfleigh Kennels, Indianapolis, photographs were obtained of the young English bulldog Thornlea Knut, a dog whose pedigree has a distinct purple tint. Thornlea Knut is one of the best bulldogs of all times. He is the son of the famous champions Failsworth White King and Thornlea Princess. In his line are the Champions Thornlea Prince, White Marquis, Heywood Duchess, Prince Albert, Woodend Thaddens, Silent Duchess and Silent Knight of Hollybrook—all names to conjure with in the dog world.

Bandit Victim Was E.-B. Man.

The Mexican situation was brought home to the implement industry by the recent kidnapping of William O. Jenkins, American consular agent at Puebla, Mexico, which has been the subject of nation-wide interest. Mr. Jenkins was abducted by bandits on Oct. 19 and released only after the payment of $150,000 by his friends.

Mr. Jenkins is one of the largest E.-B. dealers in Mexico, and the company actively assisted in the efforts to obtain relief for him. In addition to his activities as an implement dealer Mr. Jenkins owns and operates a cotton mill, a hosiery mill, a foundry and a plantation of several thousand acres at Puebla. He has had former encounters with Mexican bandits, and has had to pay large sums of money for the privilege of operating his plantation peaceably.

"A Great Help to Her Father"

By George F. Massey
Editor

SHE sells tractors and trucks. She knows repairs. She keeps the books. She runs the typewriter. She is 22 years old. And she is—well, let the picture on this page speak for itself. Did you ever see a likeness easier to look at?

So is Miss Business—exactly that!—Miss Esther Busness, daughter of Nels O. Busness, the dealer of Fort Dodge, Ia. In business hours she is all business.

Yes, she sells tractors. She speaks the tractor sales language just as naturally as any mere salesman ever spoke it.

It has already been said that Miss Busness knows repairs. She does. She is perfectly at home among the pawls and shanks and hub-caps and cams and rachets and so forth. When a farmer calls for a part she can take the electric light bulb in its little cage at the end of a wire and find the repair in the bins a shade quicker than anybody else about the establishment.

Miss Busness was a high-school girl when it came over her that she wanted to be a business woman. "But," said her parents fond, "you want to finish high-school." She thought that over for a while. It didn't meet with her 100-percent acquiescence. At length her persuasiveness prevailed. She left school saying that she could learn as much in the store as she could in the recitation-hall and she has just about proved her case.

Taking a few lessons in bookkeeping from an expert, she soon worked out a system that would apply peculiarly to her father's business and it has been functioning admirably ever since. Under all the circumstances, some of which have been presented herewith, can you blame Nels O. Busness for entertaining a certain enthusiasm wherever his daughter is concerned?

It's a little old-fashioned, of course, to talk about the "new woman." She arrived a good while ago. It took her a number of years of hard work to get the vote, but she has it now, and the man of the family is trying to act as though he wanted her to have it all the time. To his line of talk, contrary to her instincts, she is not talking back. But she is taking it with a liberal dash of salt.

Miss Esther Busness belongs to the new school of womanhood. She was born with a liking for business. Why shouldn't she make a business woman of herself? There was no negative answer. She could and she did.

MISS ESTHER BUSNESS

CALLING NAMES AGAIN

NO good purpose can be served in adding to the misunderstanding that that still exists between the farming and the farm equipment industries. In the trade there is a widespread feeling that everything should be done to promote their community of interests.

There are elements among the farmers, however, which prefer to look upon themselves as the victims of everybody to whom they sell their products or from who they buy their supplies. This is a backward element of course. It used to be that a considerable portion of the farm press catered to this weakness. But the leaders among the agricultural papers have long since seen that the implement manufacturer and dealer are no less servants of agriculture than the farmer himself.

In a recent issue of Campbell's Scientific Farmer appears an editorial tirade against the implement dealers and manufacturers. With uncontrollable fury it refers to them as "the worst class of grafters" and "pirates who are roaming the high sea of commerce flying the black flag." Those who know this industry will of course smile at the lurid picture, but such unreasonable and vitriolic propaganda is surely not calculated to promote the health of either agriculture or this trade which serves it.

The whole attack is based upon a price of $310 alleged to have been asked by one dealer for a six-foot binder. Certainly that is high, but there may be special circumstances involved in the incident which would throw an explanatory light upon it, although it is difficult to guess at what they might be.

But the editorial in question cites only the one case of "profiteering," of such it is. The attack admits that the average price charged for a six-foot binder in that territory by the average dealer was $235 before the 1919 harvest. As every dealer knows, that figure is too close to permit anything more than a scant profit, to say nothing of a "profiteer's" margin.

A good deal is said nowadays about free speech and a free press. Some of the samples of the freedom of this kind that we have seen of late constitute scarcely the best of recommendations. The wild-eyed orator and the wild-eyed editor is privileged beyond let or hindrance to open his mouth or his typewriter and let it say what it pleases.

Such a condition is an extremely unwholesome one at any time; especially during this period of post bellum readjustment. A large share of our present economic ills is attributable to unwarranted attacks of this unsavory sort.

ONLY by its output can you tell one of these new-fangled washing-machines from a victrola.

The Work of Farm Women.

Nebraska farm women are not joining in the crusade for shorter hours. A United States Department of Agriculture survey of 350 Nebraska farm homes shows the following facts: Farm women are working an acerage of more than thirteen hours a day during the summer months. Only 30 percent of the farm homes have running water. Sixty-five percent of the women carry water an average of fifty-five feet. Only 17 percent of the farm homes are lighted by electricity or gas. Ninety-nine percent of the farm women are doing the family washing and ironing, and 28 percent of them are still washing with a board and tub, and only 20 percent have gas or electric irons. With less than 3½ percent of the farm women work in the field, 23 percent of them help care for live stock and 42 percent help with the milking. Ninety-eight percent of them bake their own bread and 97 percent do the family sewing. Fifteen percent of the farm houses have bath tubs. Kerosene stoves are in 77 percent of the homes. Women produce 95 percent of the $60,000,000 worth of poultry products raised in the state annually.

The Cook Paint & Varnish Co. of North Kansas City, Missouri, made a rush shipment of their products to their branch store in Omaha, Nebraska, by an airplane. A strike by railroad switchmen prevented using the normal method of shipping. The 90 hp Curtis Army plane that was used had been painted gold and silver (with Cook paint, of course) and required one minute less than two hours to deliver the cargo to Omaha.

Missouri, the first state lying entirely west of the Mississippi to become a state, celebrated 100 years of statehood in 1921. Small Pox vaccination reduced deaths caused by the disease significantly. Sales and collections for earlier sales were lower for most companies in the farm equipment industry during 1921. Extension of credit to the farmers, so they could buy new equipment, was seen by some to be the best way for the industry to recover. The manufacturers and the dealers had stretched payments as far as possible, but additional financing was required. The government's War Finance Corporation was blamed for slow, ineffective help with financing and many were concerned about the War Department's dumping of unused military merchandise on an already glutted market. Efforts by some manufacturers to stop the downward price spiral caused by bankruptcies and dumping were considered "price fixing" by the government.

Reorganizations, mergers and other activities of the manufacturers and dealers usually was a result of re-

Makes Ironing a Pleasure.
Weight ---
7 lbs.

duced profits. Allis-Chalmers Mfg. Co. and the Moline Plow Co. were reorganized. Later, the Moline Plow Co. purchased (or merged with) the Root & Vandervoot Engineering Co. Durant Motors, Inc. was begun by the much respected W.C. Durant. The Austin Motor Co., Ltd., manufacturers of automobiles and trucks in England, was in financial trouble and appointed a receiver. A tractor cab was standard equipment on the Savage tractor sold by the Savage Harvester Co. of Denver, Colorado. The Simplicity Manufacturing Co. was started by the former sales manager of the Turner Manufacturing Co., William J. Niederkorn and a partner Francis Bloodgood. The two men began by purchasing the remaining inventory and the "Simplicity" name formerly owned by the Turner Mfg. Co. George T. Briggs was appointed the

general sales manager of Wheeler-Schebler Carburetor Co. of Indianapolis, Indiana. The Delco Light Co. and the Frigidaire Corporation, were consolidated by the General Motors Corp.

The recent consolidation of rural schools provided the opportunity for manufacturers to build special bodies for trucks, thus converting them to "School Busses." These passenger trucks were similar to bodies already being made, but in much smaller quantities. The roads were being improved gradually and at great cost, only to be torn up by tractors with spade lugs. State legislatures began to take action against tractors operating on hard surfaced roads. The Russell Grader Mfg. Co. of Minneapolis, Minnesota, in affiliation with the Allis-Chalmers Mfg. Co., began producing the Russell Motor Hi-Way Patrol grader.

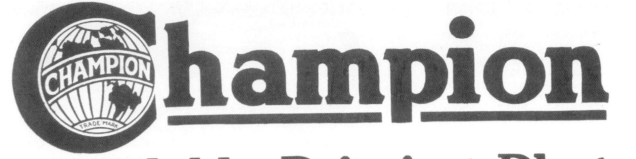

Champion
Dependable Priming Plugs

Equip Your Engine So It Sure-Fires When Cold

DO not wait until you are in the grip of cold weather. Get a set of Champion Dependable Priming Plugs now and have your engine ready for quick starting in winter.

Champion Dependable Priming Plugs sure-fire cold motors because, with the priming cup right in the plug, the gasoline trickles down the core of the plug and drips from the sparking point where the spark jumps and is the hottest.

They are imperative in cold weather for the hundreds of thousands of cars that do not have priming cups, and are infinitely better for those that do, because priming cups let the gas in too far from the spark plug. Every car can be easily equipped with these plugs in a few minutes.

Get a set of Champion Dependable Priming Plugs from your dealer, and insure easy starting in cold weather. Price $1.50 each.

Champion Spark Plug Company, Toledo, Ohio
Champion Spark Plug Co. of Canada, Limited, Windsor, Ontario.

Invest in a Sioux Tool Set

It Pays 100% Interest

The quick, accurate, efficient Sioux Way of Valve Seating saves hours of time on every Valve Grinding job. This means a big increase in profits, so that a few jobs pay for the set. And the tools are still in perfect shape after they have paid for themselves several times over, **because we keep them ground and sharpened -----** free of charge. Isn't that at least 100% on your investment?

Sioux Tools are superior on working principle, material and workmanship — each one a time saver and a profit-booster. This complete set $28.50.

Sold by All Live Jobbers

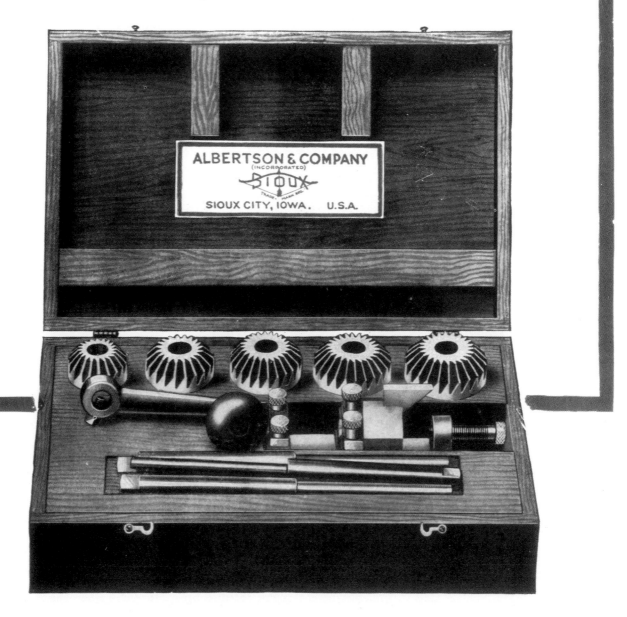

"Combine" vs. Labor Shortage

By J. M. Collins

THE recently introduced small harvester-thresher, known among farmers as the "combine," apparently is the salvation of the big winter wheat farms of the Middle West. That is the belief of N. M. Schlyer of Schlyer & Arnhold, implement dealers of Hays, Kan. Believing that, Mr. Schlyer, being an implement dealer, is in the van pushing the "combine." He's selling the machine just as fast as he can get it from the International's branch house at Salina. This year Schlyer & Arnhold are placing fourteen among their customers and would sell many more could they be obtained.

However, Mr. Schlyer refuses to take any credit for salesmanship.

"They sell themselves," he explains succinctly. "Every machine placed in a community means several more. It's a question of obtaining the 'combine,' not selling it."

It is probable no machine since the reaper has had such instant popularity among wheat farmers. In 1918 three harvester-threshers were placed in central and western Kansas. Last year the Salina distribution house sent out fifty-nine and this year, according to C. A. Morrison, branch manager, Salina will distribute between four and five hundred while the Hutchinson branch will sell a thousand.

All that has been done without the manufacturers pushing the machine. Production having always been behind demand, no big advertising campaign has been started for the new implement.

"The labor saving element is one of the strongest arguments the harvester-thresher has with farmers," Mr. Schlyer says. "In our section eight men are required for a header crew, and sometimes they're mighty hard to find, to say nothing of the expense of employing them. The harvester-thresher uses only four men. Harvesting and threshing is done in the same operation. There is no repetition of harvesting labor costs through threshing at a later period. There is no loss due to several handlings of the grain."

The weather last year had a great deal to do with booming the "combine" in western Kansas this season. Wheat lodged badly in many sections last summer and neither header nor binder could cut it efficiently. But, according to Mr. Schlyer and farmers, owners of the machine, the harvester-thresher "walked" into the fallen wheat and made thousands of dollars over its initial cost by saving this grain.

For instance, after B. J. Hammond, a farmer at Lake City, Kan., had cut and threshed 470 acres of standing wheat at a total cost of $807 and in addition had threshed 60 acres out of the shock, he sold his machine to R. E. McGrath. Mr. McGrath had 100 acres which he had given up cutting with a header or binder. With the "combine" he **saved about 1,300 bushels off the 100 acres.**

James Buffington of Delphos was another man who saw his wheat go down—300 acres of it. He purchased a harvester-thresher from a farmer near Minneapolis, Kas., paying the market price for it, and saved his crop. Then, upon the pleas of his neighbors, he went into their fields for one-half the crop and worked until Aug. 1, saving grain and putting the land in better shape than if the farmers had been forced to let the lodged wheat remain.

At Larned J. J. Jones says he had 320 acres of wheat last year, with straw five feet long, which was down after a rain, foiling both binder and header. His "combine," he says, threshed out between five and six thousand bushels with very little loss. Louis Bertrand of Oakley worked 102 hours to cut and thresh 300 acres, at an actual money expense for oil, gasoline and kerosene, including his tractor, of two cents a bushel. Allowing $7 a day wages for himself and his two boys and eighty cents an hour to each of two men and teams to haul away the grain, the total cost per acre was $2.20, which included all expense from the field into the car six miles away. Mr. Bertrand figures the cost of heading, stacking and threshing at prevailing rates would have equaled the cost of the machine—$1,885—and besides he believes he saved 300 bushels of grain which would have been lost.

"It knocks the wheat out of the straw and cleans the grain as well as any thresher I ever saw," Mr. Bertrand testifies. "I was counting a few days ago—making allowance for everything, wages of operators, fuel and oil, two teams to take care of and bin the grain, wear and tear, allowing fifteen years for the life of the machine, interest on money invested—and found that the total would be about $2.25 per acre on a 300-acre cut annually."

Mr. Bertrand used a Titan 10-20 to pull his "combine." Power for the harvesting-thresher is furnished by the machine itself in a 4-cylinder, 16 horse power motor, the same motor International uses in its trucks. Quite a few men pull the harvester-thresher with horses, twelve being required. W. O. Smith of Hoisington placed the use of his twelve horses in harvesting-threshing 450 acres at $966. He says the combined machine, in handling his wheat, effected a saving of $1,954.

The "combine" spreads straw back on the land as fast as the wheat is cut, an important fertility feature. Some wheat farmers with live stock, however, desire some straw. In that case they use the header or binder in the usual season, cutting enough to thresh out of the shock with the "combine" stationary after the rest of the crop has been handled. Attachments are provided at additional expense, with which the "combine" can be made into an "individual separator."

Two "combines" are manufactured by International, the Deering and the McCormick. They differ very little. Both cut a 9-foot swath and with both a 3-foot extension can be used, providing a 12-foot cut in light grain. With the 9-foot machine from fifteen to twenty acres a day can be cut, the manufacturers say. Both the Deering and the McCormick are regularly equipped with auxiliary engine, although they can be arranged to derive power from the wheels. Farmers, however, invariably prefer the auxiliary engine type.

A bagging platform is provided with either machine, or a wagon loader can be used. A pulley, straw carrier and feed table for use as a stationary thresher are extra equipment.

Hitches are supplied extra for use with eight, ten or twelve horses. A tractor with 10 hp. drawbar pull will handle the machine, equipped with auxiliary engine, if the grade is fairly level and the soil is firm. A light draft machine is made possible through the use of thirty-seven roller bearings.

A feature of the "combine" is that the straw walkers and cleaning riddles are in sections, so that should the machine tilt on uneven ground, straw and grain still would be evenly distributed. The separating process begins at the grates under the cylinder and it is claimed that by the time the grain leaves the beater 75 percent of the separation has been accomplished. The separation, however, is continuous from the cylinder to the bagger. Tailings go through the cylinder again. Undersized grains and weed seeds are delivered to a refuse sack.

The separating capacity is unusually large for the size of the machine. That is due to the fact that there is an extra chaffer screen which separates a very large part of the coarser materials from the grain, instead of leaving all the work to be done by the cleaner riddle. Cylinder speed is 950 revolutions per minute.

The machine weighs approximately 5,200 pounds, of which the auxiliary engine, rigidly supported above the main axle and directly connected to the cylinder shaft, weighs 1,158 pounds. The entire machine is eighteen feet wide.

Ease of operation is claimed for the "combine." The tilting lever is placed within easy reach of the operator. Large springs help raise the platform. It is an easy matter to change the height of cut to meet different heights of grain.

As the grain must be field ripened thoroughly for successful operation

THE HI-WAY PATROL COMBINES TRACTOR AND GRADER

of the harvester-thresher, some have feared it would shatter. Mary Best of Medicine Lodge, however, says not a cracked berry was found in the yield from her 250 acres. In central and Western Kansas, according to other farmers, no difficulty will be experienced in allowing the wheat to become dead ripe when the "combine" is used.

Whether grain threshed directly from the standing head would "keep" is another fear of some observers, who have pointed out the theory wheat should be allowed to "sweat" in the shock or stack. But Jacob Brull and George Bellman, farmers of Ellis County who used "combines" purchased from Schlyer & Arnhold last year, say they have been forced to hold wheat from last year because of the car shortage and it has not deteriorated so far as they can see. There is little "sweating," they say, when grain is cut with the

"combine" as it is riper than that cut with header or binder.

One rather grave situation is being forced upon the "combine" sections through the use of the new machine coupled with the car shortage. That is from lack of storage. Mr. Schlyer says he fears the possibility of wheat being dumped upon the ground when farmers who haven't provided bin room find they cannot get cars. But he's working, as a true friend of his customers, to bring home to them the importance of providing granaries, either through purchasing manufactured ones, building their own or putting sheds and chicken houses in shape.

"But we'll weather the crisis," he says optimistically. "With money and labor saving possibilities in the harvester-thresher and with our crops running 20 to 30 bushels to the acre, watch out for western Kansas."

PUPILS IN CONSOLIDATED SCHOOL DISTRICTS ARE TRANSPORTED IN PASSENGER TRUCKS

Incorporate for Farming

St. Louis, Mo., Nov. 17.—What is held to be the first instance of incorporated farming in the United States has been inaugurated by a company of St. Louisans. The company is operating under the name Riverview Ranch, the property being 1,100 acres of farm land in Jersey County, Illinois. It proposes to raise hogs for the market, being easily accessible by river to the stockyards in East St. Louis. The new enterprise is also held to be the first instance in the farm field of a concentration on one species of product.

Highway Law to Be Tested

Dallas, Tex., Mar. 8.—A case to test the constitutionality of the highway law, which limits the capacity of trucks to 4,000 pounds, was filed here against J. H. Faison, operator of a freight truck line.

On a Chautauqua Ticket

LYONS, KAN., Sept. 21.—Mel Taylor of Taylor & Sons recently wrote "Buick Valve-in-Head" 1,225 times on a chautauqua ticket three inches long and an inch and a half wide, winning the $100 discount offered in a contest conducted by the Buick agency. A local paper calls Mr. Taylor's chirographic exploit "a beautiful example of copper-plate script."

Johnson Motor Products

SOUTH BEND, IND., Nov. 22.—The Johnson Motor Co. has taken over the activities of the Johnson Motor Wheel Co. and of the Quick Action Ignition Co., and is now building the Johnson motor wheel, the Johnson outboard motor and the Quick Action flywheel magneto. Charles Kratsch is manager of the Johnson Motor Co. and P. A. Tanner, formerly with the Splitdorf organization, is assistant general manager.

Maytag Adopts Quick Action

NEWTON, IA., Nov. 17.—The Maytag Co. has adopted as standard equipment on their Multi-Motor washing machine the Quick Action magneto, manufactured by the Johnson Motor Co. of South Bend, Ind. The Quick Action magneto, which the Maytag Co. has been using as optional equipment for some time, replaces the former battery equipment.

This picture shows the complete operating mechanism

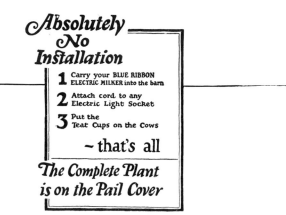

Absolutely No Installation

1 Carry your BLUE RIBBON ELECTRIC MILKER into the barn

2 Attach cord to any Electric Light Socket

3 Put the Teat Cups on the Cows

~ that's all

The Complete Plant is on the Pail Cover

Why You Can Sell
This *Portable Electric* Milker

A Farmer's Milker. The farmer knows that *this* is the milker which *he* can use profitably. A demonstration is usually a sale.

Positive Pulsator. The Blue Ribbon Pulsator, worm-driven direct from the motor, is the only one of its kind. Pulsations are always exactly the same, regular and uniform.

Easily Portable. Light and compact. Boys, women and girls can easily handle and operate The Blue Ribbon Electric Milker.

No Installation. Electric cord which comes with milker is the only "installation."

Simplest to Clean. Simplified teat cups have only two parts—outer shell and the one piece rubber "inflation" which also forms part of the milk tube. Not a corner or crevice to make cleaning hard.

Vacuum Control. Vacuum instantly adjustable. Made just right to suit hard or easy milking cows, in a second's time.

Dependable Power. The Blue Ribbon Electric Milker runs from any standard farm light plant or central station current.

Low Price. The Blue Ribbon Electric Milker is priced within reach of nearly every farmer.

The BLUE RIBBON ELECTRIC MILKER

MILK with Electricity

No Field Service
on the Blue Ribbon Electric Milker

The Blue Ribbon Electric Milker is so simple that the Complete Plant is on the Pail Cover, so sturdy that it almost never gets out of order. In those rare cases when trouble does arise, the farmer simply brings his machine top to you, swaps for a service top, and is home again in a hurry—all ready for the next milking. This plan keeps the milker in constant operation for your customers. It saves you from long, expensive trips into the country making adjustments and repairs, so you can devote your whole time to selling. It is a money-making proposition for you both.

Free Demonstration *on Their Own* Cows Helps Make Sales to Farmers

This exclusive Blue Ribbon feature gives our dealers a big sales advantage. The Blue Ribbon Electric Milker requires absolutely no installation. You can walk right into a farmer's barn at milking time, attach the cord to the electric light socket, do his milking, so that he can **see** just how much time and work this wonderful little machine will save for **him**.

He'll want it just as hundreds of other farmers have wanted it. The low price will clinch the deal. You can often make the sale **in a single call.**

The Blue Ribbon Electric Milker is profitable for small herds as well as large ones. It opens up an entirely new field. There are dozens of prospects right in your vicinity. A powerful advertising campaign is bringing The Blue Ribbon Electric Milker to their attention. There is plenty of business for the men who are willing to go out after it. Such dealers will find our proposition very profitable.

Write for booklet and territory today

ELECTRIC MILKER CORPORATION
706 Tower Building :: CHICAGO

LE ROI POWR-UNIT

THE farmer is tired of a multiplicity of engines. He needs an all-purpose power plant, powerful enough for the heaviest belt work, economical enough for the smallest power chore.

The demand has produced the machine—LE ROI 4-cylinder POWR-UNIT. Provides a steady "pulse of power" for any belt work job on the farm. Tractor-type construction and automobile controls. Develops 15 H. P. *Weighs 850 lbs.*

LE ROI motors lead as power plants on tractors, motor cultivators, trucks, autos, contractors' equipment. LE ROI POWR-UNIT is built to enhance that reputation.

LE ROI COMPANY, Mitchell St. and 60th Ave., Milwaukee, Wis.

HORSEPOWER—15.65 H. P. Standard rating.

EQUIPMENT—Magneto with impulse coupling, Carburetor, Air Cleaner, Fuel Tank, Speed Governor, Clutch, Two Belt Pulleys mounted on anti-friction bearings, Radiator and Fan.

Dealers who wish particulars of our attractive selling franchise are invited to communicate with us at once.

SPREADER SALES TALK--No. 2

The Distributor

The Distributor is an extremely important feature on any manure spreader. In fact it **is** the spreader. Without it the machine becomes a mere unloader.

The work of the distributor is three-fold. First it must spread the manure **widely** so as to cover the greatest possible amount of ground each time the machine is driven across the field.

Next it must spread the manure **evenly**. Manure dropped patchily—thick in some spots and thin in others—can never give as good results as when it covers the ground in a thin, absolutely even blanket.

Thirdly, after the beater cylinders have thrown the manure to the distributor, the revolving blades of the latter must, by their action, assist in breaking up any lumps or clods that may remain unbroken.

There are eight blades on the Distributor of the C2 New Idea or No. 26 NISCO SPREADERS and you will note the peculiar spacing. The shape and spacing of these blades is patented.

They are spaced successively around the shaft, so no one blade is in the path of the one following. The outer edge of the blades describes a curve eccentric to the revolving shaft and this causes the material handled to slide to the outer face of the blade, where the centrifugal and lateral motion is greatest, giving a wider spread and preventing the spasmodic dropping of pulverized material—a trouble largely prevalent in other machines when handling manure difficult to pulverize and spread.

The Distributor on the NEW IDEA and NISCO SPREADERS fans the manure out into a path seven or more feet wide—often wider—at least six to twelve inches on each side of the wheel tracks. It entirely covers the wheel tracks and you do not have to drive over the manure after it has been spread. It spreads the manure with an evenness that has never been excelled and it is specially designed to break up and pulverize any lumps or clods that may have escaped the beaters.

When the NEW IDEA SPREADER was invented, the Distributor **was** the "New Idea." Prior to that time all so-called manure spreaders were entirely of the crude "unloader" type.

It is a tribute to the inventor to say that practically **all** of the present day manufacturers of spreaders have attempted to imitate the New Idea Distributor, to a greater or less extent.

We are the pioneers in building spreaders with distributors. Our constant experience of over twenty years has been utilized to the utmost in designing the shape and spacing of the distributor blades on the New Idea and NISCO SPREADERS.

We claim that this is the only real Distributor—the only one that will do the work as it should be done.

The New Idea Spreader Co., Coldwater, Ohio

Harrisburg, Pa.	Chicago, Ill.	Kansas City, Mo.
Syracuse, N. Y.	Jackson, Mich.	Waterloo, Ia.
Columbus, O.	Minneapolis, Minn.	Peoria, Ill.
Indianapolis, Ind.	Omaha, Neb.	St. Louis, Mo.

1919-1923

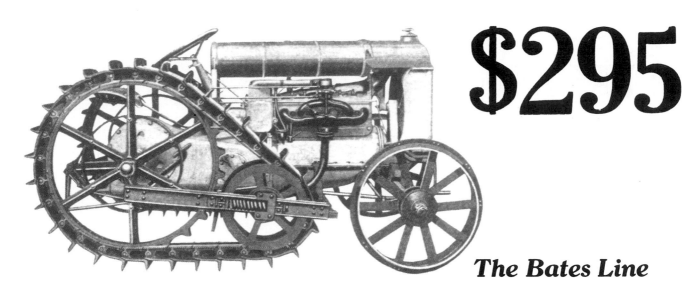

$295

The Bates Line

Our Business is Good

SINCE making our announcement of Bates Steel Crawlers for Fordson Tractors, wide awake distributors and dealers from every part of the country have been wiring or writing for territory.

Our large dealer commission and the granting of exclusive territory to accepted dealers is very popular.

Since the recent reduction in the Fordson Tractor, the Market for this Steel Crawler is increasing every day.

Road Contractors use them for work with Wheel Scrapers, Building Contractors for excavating. Counties and Townships for dragging and maintaining their dirt roads and Farmers for their Spring work where the Crawlers make their Fordson accomplish twice the work.

Many distributors and dealers are using this Fordson Crawler to pay the expense of their present selling organization.

Territory going fast.

Wire or write for full information.
Catalog and full details on request.

The Bates Steel Crawlers make Fordsons disc faster.

The Model "G" Bates Steel Mule for Big Farmers and Contractors.

The Bates Steel Mule Speeds up Farm work.

The Model "I" Bates Steel Mule for County Road work.

Bates Machine and Tractor Co.

Established 1883

220 Benton Street,   **Joliet, Illinois**

(J-105)

"The Automobile Club of France accepted an offer of the authorities of LeMans to hold the French Grand Prix Race for 183 cu. in. cars at that place." The race was held at LeMans on July 23, 1921. It was decided by the National Implement and Vehicle Association that changing to the Metric system would lead to disorganization and confusion. The Society of Automotive Engineers boasted of "over 5,000 members resident throughout the world." The S.A.E. had already "established over two hundred separate and distinct standards relating to materials and construction of automotive apparatus." Reginald Heinzelmann, inventor of the Timken roller bearing, died at the age of 45. Heinzelmann sold to Henry Timken, his uncle, the rights to the bearing for $60,000. E.R. Koontz of Richards, Missouri, invented and perfected the "Self-Tier" for the baler. The new Ann Arbor Machine Co. baler incorporated the tieing mechanism that had been fully tested the previous two years. The self

tier made the baler a one-man machine.

In 1922, the Fordson tractor price was cut an additional $230 to $395. Three years earlier, the price had been $885. Other tractor manufacturers were slashing prices to similar levels and some were including tillage equipment to induce sales. The Bulldog Tractor Corp., the Franklin Tractor Co., the Illinois Tractor Co., the Indiana Silo & Tractor Co. and the Oshkosh Tractor Co. liquidated operations. The New Moline Plow Co., near bankruptcy, ceased production and the Dowagiac Drill Co. was dissolved. The General Ordnance Co. of New York, New York, entered receivership and the General Motors Corp. left the competitive farm equipment market. Early in the year, General Motors Acceptance Corporation financed both wholesale and retail purchasing for their Sampson dealers. The Sampson plant was later closed and converted to a Chevrolet assembly plant. The Trackson Co. began in Milwaukee, Wisconsin and Harry W. Bolens bought

the property of the Globe Metal Products Co. at a receiver sale and formed the Wisconsin Castings Co. International Harvester tested 20 prototype "Farmall" tractors in actual field conditions. Wireless corn planters, invented by A.C. Croft, began to be manufactured in Wagner, South Dakota.

Even with coal mines shut down and many factories closed, it was difficult to find men to work in the harvest fields. Farmers were paying $3.50 and three to five meals per day for field hands, but the shortage was critical in some areas.

THE RIGHT TIRE FOR THE TRACTOR

Copyright 19__ by The Goodyear Tire & Rubber Co., Inc.

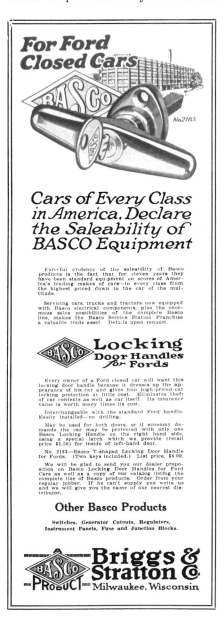

Some Mighty Attractive Business Awaits You on These Improved Machines

The Massey-Harris Reaper-Thresher

Cuts and threshes the grain at one operation at far less cost and with less waste than it can be harvested with either a binder or a header and then threshed with a grain separator. It's the culmination of many years of study and field observation on the part of Massey-Harris Co., Ltd., who pioneered the development of the small type of Reaper-Thresher and has built into it only such principles as time and experience have proven to be correct. Already has it won its way to leadership in Australia, South America, Spain, North Africa and other countries where machines of this type can be used and because it has, it is now offered to you with every assurance of its efficiency and success. It is bound to be in great demand in this territory, and the dealer who ties up with it at this time **will cash in big.** And just as this Reaper-Thresher is destined to play an important part in the harvesting operations of this country during the coming years so is

The Massey-Harris Field Cultivator

destined to become an important factor in the operation of tilling the soil and eradicating weed growths. For the destruction of thistle and quack grass, the summer fallowing of land, the breaking of wheat ground, etc., this cultivator is unsurpassed. It is the equivalent of a disc harrow and a spring tooth harrow combined, and will dig up hard ground that neither of those implements could touch. Sizes for either horses

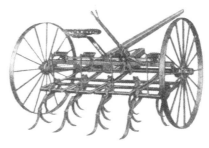

or tractors are built and both hand and power lifts are available. In the full sense of the words **it is improved farm equipment.**

Both of these machines are on display at our Kansas City Branch, where you are cordially invited to call at any time. See them and the other machines of the line during Convention Week and ask about the Massey-Harris Contract. It's a real asset to any wide-awake dealer who is ready to forge ahead.

Massey-Harris Machines and Implements are a warranted product backed by 70 years' experience in implement building. This is of what the line consists—

Mowers	Spring Tooth
Reapers	Harrows
Grain Binders	Culti-Packers
Headers	Field Cultivators
Corn Binders	Corn Cultivators
Self-Dump Rakes	Manure Spreaders
Side Delivery Rakes	Grain Drills
Hay Tedders	Reaper-Threshers
Hay Loaders	Cream Separators
Disc Harrows	Canvas Covers
Spike Tooth	Binding Twine, etc
Harrows	

Massey-Harris Harvester Company, Inc.

Builders of Farm Machines and Implements Since 1850
Head Office and Factory, Batavia, N. Y.
Southwestern Branch, 1306 West 12th St., Kansas City, Mo.

"Tuning In" With a New Trade

By Thomas Whipple

IT'S time for the hardware and implement dealer of the west to "tune in" with the volume of wireless telephone, or radio, equipment that is being "broadcast" from San Diego to Halifax. Clubs, churches, lodges and individuals are installing receiving sets that enable thousands of persons to sit in their own homes or in a neighborhood auditorium of some kind and hear music, monologues, speeches, weather forecasts, market reports and news bulletins. Many a popular "craze" has seized the people of this country from time to time, but it is safe to say that they are giving their interest to the development, use and enjoyent of wireless telephony with greater enthusiasm than they ever before gave to anything of the kind.

As a matter of fact, that foregoing sentence is not accurate, for there isn't anything else of the kind. The radiophone is one more successful challenge to the ancient saw that there is "nothing new under the sun." It is silly to argue that the ether, which carries the wireless impulse, and the force that exerts the impulse, have always existed. To all intents and purposes, so far as the average individual is concerned, the wireless telephone is a good deal newer than fresh paint. To any intelligent being it is a miracle of science and he who marvels merely displays his intelligence.

Metropolitan newspapers from one side of the Continent to the

FARM FAMILIES ARE NOW IN THE WIDENING CIRCLE OF RADIO "FANS"

other are broadcasting all sorts of entertainments and news. Not many days ago a Kansas City paper received a dispatch that its wireless entertainment had been enjoyed 'way up in Alberta. The development of the radiophone is so important in the eyes of the United States Department of Agriculture that steps are being taken to encourage it officially. W. A. Wheeler of the Department says:

"There are more than thirty-two million persons on farms, comprising nearly one-third the total population of the United States. Most of these are located where they are virtually cut off from immediate contact with the outside world. The radio is the only means of getting to them quickly at small cost the economic information necessary in the conduct of their business. . . . And anything in the way of entertainment that will afford the farmer even a slight diversion from his daily labors will be of great benefit to the nation."

All of which hooks up directly with the business of most hardware and implement men. The larger part of their business is with the farmer and his family; at the same time the nature of the radio is such that it is equally interesting to town and city people. A trade paper man visiting the west recently said, "There are seven radio outfits installed in the New York apartments where we live."

By far the best radio customer and booster is the boy with mechanical—especially electro-mechanical—leanings. This is the chap for the hardware man who expects to get very far in the sale of radio equipment to cultivate.

Tractor Prices Compared

KANSAS CITY, Mar. 1.—Following is a comparison of tractor list prices f. o. b. factory on March 1, 1922, with the peak prices of the various models:

Tractor	Peak Price	Mar. 1 Price
Advance-Rumely 12-20	$1,750	$1,085
Advance-Rumely 16-30	2,665	1,750
Advance-Rumely 20-40	3,650	2,550
Advance-Rumely 30-60	5,250	3,775
Allis-Chalmers 15-25	1,495	1,310
Allis-Chalmers 20-35	2,150	1,885
Aultman & Taylor 15-30	2,600	1,900
Aultman & Taylor 22-45	4,300	2,800
Aultman & Taylor 30-60	5,900	4,000
Avery 6 cylinder	915	545
Avery 8-16	915	445
Avery 12-20	1,625	1,360
*Avery Track-Runner	*1,625	*1,625
Avery 12-25	1,385	625
Avery 14-28	1,830	1,415
Avery 18-36	2,450	1,910
Avery 25-50	3,375	3,000
Avery 45-65	4,300	3,865
Bates F. Crawler	2,350	1,785
Bates H. Wheel	1,785	1,275
Wallis	1,795	i995
Case 10-18	1,250	p700
Case 16-27	1,865	1,420
Case 22-40	3,350	2,550
Cletrac F.	845	795
Cletrac W.	1,585	1,345
Waterloo Boy	1,400	675
Emerson-Brantingham 9-16	1,125	600
Emerson-Brantingham 12-20Q	1,395	750
Emerson-Brantingham 12-20AA	1,700	1,095
Emerson-Brantingham 16-32D	2,575	1,820
Fordson	885	395
Hart-Parr 20	1,195	945
Hart-Parr 30	1,595	1,295
*Caterpillar T35	*2,350	*2,350
Caterpillar 5 Ton	4,500	3,975
Caterpillar 10 Ton	7,000	6,050
Indiana	1,100	675
International 8-16	1,150	p670
Titan 10-20	1,295	p700
Twin City 12-20	1,760	1,395
Twin City 20-35	3,175	2,950
Twin City 40-65	5,750	4,750
Minneapolis 12-15	1,200	800
Minneapolis 17-30	2,000	1,600
Minneapolis 22-44	3,300	2,650
Minneapolis 35-70	4,900	3,850
Moline	1,395	970
Nichols & Shepard 20-42	3,100	2,650
Nichols & Shepard 25-50	3,460	3,000
Nichols & Shepard 35-70	4,160	3,650
Heider D.	1,270	870
Heider C.	1,570	905
Samson	1,075	445

I—Price includes power implement.
P—Price includes plow or disk harrow.
*—New model, price just announced.

A Power-Driven McCormick

A POWER-DRIVEN grain binder, which takes its power direct from the tractor, is being put out this year by the International Harvester Co., under the McCormick trade name. It is a 10-foot cut machine and is intended for use with the International 8-16 hp. tractor.

The binder is practically the same as the regular McCormick binder excepting that it is equipped with the power drive mechanism. Power is transmitted direct from the tractor to the binder by a revolving shaft. Tractor and binder must be ordered together. The International 8-16 hp. tractor, when ordered with the binder, is equipped with a power take-off transmission. A slip clutch is provided which can be adjusted easily to transmit the right amount of power to the binder.

The McCormick power-driven binder has two sets of control levers. One of them is located within easy reach of the tractor seat. For instance, the levers for controlling the reel, tilting the machine, shifting the binding attachment, may be easily reached by the operator. The bundle carrier trip is operated by the tractor driver with ease. The other set of levers are for use in extremely bad conditions where it is necessary to put a man on the binder seat.

It is supplied with transport truck, one-man binder control tractor hitch, and stub tongue. The stub tongue can be attached to the binder platform the same as on the regular McCormick binders for transporting through narrow gates and along the road.

Inasmuch as the binder mechanism is run by the engine of the tractor, the grain can be cut in wet spots wherever it is possible to travel with the tractor. This outfit has cut grain successfully where horses could not go.

A 10-foot swath is an unusually wide cut and enables one man to cut from 25 to 75% more than with the ordinary binder. The extra width of cut, the more rapid and uniform rate of travel and being able to work the tractor more hours a day, make this possible. Not constructed to be drawn by horses.

REVOLVING SHAFT, A, TRANSMITS POWER TO THE BINDER MECHANISM

A. B. C. Ironer Is All-Electric

A NEW, 26-inch, electrically heated and operated ironer is being added to the A. B. C. line of electric washer and ironers put out by Altorfer Bros. Co. of Peoria, Ill. The new ironer will be known as the A. B. C. All-Electric and will retail at $145.00, and is for the average family.

An important feature of this Ironer is the heating element which takes only 1500 watts maximum; 1350 watts average. This means that expensive, special wiring necessary for other ironers is done away with. The only wiring necessary is to run two No. 12 wires from the meter box to the outlet plug for the heating elements. This should not cost more than $5 or $10.

The low wattage, together with the convenient size and method of operation of the A. B. C. All-Electric is expected to make it as acceptable to the housewife as a washer. It requires only 25-inches by 33-inches of floor space and can be used in the kitchen, on the back porch, or anywhere.

The All-Electric is equipped with an open end roll for such pieces as cannot be fed through flat. A full width foot lever gives instant control of the shoe from any position, leaving both hands and full attention for the work. The shoe, under instead of over the roll, allows steam to escape and facilitates feeding.

Economy of electricity is effected by a three point rheostat on the heating element. After the shoe is brought to the ironing heat, it can be kept uniformly hot by reducing the current first one-third and then two-thirds. An automatic motor switch is kicked on every time the shoe is let into the ironing position. This minimizes the possibility of scorching the pad by keeping the roll turning whenever it is in contact with the shoe. Power from the over-sized motor is transmitted through a belt-worm-chain drive that is noiseless, trouble proof and flexible enough to prevent motor burn-outs.

Double Gasoline Mileage

St. Louis, Feb. 7.—Julius H. Barnes, president of the United States Chamber of Commerce, while in St. Louis stated that the General Motors Corporation had worked out at its Dayton laboratories a formula for liquid with a lead base, which, when mixed with gasoline, will double the mileage obtained from gasoline by motor cars.

He was told that the mixture for gasoline will be available at filling stations to motorists in sufficient quantities this year. He pointed out that the discovery would have the effect of doubling our oil resources.

To Market a New Gasoline

New York, April 16.—The General Motors Chemical Co. has been incorporated for the marketing of a new type gasoline which is known as a modified gasoline, "containing an anti-knock compound developed after many years of experiment" by the General Motors Research Corporation, Dayton, O.

In 1923, combines continued to be popular. The new combined threshers were usually owned by one farmer, who used it on his crop and the crops of his neighbors. The job of the professional threshermen was considered threatened.

John Deere introduced the model "D" tractor which continued in production until 1953. The Winnebago Tractor Co. of Dixon, Illinois, merged with the Clipper Lawn Mower Co. Production capacities for the Winnebago Chief tractors.

Bement differentials and Clipper lawnmowers were all to be increased.

Recently patented "V" belts were beginning to be used to drive accessories such as the water pump and fan on gasoline engines.

Thresher Has Novel Design
Wichita Inventors Bring Out Separator Making Use of Centrifugal Force

DEPARTING almost entirely from the accepted design of threshing machines on the market today, H. L. Strong of the Strong Trading Co., Wichita, Kan., has brought out a separator built on the "centrifugal force" principle, which is known as the New Centrifugal thresher. This machine is not yet on the market, as Mr. Strong has declared himself unwilling to undertake the manufacturing of it without the proper equipment and resources.

The machine, designed by T. E. Forster, was built in Wichita machine shops in January of this year and was first shown and operated the following month. Since that time the designers have tested it in field work under all conditions and have just released the detailed information of the design with photographs.

The outstanding departure from standard design is in the threshing chamber which is a steel "boiler" into which the feeder brings the grain. This chamber is tapered slightly at the far end where the suction fan takes the straw to the stacker. The chamber is constructed of 14 gauge steel, riveted, and is 8 feet long and 32 inches in diameter.

Through the chamber operates the main propeller threshing shaft, a 3-inch cold rolled steel shaft which is 11 feet long, running horizontally through the center of the threshing chamber and also through the fan fan. The drive-pulley end of the shaft is located directly below the feeder and band-cutter.

On the main threshing shaft are a number of specially shaped steel propeller arms called "threshing arms." These are not designed as beaters, the designers say, as there is no resistance to cause beating. These arms housing, operating the large suction are bolted to the main shaft in a spiral formation, and thus whirl the straw by centrifugal force while carrying it through the cylinder.

The threshed grain drops into a grain pan below, through what are known as "adjustable deflectors." In this pan below is an auxiliary threshing shaft, a 2-inch cold rolled steel shaft, 15 feet long, running horizontally through the 10-inch tube-pan below the main threshing chamber. Small arms are fastened to the auxiliary shaft spirally. These aid in the threshing, while also scouring and polishing the grain. The arms are also designed to serve as conveyors, taking the threshed and cleaned grain to the back of the machine where it is elevated to the cleaning shoe for a final separating from the chaff.

The suction fan, discharge stacker and fly wheel are at the back of the threshing chamber. The fan is 32 inches in diameter and is so constructed as to aid the centrifugal force of the main propeller arms in separating the grain from the straw. The fan also operates as the stacker fan and serves as a balance wheel for the entire machine.

From the auxiliary cleaner the grain and chaff are taken to the cleaner by an elevator-conveyor of endless chain and steel buckets. From here the grain with the chaff reaches the cleaner shoe where it is cleaned and delivered into the loading elevator and then into the grain wagon.

The cleaning mechanism consists of a shaker shoe which carries an adjustable chaffer and screens, operated by a pair of eccentrics giving a three-quarter inch throw. A small fan screens and chaffer. Any unthreshed throws an air current under the fan chamber and then into the stacker grain not going to the loading elevator while the chaff is blown in the suction drops back into the auxiliary cleaner pipe.

Roller bearings are used on the main propeller shaft and on the auxiliary threshing shaft.

The feeder is of the chain and slat carrier type with revolving band cutter knives, feeding the grain into the threshing cylinder. The designers claim that a new type feeder governor has been designed which is a special feature and not yet ready for announcement.

The machine is also so designed, it is claimed, as to be available for use as a header thresher, driven directly from the tractor. It is declared that the New Centrifugal will thresh different grains similar to machines now on the market of the accepted standard design.

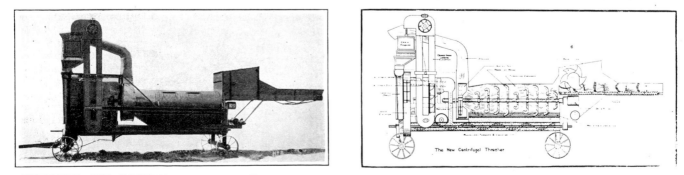

EXTERIOR AND INTERIOR VIEWS OF THE CENTRIFUGAL THRESHER RECENTLY INVENTED BY WICHITA MAN

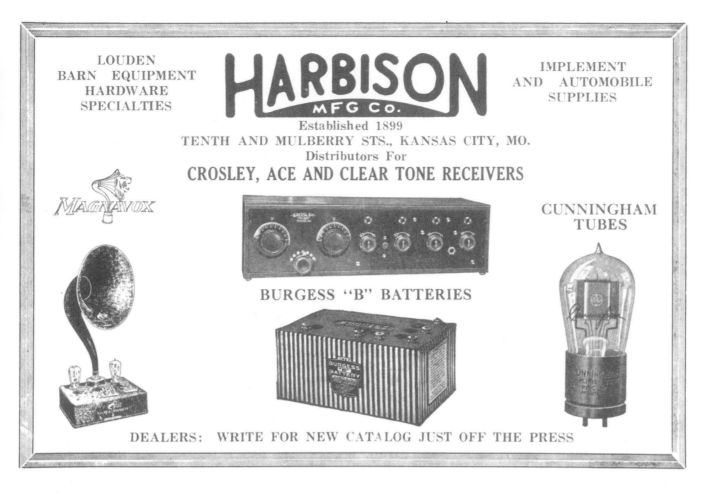
To Cover Territory in Roadsters

SOME of the Missouri and Kansas travelers for the John Deere Plow Co., Kansas Ciy, have been equipped with Ford roadsters and will cover their territories in these cars. M. J. Healey, vice-president and general manager of the house, plans to use this equipment over a period as a try-out of the automobile in contrast with train travel.

Soon he expects to accumulate enough data to settle the question of service and economy as between the two modes of transportation for traveling salesmen. They are being used only by regular travelers in territories where such practice seems to be practicable. The men to whom machines have been assigned for the present are: N. W. Ballew, Kansas City; J. R. Hall, Sarcoxie, Mo.; C. M. Hassel, Chillicothe, Mo.; R. E. Circle, St. Joseph; H. F. Manny, Wichita; C. E. Bair, Salina, Kan.; R. S. Scymanski, Norton, Kan.; E. M. Burke, Manhattan, Kan.

In uniform style, each of the roadsters has been decorated liberally and attractively with gilt letters telling the world that the man behind the wheel represents the full-line farm implement manufacturing and selling organization of John Deere. As an advertisement, at any rate, it is not believed that the train can compete with the roadster.

His Reception

Finally, there is a new story about British efficiency. To the hospital for seamen in London came lately a strange old man whose conversation was unintelligible to the attendants. So they burned his clothes, scrubbed him, shaved him, gave him a bromide and put him to bed. When he woke up the next day it was discovered he had dropped in to call on a sick friend.—

THE "FLEET," ALL READY TO GO, IN FRONT OF THE HOUSE OF DEERE, K. C.

1924-1927

COMPANIES TRY TO SATISFY
NEEDS OF THE FARMERS

Implement & Hardware Trade Journal

Member Audit Bureau of Circulations and Associated Business Papers, Inc.

1924-1927

In 1924, the Implement & Tractor Trade Journal had a slogan contest and the winning entry was "Good Equipment Makes A Good Farm Better." Many manufacturers and a large number of dealers began using this slogan to encourage the farmers to buy better and more modern farm equipment. More implement dealers began to handle radios and other items of hardware. The farm equipment was not selling as well as before and other products were necessary for the established dealers to remain in business.

The Soviet (Russian) government took over the International Harvester Co. factory at Lubertzy, near Moscow, dispossessed the company's representatives and began operating it as a government institution. There was no compensation received or promised for the estimated value of the $2,291,000 facility and the property was considered by many as stolen. "That is how 'idealism,' gone mad, works in Russia."

The Westinghouse Light Co. "fooled" plants into growing faster using electric lights. The U.S. Marines tested a new amphibious tank. The Federal Trade Commission issued a complaint that listed most of the major farm equipment manufacturers and a dealers' association suspected of "fixing prices." The Eclipse Lawn Mower Co. changed its name to Eclipse Machine Co. The Advance-Rumely Thresher Co. purchased Aultman & Taylor Machinery Co. of Mansfield, Ohio, effective on the first day of 1924. Early in the year, the Avery Mfg. Co. (previously called the Avery Planter Co.) of Peoria, Illinois, filed a petition for bankruptcy, but was reorganized as the Avery Power Machinery Co. by July. Engines designed by the defunct Midwest Engine Co. were continued by other manufacturers including Allis-Chalmers Mfg. Co., Huber Mfg. Co., J.G. Brill Co. and the Waukesha Motor Co. The Walter A Wood Mowing & Reaping Co. of Hoosick Falls, New York, was sold to the Bateman Bros. of Philadelphia, Pennsylvania. The company was renamed the Walter A. Wood Implement Co. The Champion Line was

purchased by B.F. Avery and Sons of Louisville, Kentucky, and the New Moline Plow Co. purchased the Milwaukee line from International Harvester which had to divest several lines. The first Farmall tractor was sold to an Iowa farmer. Two hundred models were released the first year. International Harvester also introduced the power drive McCormick corn picker, that was driven by the tractor's power take off shaft. Not everything was powered by gasoline or electricity, the J.I. Case Threshing Machine Co. still issued a twenty page catalog of horse drawn equipment.

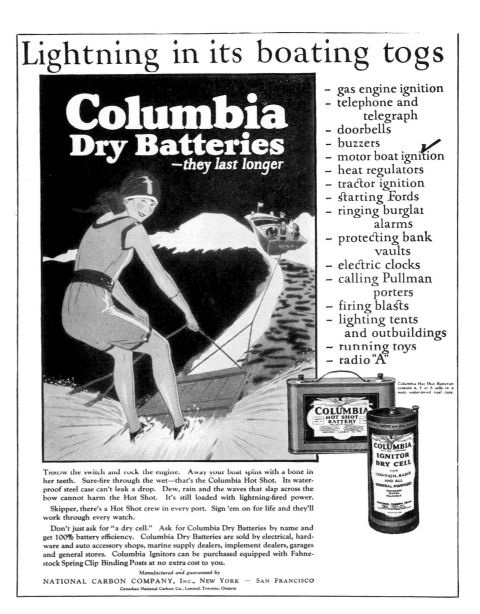

Alcohol Is Best For Radiators

LETTER circular No. 28 in revised form has just been issued by the U. S. Bureau of Standards. Copies of this may be had on application. It embodys the tests made on a number of anti-freezing mixtures and gives a table showing the percentages of alcohol, of glycerine, and of combinations of the two that are necessary to produce a solution having a given freezing point.

Alcohol is still regarded as the best material to keep automobile radiators from freezing. If wood alcohol is used care should be taken to see that it is free from acid, otherwise corrosion of the radiator and circulatory system may result.

The great drawback to the use of alcohol is the fact that it readily evaporates and has to be replaced. Glycerine does not have this drawback, but it is more expensive.

Among other anti-freeze solutions discussed are calcium chloride, honey, glucose, and kerosene.

Calcium chloride is frequently used and found to be very effective, but it had a decided corrosive action particularly on solder and aluminum.

This salt forms the basis of many of the patent antifreeze mixtures on the market. To some of these a soluble chromate is added to prevent corrosion. The bureau finds that little corrosion results when this is added except to the aluminum parts.

Another troublesome feature of calcium chloride is its tendency to cause short circuits when it gets on the spark plugs or ignition wires. When cold it takes up moisture and forms a good conducting layer where such is not desired.

It is hard to remove and the short circuits formed by it hard to find because they disappear when the engine gets hot.

Honey and glucose were found unsatisfactory because a high concentration is necessary to prevent freezing, and this results in a thick solution that does not flow freely. There is also danger of depositing sugar in the circulatory system.

Low percentage solutions do not lower the freezing point to any great extent, but they do prevent bursting the radiator. When such a solution does freeze it first turns to a slush which must first be cooled to a considerably greater extent before it turns solid.

Kerosene is not recommended because its vapors are inflammable and its high and uncertain boiling point is likely to cause serious overheating of the engine, or even to the melting of

solder in the radiator. It also has a slight solvent action on rubber.—G. H. Whiteford, Professor of Chemistry, Colorado Agricultural College.

"Bolens Broadcaster"

MUCH information about the market for garden tractors is carried in the "Bolens Broadcaster," a monthly house organ published by the Gilson Mfg. Co., Port Washington, Wis. It is the namesake of Harry W. Bolens, president of the company, and is full of interesting facts about the construction and use of the Bolens Hi-Boy and the Bolens Power Hoe, also namesakes of his.

The second number of the first volume came off the press recently. Its leading article entitled, "The Garden Tractor Age Is Here," written by L. L. Heller, farm advisor for Cook County, Illinois, discusses the practical side of small-unit power for the truck farmer and declares that "the new garden tractor will emancipate the truck grower's family from wheel-hoe slavery."

He ends his article with: "Here's a toast to the garden tractor—It keeps the boys on the farm and the girls out of the field."

Both of the Bolens models are thoroughly described in this number of the "Broadcaster" and a number of strong testimonials are printed. Dealers will be placed on the mailing list if they request it.

PISTON RING LOGIC

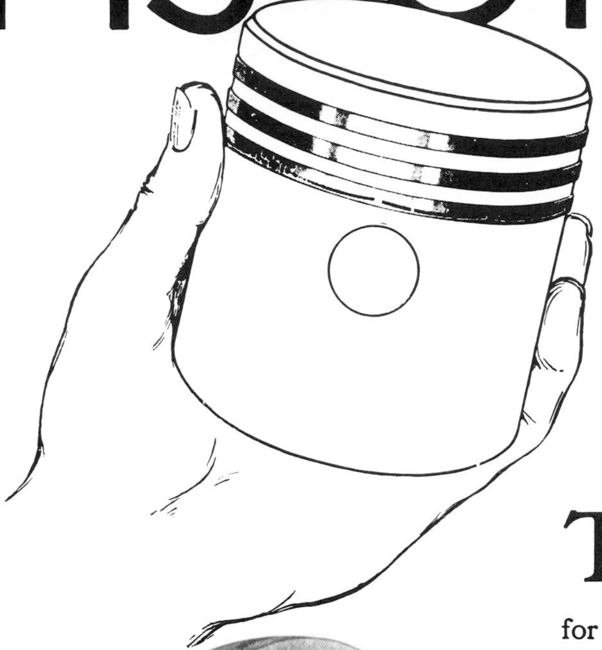

THE Quality Drainoil Ring, the latest development in oil controlling piston rings, has provided the most reliable piston ring combination for the control of oil and is the most logical in principle, when installed correctly in the lowest groove of the piston, with Quality Piston Rings in the remaining grooves.

Not until every possible test had been made to prove that the design is mechanically correct and that it would perform satisfactorily, was this ring adopted. The reputation of the Piston Ring Company is too highly prized to risk on an unproven product.

When installed, the knife-like scraping edges of oil slots cut excess oil from the cylinder walls, on downward stroke of the piston, and force it back to the crank-case through channels formed by slots in ring and drilled holes in piston grooves.

Quality Drainoil Rings are shipped four to the package in durable, attractive, non-fading orange and black boxes with all sizes and oversizes plainly marked on boxes and on rings.

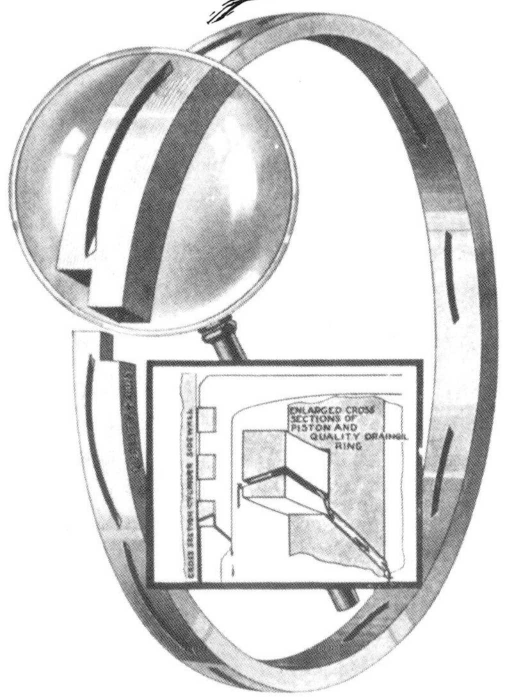

Scientifically Designed
Proper Material
Quick Seating
Lasting Resiliency
Individually Cast

Accurately Machined
Correct Wall Pressure
Easy to Install
Long Life
Turned Surface

The name QUALITY stamped in each ring is your guarantee. Accept no substitute.

The *Piston* RING COMPANY

Muskegon, Michigan

"Good equipment makes a good farmer better"

For One More Dab

IN and around any well-regulated rest-room should be plenty of mirrors. The ladies appreciate them. Some are "catty" enough to say the men do, too.

But all the rest-room looking-glasses furnished their customers by the members of the Aid Hardware Co., West Plains, Mo., are not sufficient to satisfy their idea of "service." So when Charles T. Aid, head and founder of the firm, erected his present three-story building, he contrived to have a two-by-five mirror set permanently into the South Street side of the structure.

The Presbyterian Church is just back of the store. Scores of worshippers pass the mirror on their way to and from this and other churches every Sunday. Many a woman pauses a half-minute in front of the looking-glass for one more dab at her nose with the faithful powder-puff. The feminine contingent of West Plains vote the mirror a real convenience.

A Hot Clue

The great detective glanced around the room with a practiced eye. The pictures were torn into shreds, the chairs were broken, the table lying on the top of the piano. A great splash of blood was on the carpet.

"Someone has been here," he said.

Hillside Combine

SPECIAL advertising matter has been placed in the hands of dealers and prospects by the J. I. Case Threshing Machine Co., Racine, Wis., on the hillside type of Case harvester-thresher. The machine is so constructed that the thresher can be maintained in a level position by means of a special device, even on a fifty percent grade, it is claimed, where it will thresh perfectly.

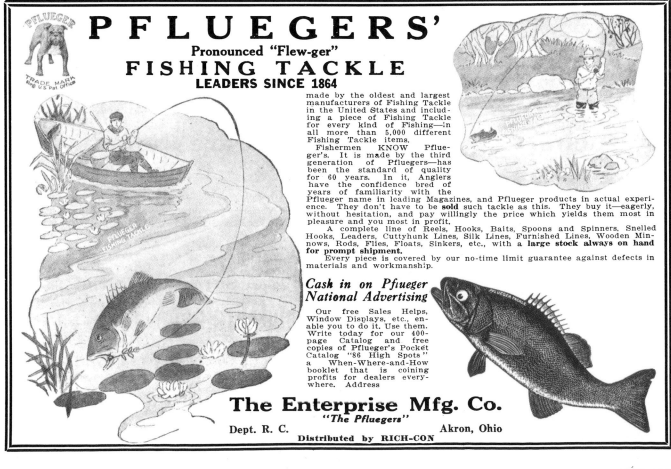

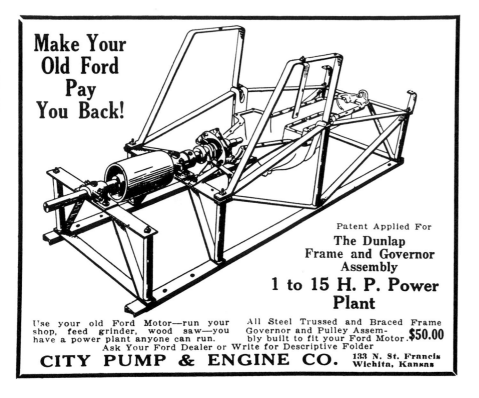

In 1925 credit was used to sell power machinery at a frantic pace. The additional equipment permitted farmers to cultivate much more ground than before, but much of the indebtedness was not properly secured. Many dealers and some manufacturers failed because of loan practices that were too trusting.

Foreign competition was never taken lightly by the American farmer. The farm equipment industry and farmers in the U.S.A. were aware that Argentina, Australia and Canada had great unplowed fields, cheaper land, cheaper labor and good transportation, all suitable for competitive wheat production. All that was missing was power equipment and farming expertise. Lower prices for farm products grown in the U.S.A. was dependent upon farmers making the most of these two advantages. The Advance-Rumley Thresher Co. began to distribute their tractor oil, called "Oil-Pull" in sealed containers to stop the trouble caused by the "oil bootlegger." The Breyer Ice Cream Co. of Philadelphia, Pennsylvania, ordered several thousand of the new Frigidaire ice cream cabinets. Kansas was the last state to adopt a state flag.

The Minneapolis Threshing Machine Co. began to sell a prairie-type combine in 1925 designed by McVicker and C.C. Cavanaugh. In March, Monarch Tractor Inc. was moved to Springfield, Illinois, and was reformed as the Monarch Tractor Corporation. The Avery Power Machinery Co. of Peoria, Illinois, was formed by the new owners of property previously owned by the bankrupt Avery Co.

The Farmall tractor sold so well in 1925 that IHC decided that it was not necessary to advertise in 1926. After all, they couldn't build the fantastic new Farmall fast enough to satisfy demand for the already popular tractor. Everything about the new tractor seemed to be on target, even the name. On April 25, the $12.5 million Caterpillar Tractor Company of Peoria, Illinois, was formed by the merger of:

C.L. Best Tractor Co. of San Leandro, California.

Holt Manufacturing of Stockton, California.

Holt Manufacturing already controlled the C.L. Best Tractor Co. and "Caterpillar" was the trade name of the Holt company, who had developed excellent manufacturing ability during WWI. The Best company contributed a very good dealer organization, that was still improving, even better tractor design and the excellent financial standing that was much needed to expand.

"We Have Used Link-Belt Chains For Fifty Years"

THE Sandwich Manufacturing Co., Sandwich, Ill., adopted Link-Belt chain 50 years ago as the result of a demonstration made in the field by William Dana Ewart, founder of Link-Belt Company.

"From that time until the present," writes Mr. C. C. Jones, their Manager, "the Ewart Manufacturing Co. and its successor, the Link-Belt Company, have had our annual contract for malleable iron chains. We have bought none elsewhere.

"In this long term of years, you have certainly furnished us enough chain to encircle the globe. Some years we have used over two miles per day, for months at a time, of plain No. 55 Ewart Link-Belt.

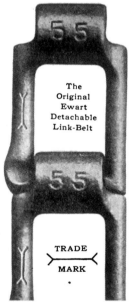

5 5

The Original Ewart Detachable Link-Belt

5 5

TRADE — MARK

"We reflect with much gratification that during all of this long period the great quantities of plain chain, and the multiplicity of attachment links that we used, have come through to us on time, and with almost the regularity of the sun. Complaints from our customers as to the quality of your products have been so few as to be really negligible."

Link-Belt chains have for half a century played a conspicuous part in the successful career of the Agricultural Implement Industry. Their performance has materially assisted in the economical preparation of millions upon millions of bushels of grain. Today, as in the early years, the famous Link-Belt trade>——<mark continues to stand for economical and reliable performance.

2105

LINK-BELT COMPANY
Leading manufacturers of Elevating, Conveying and Power Transmission Chains and Machinery

PHILADELPHIA, 2045 Hunting Park Ave.   CHICAGO, 300 W. Pershing Road   INDIANAPOLIS, 200 S. Belmont Ave.
Offices in Principal Cities

LINK-BELT

THIS YEAR LINK-BELT IS FIFTY YEARS OLD

"Three Toes" Caught

ANOTHER notorious stock killer is dead. "Three Toes," a wolf said to rival the famous Custer wolf in the extent of stock killing and the way in which he has eluded capture, was successfully trapped in Harding County, S. D., last July. The honor of his capture goes to Junior Inspector Clyde F. Briggs of the Biological Survey of the United States Department of Agriculture.

More than 150 men have tried for "Three Toes" besides the men engaged in several big drives made by stockmen. Mr. Briggs took him in 15 days' trapping, using natural wolf scent as bait. The work was conducted in cooperation with the state department of agriculture.

This wolf has been a noted killer for more than 13 years, ranging over the western half of Harding County and northwestern Butte County in South Dakota; Bowman County, N. Dak.; and southeastern Corter County, Mont.

It would be impossible to estimate his entire killings, but from available data and statements of stockmen who have lost heavily in recent years through his depredations, $50,000 is a conservative figure.

He has destroyed several thousand sheep as well as a large number of cattle. Horses have not been counted against him but he has killed many. He seldom killed only enough animals for food but almost invariably destroyed from 10 to 30 head on each visit.

Ever Meet These Famous Announcers? Anyhow, You Probably Know Them

Beginning at the lower left and reading clockwise around the microphone, they are: Lambdin Kay, "The Voice of the South," WSB, Atlanta; "Bill" Hay, KFKX, Hastings, Neb.; John Schilling, WHB, Kansas City; George Hay, WLS, Chicago; Leo Fitzpatrick, "The Merry Old Chief," WDAF, Kansas City; "The Hired Hand," WBAP, Fort Worth; "Gene" Rouse, WOAW, Omaha.

What Mr. Sarnoff Says

Why should this vast increase in radio interest on the farms continue? David Sarnoff, vice-president and general manager of the Radio Corporation of America, points out that educators who devote themselves to agricultural betterment see the following possibilities for radio with reference to what it will do for the farmer: It will help—

1—To relieve the farmer and his family from the sense of isolation which is perhaps the harshest handicap of agricultural life.

2—To broaden their social, spiritual and religious life.

3—To cope with class and sectional differences and develop greater national unity as between the farmer and other elements of our citizenship.

4—To make possible a system of agricultural colleges which will be open to all the 30,000,000 Americans who live on farms.

5—To aid in keeping the boys and girls on the farm, thus preserving for agricultural development the energies of the thousands of ambitious young men and women who are drawn away each year to urban pursuits.

6—Radio can be employed to furnish accurate time signals and weather reports to the farmer and townsman. It can broadcast warning of approaching storms and flood warnings received through radio will be of priceless value to the farmer.

7—Radio can furnish accurate news of prices and trade conditions of farm products at all principal markets within the hour and make it available to every farm home.

8—Radio, in my judgment, is destined to become one of the most effective elements in the business equipment of the farm, comparable perhaps to the great utility of the automobile.

Mr. Sarnoff says: "Above and beyond its utilitarian uses, the message that radio brings to the farmer is a message of human contact, human sympathy and culture.

Care of Refrigerator

REAL economy is always to keep the ice-chamber well filled so that a uniform temperature—preferably about forty degrees, never above sixty degrees—is uniformly maintained.

It is the greatest mistake to try to prevent the ice from melting by any such habit as covering it with papers or cloth, and never put milk, butter or other products directly on the ice. Actual contact between the air to be cooled and the ice surface is necessary.

The doors of the refrigerator should be kept tightly closed. This saves ice and helps to maintain a more even temperature. Since the cold-air currents fall, this makes the bottom compartment under the ice-chamber the coldest location of all.

That is why this is the place to set the most perishable foods, which are generally milk, butter or raw meats. The next coldest place is the lowest compartment on the opposite side, and here may be placed meats or delicate watery vegetables or berries, all most likely to rapid decay.

The racks above are less cold as we ascend, and we can remember that the top rack or location is the least cold and therefore the place for relishes, dry foods and those that have been cooked.

Warm foods should never be placed in the refrigerator. They cause the ice to melt wastefully, the temperature of the air is raised, which increases the bacteria count and the chances of the food to sour or spoil.

This is particularly true of soup, milk and semi-liquid dishes whose watery consistency is exactly the desired medium or potential "culture" in which little bacteria love to play—and grow.

That, too, is the reason why solid and dry portions of food are best laid away apart from their gravy or sauce, since the dry portion will tend to keep better and longer than a fluid.

Again, one speck of mold will, like radio waves, travel to a distant article or saucer of food and invisibly but quite surely contaminate any surrounding surface which is also moist and warm.

Thus an unnoticed piece of cheese in the rear, a portion of ham bone, a bit of pimiento or other soft food will be the unseen cause of odors and spoilage.

Nor can one expect the best refrigerator to keep sweet milk which has stood on the back steps in the sun, on the hot shelf over the range, or been in an open pitcher on the table some hours before its return to the bottle.

Frequent changes or rise in temperature in the ice-box is the most common cause for souring and spoiling of its contents.—From the Designer.

KANSAS CITY, April 24.—Reports are coming of municipal parks going in for horse-shoe pitching. Calls are being received for pitching shoes so rapidly that one jobber has been compelled to order a second shipment already this spring. "It is surprising, too, about croquet," said a buyer in one of the jobbing houses. "Last year we bought and sold over a car and a half of sets and this spring it is much livlier than last."

"Prospects for this year's trapshooting indicate the breaking of all previous records. Gun clubs are already getting into action and the demand for traps, guns, clay birds and ammunition is going to be good.

THESE DEALERS ARE ALREADY UNDER WAY

The thousands of dealers who make up the A-C Dayton distributing system are not only ready now for the big radio buying season, but **already under way on active selling work.** These dealers are equipped with "The Master Radio Salesman"—and with that as their guide, they are well started toward a fully resultful season.

"The Master Radio Salesman," prepared only after thorough study of radio selling methods, is the most forward step yet taken in radio merchandising. It is the key to success in this field—a complete, practical, workable plan for developing retail radio business—and holding it. With this work, advertising material provided to the dealers doubles in value —and A-C Dayton dealers are prepared with plenty of such selling

helps in addition to extensive factory advertising.

A-C Dayton dealers will make records this year. They have, first of all, the right merchandise. Added to that, they have not only selling help in the form of advertising material, but constructive guidance in using that material.

Remember these dealers when you go to take on a radio line. Remember that selling is your only road to profit, and remember that general advertising cannot do the whole job. With the A-C Dayton you will have these things—plus that which no other line can give you—actual, concrete, scientific selling plans, ready for you to apply. Write direct or to any A-C Dayton distributor asking to see "The Master Radio Salesman."

The A-C Electrical Manufacturing Company
DAYTON, OHIO
Manufacturers of Electrical Devices for More Than Twenty Years

The Console (only thirty-eight inches high) $185; west of Denver $190.

The Glass Set (heavy French plate glass cabinet) $125; west of Denver $130.

The Phono Set, for installation in practically any phonograph, cabinet or console, without accessories, $95; west of Denver $100.

5 Good Reasons:

1 A-C Dayton performance sells for you. The entire line is built to sell on its own merits; demonstrate and you will sell.

2 The A-C Dayton line is complete yet compact. Four models provide sufficient variety of style and allow most rapid turnover; one circuit insures equal performance from all receivers, uniform satisfaction and simplified selling problems.

3 A-C Dayton stands for complete price protection—maintenance of list through carefully selected outlets and full protection against price slashes, stock dumping and bargain basement sales.

4 A-C Dayton price is honest price—sufficient to permit good workmanship and good materials—yet no more than is necessary for fully satisfying performance.

5 A-C Dayton dealer-assistance is **real** assistance. "The Master Radio Salesman" shows you how to sell, and points the way to the fullest realization on all A-C Dayton superiorities.

A-C DAYTON RADIO

MEMBER RMA

For the Man Who Believes His Own Ears

Tractor Testing and Rating

A. S. A. E. Adopts Code—Manufacturers Approve

FOR several years the engineers of the tractor industry made real efforts to formulate a standard method or procedure for the testing and rating of farm tractors. The fact that tractor development was still in its early stages naturally made this objective difficult of accomplishment. The efforts of the engineers, however, seemed to have been crowned with success when the American Society of Agricultural Engineers, on the recommendation of its committee on tractor testing and rating, the membership of which was comprised of agricultural engineers from the tractor industry and the state agricultural colleges, adopted what is known as the "A. S. A. E. Tractor Testing and Rating Code."

Recommends Minor Revisions

Early this year the committee on tractor testing and rating recommended some minor revisions in the code, which were approved by the council of the society. At its meeting last April the tractor and thresher department of the National Association of Farm Equipment Manufacturers approved the A. S. A. E. code and recommended its general adoption by the tractor industry. The code has also been approved by the standards committee of the Society of Automotive Engineers and by that organization as a whole at its annual meeting held in June of this year.

With the approval of the A. S. A. E. tractor testing and rating code by the three organizations named, the solution of the problem of a standard method of rating farm tractors seems to have been reached and it is hoped that its use by the tractor industry will become universal. The code in its latest revised form is as follows:

"The belt horsepower rating of the tractor shall not exceed ninety percent of the maximum load which the engine will maintain by belt at the brake or dynamometer for two hours at rated engine rpm., the test to be carried out as specified herein.

Test Will be Repeated

"The drawbar rating of the tractor shall not exceed eighty percent of the maximum drawbar horsepower developed at a rate of travel recommended for the ordinary operation of the tractor, under conditions of testing as specified herein.

"The following rating tests are to be conducted in the order given on three or more tractors picked at random from factory stock run by the engineer or engineers conducting the test; the averages of all tests are to be used in determining the results:

"Test A (Limbering-up Run)—Before a test is undertaken it is important that the tractor shall have been in operation for a sufficient length of time to attain proper operating conditions throughout so that the results of the test shall express the true working performance.

"The tractor or tractors to be tested shall therefore be submitted to 'limbering-up' runs on the drawbar of twelve or more hours duration. Drawbar loads of approximately one-third, two-thirds, and full load shall be pulled by the tractor during the runs, each load being drawn for approximately an equal length of time, the lighter loads being used first.

When Test Shall be Made

"Test B (Maximum Brake Horsepower Test)—The tractor engine is to be tested in the belt with the governor set to give full opening of governor valve, and the carburetor set to give maximum power at rated engine crankshaft speed. The rated speed is that which the manufacturer recommends for the engine under normal load. The test shall begin after the temperature of the cooling fluid and other operating conditions have become practically constant. The duration of this test shall be two hours of continuous running with no change in engine adjustment.

"If the speed should change sufficiently during the test to indicate that the operating conditions had not become constant when the test was started, the test will be repeated with the necessary change in load. The term 'load' as used in this connection means pounds on dynamometer or brake scale. Minor changes in load to be made to maintain rated speed and the average load and average speed for the period shall be used in computing the horsepower.

"All belt horsepower tests must be made with an electric dynamometer, or with an accurately tested Prony brake or other accurate power-measuring device. Correction shall be made for temperature and altitude effect on horsepower output. Standard conditions of barometric pressure of 28.6" Hg. and a temperature of 70° F., or 530° abs. T. shall be used.

Formula to be Used

"The following correction formula 1 shall be used: $B\text{-}Hpc = B\text{-}Hpo \times (Ps \div Po) \times (To \div Ts)$.

"$B\text{-}Hpc$=corrected brake horsepower;

"$B\text{-}Hpo$=observed brake horsepower;

"Po=observed barometric pressure in inches of mercury;

"Ps=standard barometric pressure in inches of mercury;

"To=observed absolute temperature in degrees Fahrenheit.

"Ts=standard absolute temperature in degrees Fahrenheit.

"Test C (Maximum Drawbar Horsepower)—After the tractor has attained proper working conditions it shall be subjected to a series of drawbar tests.

"In the correction, formula 70° F. is used instead of 60° F., and 28.6" Hg. instead of 30", since these figures conform more nearly to the conditions as found throughout the tractor region of the Middle West. If a rating is made on corrections made at 30" Hg. and 60° F, the rating may be higher than the power actually delivered in a test.

"If the ordinary accepted standards of 30" Hg. and 60° F. are insisted on, then the only alternative will be to drop the percentage figures back to 80 percent for the belt or brake horsepower and 75 percent for drawbar horsepower.

Loads Will Vary

"The loads shall be successfully increased or decreased to a point where the tractor can sustain a constant pull for a distance of not less than 1,000 feet with the average engine rpm. maintained within 5 percent of its rated full load speed and a wheel slippage of not more than 10 per cent, the wheel slippage to be secured as follows:

"The tractor shall be driven without any drawbar load for a distance sufficient to give 10 revolutions of the drivewheels. This distance is then accurately measured and is used as the basis for computing wheel slippage. The number of revolutions of all drivers shall be counted for the entire distance through which the load is pulled.

"The drawbar pull shall be measured by means of an accurately calibrated draft dynamometer or draft measuring device placed between the tractor and the load. The actual distance traveled shall be used in calculating the horsepower, no allowance being made for wheel slippage. During this test the tractor shall be run in the gear recommended for plowing under favorable conditions.

"These tests will be made on the lowest commercially available grade of fuel which the manufacturer recommends for the tractor.

Testing Procedure

"Nature of tests—The following tests are to be conducted on one or more tractors picked at random from factory stock run; records of fuel consumption are to be taken during all tests except Test A:

"Test A (Limbering-up Run)—Before a test is undertaken it is important that the tractor shall have been in operation for a sufficient length of time to attain proper operating conditions throughout so that the results of the test shall express the true working performance.

"The tractor or tractors to be tested shall therefore be submitted to a 'limbering-up' run on the drawbar of twelve or more hours. Drawbar loads of approximately one-third, two-thirds, and full load shall be pulled by the tractor during the run, each load being drawn for approximately an equal length of time, the lighter loads being used first.

"Test B (Maximum Brake Horsepower Test)—The engine is to be tested in the belt with the governor set to give full opening of governor valve, and the carburetor set to give maximum power at rated speed. (The rated speed is that which the manufacturer recommends for the engine under load.) The test shall begin after the temperature of the cooling fluid and other operating conditions have become practically constant. The duration of this test shall be two hours of continuous running with no change in load or engine adjustments. (The term 'load' as used in this code, in connection with brake tests, means pounds on dynamometer or brake scale.)

"If the speed should change during the test enough to indicate that conditions had not become constant when the test was started, the test will be repeated with the necessary change in load. If, however, the speed should tend to increase only slightly during the progress of this test sufficient load shall be added to maintain the speed constant and the average load for the period shall be used in computing the horsepower.

"Test C (Rated Brake Horsepower Test) —The engine is to be tested in the belt at rated speed with carburetor to be adjusted for best economy. The load is to be such as to give not more than eighty percent of the horsepower obtained in Test B. The test shall begin after the temperature of the cooling fluid has become constant and shall continue for two hours continuous running with no change in load or engine adjustment.

"Test D (Varying Load Tests)—The engine is to be tested in the belt with all adjustments as in Test C with no stops. The total running time shall be one hour and ten minutes, divided into seven ten-minute intervals as follows:

"(a) 10 minutes at load as in Test C.
"(b) 10 minutes at maximum load.
"(c) 10 minutes at no load.
"(d) 10 minutes at one-fourth load.
"(e) 10 minutes at one-half load.
"(f) 10 minutes at three-fourths load.
"(g) 10 minutes surging loads varying suddenly from maximum load to no load and other varying loads between these extremes.

"All belt horsepower tests must be made with an electric prony brake or other accurate power-measuring device.

"Drawbar tests are to be conducted after the brake tests are completed. The drawbar tests are to be conducted on as near level ground as is available. The ground shall be firm and of such a nature as to provide sufficient tractive resistance to permit the tractor to exert its full power. The loads shall be successively increased or decreased to a point where the tractor can sustain a constant pull for a distance of at least 1,000 feet with the average crankshaft revolutions per minute maintained within 5 percent of its rated full load speed and a wheel slippage of not more than 10 percent. The wheel slippage is to be determined as follows:

"The tractor shall be driven without any drawbar load for a distance sufficient to give 20 revolutions of the drivewheels. The distance is then accurately measured and is used as the basis for computing the wheel slippage. The number of revolutions of all drivers shall be counted for the entire distance through which the load is pulled.

"Other drawbar tests shall be carried out in a similar manner as above, except the determination of the amount of load. This can be determined before such tests are begun by using any desired portion of the rated load. The rated load shall be not more than eighty percent of the maximum load.

"If fuel consumption records are desired on the drawbar test, then the test shall continue for at least two hours with no stops. The amount of fuel shall be determined by starting with the fuel tank full. At the conclusion of the test the tank shall be refilled, carefully weighing the amount required to refill and converting it to United States standard gallons at 60° F. Draft and slippage records shall be taken, as for the maximum test, at 30-minute intervals during the test.

"The drawbar pull is to be measured by means of an accurately calibrated draft dynamometer or draft measuring device, properly placed between the tractor and the load being drawn.

"Miscellaneous tests may include any desired tests not included in the given outline. Such tests shall be so conducted that the results will be accurate and reliable.

"All tests will be made on the lowest commercially available grade of fuel which the manufacturer recommends for his particular tractor. All fuel used shall be purchased on the open market and shall consist of the low grades of such fuel commonly sold in the locality, that is, if the tractor is to operate on gasoline, the lowest grade of such fuel commonly sold in the community shall be used. The same is true of kerosene and distillate.

"The lubricants used in these tests shall be such as are regularly recommended by the manufacturer for use with the tractor.

"The belt or belts used in these tests shall be such as the manufacturer recommends for use with the tractor in ordinary operation. No allowance will be made for belt losses.

"Lubricants specified by the manufacturers are the kinds and grades of lubricants to be used in the different parts of the tractor during the tests."

Christmas Seals

"HOW far that little candle throws his beams; so shines a good deed in a naughty world." When Shakespeare wrote these lines in 1598, the duration of human life averaged 33.5 years.

Since then, 21 years have been added to the normal span which is now about 55 years. About seven of these 21 years—or one-third of the total gain—has been added in the 18 years since the birth of the Christmas Seal.

Life Savers

The design of the seal this year is most appropriately the lighted candle symbolizing knowledge, truth and service, throwing beams of light into the ill health crannies of the world.

The candle was lighted in America 18 years ago when the idea of the Seal which came as a vision to the Danish postal clerk was accepted by the National Tuberculosis Association.

Since that time, there has not been a moment the beams of the candle have been darkened until today the Tuberculosis movement permeates every city and hamlet in the land, saving each year 100,000 lives; and, yet, there are still 100,000 lives lost each year from tuberculosis.

High Water Mark

A tourist passing through a small country town noticed a post on which was marked the height to which the river had risen during a recent flood.

"Do you mean to say," he asked an inhabitant, "that the river rose as high as six feet?"

"Oh, no," was the reply, "but the children used to rub off the original mark, so the mayor ordered it to be put higher up out of their reach."

Courtesy Underwood & Underwood

WHEN CALVIN COOLIDGE DELIVERED HIS INAUGURAL ADDRESS THERE WAS A ROW of Microphones Directly in Front of Him. Here He Is Speaking to His Countrymen All Over the Land. What a Change Radio Has Wrought in the Speed With Which Information Is Transmitted! It Will Be a Pleasure to Take Part in Its Future Development. Judging the Future By the Past, Radio Is As Certain of Becoming a Utility of Common Use As Were the Telegraph and Telephone.

Favors "Combine"

NORWICH, KAN.—To the Implement & Hardware Trade Journal: We noticed an account in your issue for Aug. 22 relative to the "evolution" in the harvesting business, calling particular attention to the combines. We also noted where a dealer had sold 25 binders.

As I have heard arguments pro and con on the success of the combines, I will say I am a full-fledged combine dealer, selling 35 McCormick-Deering combines this year. I claim this is the largest business done in the Wichita territory.

The harvester-thresher, or combine, as we term it, I introduced by selling the first one in this district in 1920. At the time the sale was made, rumors went around that I had made a mighty poor sale. In fact, I was almost tempted to cancel the order. But I made up my mind that if everything was all properly set up and operated, it would be a money-maker at the lowest yieldage.

The boys I sold the machine to were in debt approximately around $8,000, with nothing else but a determination to make a success. They cut an average of 400 acres of wheat every year, with no expense outside their gas and oil and repairs.

They cleaned up the old debt, and today they established their credit in such a manner that they bought 320 acres of land. I sold them their second machine, which is another McCormick-Deering like the first one. Their old one was sold for $800.

Our biggest argument we had to overcome was that the wheat simply had to be left standing too long in the field. Now this is true; but, on the other hand, how does the thresherman thresh wheat? He waits until it gets thoroughly dried out. Suppose it happens to rain about the time he is ready to thresh? It very seldom rains when it is in the shock.

We have had instances this year where a high wind raced through a strip of country and tore all the shocks and headed stacks to pieces. One man raked his wheat after a storm and got three bushels an acre after

threshing the shocks off of it. There are a number of good argumentative points that come up against the machine, but neverthe-

less, when they pay their threshing bill, which is a big item, in this neighborhood, farmers are usually ready to buy a combine.

A dealer had remarked to me: "How in the devil do you think these farmers are going to pay for a combine when they can't pay for a pull binder? I think it is foolish to sell a farmer who can't afford one." A combine will pay for itself quicker than any machine he can buy.

Speaking from the financial angle of the matter, take the popular priced machine of $1700. The safest investment only nets you from four to five percent, which on the combine will bring in from sixty-eight to eighty-five dollars. The farmer with an ordinary crop of about 250 acres knows that the cost of handling it will cost from eight to nine hundred dollars. This, on wheat averaging 20 bushels and less, means a 50 percent gain on his investment. These machines, with proper care, will last the farmer from eight to ten years. Does it pay to buy a harvester-thresher? Give me the field where there are combines for business. In my experience this machine beats them all.

Trusting this will be of interest to you, we are sure dealers who are still in doubt about combines will have a chance to realize what other dealers are doing. Thank you for the information regarding certain repairs, on which we are unable to find the proper manufacturers.

> JOHN H. W. BOMHOLT,
> Manager, Wulf Bros. & Wilkie.

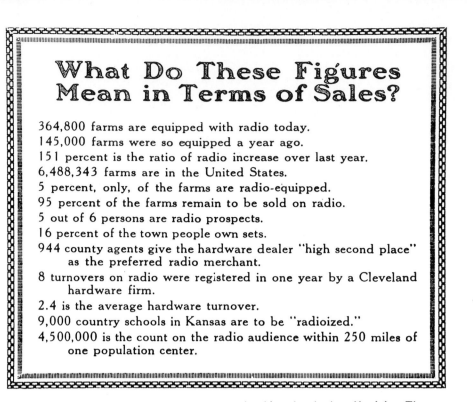

At last—
Alemite
on
Farm Implements
The Bassick Manufacturing Co.,
2696 N. Crawford Ave.,   Chicago, Ill.

Announcing the
FARMALL
The New General-Purpose Tractor that PLANTS and CULTIVATES, Too!

THE Harvester Company now offers the trade a remarkable new development—the McCormick-Deering FARMALL Tractor. As the name implies, the FARMALL handles all the usual farm power jobs with complete success—*drawbar, belt,* and *power take-off* —and in addition fits the need for a tractor that will *plant and cultivate corn, cotton, and other row crops.*

The FARMALL can be used with much of the regular horse-drawn equipment already on the farm, such as harrows, drills, planters, mowers, binders, etc. In addition, it can be fitted with the FARMALL 2-row cultivator, FARMALL mower, tractor-binder, etc. It will plant 25 to 50 acres of corn a day, depending on whether 2 or 4-row planter is used. Its steady power and ease of control insure straight rows and unusually perfect cross checking. With the FARMALL 2-row cultivator one man can cultivate 15 to 25 acres a day, depending on speed and conditions.

At all other farm power operations the FARMALL challenges comparison with other tractors of its size. Its power, ease of control, and clear view ahead adapt it perfectly to drawbar work—plowing, tilling, list-ing, seeding, cutting grain and corn, loading and hauling hay, corn picking, etc. The FARMALL brings special advantages to hay making. When equipped with the simple 7-foot FARMALL power-drive mower, it will cut 20 to 25 acres a day. Its steady power (governor controlled), and wide, properly located belt pulley, fit it equally for threshing, silo filling, grinding, and all other belt work.

* * *

It has been many years since the trade has been given an outfit so full of convincing, selling, order-getting demonstration possibilities as the new McCormick-Deering FARMALL. The dealer with an eye for sales volume will arrange at once for well-attended public demonstrations. He will call on his nearby Harvester branch house for cooperation—and he will get it! One thing is sure—the man who *sells* or *handles* this FARMALL Tractor outfit in planting or cultivating is going to be strongly impressed by it. Because of the wide range of work done successfully by the FARMALL its demonstration possibilities are almost endless. For a big power farming equipment year, work these possibilities for all they're worth!

INTERNATIONAL HARVESTER COMPANY
606 So. Michigan Ave.   of America   Chicago, Ill.
(*Incorporated*)

In 1926, Western Harvester of Stockton, California was formed as a subsidiary of the Caterpillar Tractor Co. and Ferguson-Sherman, Inc. was formed to manufacture and sell Ferguson plows built especially for the Fordson tractor. The four principals of the new company were Harry Ferguson, E.C. Sherman, George B. Sherman and Capt. J. Lloyd Williams. The Waterloo Gasoline Engine Co., owned by John Deere since 1918, was renamed the John Deere Tractor Co. The McCormick-Deering all steel threshers of 1926 were equipped with ball bearings to eliminate the need for frequent oiling and adjustment.

"Kleenbore" was selected, by the Remington Arms Co., from thousands of names submitted in a contest to identify their priming mixture for shells that made cleaning the gun barrel unnecessary. The new name was submitted by W.A. Robbins of Jonesville, Louisiana, and Nelson Eugene Starr of Goshen, Indiana. According to Deutche Zeitung, a German firm discovered a solder suitable for "welding" aluminum. The new solder would be especially valuable to the aircraft industry and was said to be composed of seven different metals and have a melting point of 300 degrees F. The new Coleman lanterns had a built-in pump. The Hobart Mfg. Co., the Ohio maker of electric food preparing equipment, bought the Crescent Washing Machine Co., makers of dish washing and metal washing machines.

Mouldable Wood

THE appearance in the hardware market of a mouldable wood brings up a number of interesting possibilities. The new product is called Plastic Wood and is manufactured by the Addison-Leslie Co., of Canton, Mass. Tests have proven that Plastic Wood is true to its two-fold slogan: "Handles like putty—hardens into wood."

As it comes from the can it may be moulded in the fingers or pushed into holes and cracks. It will adhere to metal, tile and glass as well as to wooden surfaces.

In its hardened form, Plastic Wood can be carved, planed, sand-papered and turned on a lathe. It will take paint, stain and varnish exactly as any other wood—in fact it is wood differing only from what we are used to in that it has no grain. Nails and screws will not split it.

The manifold uses of such a product are obvious. Furniture manufacturers, cabinet workers, boat builders, pattern makers, automobile body builders, etc.—these trades by it in bulk.

And now the householders have a chance to use it. It has been put up in one-quarter pound cans and one-pound cans and is sold by hardware stores for their convenience.

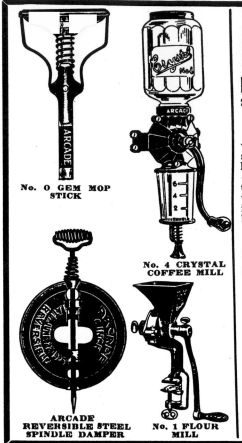

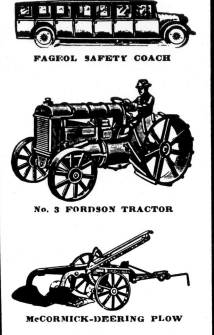

Our Petroleum Supply

Oil for Many Years to Come--
Then Some Other Source of Power

By W. F. Schaphorst, M. E.

Newark, N. J.

ENGINEERS are now seriously looking into the future and are estimating the probable demand for gasoline and fuel oil. The American Petroleum Institute Committee on Conservation has made the following estimates of the population of the United States:

Year	Poulation
1930	122,397,000
1940	136,318,000
1950	148,678,000

In 1924 the number of automotive engines, including automobiles motorcycles, airplanes, etc., was 13,028,000; in 1930 it is estimated that the number will be 31,018,000; in 1940, 39,755,000 and in 1950, 44,990,000.

Thus it is figured that the number of engines will be more than tripled, yet the committee is of the opinion that with the improvements in refining, which are being rapidly made, and with the improvements that may be made in the economy of automotive engines no more than 500,000,000 barrels of petroleum will be needed in 1950 instead of the 643,966,000 used in 1924.

Regarding the possibility of improving automotive engines C. F. Kettering, president of the Central Motors Research Corporation, is credited with the following statement:

"We believe that it is possible to make automobiles go twice as far per gallon of gasoline used. The present internal commustion engine and automobile only transform an average of 5 percent of the energy originally in the gasoline into useful work.

"It is possible at present to transform 10 percent of this energy into useful work, and this will be common practice in the future."

A similar statement was made by H. L. Horning, president of the Society of Automotive Engineers, addressing the American Petroleum Institute in December, 1923.

As for the number of years our oil supply will continue before exhaustion, that of course is problematical. At the present time the supply is more than ample and it is assumed that there are many undiscovered oil fields in existence in the United States which will sooner or later be found.

At any rate, we are assured that we have nothing to worry about so far as oil supply is concerned for 25 more years. When 1950 arrives it is quite probable that water power from rivers, water falls and tides, wind power, sun power, etc., will be more economically utilized than at the present time.

Before many years the problem of supreme importance to the human race will be that of heat and energy. Until we can be reasonably certain that there will always be a generous supply for everybody, fuels of all kinds should be conserved.

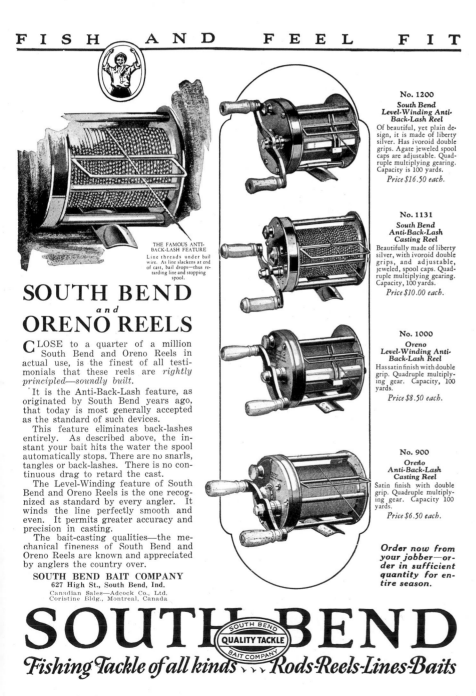
Building Small Combines

INDEPENDENCE, Mo., May 19.—P. J. Hansen, factory superintendent of the Gleaner Combine Harvester Corporation here, says the factory has been turning out combines at the rate of 50 a day. The factory is working 350 men, double shift. Mr. Hansen formerly was general superintendent for Bucher & Gibbs, implement manufacturers, Canton, O. He was also connected with the production department of the J. I. Case Plow Works, Inc., Racine, Wis., before taking up his residence here.

The Gleaner corporation is making two types of combine. One is especially for the Fordson tractor and the other is for any type of small tractor. Both combines convey the wheat to the cylinder on the auger principle. These small combines are a one-man proposition, the operator of the tractor raising and lowering the sickle by means of a lever to the side of and behind him.

The Gleaner offices are in the Land Bank Bldg., Kansas City. The names of the two combines are the Gleaner and the Baldwin. With the Gleaner the power of the tractor carries the machine and operates the mechanism. On the Baldwin, the machine is pulled by the tractor and the combine machinery is operated by a gasoline motor. Both are equipped with Hyatt, Timken and New Departure bearings.

In 1927, Charles A. Lindbergh was the first to fly a monoplane solo, non-stop from New York to Paris. Implement dealer, J.B.R. Vaughn of Wray, Colorado, had earlier employed Captain Lindbergh as a stunt flyer for a group called "Mil Hi Flying Corps." Lindbergh had been the first to operate a tractor and to use a milking machine in Morrison County, Minnesota. Tennis balls named "Lively" were beginning to be "canned in air" by the Pennsylvania Rubber Co. of Jeanette, Pennsylvania, and the Lincoln Electric Co. of Cleveland, Ohio, began to sell "dipped" welding rod called "Stable-Arc." On May 10, 1927, it became illegal to send "one-hand weapons through the mails" according to the Post Master General. Allis Chalmers purchased Pittsburgh Transformer Co. of Pittsburgh, Pennsylvania. The American Society of Agricultural Engineers adopted standards for Power Take-Off drives. Lack of standards before this time limited success of implements driven by a tractor's power.

Rumely Gets Toro Machine

MINNEAPOLIS, Oct. 10.— J. S. Clapper, president of the Toro Mfg. Co., announces that the company has consummated a deal with Advance-Rumely Thresher Co., La-Porte, Ind., for the assignment of all patents, and the exclusive rights to manufacture and market the Toro combination tractor and row-crop cultivating machine for agricultural use.

The Toro Mfg. Co. will continue the production of its present line of golf course and city park maintenance machinery, which is not effected by the deal with the Advance-Rumely company for the farm machine.

JAPANESE USE AMERICAN ENGINES

"The thrifty Japanese hold the rice up to the cylinder of the thresher, which is very slow as compared to those used in America, and thresh only the grain or head off. The straw is then carefully piled away for use later in making matting, brooms, etc.

"A typical salesroom for the display of our engines in Japan is shown in another picture. One will note that a few of the attendants are dressed in American style clothes.

"However, in many of the stores the merchants and clerks wear the gentleman's kimons so common in Japan. This particular store is located on Ginza Street, one of the principal thoroughfares in Tokio."

Mr. Madden says most of the Japanese warehouses are built of galvanized iron, which has become a popular building material since the earthquake.

It is frequently used to make large storage places, fences, etc., and so far seems well able to resist the earth tremors which are a common occurrence in Japan.

Supply the Japs

THE United States is the chief source of Japanese electrical imports, having supplied a value more than three times greater than either the United Kingdom or Germany during 1924.

American exporters maintained their leadership in shipping all classes of electrical goods to Japan except insulated wire. Germany took second place as an electrical exporter to Japan in 1924, having a total value slightly larger than that of the United Kingdom.

The greatest progress in sales of German equipment was that of small apparatus, notably meters and insulated wire. The import of heavy equipment and telephone and telegraph apparatus was predominantly from the United States and the United Kingdom.

Occident vs. Orient

AS an illustration of the difference between occidental machine methods and oriental "cheap" labor in the field of agriculture, Arthur Huntington, of Cedar Rapids, pointed out to the northern central division meeting of the Chamber of Commerce of the United States that it takes four Chinamen in China to raise an acre of rice while one Chinaman in California can raise 3,000 bushels.

"I was in California last year," he said, "attending a meeting of the American Society of Agricultural Engineers. After the meeting we were shown California. We saw much rice, and much of the labor was oriental.

"We asked what they paid for oriental laborers and were told $6.00 per day and 'found' or the equivalent of $8.00. We asked where they found a market and were told 'in China and Japan,' where, by the way, rice field labor is from ten cents to fifteen cents per day.

"A little further inquiry brought the information that the Chinese Agricultural Association had a commission in San Francisco protesting against Americans flooding China with cheap rice, thus destroying their agriculture."

Raw-Meat Climax

KANSAS CITY, Jan. 24.—Starting with a lively exchange of biffs by light-weight boxers and working surely up to a raw-meat climax in the last act when Billy Edwards, world light heavyweight champion, and Jack Kogut struggled desperately and without make-believe for mat mastery, the Athletic Carnival, staged by the Kansas City Implement, Hardware and Tractor Club last Wednesday night in the American Royal Arena for the entertainment of all Western convention visitors, may be written down as a success unqualified.

Ask anybody who was there. Unless the witness interviewed happens not to care for the kind of "blood and sand" exhibited at this pugilistic and palestric miscellany, the answer is bound to be a pious, "Amen, Brother!"

Though there was plenty of room to spare in the big arena, seven or eight thousand people, men and women, enjoyed the festivities.

When Kogut finally succumbed to Edwards' chiropractic headlock, a large fraction of the audience stood on chairs and booed the hard-working victor. The crowd had to have a favorite.

But it was all part of the noise and spectacle. That's what ran up the sum total of enjoyment. And the required dash of comedy was tossed into the pot at exactly the right moment to make it taste best when "Teddy Bros," an eccentric wrestling team from Omaha, made up as aged Rubes, butted into the proceedings and made the patrons of the performance howl with delight. "Teddy Bros" are going to be a tender memory with most of them for several years to come.

Some clever boxing was seen in the first two matches in the 105-pound and 160-pound classes, each going four rounds. Then McWilliams and Hawley, 160 pounds apiece, showed both anger and steam and gave the spectators a foretaste of the hot action that was to come.

Durrell Black and Tommie Maroon, 122 pounds, socked each other skillfully and industriously for four more rounds. They tried hard for a knockout, but couldn't quite reach it.

But in the next match Fred Urick treated Eddie Heilman to a K. O. in the second round.

Some Tight Scrapping

Both demonstrated what piston-like blows can do when well placed. Each weighed 142 pounds.

Carl Kenny and Charlie Meyers, 133 pounds, then attacked each other and kept it up four rounds, ending in a tight draw.

At the finish of the second round Meyers failed to hear the bell. In the third he hit the husky referee who didn't seem to mind it.

After some more music by the band two brawny wrestlers entered the ring, took off their bath-robes and got to work.

Signs of temper. One, cast for the villain in the piece, began slugging. The onlookers began booing.

A ringside spectator, with passion for justice in heart, was about to climb through the ropes and settle it himself when the unfair wretch was disqualified.

The Edwards-Kogut match looked just like a wrestling ought to look for those who like to see it rough and earnest. Yes, it was a splendidly compiled pocket edition of the world war.

Billy took the first fall in thirteen minutes with his copyrighted chiropractic headlock, which the audience refused to endorse.

They said it with boos. His knuckle action didn't seem fair, although nothing was barred but the strangle-hold.

The second fall didn't go so long and Jack, much to the hectic joy of the majority, conquered on his feet, with the champ on his tummy, at the end of five minutes and ten seconds.

It was a queer hold that hadn't been seen in these parts. It had to have a name, something quaintly fraught with advertising possibilities.

After a conference of the experts, it was announced as the "Boston crab hold."

Everybody was happy until Mr. Edwards applied his chiropractic headlock to Mr. Kogut at the terminus of the third fall, thus removing the championship from the danger most of the spectators thought it was in.

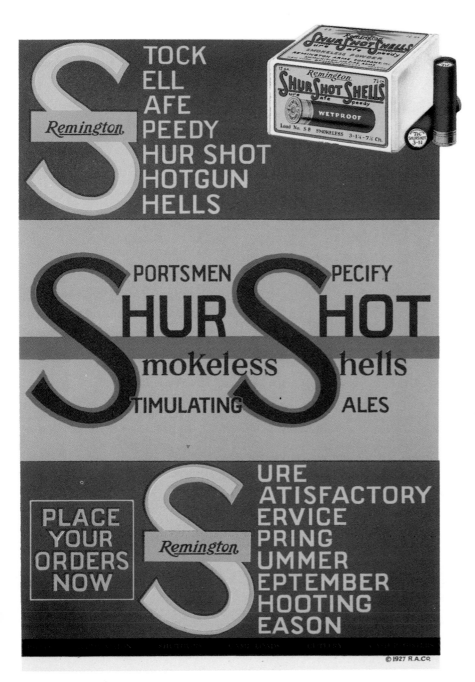

New Cotton Picker

Spindle Type Harvester Operated by Tractor, Stripper for Upland Crop, and Cleaner, Will Effect Saving in Production

THE International Harvester Co. has developed, in connection with its Farmall tractor as the power plant, what it believes will prove to be entirely practical machines for mechanically harvesting cotton.

These machines should materially reduce the cost of production. They are also expected to improve the grade of cotton produced, and under some conditions will substantially increase the amount gathered per acre.

In recent years the company has redoubled its efforts to produce machines that would successfully harvest cotton in the lowlands of the South as well as the uplands of the southwest where soil and climate conditions are vastly different. The result of these efforts is three machines for harvesting cotton:

A picker of the spindle type for the lowlands and other sections of the Old South where the entire crop cannot be picked at one time due to a long season and uneven ripening.

A stripper or boller for harvesting upland cotton, as it is usually called, which is cotton that matures quickly and ripens evenly.

A cotton cleaner for cleaning stripped cotton and bolls.

The picker is still looked upon by the Harvester company as semi-experimental. Only a limited number have been produced this year and these have been placed in various sections of the South where they will be operated during the fall and winter and results carefully watched.

The machine is what is known as the spindle type, having two picking cylinders set vertically, each carrying a large number of spindles which work horizontally and are close enough together for at least one of them to come in contact with every open boll on the plant.

The two picking cylinders and doffers are suspended by pendulum and spring floating action. The picking mechanism floats in all directions and thus adapts itself to the variations in the cotton row and the ground, thereby obviating the need of accurate guiding of the wheels of the machine and eliminating danger of breakage of spindles and injury to the cotton plants.

As the picker is pulled along over the cotton row by the tractor, two large gathering shoes on the front of the machine pick up the spreading branches of the plant and place them in a position for the picking spindles on the two cylinders. The spindles revolve rapidly, at the same time moving backward on the cylinder in a horizontal position at exactly the same speed that the picker moves forward.

The cotton in the open bolls winds around the spindles and is carried back to the doffers where the cotton is released by a quick reverse action of the spindle as it passes between two sections of the doffer.

The McCormick-Deering Cotton Picker As It Looks Behind a Farmall Tractor

Each section of the doffer is equipped with a small set of brushes on the upper and lower sides. Each spindle passes between the two sets of these brushes which clean the spindle of cotton at the time when the reverse action of the spindle takes place.

The next operation is to separate the cotton from dirt and trash, which is accomplished by means of a revolving disk cleaner. From this centrifugal cleaner the cotton passes on to an elevator where a cylinder and belt cleaning device continue the cleaning action. The cotton then passes into one of two large gathering bags at the rear of the machine. These gathering bags are removed when full and replaced with empty ones.

The outfit is operated by two men, one guiding the tractor and the other controlling the cotton picker. The machine will pick from two to five bales of cotton a day, which is equivalent to what two pickers could gather in from eight to fifteen days. On the basis of present wages paid to hand pickers, the machine will save from $10 to $15 a bale over hand picking. Mechanically harvested cotton will, as a rule, grade higher than the average hand-picked cotton.

Whereas the picker gathers only the ripe cotton or open bolls, the stripper or boller is designed to gather the ripe cotton and also the unopened bolls in one operation. It is a very simple machine in comparison with the picker and can be operated either by tractor or horse power. The stripper has a pair of long dividers, similar to the dividers on corn binder, which pass one on each side of the cotton row. These dividers are adjustable up and down and are operated close to the ground.

As the machine is drawn forward, the dividers guide the cotton plants between two stripping chains immediately back of the dividers. A series of stripping fingers on these chains strip the bolls from the plant and deliver them into a gathering box at the rear. Spring bars and leaf springs known as picker fingers are located immediately below the stripping chains. These fingers gather and retain any loose cotton that is not in the bolls.

When the gathering box is filled with bolls, it is dumped on a large piece of canvas located at convenient points in the field where a cleaning machine may be located.

The stripper, operated by two men, is capable of gathering from two to five bales of cotton a day, depending upon the yield and condition of the field.

The cotton cleaner used in connection with the stripper is a self-contained power-operated stationary machine to which the bolls and stripped cotton are brought after being dumped from the stripper. The cleaner is of the spindle type and consists of a revolving drum having spindles of the same type as the cotton picker previously described.

The stripped cotton is delivered into a boll-breaking device which opens the closed or unripe bolls without injuring the fibre. The cotton is next passed on to the spindle drum where the spindles pick up only the lint, all other material being ejected. The cotton is removed from the spindle by doffers in the same manner as on the picker, and the lint is pneumatically conveyed to a wagon or cotton house where it is then ready for the gin.

Thousands of bales of cotton are left in the field each year in the unopened bolls or hopper and then passed through a feeder which delivers the bolls at a uniform rate to "bollies." With the cleaner and the stripper, which may be taken into the field late in the fall following hand pickers or the mechanical cotton picker, the cotton in these "bollies" can be harvested, adding considerably to the grower's income.

The International Harvester organization has actively assisted many inventors working on mechanical cotton harvesters during the last thirty years. Ever since the World War, when hand pickers became scarce and high priced, unusual efforts have been put forth by the Harvester experimental department in the development of thoroughly successful cotton harvesters. It has built and conducted experiments with suction, spindle, and stripper types of machines. A vast amount of time and money has been devoted to this work.

Efforts to develop a successful cotton harvester date back as far as 1850 when the first patent was taken out. In the late nineties, when Angus Campbell was in the experimental department of the Deering Harvester Co., he worked out many of the features of what later was known as the Price-Campbell cotton picker.

W. H. Turner was another inventor who worked along similar lines as Mr. Campbell. Later John F. Appleby, inventor of the Appleby knotter for twine binders, turned his attention to a cotton harvester and developed a machine with spindles entering the plant vertically.

The Price-Campbell machine would pick the mature cotton, but it was claimed that it left too large a proportion of the cotton unpicked and also damaged the unopened bolls and the oncoming crop in sections where the season is long and the leaves remain green on the plant.

Some of the Appleby machines were built and sold in the upland cotton sections, where they were fairly successful in picking the entire crop at one time, but this machine had the same shortcomings as other pickers when working in lowland cotton.

Has Parking Space

A. G. HOGE, a hardware dealer in Oklahoma City, Okla., sells builder's hardware to a great many architects and contractors. Builders hardware is something that can't be picked out in a hurry; customers like to drive up to the store, park their cars, and take their time about selecting hardware for a new house.

So Mr. Hoge moved his store several blocks farther out in the city, where his trade could park as long as it pleased. There is no time limit—a man can leave his car in one spot all day if he wants to. This is a great attraction to car-driving people. They like to feel that they can leave the old bus at the curb as long as they like without finding a tag attached to the steering wheel when they return.

New Pick-Up for Combine

I NVENTION of a new pick-up feeder now makes it possible to use a combined harvester not only for harvesting standing grain, but also for handling bundle grain. With the pick-up feeder attached to a combined harvester, the shocks or stools can be picked up, the bundles cut, the grain fed evenly into the separator, threshed, cleaned and delivered ready for market.

The big, expensive harvesting crew is eliminated—the women-folks no longer face the nightmare of feeding an army of men; the grain-wasting handlings to get the bundles to the threshing outfit are eliminated; the threshing outfit itself becomes a thing of the past; gigantic straw stacks are no longer necessary, while straw can be saved, or scattered and burned or plowed under.

Equipped with a combine, standard header and pick-up feeder, the grain grower can combine all his crop in seasons when conditions permit, or can bind it all in an unfavorable season. Or he can bind a portion of his crop and handle the rest of it by the straight combine method. He is prepared to meet bad-weather emergencies or is equipped to undertake a broader range of custom work.

ANOTHER OBSTACLE HAS BEEN OVERCOME AND THE COMBINE TAKES A FIRMER AND BROADER HOLD ON AMERican Agriculture. Grain May Be Cut Rather Green With a Header and Run off in Windrows. The Combine Attachment Picks the Grain From the Windrow and Threshes It. The Crop is Machine Handled Throughout and There Is, Therefore, Very Little Shattered Grain. This Attachment So Widens the Scope of the Combine That It Is Now Practical for the Harvesting of Sweet Clover and Other Seeds That Need to Dry Before Being Threshed.

A Hardware Man's Widow Wins

MRS. Nellie E. Trego, president and sole owner of the Trego Radio Mfg. Co., Kansas City, is one woman who has broken into the ranks of big business without sacrificing one whit of her womanliness.

In her belief, such sacrifice is neither desirable nor necessary. Her business runs into the millions annually and she has built it up from a little short of nothing within a few years.

Mrs. Trego's husband was a hardware dealer in Humboldt, Kan. When he died she had to make a living for herself and one-year old daughter. She went to Kansas City and attended business college.

After learning stenography she began earning nine dollars a week as an employee of A. J. Stephens, a well-known Kansas City tire man.

Her appetite for "knowing what it was all about" and an ability to execute details soon made her invaluable to her employer.

Soon he was able to go out into the field, knowing that she would manage well at headquarters. Her salary grew in proportion to her knowledge and ability.

Ambitious to direct larger interests, Mr. Stephens organized a tire manufacturing company and made Mrs. Trego treasurer. She held a small block of stock. This investment she lost in 1921 when uncontrollable business reverses were encountered.

Faith in His Ability

She still retained ownership in her home intact, however. Mrs. Trego believed implicitly in Mr. Stephens' executive ability and offered to lend him money on her home that his business might be restored.

He refused. Then she decided to enter business independently. Borrowing $3,000 on her home, she organized the Trego Radio Mfg. Co.

About the first thing she did was to make Mr. Stephens manager, paying him in both salary and profits.

As manager Mr. Stephens spent $2,500 of the $3,000 in well-directed advertising. The remainder was invested in materials and running expenses.

This bold policy would have set the average conservative aghast. Nevertheless, by mingling plenty of care with courage, the company progressed rapidly.

More than a hundred thousand radio sets have been sold by the company since that slender beginning.

But Mr. Stephens was a tire man. He wanted to get back into the manufacture of automobile tires. An opportunity developed for him to regain the old plant.

With Mrs. Trego and three Kansas City business men who had unbounded confidence in him he reorganized his former rubber company.

Gradually the company accumulated a capital of a half million dollars. Mrs. Trego and Mr. Stephens bought the other interests involved. They now have complete control and ownership.

They Cover Much Ground

The radio and tire plants together cover the full block from Chestnut street to Kansas Avenue on East Fourteenth Street and reaches along Chestnut for nearly a block.

Though, naturally, she doesn't boast of it, Mrs. Trego's personal interests must be worth something like a quarter of a million dollars. She owns valuable Texas oil leases and an interest in an Omaha rubber company.

Mrs. Trego is much interested in the success of women who have entered business. Recently she spoke before the convention of Business and Professional Women's Clubs in Oakland, Cal.

It was there that she gave utterance to her belief in the preservation of femininity among business women. She "stood up" for those of her sex who insist upon being womanly women even though successful in business.

"Powderless noses, flat shoes, sailor hats, mannish clothes and manners and rougeless cheeks," were not necessarily marks of success for feminine executives, she told her audience.

"For a woman to succeed in the business world it is necessary that she look her feminine best. There is no reason to imitate men in clothes or manner. It is her ability that counts."

The Trego Radio Mfg. Co., which is the expanded personality of this resourceful and courageous woman, continues to grow vigorously.

Just now Mrs. Trego and her loyal staff are pushing their latest model on the market. They call it the Tregofonic.

EVEN in her attitude about women in business Mrs. Nellie E. Trego, head of the Trego Radio Mfg. Co., shows a brand of common sense which is not so common as it ought to be. She cultivates not a single "manly" mannerism. She powders her nose just as consistently as she polishes her mind and takes no special credit for either commendable habit.

Are Skating Rinks Returning?

Rolling to Music Popular at Warrensburg, Mo., As Well As at Ann Arbor, Mich. -- Prominent Educator Believes That Schools Should Help Direct Leisure Time

HAS the popular skating rink of twenty years ago returned to bring additional profit to the hardware trade as it did then? For some time there have been rumblings about how the coeds of Ann Arbor were taking to the roller skate, but now appear signs closer home—at Warrensburg, Mo., where every afternoon and night the skating rink of A. L. Mahaffey is filled with the young of both sexes who roll, swerve and glide in delightful rythm as a large player piano booms out the popular airs. They go 'round singly, doubly and three abreast.

Scenes like this recall the status of the roller skate some twenty years ago. At that time many of the very small towns had rinks that did a thriving business. For example, Oakley, Kan., one of the small towns of that state, out in the corner of Logan County, had one of the biggest skating rinks in the whole state and among the largest in the country. In its heydey, it attracted such celebrated skaters as Ole Overfeldt, noted stilt skater from Europe.

May Be Coming Back

That was before concrete highways and sidewalks, in conjunction with the craze for dancing, took the skating out of the rinks. The lovers of music went then to the dances and those who still liked the rollers went to the sidewalks and concrete highways.

Is it possible that we are about to witness the return of the roller skater who prefers to roll and cut graceful curves to music rather than keep time by the shuffle of feet? Certainly the signs of the times are sufficiently indicative to merit observation. It is said, at Warrensburg, that roller skating to music is especially appealing to students of all grades from the common school to the college. A visit to Mr. Mahaffey's rink proves it. There are all sizes, from the grade schools to students and instructors from Central State Teachers' College located in the edge of town.

More Time for Play

"In these days of modern production and distribution, the problem of leisure time, and what to do with it becomes important and it is the business of the schools to help solve that problem.

"In my opinion, under the productive mechanism of the present time, it is possible to create enough wealth to sustain one in four hours of actual work each day. What are we going to do with the rest of those hours? They are going to be spent to some extent in recreative activities, and it is largely the business of the educational machinery to see that such activities are directed toward proper and wholesome ends. The problem of leisure time may be expected to become more acute with the development of productive facilities in both the rural and urban centers.

Education Anticipates Change

Mr. Mahaffey is an old alumnus of the Warrensburg institution. He conducts a confectionery and operates the skating rink as a side issue. Here are the facts and figures on his venture:

He makes a charge of 25 cents for one hour and a half of skating; week days the rink is open from 2:30 to 5:30. In the evening from 8:00 to 10:45. On Saturday nights the young people skate until midnight if they desire.

Mr. Mahaffey's Outlay

The initial outlay of capital was the purchase of 100 pairs of skates. The player piano is rented for $20 a month. Electricity for the piano and lights amounts to $5 a month. Three boys are hired to assist in putting on skates, sweeping out, etc. They get $2.50 a week each. Rent on the building is $75 a month.

The building is too small, Mr. Mahaffey says. It is about sixty by eighty feet. The returns amount to all the way from $10 to $40 a day. Forty dollars was the largest day's business. Mr. Mahaffey believes, however, that winter time will show a large increase in skating, as he has the competition of the swimming pools in summer, which is a popular sport among students, as well as others, during the summer season.

1928-1932

GIANTS ARE FORMED

JOHN DEERE COMBINE

The Grain-Saving Combine
That's Easier to Operate

WHETHER or not your prospects are expert threshermen, you will have, in the John Deere Combine, features that will meet their every need. Into it are built proved mechanical features that mean clean, thorough threshing and the simplicity of operation that is wanted both by expert operators and by farmers who are not experienced in the handling of threshing machines. Put the John Deere on display NOW. Take advantage of the farmer preference that is bound to grow with the sale of each John Deere Combine. Offer your combine prospects these features—

1. Its simplicity and the convenience of its controls make the John Deere easy to operate. All main controls—the motor, the main drive clutch, platform and reel elevation, the re-cleaning sieve and the wind on the lower shoe—are controlled from the operator's platform. One man has instant and complete control of his machine within easy reach. From his position on the platform, he can see the vital parts of the combine and watch its work.

2. Cutting, elevating, separating and cleaning units of the John Deere are constructed with a great margin of safety for saving grain under varying conditions. In each operation from cutting the grain to emptying the grain tank, the John Deere is a grain-saver. The method of re-cleaning employed is especially important to the delivery of clean grain.

3. Typical of all John Deere products, the John Deere Combine is built with a wide safety-margin of strength. It has strength to stand the severe strains of working in rough fields. Every part is built to last—only the best materials are used in its construction. It is designed to give the maximum length of service with a minimum of repair expense.

4. Draft and wear are reduced by the use of roller and ball bearings at all important points of friction. The lighter-, smoother-running advantage that results means that the John Deere, and the motor that operates it, will last longer and give better service.

5. Oiling the John Deere is a simple operation that requires little time. Every bearing can be oiled from the outside of the machine with a high-pressure grease gun. This high-pressure system makes thorough oiling a swift and easy job and longer life results.

6. The motor that operates the John Deere has a surplus of power, insuring the steady, even speed so essential to good threshing. It is of four-cylinder type and modern design throughout.

7. The 65-bushel grain tank on the John Deere is emptied in less than one minute. This is done by simply turning a crank that raises two doors at the bottom of the tank. The saving in time due to this feature alone means many bushels added to the day's run.

Write us TODAY for more complete details about The John Deere Combine

John Deere Plow Company
Kansas City, Mo.

The John Deere No. 2 Combine, shown here, is built in 12- and 16-ft. sizes. Its separator is 24″ x 40″. The capacity of its grain tank is 65 bushels.

Implement & Tractor Trade Journal

Established~1886   Published Fortnightly

1928-1932

Farmers in the Southern Plains bought large quantities of machinery during the time between 1925 and 1930. Value of just the new equipment in a 27 county area increased from $16.1 million to $36.0 million during this time. Because of the cost, tractors and combines were usually purchased on credit.

Allis Chalmers, J.I. Case, Caterpillar, John Deere, International Harvester, Massey-Harris, Minneapolis-Moline, New Idea, Oliver and others clearly became forces in the manufacturing muscle of America.

The most popular combines of 1928 cut 16 or 20 feet wide. Allis Chalmers purchased the Monarch Tractor Inc. of Springfield, Illinois. Two Monarch models, the "50" and the "75", were

especially popular. In July, the J.I. Case Threshing Machine Co. purchased the Emerson-Brantingham Co. of Rockford, Illinois. In August, the J.I. Case Co. was formed from the J.I. Case Threshing Machine Co. of Racine, Wisconsin. An all metal biplane designed by Waverly Stearman, brother of Lloyd Stearman, was made by the Butler Manufacturing Co. Butler steel buildings were often used as airplane hangers. The Caterpillar Tractor Co. purchased the Russell Grader Mfg. Co. of Minneapolis, Minnesota. The Sheffield Steel Corp., formerly the Kansas City Bolt & Nut Co., expanded production to include nails, barb-wire, field fence, poultry netting, bale ties and smooth wire. John Deere introduced the "GP" model, which

continued in production until 1935, and Ford suspended attempts to sell Fordson tractors in the United States. Later, some Fordson tractors were imported from Cork, Ireland, by O.J. Watson. International Harvester began to sell track laying crawler tractors called "Trackson." Early in 1928, Massey-Harris Co., Ltd., of Toronto, Canada, purchased the J.I. Case Plow Works, manufacturers of the Wallis tractor, located in Racine, Wisconsin. Massey-Harris sold the "J.I. Case" name to the J.I. Case Threshing Machine Co., thus saving much of the confusion of the similarly named products and recovering some of the purchase cost. Advance-Rumley "DoAll" tractors were announced on April 13.

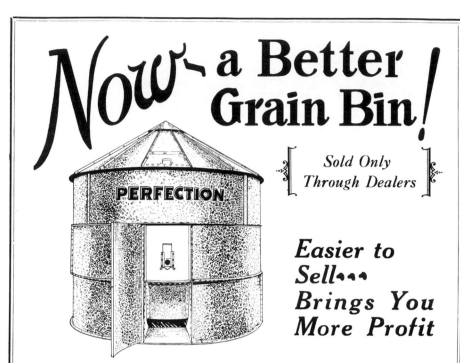

IN FUR

"How are you going to vote this fall?" asked her dearest friend.

"In a new fur coat if I can work John for it," she confided.

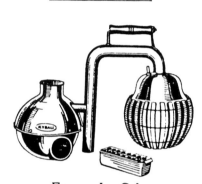

Freezes Ice Cubes

Electricity Not Needed

Call It "Icyball"

IN pacing on the marktt the Crosley Icyball, a new step has been taken in the field of refrigeration. Says the Crosley Radio Corporation, Cincinnati, O., in the following unusual announcement:

"At the amazing low cost of two cents daily, Icyball will keep a refrigerator cold from twenty-four to thirty-six hours. At the same time it will freeze ice cubes, the same as other devices whose cost of opration is many tims as much.

"Icyball may be used to advantage anywhere, the city as well as the country. Electricity is not needed to operate Icyball. A cook stove is the only thing necessary to assure complete satisfactory results with this new device.

"To those living in rural districts or in homes not supplied with electricity, Icyball will be a Godsend.

"Every housewife who has trouble in keeping a plentiful supply of ice in her home, dreads the coming of summer. To her it means the waste of many dollars in spoiled foods and the many other inconveniences and distractions caused by high temperatures. With Icyball, all of these troubles are eliminated.

"Icyball will freeze ice cubes within from three to four hours and as many cube trays as one desires may be used. This gives assurance of ice cubes for ice water at all times.

"As the temperature of the Icyball cabinet remains below freezing, foods generally classed as perishable may be kept pure and usable at all times.

"Operation of Icyball is simplicity itself In a few minutes even a child can replenish the refrigeration cabinet with a temperature sufficient to meet all purposes for the preservation of foods and the supplying of ice and cold drinks.

"The upkeep of an Icyball is negligible. It requires no service. There is nothing to get out of order. It is not necessary to replace the liquid with which Icyball is charged.

"Used according to instructions, there is no danger attached to Icyball and even if misused there can be no danger. Icyball has solved the refrigeration problem.

"Icyball is being manufactured by the Crosley Radio Corporation of Cincinnati which plans production on a large scale. Despite the fact that Icyball has been introduced into only a few communities hundreds have been sold. The manufacturer thoroughly tested and tried it out before placing it upon the market."

Gleaner Expansion Plans

KANSAS CITY, Oct. 8.—Plans are under way to expend $150,000 on the early expansion of the Independence (Mo.) plant of the Gleaner Combine Harvester Corporation.

Work is expected to begin soon. President S. H. Hale of the corporation announces that 60,000 square feet of factory floor space are to be added, together with more manufacturing equipment.

The factory is 527 feet long. Three hundred feet are to be added to its length. Capacity of the machine shop will be doubled. "Production line" methods are followed in the manufacture of the company's combines.

To the regular output of combine harvesters the manufacture of a new power feed mill, the "Gleaco," has been added. It is planned to make about two thousand mills for next year.

Upwards of three thousand combines were made and marketed last year. T. J. Turley, general sales manager, says that this may be doubled.

At a recent meeting of the board of directors a stock dividend of 200 percent was declared. At the same time the expansion plans were made known.

George H. Burr & Co., investment bankers, have bought a large share of the unissued stock. Proceeds will revert to the corporation's treasury for use in expansion.

Outstanding shares of stock now number 72,000; 100,000 are authorized. The Gleaner business was started five years ago. It is now one of the largest manufacturing enterprises in the southwest.

Six hundred men were employed last year. This force is to be increased to eight hundred, with an annual payroll approximating half a million dollars.

Leaves Bequest to Veterans

CHICAGO, Oct. 11.—Approximately 1,200 men and women who served the United States in the World War, or their children, will receive free tuition during the coming school year in 60 universities and colleges in all parts of the country, under provisions made by the late LaVerne Noyes, Chicago capitalist and pioneer steel windmill manufacturer, it is announced by trustees of the Noyes estate.

Of this number, 700 scholarships are to be furnished from earnings of the Aermotor Co. of Chicago, while the other 500 are available from a $2,000,000 endowment established by Mr. Noyes at the University of Chicago, a year before his death.

Some 7,500 qualified applicants have availed themselves of scholarships since establishment of the two foundations.

Deeply touched by the spirit of the men and women who offered the supreme sacrifice for their country, Mr. Noyes cast about for some means of expressing his appreciation.

To that end he established the foundation at the University of Chicago. On his death in July of 1919 it was found he had bequeathed the income from his business to extend the scope of his educational foundation. The trustees of the estate, who operate the Aermotor company, supervise the expenditure of the funds set aside each year for the extension work.

Any American citizen, regardless of race, color or creed, who served in the armed forces of the United States 60 days prior to the signing of the armistice and was honorably discharged, is eligible. His children also are eligible, as are the children of men who died in the service. Women who served as nurses in the army and navy, and their children are included.

Mr. Noyes was born at Genoa, N. Y., in 1849. His parents moved to Springville, Ia., in 1854. Mr. Noyes won his degree of B. S. at Iowa State College in 1872. In 1915 his alma mater conferred on him the honorary degree of Doctor of Engineering "in recognition of his eminent success in the field of engineering and his interest in the promotion of higher education."

In 1887 Mr. Noyes became interested in windmills, the field of his greatest success. He undertook the development and promotion of a new steel windmill which had just been invented. The new wind-power motor he called the "Aermotor" and it was manufactured by the Aermotor company, of which Mr. Noyes was at all times the head.

In 1913 he improved the mechanism of the Aermotor so that its working parts are at all times flooded with oil. To this type he gave the name "Auto-oiled Aermotor," as it has an inclosed gear case forming an oil reservoir that insures automatic lubrication.

Handed Him One

Brown was an easy-going old fellow. He believed in taking things as they came. Not so Mrs. Brown.

"Don't you think," she remarked one Sunday afternoon, "that we should be considering Mary's future? It's time she was married; she is already 34."

"Oh, I shouldn't worry!" replied old Brown. "Let her wait until the right sort of man comes along."

"Why wait?" returned Mrs. Brown. "I didn't!"

Radio Fans: Tune in Tuesday on "The Cook Painter Boys," Station WDAF, The Kansas City Star, from 9:30 to 10:00 p.m. Central Time.

Spring Days bring big sales for Cook Dealers!

When Sportsmen find that they can buy....

Winchester Staynless Center Fire Cartridges, it will make a world of difference in your sales. Last year they bought and tested out the Staynless .22 Rim Fire Cartridges. They found, as they had hoped, that nothing of the old-time Winchester accuracy and dependability had been lost in making this new type of non-corrosive cartridge. They found a cartridge which could neither rust nor pit their rifle barrels. When they find this same improvement available in

Winchester Center Fire Cartridges, there will be no limit to the new demand.

With the first success of small-bore Staynless Cartridges, in fact, sportsmen asked for Staynless Cartridges of larger size. It was mighty hard to make them with no sacrifice of former shooting qualities but now that this has been done and they are here—you will find it mighty easy to make sales.

WINCHESTER STAYNLESS CENTER FIRE CARTRIDGES

You now can offer Winchester Center Fire Cartridges in the following popular calibers:—25-20 Win. Lead; .25-20 Win. S.P.; .25-20 Win. Superspeed; .25-35 Win. S.P.; .30 Win. S.P.; .30 Win. Superspeed; .32 Win. Lead; .32 Win. Superspeed; .32 Win. Special S.P.; .32 Smith and Wesson Lead; .32 Smith and Wesson Long Lead; .32 Colt N.P.; .303 British Soft Point; .32-40 Win. S.P.; .38 Short Colt; .38 Colt N.P.; .38 Colt Special; .38 Smith and Wesson Lead; .38 Smith and Wesson Special Lead; .38 Long Colt; .45 Colt; .45 Auto Colt F.P.; .405 Win. F.P.; .405 Win. S.P. We are extending this Staynless Center Fire line as rapidly as possible to include all center fire sizes. Additional calibers will replace the present cartridges as they come into production. Soon, therefore, you will have a complete line of Staynless Center Fires which is sure to take the sporting world by storm. This is surely "a word to the wise."

WINCHESTER LESTAYN RIM FIRE CARTRIDGES [LESMOK]

Another new departure—the complete line of .22 Rim Fire Cartridges loaded with Lesmok powder for greatest accuracy in target shooting, is now offered with this same non-corrosive priming mixture—a combination which reduces rusting and pitting to the minimum. (No change, however, will be made in the super-accurate target cartridges—Winchester Precision 75 and Precision 200.)

WINCHESTER STAYNLESS RIM FIRE CARTRIDGES [POWDER]

The Winchester Staynless .22 Rim Fire Smokeless Cartridges, which have everywhere proved so popular with shooters, are now available in a complete line. These cartridges have laid a firm foundation for your 1928 success with Winchester Staynless Center Fires.

BE FIRST TO SHOW THE WINCHESTER STAYNLESS LINE

This line will be publicly announced in February issues of the Sporting Magazines. Get in touch with your jobber today. Be first to stock and show the Winchester Staynless line and to cash in on the prestige of the Winchester name.

WINCHESTER REPEATING ARMS CO., New Haven, Conn., U. S. A.

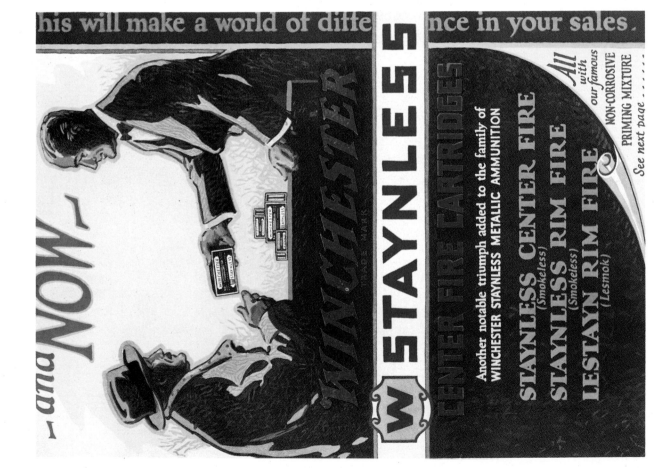

his will make a world of difference in your sales.

—and NOW—

WINCHESTER
STAYNLESS

CENTER FIRE CARTRIDGES

Another notable triumph added to the family of
WINCHESTER STAYNLESS METALLIC AMMUNITION

STAYNLESS CENTER FIRE *(Smokeless)*

STAYNLESS RIM FIRE *(Smokeless)*

LESTAYN RIM FIRE *(Lesmok)*

All with our famous NON-CORROSIVE PRIMING MIXTURE

See next page......

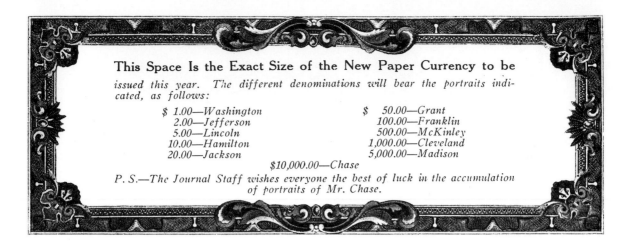
Acres in Hours With Insecticide Duster Service

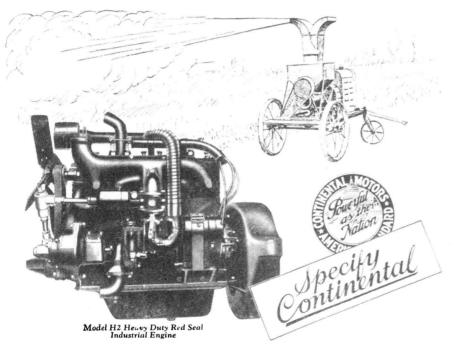
The average farm income, according to some sources, during 1929 was $387.00. Land prices had declined to the point that land had the same value as in 1917. The size of the paper currency also became smaller. The Wall Street Crash of October was only one indication of the difficult times that followed, but most farm equipment manufacturers set record company profits in 1929. Overall production and sales of farm equipment had been better only in the boom year of 1920. Exports were up, but it was noted that German and British tractors and other farm equipment had just about shut off the Canadian and Russian markets. The farm equipment dealer of 1929 ran a main street business requiring better and more attractive retail establishments. Many dealers improved their buildings and expanded to carry items for the whole family.

Hopeless

Host—"Have a cigar?"

Guest—"No, thanks, I don't smoke."

Host—"Have a drink."

Guest—"I really don't drink."

Host—"Maybe you'd like to go in the ballroom and dance a bit."

Guest—"Sorry, but I don't dance either."

Host (desperately)—"Well, my wife has a ball of yarn and some needles; perhaps you'd like to knit a little."—Kansas City Star.

Picked a Flaw

An alumnae of a nearby college was entertaining her former history professor at dinner, much to the interest of her three small sons. Jack, the eldest, kept eyeing the professor closely, and finally, after dinner, he asked, "Mother, you say Professor Smith was your history professor?"

"Yes, sonny, and a splendid one, too!"

"Well," pronounced the 6-year-old observer, "he has a hole in his sock."—Kansas City Star.

President Herbert Hoover established the Federal Farm Board and chose IHC's Alexander Legge to serve as chairman. By the time farm relief was available, the farmers who needed financial help had either recovered or were already destroyed. Overproducing lowered prices of crops, but the overall farm outlook was optimistic.

Allis Chalmers purchased the La Crosse Plow Co. of La Crosse, Wisconsin, during 1929 and also began producing the United tractor for the United Tractor and Equipment Association. The association was made up of manufacturers who had made equipment primarily for Ford tractors. John Deere introduced the "GP" tractor and the Foote Bros. Gear & Machine Co. took over Bates Machine and Tractor Co. operations. The new Belle City corn picker-husker was convertible from pull-type to tractor-mounted. Oliver announced a new potato combined digger-harvester. A new company started by W.C. Durant, the American Cotton Picker Corp., began manufacturing a new cotton picker in St. Louis, Missouri.

Minneapolis-Moline Power Implement Co. of Minneapolis, Minnesota, was formed by the merger of:

Minneapolis Steel and Machinery Co. (Twin City Tractor) of Minneapolis, Minnesota.

Minneapolis Threshing Machine Co. of Hopkins, Minnesota.

Moline Implement Co. (New Moline Plow Co.) of Moline, Illinois.

Early in April of 1929, the Oliver Farm Equipment Co. was formed by a merger of:

Hart-Parr Co. of Charles City, Iowa.

Nichols and Shephard of Battle Creek, Michigan.

Oliver Chilled Plow Co.

Later in the month, the American Seeding Machine Co. of Springfield, Ohio, merged with the newly formed Oliver Farm Equipment Co. The American Seeding Machine Co. was formed earlier by the consolidation of several smaller companies including:

Bickford & Huffman of Macedon, New York.

Brennan & Co. of Louisville, Kentucky.

Empire Drill Co. of Shortsville, New York.

A.C. Evans Co. of Springfield, Ohio.

Hoosier Drill Co. of Richmond, Indiana.

P.O. Mast Co. of Springfield, Ohio.

Superior Drill Co. of Springfield, Ohio.

On August 9, 1929, the Minnesota-Nilson Corp. filed articles of dissolution. Deere and Company purchased the Wagner-Langemo Co. of Minneapolis, Minnesota, manufacturer of threshing machines.

In 1930, steps were taken, by the directors of the Board of Trade, to stop the Russian government from selling wheat in the Chicago market. It was believed that the Russian grain lowered the price American farmers were paid. The New Idea Spreader Co., manufacturers of spreaders, husker-shredders, two-row corn pickers and transplanters purchased the Sandwich Mfg. Co. of Sandwich, Illinois. The Sandwich company, made hay rakes, hay loaders, portable elevators, all sizes of corn shellers and portable gas engines ranging from 1.5 to 10 horsepower. The A.B. Farquhar Co., Ltd. of New York bought the Iron Age line of implements, except for the garden tools. The Stover Manufacturing & Engine Co. was purchased by locals in Freeport, Illinois, from the Chicago owners. The Electric Wheel Co. stopped production of tractors, but continued production of wheels for wagons, tractors and other equipment. The New Way Motor Co. of Lansing, Michigan, was dissolved by court order. The Lauson Corp. was reorganized to take over the assets and liabilities of the John Lauson Co., then introduced the new model "RA." The C.H. Turner Mfg. Co. of Statesville, North Carolina, was reorganized as Turner Mfg. Co. in September after being in the hands of receivers since February.

NEW IDEA
Announces
that It has Purchased the

SANDWICH
MANUFACTURING
COMPANY
of Sandwich, Illinois

Together with the entire line of
SANDWICH FARM MACHINERY

including Hay Rakes, Hay Loaders, Portable Elevators, all sizes of Corn Shellers and Portable Gasoline Engines of from 1 1-2 to 10 H. P.

SANDWICH-NEW IDEA GASOLINE ENGINES
Portable models, 1½ to 10 H.P.

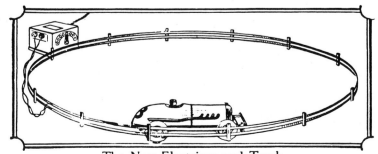

The New Electricar and Track

May Increase Poultry Demand

Two new methods of marketing chickens to the consumer are expected to result in an increased demand. By one of these methods, at least two packing companies are marketing whole chickens in cans. The birds are inspected for condition by U. S. Department of Agriculture representatives, and are then dressed, cooked and canned. By another method, inspected chickens are being full-drawn, hard chilled and placed in individual containers at the packing plants. They are thus sold ready to cook.

These new processes are brought into being in the hope that they will increase poultry consumption. The housewife is saved disagreeable labor in their preparation. The canning method will also make it possible to carry chicken in all stores which handle canned goods, making more business for grocers and less for butchers, adding another factor in breaking down the distinction between grocery stores and meat shops.

Announces Toy

A NEW and different electric toy, the "Electricar," has just been announced to the trade by the Kokomo Stamped Metal Co., Kokomo, Ind., manufacturers of the Chieftain balloon wheeled roller skates.

The Electricar is a miniature of the latest and fastest type of real racing cars. The toy operates by electricity through a step-down transformer from the light socket.

Powered by a real commercial electric motor, the Electricar is capable of extraordinary speed and endurance. It runs on the inside or outside of a special 20 foot flexible spring steel fence. The fence can be placed in any position or shape. A patented steering arrangement keeps the Electricar to the fence regardless of shape. The toy is lustre enameled in three colors, has nickel radiator, genuine molded rubber tires, and is sturdily built.

The Electricar, complete with track, is attractively boxed for display.

The Electrcar is expected to have a genuine sales punch, as it is a really new idea in propelled toys and has many features not found in toys of this nature being offered to the trade. It will be a welcome addition to the Kokomo toy line.

Summer offers *bigger profits*
with this

Insulated
TAPPAN
GAS RANGE

SUMMER is one of your best range merchandising seasons because summer increases the demand for the Insulated range—the volume builder of the range business. Women are busy baking and canning—more than ever conscious of the heat leaks and gas waste of their old stoves. Add Tappan's exclusive sales-clinching features—the Smokeless Broiler and Chromium-lined Oven—to Insulation's persuasive summer appeal, and you've opened the door to profit.

Tappan has prepared a complete merchandising campaign to help you *pull* in business. Forceful newspaper ads—one of the most novel display pieces ever designed for a gas range—and a complete assortment of consumer sales literature. Act now, write for detail—and cash in on Insulation profits!

The
TAPPAN STOVE CO.
Mansfield, Ohio
Western Office:
665 Howard St., San Francisco
California

TAPPAN
GAS RANGE

An Immediate Opportunity

There remain a few choice dealer territories in Kansas and Western Missouri, on the Improved Fordson Tractor, the tractor that will sweep the country with its sales in 1930.

The Improved Fordson is the result of seventeen years of study and experience by the world's greatest industrial organization. It is a larger tractor, has Bosch "high tension" ignition, larger radiator, with water pump and a plow speed of 3⅛ miles per hour.

Thousands of farmers in this territory will cut production costs and increase their profits this year with the Improved Fordson.

Scores of Implement dealers will share in the dealer profits to be made through the marketing of these modern tractors.

The demand has already been built by national advertising, the quality is known wherever the name of Ford is known and here is your opportunity.

Lauson Vertical Engine

Announcing the New Air Cooled Model Built for Continuous Service

MEETING the increasing demand for a motor built on automotive lines, yet not too heavy in weight, the Lauson Vertical air cooled engine, recently put on the market, is a substantial model designed for continuous service.

Neat, compact, self-contained construction makes the new engine readily applicable to portable units. It also meets the demand of stationary power users for a motor of the higher class design that can be installed on water or light unit in a residence, dairy, or power house, or wherever smooth, quiet operation coupled with cleanliness is appreciated. The unit can also be furnished in tank or radiator cooled job.

Years of experience in the manufacturing of engine and tractor governors for commercial purposes have been incorporated into the design of the governor on this vertical engine. The drive is correct; spiral gears from the cam shaft; bronze bearings are used; flyballs are mounted without the use of complicated parts. It is a quick-acting sensitive governor, maintaining a constant speed that can be quickly changed by moving the speed lever.

Mounted in high grade heavy duty roller bearings which are lubricated from the engine crankcase, the crankshaft, 1⅜ inches in diameter, is balanced with counterweights which reduce vibration to the minimum.

The camshaft is one piece, drop-forged, hardened and ground, and is mounted in large bronze bearings. Such a camshaft will show very little wear and will maintain correct valve action. The connecting rod is interchangeable with Ford model T rods. Pistons are of grey iron, especially designed for high speed performance and equipped with three rings.

An oil reservoir is formed by the lower base of the engine, from which the oil is raised by a submerged plunger pump driven by an eccentric on the camshaft. Distribution to all vital parts of the engine is by splash system.

High tension Wico magneto, specially designed for high speed motors, is driven by helical cut gears from the camshaft and operates at the camshaft speed. The spark plug is mounted in head, directly in combustion chamber.

Ample cooling capacity is provided by means of a special construction fan in the engine flywheel. Cold air, drawn in over the cylinder head, cools the valves. The air passes around the cylinder through large cooling fins, with which it is provided and is then expelled through the flywheel fan. Such constructon keeps the head and valves cool at all times.

Facilitating simple cranking, the engine has the same hot spark when cranking as when running at normal speed, due to magneto construction.

The cylinder is ground by the Hutto method, the method used by 85 percent of auto and airplane engine builders.

Complete specifications of the Lauson vertical air cooled engine are as follows: Number of cylinders, 1; bore, 2¾ inches; stroke, 3¼ inches; cycle, 4; speed, 900 to 1800 r.p.m.; horse power, 1½ h.p. at 12000 r.p.m.; ignition, Wico type B magneto; fuel, gasoline; carburetor, Tillotson; flywheel diameter, 12 inches; flywheel face, 2¾ inches; crankshaft bearings, 1⅜ inches; crank pin, 1¼ inches; governor, vertical flyball type, gear drive, enclosed in dust proof housing, speed adjustable.

Height over all 24½ inches; width over all 21⅜ inches; pulley, 4⅝ inches in diameter, 3 inch face, fibre pulley; net weight, 160 pounds; domestic shipping weight, 200 pounds; export shipping weight, 260 pounds; boxed for export, 10 by 19 by 30 inches.

An Allis-Chalmers Monarch "75" crawler tractor was equipped with a Diesel engine and first tests indicated that it was very economical to operate. The men testing the Diesel believed "that the next 10 years will be an era of Diesel-driven tractors." It was thought by some that the price of Diesel fuel would increase as soon as it became more popular and some other hidden costs would make the Diesel actually more costly. J.I. Case introduced the C and CC tractor models, Massey-Harris Co. introduced a four wheel drive general purpose tractor and Oliver introduced the new Hart-Parr "Row Crop" tractor. Stockholders of the Steinke Bros. Mfg. Co. of Peoria, Illinois, changed the company's name to Little Giant Products, Inc. Experiments were being conducted to broaden the use of wheat combines for harvesting other crops including corn, oats, sorghums and sunflowers. Two men were required on the baler and another to drive the tractor, but the combine unit would pick up windrowed hay or straw and compress the crop into bales. Soybeans, long considered a "pinch-hitter crop," increased in popularity as a crop that fit well into the rotation with other crops.

The self-propelled Sunshine combine, made by the H.V. McKay Co. of Sunshine, Australia, was introduced by the U.S. distributor, the Ohio Cultivator Co. of Bellevue, Ohio. The horseless and tractorless harvester was invented in 1924 by Hugh Victor McKay and H.S. Taylor. Massey-Harris Co., Ltd. of Canada, purchased interest in H.V. McKay Proprietary, Ltd. of Australia.

Late in 1930, Allis-Chalmers began selling a "baby" combine that was only 5 feet long and weighed only 3000 lbs. Allis Chalmers also introduced the "UC" All Crop tractor. The Ann Arbor Machine Co. designed a traveling, tractor operated hay baler. The Advance-Rumely "6" tractor was announced in December and was equipped with a six cylinder engine. Twenty-five years earlier, it was noted that a single cylinder was enough for automobiles because "one cylinder engines were providing all of the trouble needed."

The Tractor Division of Allis Chalmers Manufacturing Co. purchased the assets of Advance-Rumely Corp. of La Porte, Indiana. Stockholders of Advance-Rumely ratified the merger by voting on May 29, 1931. The Avery Power Machinery Co. was reorganized again, the Associated Mfrs. Corp. was assigned a receiver, the United Engine Co. filed for voluntary bankruptcy and the Gleaner Combine Harvester Corp. reorganized as a result of financial problems. It was very hard for some to see recovery from the depression, but some indicators showed improvements.

After more than five years experimenting, testing and demonstrating, Caterpillar strted production of a diesel powered tractor.

Manufacturers tried to sell farmers on not decreasing acreage, but to increase yields and lower production costs to enable American goods to compete in the world markets. Manufacturers and dealers knew this required new and better equipment, but farmers thought that reducing costs meant not spending money on new equipment. Equipment sold around the boom year of 1920 was wearing out and more than baling wire was needed to repair it. Dealers with experienced repairmen with the right tools, parts and knowledge were needed to repair the old machinery. A slogan used was "Dealers' Service Makes Good Equipment Better." Some people who had lived in the cities until they were out of work, returned to farms. Intervening years had changed farming too much for some, but others were receptive to modern farming methods and succeeded. Running water was still thought to be the most needed modernization for the farm family.

The Allis Chalmers tractor of 1932 was available on rubber tires, but soon most other makes offered "air tires" made by Firestone and other tire companies. Allis Chalmers purchased the American Brown-Boveri Co. and the Conduit Mfg. Co., both of Boston, Massachusetts and the Birdsell huller line from the Birdsell Mfg. Co. of South Bend, Indiana. After another reorganization, the Gleaner Combine Harvester Corp. plant and assets were sold to a new company called the Gleaner Harvester Corp. Assets of the Kentucky Wagon Co. of Louisville were sold on July 20 to a group starting the Kentucky Mfg. Co.

County agents of the U.S. Department of Agriculture were blamed for "propaganda which prejudiced farmers against purchase and use of tractors." It seems that a number of agents were influenced by members of the Horse Association of America who were working to support their cause. Though the agents were actually independent, it was felt they represented Washington.

The Piston Ring Co. of Muskegon, Michigan, changed its name to Sealed Power Corp. effective July 15, 1932.

Dust storms began late in January of 1932 as a duster swept across the panhandle of Texas. Net farm income averaged about $230.00 in 1932. "Bunc" was used by a pessimistic buyer to describe many advertising claims. Live stock were thought to be safer than grain crop farming by some and garden tractors were easier to sell than large equipment.

A Halt in Foreclosures

New York Life Insurance Co. has suspended foreclosures in Iowa, pending action by the Iowa Legislature. This company reports that its loans on farm lands in that state amounted to $1,837,916 as of Dec. 31, 1931.

Prudential Insurance Co has announced a nationwide suspension of foreclosures on owner operated farms throughout the United States, which involve 37,000 farms and amount to approximately $200,000,000 in the United States and Canada.

Mutual Benefit Life Insurance Co. has announced that it will comply with the request of Gov. Clyde Herring of Iowa and suspend all future foreclosures in the state. This decision involves about $5,000,000 which is in foreclosure litigation.

Gov. C. W. Bryan of Nebraska has appointed a board of conciliation to aid in bringing about fair and equitable settlements between debtors and creditors in that state.

"Our Gang" Inspects the McCormick-Deering Ball-Bearing Cream Separator, and Pronounces it "Okay."

Farina, the little colored gangster, just had to show everyone how easy it was to turn. Chubby little Wheezer thought the big bowl just his size, and it was—while little Mary Ann Jackson demonstrated just how nice and comfortable a seat the adjustable pail shelf could be. The youngsters are the famous Hal Roach protegés.

We're Living in a Strange Era

VISITING not long ago an eastern Kansas community with which there has been personal association over a forty-year period, we could not help but note the evidence of agricultural decadence on every hand. It was a community generously endowed by nature with every essential for a prosperous agriculture, and of such it boasted in earlier days. There has been great dissipation of these natural agricultural resources.

The spirit of the pioneers who transformed a wilderness into a beautiful countryside has vanished. The comfortable and spacious farm homes which they erected, the large barns and other farm improvements are crumbling, showing a lack of paint and other ordinary maintenance. Once neatly-trimmed hedges are soaring skyward like California redwoods. Fences are dilapidated. Pastures of native prairie grass are overgrown with weeds while mowing machines are rusting in fence corners. Fields that once produced bounteous crops are now gullied to the point where their fertility can be restored only by heroic engineering. The manure spreader has disappeared as an implement of farm operation. Fields in which corn failed to make a stand last spring have since been untouched.

A thriving town of four thousand persons does not boast of a retail implement store. In the forty-year period the mortgaged indebtedness of the farms has increased from a purely nominal figure to several hundreds of thousands of dollars. The farms are still inhabited, but there is none of the enterprise of former years. Modern farm equipment has eliminated the drudgery of the nineties, but even with this relief few farmers are working. Some are obsessed with hopelessness, others with lassitude, and still others with strange economic ideas.

Their fathers believed in the theory that "God helps those who help themselves." A few still survive who believe in the combination of the Almighty and work, but the great majority incline to a belief that the government will relieve those who don't work.

It may be a mere coincidence, but we can recall a few chapters from early youth when meetings were held in the country schoolhouses, and there came from these a lot of talk about "hell-raising" instead of corn raising. We can recall that the subjects of conversation at the impromptu Saturday afternoon group meetings of the farmers in town suddenly changed from millet, kafir, the virtues of different dairy breeds, etc., to something closely resembling politics. There was considerable ecstacy in the community when occasionally some political idol was toppled from his throne.

We do not know that there was anything more than a coincidental relationship, but in our more mature years we became conscious that every man in politics made it a point to find out what the farmers wanted and then to "come out for it."

One must admit, of course, that the economic lot of the farmer is not bright. The same is true of most any other industry. But it is equally apparent that in every industry the man who works is still better off than the man who merely bemoans his lot and does nothing to correct it.

Some day perhaps some brain trust will be installed in the high places which will place a different interpretation on agricultural relief, which will distribute benefit payments only in direct proportion to the farmers' own efforts toward self-relief, bonuses for the improvement of soil fertility, for terracing farms, for improvements in live stock, etc. There may perhaps some day develop a program which will make agriculture basically and permanently sound for those willing to work.

Meanwhile, in this strange world of the present—what next?

HOWARD E. EVERETT.

A New Battery Terminal

A new type of battery terminal, which is said to overcome permanently the annoying troubles of the garage man and motorist due to corrosion and consequent bad contacts is being manufactured by Louis Gross, 3739 West 47th Place, Cleveland, O. All parts exposed to corrosive action are of lead, unmixed with tin or zinc and are therefore acid-resisting. As an additional protection against bad contact, the connecting cable and lug are secured by special lead burning under the Gross formula, no solder being used.

To connect up the battery, the terminal is slipped over the battery post. The "U" shaped clamp, (made from hard acid-resisting lead alloy) is then forced down over the projecting ends of the clip. The compressing action of the tapered slot in the clamp assures a thorough contact without the aid of bolts or spring washers; the lead-to-lead contact giving unusually effective resistance to jarring loose from vibration. The terminal may be readily removed from the battery post when desired merely by levering off the clamp with a screw driver.

1933-1935

BAD TIMES FOR EVERYONE

Time is always a factor when plowing must be done. What farmer wouldn't prefer air tires, when they enable the tractor to crowd an extra day or two's performance into the week?

Implement and Tractor

★ THE BUSINESS PAPER OF THE FARM EQUIPMENT INDUSTRY ★

1933-1935

The whole country was having financial problems. Farmers in the "Dust Bowl" had to fight weather and low grain prices at the same time. The "Dust Bowl" that started in 1932, continued until 1940 and affected not only wheat production, but the lives of nearly everyone in the nation. Farm Relief or Agricultural Adjustment Act began in the spring of 1933.

Two-Fold Economics of "Buy American"

"Buy American" is the battle-cry of many industries today. Back of the idea lie two sound reasons—not only one, as many people think.

The first reason is that the purchase of American products creates more hours of work for American workers. Work means wages to spend. And money in circulation again means increased confidence and credit. With the increase of credit, industries' management will be enabled to undertake larger operations, employ more people.

This theory is, of course, attacked. Its critics say, "We must balance exports with imports." But in the case of many products, no real "balance" could ever be achieved, because of the radical difference in quality between American products and their foreign competitors. The basic appeal of foreign goods is "cheapness"—and little else. Price is a potent factor today, but cheap price coupled with cheap materials and workmanship is never a bargain. American consumers have learned this. So have American merchants. And today the demand is for that quality in merchandise which makes its purchase truly economical.

There is no denying the fact that American goods, as a whole, excel all foreign substitutes. Price competition exists, but quality competition—never!

"In no line is this more evident than in binder twine," says J. S. Bradford, of the Plymouth Cordage Co. "In this market now there is a powerful trend toward quality, and away from foreign twines which looked deceptively economical be-

cause of competitive price. But the court of last resort is the farmer. When he uses inferior twine, his harvesting costs rise alarmingly. Back he comes—and has come—to demand the economy of quality. Dealers in Plymouth twine report this trend from all sections of the grain country. And Plymouth, which has upheld its quality standards for over a century, is again reaping the inevitable reward, along with its loyal dealers and distributors."

CROP PRICE PLANS FOR 1933

Both International Harvester Co. and Deere & Co. Make Arrangements for This Year

Both International Harvester Co. and Deere & Co., have announced crop price guaranty plans for 1933, similar to those which were in effect for 1932 but elaborated to extend over their entire farm machinery lines.

The basic crop prices used in the guaranties are 70 cents a bushel for No. 2 hard wheat, Chicago delivery; 45 cents a bushel for No. 2 yellow corn, Chicago; and 8½ cents a pound for middling cotton, New Orleans. These prices are the same as last year, except for corn which was 50 cents in the 1932 guaranties.

Another elaboration in the crop plans of the two companies provides for handling 80 per cent of the purchase price of all machines costing less than $150 under the guaranty plan during 1933.

The plan announced by International Harvester does not apply to repairs nor motor trucks. In announcing its plan International said:

"We believe that present crop prices will show a gradual improvement, and we also believe that if the farmer is assured a fair price for his 1933 crops, he will purchase the implements which he needs. We have decided, therefore, to give American agriculture again, as we did in 1932, the assurance of a fair price for wheat, corn and cotton on this year's purchases of equipment from us in so far as 1933 payments therefore are concerned.

"Where the purchase amounts to $150 or more and the farmer's note maturing this year represents not more than 40 per cent of the purchase price, there will be endorsed on the note the equivalent number of bushels of wheat computed at 70 cents per bushel, Chicago, for No. 2 hard wheat. If the average Chicago quotation at maturity of the note is less than 70 cents per bushel, the purchaser will be credited with the price differential multipied by the number of bushels shown on the note, provided the remainder of the note is paid in cash within ten days of maturity.

"Where the purchase amounts to less than $150 and the note represents not more than 80 per cent of the price, and the Chicago market wheat price at maturity of the note is less than 70 cents per bushel, the farmer will be credited with one-half the price differential per bushel, multiplied by the number of bushels endorsed on the note, provided the remainder of the note is paid in cash within ten days of maturity.

"The wheat price guaranty plan will be generally used throughout the country, but in territories where corn or cotton is the principal money crop and practically no wheat is grown, either the corn price guaranty will be applied at 45 cents per bushel for No. 2 yellow, Chicago, or the cotton price guaranty plan at 8½ cents a pound, New Orleans, for middling cotton.

"The company reserves the right to take delivery at the specified price of any portion of the commodity involved in any transaction, if available at the time of payment. It also reserves the right to withdraw the offer or to readjust the terms in case of governmental action materially affecting prices of the crops in question."

While Deere & Co. has not made formal announcement of the details of its plan except to its organization and dealers, it differs from that of the Harvester company only in minor particulars.

Farmers Not Revolting

WHEN two leaders of farm organizations recently told a congressional committee that a revolution is brewing in the farm areas, they spoke words which if true would be fraught with serious eventualities for our national welfare. Their statements, however, were far from the truth and constituted another example of the use of threats instead of logic.

Despite the fact that leaders presuming to speak for agriculture talk of revolution, more than six million farmers are either in the fields, plowing and seeding their ground, or actually preparing their 1933 crop programs, hoping this year will produce the long-awaited "break" in their favor.

Does this sound revolutionary?

A revolution, in the proper conception of the word, means a movement to upset the existing order of things, using violence if necessary. Present activities of the rank and file of American farmers are absolutely devoid of revolutionary semblance.

It is true that here and there throughout the farm belt, where farms are heavily mortgaged and foreclosure movements have been initiated, some mild acts of violence have been invoked to forestall such legal action. But by no stretch of the imagination can these actions be interpreted as revolutionary signs. These actions have been designed to maintain the existing order, not to foment or upset.

There are no signs that farmers are abandoning agriculture; moving into towns to be fed at public expense; indulging in violence to prevent their neighbors from planting crops; destroying farm property, or even demanding cancellation of farm debts. Until something on this order occurs we cannot talk truthfully of a farm revolt.

Six million and more farmers want the established order to prevail. They want to retain possession of their farms and their homes, and to carry on the pursuit of agriculture as they have in the past. They realize that so long as they can remain on their farms they have a chance. They have maintained their hope in the future. They do not deny their obligations, even admitting that part of them are due to their own mismanagement and speculation. They ask only a fair chance to pay them off.

Mild demonstrations in defense of farm homes have not been without their value. Mortgage holders have been impressed and are becoming more sympathetic. They are realizing that their own best interests call for a more conciliatory attitude. There are many indications that deserving farmers will be given a new deal involving longer tenure of their homes and a long-term refinancing of their indebtedness. State and national legislative bodies are seeking to lighten the tax burden upon agriculture. The working of economic cycles eventually will restore commodity prices to profitable levels. No revolution is necessary.

Four out of every five farmers in the western farm belt are thinking their way clearly through the stern realities they face. They are solving their own individual problems with a fortitude and clarity of vision sadly missing in so-called higher places. They intend to continue farming; they are buying such necessary machinery as they can afford, buying more this spring than last year; and as fast as possible they are restoring their mechanical equipment to the point where they can operate their farms profitably. In fact, the desire to buy economical equipment has never been so great as it is today.

American farmers are sane, law-abiding and intelligent-reasoning individuals. They resent the waving of a red revolutionary flag.

They need more than all else to be left alone to work their own salvation out of the soil under conditions which will make that possible. When that can be done, agriculture will resume its old position as the foundation of our economic and industrial establishment.

HOWARD E. EVERETT

Cam Piston Finishing Lathe

The gradual adoption of the out-of-round piston by the gasoline engine makers, the truck, tractor, bus and auto manufacturers, has made piston replacement work a real problem for the average repair shop and dealer. To meet this situation, the South Bend cam piston finishing lathe has recently been developed by South Bend Lathe Works, South Bend, Ind.

The manufacturers claim that it will finish all makes and types of pistons, whether cast iron, aluminum or alloy, in the former conventional round shape with straight relief, or in any of the newer oval shaped designs. Set-ups are simple and the change from finishing round pistons to cam-shaped pistons, and vice versa, requires less than two minutes. A tool tipped with tungsten carbide is used for the finishing, permitting high speed work.

Half-Ton Truck Is Announced by Harvester Company

The recent announcement of a new half-ton motor truck by International Harvester Co. marks a significant milestone in motor truck history. For years the company has been a leader in producing a full line of heavy duty and speed trucks with ratings up to three tons. With the advent of the new Model D-1, the half-ton truck, the size range is considerably enlarged, extending from the new half-ton up to a 7½-ton heavy duty unit in a variety of types. The new Model D-1 chassis is priced at $360 f. o. b. the factory.

The extensive Harvester after-sales service organization, which has been a tremendous factor in building up the company's truck business, will play a prominent role in developing popularity for the new unit. Generations of service experience have taught the Harvester organization its importance. The company's widely extended chain of service stations and branch houses, which has been established to further the sale of trucks will be available for purchasers of the Model D-1. Farmers and small-town merchants will appreciate the service back of the new model.

The Model D-1 chassis has a wheelbase of 113 in., and is powered by a six-cylinder, high compression engine, which develops 70 hp. at 3400 r.p.m. Down-draft carburetion and full pressure lubrication to main, connecting rod and camshaft bearings are important features. The 9-in. clutch is of single plate design with built-in vibration damper. The transmission has three speeds forward and reverse. Final drive is of the spiral bevel type with semi-floating axle shafts. Semi-elliptic front and rear springs are of chrome-vanadium steel and have self-adjusting shackles. Forty-spoke 18-in. wire wheels, a left front fender well, spare wheel and spare wheel carrier are standard equipment.

Special attention was devoted to the appearance of the new unit, a factor that is receiving increasing consideration by truck buyers. Two attractive types of body are available, one a de luxe sedan panel body and the other an all-steel pick-up body with coupe type of cab. The loading compartment of the panel body is 72 in. long, 46¾ in. wide and 47 in. high. The pick-up body has a loading space 66 in. long and 46¾ in. wide. Side panels are 11 in. high with 6-in. flare boards.

This is the International Harvester Co.'s new Model D-1, one-half ton truck, shown with cab and pick-up body, one of the two types now available. It is ideal for the farmer who requires light haulage and transportation and answers the purposes once filled by the spring wagon

The government stimulated the economy by buying large numbers of tractors and engines, but angered some because it bought direct from the manufacturers. The wheat harvest of 1933 in the "Dust Bowl" averaged 2 bushels per acre. The Tennessee Valley Authority began, Fate-Root-Heath Co. started production of "Plymouth" tractors in Plymouth, Ohio, and Armco began manufacturing stainless steel in 1933. Allis Chalmers introduced the "W" model which became the "WC" the following year and was continued until 1948 when the similar, but different, "WD" model was introduced. The "W" was the first tractor on pneumatic tires tested in Nebraska. Several manufacturers, including Hercules, announced the introduction of a Diesel engine.

Hercules Has Diesel Engine

Hercules Motor Corp., Canton, Ohio, announces its development of a 6-cylinder, solid injection, high compression, full Diesel engine for automotive, general industrial, agricultural, oil field and marine purposes. The engine is known as the Hercules DX1, has a bore of 5 in. with a 6-in. stroke and is designed to operate at speeds up to 1000 r.p.m. Its design and constructon follow the usual Hercules practice, combining symmetry and compactness with accessibility and durability. In general the external housing dimensions have been made interchangeable with the Hercules HX series of gasoline engines.

In performance, it is claimed the DX1 will develop from 100 hp. at 1000 r.p.m. to 177 hp. at 1800 r.p.m. and 188 hp. at 2000 r.p.m. Fuel consumption is .383 lb. per-brake hp. at 1000 r.p.m., .430 lb. at 1800 r.p.m. and .46 at 2000 r.p.m.

The Real Farm Crisis!

ESTIMATES of 1932 farm equipment sales from various reliable and competent sources indicate average purchases per farm of not more than sixteen or seventeen dollars. This amounts to but a few pennies per acre. Over a long period of years the average machinery needs of agriculture have been established at about sixty dollars per farm, and purchases have frequently approximated seventy for each of the more than six million farms per year.

This follows an abnormally light volume in 1931, and the poor condition of present farm equipment is further intensified by the lack of repair business during the last two years.

The inevitable results of this two-year period of reduced buying are already manifest in many sections. Competent advisors report that it will be impossible this spring to obtain a good stand of corn in one of the banner corn counties of Kansas because of the deplorable condition of planting equipment. This condition prevails in greater or lesser degree throughout the entire Corn Belt. Reduced yields are imminent unless farmers can buy new equipment. The entire Southwest faces the possibility of the shortest wheat crop in history, partly because of unfavorable soil and weather conditions, but more especially because the farmers' equipment has not been in condition to circumvent these conditions.

The breakdown of machinery is the real crisis which agriculture faces this year. It is more important than higher markets, for what will good prices avail the farmer who cannot produce something to sell?

Throughout all the crop producing area of the United States we still hear the cry—too frequently from bankers who should know better—that the farmer has gone "broke" buying machinery. The farmer has never invested more than five per cent of his income in mechanical equipment, and in the last two years has spent but little more than two per cent. Today, he is going "broke" for lack of machinery.

At a time when the whole nation realizes the need of rehabilitating industrial establishments, and ample funds are available for that purpose, similar assistance is being denied agriculture.

Two conditions must be realized in this emergency: (1) Farmers in greatest need of new equipment are least able to buy, and must have assistance, and, (2) that assistance should come from the local bank.

At the beginning of 1932 the farm equipment industry was acting as local banker to the farmer to the extent of approximately $300,000,000. This amount has probably been reduced ten per cent during the year by a vigorous, though not ruthless, collection program. Yet, in view of reduced returns from sales the remaining $270,000,000 of receivables represents a greater burden for the industry.

The local banker has a more vital interest than ever in the adequate equipment of local agriculture. The major portion of local bank loans are first mortgages on farms. These mortgages are good only as the farmer is able to meet the interest. Foreclosures are possible, but contribute nothing to bank solvency as they only increase frozen assets. The banker's salvation lies entirely in the ability of the mortgagee farmer to produce marketable crops and pay his interest and taxes.

How can this be done with the farmer's equipment in a pitiable state of efficiency—virtually broken down?

HOWARD E. EVERETT

Oldfield to Seek Tractor Speed Record at State Fairs

The veteran Barney Oldfield, the best known automobile racing driver in the United States, will return to the track at a number of state fairs this summer and fall after a 15-year period of retirement. He has signed contracts with several state fair managements to drive a Model U Allis-Chalmers tractor in an effort to establish a new speed record for that tractor which did 35.405 miles an hour recently on the Milwaukee speedway.

Oldfield is scheduled to appear at the Illinois State Fair in Springfield, August 25. Other appearances scheduled to date include the following: Ohio State Fair, Columbus, August 29; Iowa State Fair, Des Moines, August 31; Indiana State Fair, Indianapolis, Sept. 2; Nebraska State Fair, Lincoln, Sept. 4; Minnesota State Fair, St. Paul, Sept. 7, and Kansas Free Fair, Topeka, Sept. 13.

During Farm Week at the Century of Progress last week, Oldfield drove the Model U tractor on the speedway at the Fair Grounds as well as in the Century of Progress parade on Michigan Avenue. He also drove the tractor a few weeks previously in the Allis-Chalmers field day near La Porte, Ind., at which the company introduced its new Corn Belt Combine.

At the forthcoming fair, Oldfield promises to give farmers a new conception of tractor speed, which he considers prophetic for tractor field work.

Barney Oldfield with the Allis-Chalmers tractor on Firestone tires as he appears on the state fair circuit driving tractors at record speed

Development of the Hay Combine

WITH the increasing interest in the raising of legumes, particularly alfalfa, came the call for a machine that would combine two or more harvesting operations into one and at the same time prepare the crop for market with a proportionate saving in labor and costs; a machine that would enable the farmer to realize his ambition of greatly increasing his alfalfa acreage, but with less interference with the tending of his other crops during the harvesting of three or more cuttings per year; a machine to take to the crop instead of moving the bulky crop to the machine and thereby provide the only means of retaining practically all of the leaves on the stems as well as keeping the hay free from dust and dirt by eliminating the necessity of dragging it over the ground.

As an example of the farmers great appreciation of the application of combine methods to harvesting and preparing crops for market, reference is made to the comparatively recent crossing of the continent from the West to the East in the general adoption of the grain combine.

One resisting problem remained to be solved in that a vast percentage of the farmers refused to waste their straw. This proved the deciding factor as to the justification of the time, effort and expenditure that would be involved in the pioneering of an entirely new type of agricultural machine that would automatically pick up forage crops from the windrow and prepare them for market in one operation, for in the salvaging of this straw it is only necessary to disconnect the drive to the straw spreader of the grain combine to permit the straw to be dropped on the stubble in perfect windrows.

This machine, now perfected, provides a practical means of salvaging straw left in the field by the grain combine at negligible cost and at the same time leaves it in the bale ready for market or for the most economical storage and convenience in handling.

In developing a pick-up baler, it is highly essential that the baling press unit possess the maximum in capacity and strength with minimum weight. It must be free of mechanisms which tend to produce vibration or throw it out of balance in operation. It should partially balance on a single axle with the forward end resting on the draw bar of the tractor in order to provide the shortest possible turning radius as well as to allow the pick up unit to follow closely the contour of the ground. The frame which serves as the foundation of the entire machine must be amply trussed to guard against sagging as a result of the extra duties imposed upon it as a traveling field machine.

Even though the pick up baler has an advantage over other types of traveling field machines in that it ordinarily is operated on comparatively solid hay meadows, the problem of keeping total weight

to a minimum is of utmost importance. Power must be provided for operating the machine either by means of a separate engine, or it must be transmitted from the tractor. A separate engine adds weight, cost, fuel consumption, maintenance and depreciation. As but 8 hp. is required for operating the machine, we were quick to take advantage of the rear power take-off facilities now provided on practically all American built tractors, and thereby pioneered the power take-off baling press as well as the hay and straw combine.

The result is a complete machine that includes picking up and power transmission facilities which weighs 300 lbs. less than our heavy duty belt power baling press which was developed many years ago for steam tractor power. As a further comparison, its total gross weight operating in the field is more than 1,000 lbs. less than the average of 10 and 12-ft. grain combines.

The balancing point of the conventional belt power baler being so far forward of the center of the frame presents somewhat of a problem in mounting on a single axle, especially in view of the fact that additional weight must be added still ahead in order to provide the means of transmitting power from the power take-off of the tractor. This problem was solved to a great extent by eliminating the belt shaft of the baling press, adhering to the same gear centers and attaching the drive pinion to one end of the cross shaft of the power take-off transmission and the balance wheel to the other, and at the same time eliminating the necessity of belts and pulleys. In addition to the important reduction of weight and number of parts the only shafts on the machine revolving at a speed greater than 220 r.p.m. are carried on Timken roller bearings.

In order to keep angularity of the universal joints of the drive shaft to a minimum, the height of the power take-off transmission should be in keeping with the average height of the power take-off outlets of the various tractors.

The principal problems encountered in the actual picking up of the windrow and conveying it into the hopper of the baling press unit in sufficient volume were those of carry-under, winding, breaking the hay stream for dropping the divider block with a minimum loss of time and providing a positive means of starting again the flow of hay into the hopper of the baler after dropping the block. These problems were enlarged because of the necessity of handling different kinds of hay and straw with equal efficiency, and under all conditions. Numerous features or mechanisms were tested which worked beautifully in the handling of one crop, but proved inadequate in another. The types of grain combine pick-up units were thoroughly tested but because of the broader requirements for the clean pick-

ing up of hay and straw we eventually saw fit to develop our own pick-up.

Wider Than the Windrow

The pick-up should be somewhat wider than the actual spread of the windrow on the ground to aid the tractor driver in maintaining a straight course and still pick up all of the uneven edges. The pick-up on the Ann Arbor hay and straw combine is 54 in. wide, but it is capable of drawing in the scattered edges as far as 10 inches on either side because of the absence of any part in line with the ends of the pick-up cylinder, therefore giving in reality a picking up width of 74 in.

The first possibility of lodging or winding is immediately after the crop has been lifted over the pick-up cylinder and as it is being passed on to the elevating conveyor. Next, at the top of the elevating conveyor are the same problems on delivering the crop into the cross conveyor with the additional problem of avoiding carry-under. All of the various methods of conveying were thoroughly tried out including slats, canvases, etc., and a series of drag chains equipped with special tine links proved by far the most efficient for both the elevating conveyor and cross conveyor units.

A problem of no small consequence was that of providing a force feeding unit on which depended the success or failure of the machine as an absolutely automatic machine that would require no manual assistance except for wiring the bales. This unit must be equally efficient in working

ANN ARBOR HAY & STRAW COMBINE

OWNED BY THE ILLINOIS CANNING CO.

HOOPESTON ILLINOIS

OVER 3 TONS 17×22 BALES EVERY 45 MIN.

Some views showing the utility of the Ann Arbor hay and straw baler as used by a canning company in Illinois where large scale baling is practiced

either forward or backward, and on it depends the quick separation of the hay stream for dropping the divider block as well as the positive starting of the hay again after the block is in place. Tines 5 in. long literally reach out and pull the hay or straw in from above, semi-compress it as it passes into the hopper of the baler so that it will be kept well under the bottom of the feeder head and then strip straight out. Some lost sleep and headaches resulted along with a lot of waste paper from the drawing board, but the outcome was a highly efficient device which accomplished all of these requirements with no possibility of winding.

The Method of Drive

Should the pick-up be ground driven or power driven? Early experiments were mainly confined to the ground drive off the right transport wheel and through a counter shaft necessitating additional sprockets and other parts, together with several extra feet of drive chain. The idea was to recommend the removal of the combined pick-up and elevating conveyor unit for stationary baling (which can be accomplished by the removal of three bolts or pins), and pitch directly into the cross feed unit. It was soon discovered that the large area of the elevating conveyor offered outstanding advantages for receiving the hay or straw pitched or merely shoved on to it from stacks. In feeding from near the ground with the pick-up power driven it is only necessary to push the hay or straw against or over the end of the pick-up cylinder. The adoption of the power drive added very materially to the capacity of the machine in stationary baling and eliminated the need for at least one man. Side draft was also eliminated by relieving the transport wheel of the added rolling resistance involved in driving the pick-up.

The combining of the power drive for the pick-up into a single unit with the drive already provided for the cross conveyor was accomplished in a very simple and unique manner. It has the following duties to perform, all by means of a single control lever located in a position most convenient to the operator.

1. While picking up windrows the cross conveyor must be stopped and started at will in both a forward and reverse direction, but the pick-up must run continuously in the forward direction only.

2. For convenience in stationary baling, a small cam may be shifted permitting the pick-up to automatically stop and start with the cross conveyor, but the pick-up remains inoperative while the cross conveyor is operating in the reverse direction.

3. A backing up clutch is provided so that the pick-up will not turn backwards under any condition.

Safety Clutches Important

Safety clutches are provided at all three points of power delivery which insure against sufficient power being applied to spring or break any parts. An over-running clutch is also built into the power

Good Season Looms for Hay Tools

Prices for all kinds of hay are advancing, less spectacularly than the grain, but none the less certain. Legume fields have depreciated during the depression, farmers lacking both incentive and finances for new planting. Consequently a hay shortage with high prices is not improbable. The dealer therefore will have an exceptional opportunity to sell new efficient equipment. The upper view shows the Case pick-up baler together with visible results of its work. The left center view shows a Case mower doing a clean, fast job of cutting quality hay. At the right center the windrows are being inverted for drying after a shower with a Case side delivery rake. The lower view shows a Case hay loader elevating hay from the windrow.

shaft connecting the drive from the tractor to guard against the momentum of the balance wheel of the baler unit transmitting power through the tractor on closing the throttle.

One and one-eighth inch shafting is used throughout the entire machine, exclusive of the baling press unit. These shafts operate at speeds of from 35 r.p.m. to a maximum of 165 r.p.m. making it possible for all 15 cast iron self aligning bearings to be interchangeable.

It is possible for the running boards for the wire tiers to raise their entire length to a clearance of 17 in. for transporting over uneven ground and irrigated fields.

No disassembling or adjusting is required for transporting from one field to aonther.

Wind guards are essential in preventing the hay or straw from being scattered on windy days, especially as it is being conveyed up the elevating conveyor.

Alemite push type lubrication is provided throughout.

Many tonnage records have been made that far exceeded our own claims and expectations and we now feel it is conservative to state that the Ann Arbor hay and straw combine is capable of baling double the tonage with one-half the number of men as compared with any other method of baling from the windrow.

Faster Air Plane Service

The time required for moving air mail, express and passengers from Kansas City to many points was radically reduced Aug. 15. The United Air Lines is now operating a fleet of multi-motored passenger transports, which enables it to reach Chicago in 2½ hours, New York in 7¾ hours and San Francisco overnight, with corresponding reductions of the time to other points.

A Vision of Speed in Farming

*Barney Oldfield's Tractor Racing May Be More Than a Stunt;
Anyhow, He Thrills Crowds, and Answers Question "What of It?"*

*Barney Oldfield with the Allis-Chalmers tractor on Firestone tires as he appears on the
state fair circuit driving tractors at record speed*

REGULAR farm tractors circling state fair race courses as speeds which exceed 40 miles an hour on the straightaways at least; farmers rushing from the live stock exhibits to crowd the railings or taking seats in the grandstands for the first time in their lives; city folks getting their fill of thrills from the speed contests of the iron horses; the veteran Barney Oldfield expressing from the P. A. system his hope of being the first to drive tractors a mile in less than a minute, just as he was the first to drive a car a mile a minute and later was the first to exceed two miles a minute—such are the strange doings taking place at many state fairs this season.

One could not have attended any of the state fairs this summer and fall, from Minnesota to Texas or from Ohio to Kansas, without realizing that Allis-Chalmers and Firestone have injected a new and commanding factor of interest into that old institution known as the state fair. There have been something more than horse races, something more than the carnival of midway attractions, the farm machinery exhibits, live stock displays, etc., all of which have been getting pretty stale for a generation or more. There is new life around the fair grounds.

But what of it?

This question is asked quite frequently.

Let Barney Oldfield answer it. "The glory of winning an automobile race is a mere incident," Barney says. "The real value of a race is that it shows mechanical weaknesses and dictates where mechanical improvements are needed. The grueling test of a few hundred miles on a speedway has done more to improve car construction than any other one thing. It was back in 1901 that I first drove a car at 60 miles an hour. At that time every car owner was satisfied if he could get 25 miles an hour from his car. But try to sell a man a car today that won't maintain a 60-mile speed on the roads for an indefinite time. It can't be done. Automobile speed contests made it possible for the engineers to build better cars, and for the tire makers to build better tires.

"No one ever thought a few years ago that thousands of trucks with tons of load would be going along the highways at 60 miles an hour. Try to pass them on the roads today and you'll get an idea of their regular running speeds.

"Of course, no one today can see tractors hauling loads along the highways on air tires at 60 miles an hour, or even 40. But what has happened in the past shows what can be expected in the future. In this mechanical world of ours, we must keep looking ahead, making ready in the present for what we may reasonably expect in the future."

Let's look in on the Topeka fair for a moment. The fair management features Barney Oldfield with his Allis-Chalmers tractor and its Firestone air tires as one of the leading attractions, for Wednesday. Barney arrives Tuesday. He drives his tractor from the Fair Grounds up Topeka Boulevard to the State House, where he stops to pay his respects to Governor Landon. He tells the governor that a new speed age is dawning in agriculture, which is of tremendous interest to the chief executive of an agricultural commonwealth, and who begins to vizualize a day when perhaps agriculture can observe NRA hours. Then on to the City Hall for a little chat with Mayor Ketchum.

Tuesday night the clouds break loose and the track is drenched. Wednesday afternoon it is heavy with mud. But the tractor speed event takes place just the same, and Barney in his race against time does a little better than 38 miles an hour. Surely, he could have done much better had the track been faster. A hurried conference is held, and the crowd is informed that Barney has agreed to appear again on Friday's program.

Friday's track is ideal. Between the heats of a trotting race, three tractors come on the track. The three are exactly the same. Two of them are to be driven by ambitious Allis-Chalmers service men —Frank Roberts and Carl Hill—the other by Barney. One lap to get up momentum and they are off to a five-lap race. Maybe the younger men are a little quicker on the getaway, anyhow they get off in the lead, which they hold well for three laps. Then experience begins to manifest itself and gradually the veteran Barney, forced away from the inside, begins to overtake his competitors, passing finally into the lead just before heading into the last home stretch.

The race thrills. The tractors virtually speak speed. They have traveled 50 per cent faster than the nags who preceded them, and their speed remains in mind as fast in contrast to all the horse events which featured the remaining race program of the day.

But Barney himself isn't through. Another lap around the track and the P. A. system informs the crowds that the next lap will be an effort to drive a tractor faster than one has ever been driven before. It is a test of speed. The tractor hits into the dust at the banks, but its air tires don't skid and there is none of the sliding so common in auto racing. Over the straightaway on the far side the tractor is virtually "burning up" the track. Then into the second bank and down the home stretch the tractor comes, undoubtedly delivering its last ounce of power, but making less noise about it than one of Harry Miller's racers on the Indianapolis Speedway.

A new record has been set—39.77 miles an hour—electrically timed.

Corn Belt Combine Makes Debut

*Hundreds Witness Showing of New Allis-Chalmers Product,
Which Runs on Rubber and Harvests at Five Miles an Hour*

In this birdseye view of the Corn Belt combine's first demonstration setting, the new small combine is seen in the foreground. Barney Oldfield, veteran racing driver, is on the tractor seat starting his first trip across the field. Following in the rear is a 21-year old binder drawn by three horses. An A-C Rumely thresher is working in the background to the right, while to the left is seen a part of the crowd which was on hand.

Debt Machinery Doesn't Click

SOME rather elaborate machinery has been established by Congress for extending financial relief to debt-embarrassed farmers. The Emergency Farm Mortgage Administration has been set up. The Farm Credit Administration has been organized to coordinate various relief efforts. Both of these are in addition to the already established Federal Land Banks and Regional Agricultural Credit Banks. It would seem that the base for the government loan structure should be sufficiently broad to include every deserving farmer.

But, the machinery isn't "clicking."

At least, such are the reports emanating from the farm regions of those states where the agricultural loans are most needed. Its processes are too slow. It is freely being charged that more money has been taken from the farmers in the form of fees required for application than has been disbursed in loans. Whether or not such charges are true, there is nothing in the various legislation either in intent or literal interpretation which should make it hard for a deserving farmer to obtain a necessary loan.

The general opinion is that the failure of the various agencies to function more expeditiously is due not to inherent defect in the laws but in their administration. In some instances the administration of the relief measures has fallen into the hands of bankers, or former bankers, who traditionally at least are against government competition with private banking. There is said to be too much of the attitude of the hard-boiled banker in the high places of relief administration.

It also is being freely charged that a lack of intelligence is too dominant at the farmer's local source of contact. The same local representative acts for both the Federal Land Bank and the Emergency Farm Mortgage Administration. Many instances have been reported of these representative accepting the application fee of the farmer, then sending his application to the land bank instead of the emergency mortgage headquarters.

Another common grievance among the farmers who have applied for loans and have not yet received relief is in the length of time involved between filing the application and the appraisal. This certainly is unfair to the farmer who has listed certain assets as security, and who may be forced to dispose of the assets to satisfy an obligation before the delayed appraisal is made.

Much fault also is being found with the nature of the appraisals. The standards required by the government are quite liberal and are based on a long term basis instead of the depression levels. Yet many boards of appraisal are listing crop securities at ridiculously low prices. It has been clearly specified in the Emergency Farm Mortgage Act of 1933 that loans may be applied to payments of machinery indebtedness, yet in many instances loans for this purpose are being rejected even in the face of ample security.

The attempt of the administration in Washington to extend debt aid to distressed farmers is entirely laudable. It contemplates no catches, no impossible restrictions. It does not demand its pound of flesh from those who may become its debtors. It seeks no invasion of commercial banking, but rather to protect the stake which private capital now has invested in agriculture.

The code may reestablish employment and public buying power, but this will avail little to the farmer around whom the spectre of foreclosure continues to hover. The facilities for debt relief are all that could be desired. Will this worthy objective be defeated by unintelligent and dilatory administration?

HOWARD E. EVERETT.

Some Pitfalls in Farm Relief

A SERIOUS danger exists that in our desire to do something for agriculture we may attempt too much and thus commit the nation to a policy more uneconomic than the conditions it seeks to correct. For the most part the troubles now besetting agriculture are of non-agricultural origin, and should be so regarded. They include the loss of foreign markets, under-consumption at home, excessive taxes, weaknesses in our banking system, high transportation charges, mortgage indebtedness, etc. One of the most hopeful signs of the day is that remedial action is being directed toward the solution of these problems, now retarding agriculture and all other industry.

The relief measure now before Congress has been quite thoroughly discussed by all affected interests. It has many features which recommend it, as well as some valid grounds for opposition. Perhaps the most significant fact developed from the discussions is the apathy, the lack of acclaim for it, on the part of the farmer. The farmer seems to hope that it will work, but is more dubious than hopeful.

Two serious weaknesses have been outstanding in nearly all remedial proposals of the last few years—the utter disregard of both the human element in agriculture and the possibility of lowering production costs.

All relief programs seem to presuppose that farmers are in great distress; that all who till the soil are economic derelicts who must be reclaimed through some governmental action. No such collectivized classification can be applied to American farmers so long as the figures of the government itself show seventy-five percent of them in satisfactory financial circumstances. Granting that farm prices are too low, are we to admit that it is impossible to reduce production costs through efficient and inexpensive machinery the same as other industries have done and are still doing during this depression?

We hear much these days of marginal and submarginal lands; the need for removing the unprofitable acreage from our agricultural domain, and from ruinous competitive production with better soils. Why not apply the same analysis to the human content of agriculture? Why try to bolster up agriculture so that it will be profitable for the submarginal farmer? Such procedure means only more public taxes.

It is well known that some farmers are always foredoomed to failure through their inefficient methods; that some farmers operate unprofitably even during most prosperous times; that some could not succeed even with dollar wheat or twelve-dollar hogs. A man who cannot make a success at anything else is not essentially qualified to be a successful farmer, even though some professional farm relievers would place more men on the farms.

The new administration has taken commendable action in weeding out the weak banks—the submarginal. It is trying to strengthen the marginal ones, but if they cannot make the grade they too must go. Competition eliminates the weak from the industrial and mercantile fields. Government does not intervene to save failing manufacturers or failing dealers. Why then this preferential interest in the chronic failing type of farmer? Should agriculture be made safe for him?

Continued artificial support of incompetent farmers will but delay the day of agricultural prosperity, by building up surpluses produced at undue cost and depriving the most intelligent of their economic right to profits. Let's not arrest a system which rewards individual initiative and enterprise. Let's not deprive the progressive farmer of his right to apply mind and machines in meeting the challenge of lower production costs.

HOWARD E. EVERETT.

Power from Ford Cars

Conversion of Ford Model A cars or Model AA trucks into portable belt power units, with 20 hp. available at about the cost of 1½ hp., is made possible through the Badger belt power takeoff made by Heus Mfg. Co., New Holstein, Wis. The takeoff attached to the front of the car or truck by the bumper bolts, which makes it easily attachable or detachable.

Ford power unit, showing method of attaching to front of Model A car

Once installed it can be cranked in or out of gear in 10 seconds.

This new unit includes a set of 4-in. bevel gears and is mounted on four Timken roller bearings. It carries a pulley of 6-in. diameter and 6-in. face, which operates at the same speed as the engine. In attaching the crank ratchet of the Ford is removed and is inserted at the front end of the Badger unit, by which means the unit is cranked in or out of gear with the lug end of the Ford crank. The manufacturers recommend the unit for all farm operations where 20 hp. or less belt power is required and where the amount of work to be done does not justify the investment in a permanent power unit. A 6-blade fan is provided for summer cooling, which supplements the Ford fan. Fuel requirements are said to be about a gallon an hour. The Badger unit is made in two models and sells as low as $18.50.

Farming to *Profit* or *Subsist?*

Use of Cost Reducing Equipment Will Determine Future Status of the Individual Farmer More Than Ever Before

THE "back to the farm" movement constitutes one of the most definite trends of the depression period. Bureau of Census authorities have estimated that a million and a half persons will have returned to the farms by the end of the three-year period, 1930-32. This will be sufficient to offset entirely the population movement from farms to cities during the 1920-29 period. If continued the time is not far distant when the percentage of total population residing on farms will be higher than that of the pre-war period.

Such a pronounced change cannot fail to exert some influence, temporary at least, upon our economic and social establishment. Our main concern is in its influence over the future of the farm equipment industry.

Self-evident results of such a population tendency are the creation of more producers of agricultural products together with a decreased market for the products of the farm. But a more thorough analysis of the reasons for the trek back to the farms indicates that it is a move for economic self-preservation, that these new farmers are more interetsed in mere subsistence for the time being and that their ultimate vocation will be determined by future economic trends. In the meantime rather than contributing materially to marketable farm production their main import is in the reduced market afforded by their farm tenure.

Opens a Larger Market

Their temporary influence on the farm equipment trade should be favorable, as they afford a temporarily larger market for the local dealer, particularly for used and reconditioned equipment. Quite likely a considerable part of this returned farm population will become permanent. Some of it will return to city employment as soon as that becomes available. That which remains may replace some of the present farm population on marginal lands. From practically every angle, the trend back to the farms can be considered largely temporary.

As nothing can be considered permanent in a world of changing economic conditions, one can expect constantly changing ratios between rural and urban population. There will always be need for farmers in larger number than those following any other endeavor, as the needs for food and clothing are paramount to all others. There will doubtless be more than six million farmers in the United States for a great many years to come. Any radical decrease below this number can come only with an increased use of cost-reducing farm machinery.

No matter how prosperous agriculture may eventually become there will always be a large number who are barely subsisting, with some who are actually farming at a loss. The same is true in all vocations in good times and bad. There are grocers and other tradesmen who merely subsist and at the end of the year have nothing more to show for their efforts than the fact that they still live and have title to their business. The same is true of professional people, manufacturers, distributors and others.

The only opportunity for the farmer to rise above a state of mere subsistence is through efficient farming methods, as represented by improved cost reducing machinery. This statement is true regardless of the economic high or low levels prevailing. Studies in costs of producing major crops conducted by the U. S. Department of Agriculture show that many farmers cannot break even with market prices three times those prevailing today. Records of the same kind compiled through the National Association of Farm Equipment Manufacturers, based on the use of modern cost-reducing machinery, show that profits are possible at little more than present levels.

Ultimate prosperity for agriculture depends upon the ability of the individual farmer to rise above a mere subsistence level through the use of proper equipment. Regardless of present price levels for farm products, labor still remains the principal factor in the cost of producing crops. The ratio of the cheap farm wages of today to the selling price of farm commodities is much higher than in recent years, deceptive though the lure of cheap labor may be. The capacity of one man on the farm, exerted through power and machinery, is still the one standard through which farm production costs can be reduced.

Some Will Never Prosper

Gradually the price levels for farm crops will return to higher levels, perhaps through the working of natural laws, also perhaps stimulated by legislative methods. With each increase in market prices, there will be an increased number of farmers able to farm profitably. But at the highest levels there will still be a large percentage unable to make the proverbial ends meet.

The agricultural future of this sector of the farm population is not bright. They may eventually be forced off the farms. Until they are their production will only serve to increase visible supply and depress price. But when a more prosperous time comes again, the economic producers will be buying the products of industry in greater number, creating additional industrial employment, and affording a means of absorbing the surplus population of agriculture.

Shortage of Goods Is Near

The present supply of farm equipment, for instance, would be quickly exhausted by merely a slight upward spurt in farm buying. Already two leading manufacturers have been forced to increase fatcory production in readiness for spring trade. The purchase of only a small amount of equipment will result in immediate factory employment. Only a few repetitions of such instances would greatly change the present economic set-up.

In any future consideration of agriculture it must be emphasized that under any economic consideration, the well-equipped farmer has the advantage. A wider spread use of equipment will have more influence upon the future success of agriculture than the number of farmers involved. Agriculture of the future will be a survival of the fittest, with mechanical equipment largely determining the fitness. In the next period of agricultural prosperity it is not improbable that we shall see still greater decreases of farm population than we have yet known. It will be possible for 20 instead of 25 percent of our population to take care of all our agricultural requirements; and it is within their power to do it profitably under a below-average scale of prevailing farm crop prices.

William Ford in Bankruptcy

As a result of a voluntary petition in bankruptcy, the Guardian Trust Co., Detroit, has been made trustee for William Ford, also of that city. This action is the result of a previous receivership for William Ford & Co., Highland Park, Mich., which formerly distributed a number of lines of agricultural and industrial machinery for use with Fordson tractors throughout Michigan and northern Indiana. Liabilities of William Ford were listed at $412,900 with no assets. Mr. Ford is a younger brother of Henry Ford, the motor car manufacturer, and William Ford's attorneys stated that Henry Ford had not been advised of the suit. William Ford, for many years, was treasurer for the Tractor Implement and Equipment Distributors, a national association of Fordson tractor equipment distributors.

The Eagle Must Be Fed!

TWO million names have been added to the country's payrolls during recent months, according to reliable statistics. For the most part these additions were justified from the business resulting from a return of confidence. More recent additions, largely attributable to the NRA, have social rather than this economic justification. The retention of these later reemployed is not warranted by returns to employers, but by the hope of these employers that concerted reemployment may eventually increase buying power sufficiently for permanent employment.

It is one of the ironies of a normal recovery that wages lag behind other advances. Recovery, obviously, would be expedited were wages to increase in horizontal accord with other advances. To step up lagging buying power in greater proportion than normal is the NRA main objective.

This ambition is laudable, and, it is unanimously hoped, practical as well. But until it attains success, employers of wage earners, just as anxious as anyone to see the program succeed, are called upon to finance its operation. This is true without exception whether the employer happens to be a manufacturing corporation, a small business owner or an implement dealer out on the prairie. They are being asked to increase payrolls and face rising costs of operation without immediate returns. Someone must feed the eagle to keep it flying. It can't long be fed from profits which haven't been made in recent years nor from reserves depleted to the vanishing point.

Business must have credit, not inordinate financing, but the sound type of credit upon which our business structure has been built. In normal times business operates upon about seven times as much credit as currency. Today we have an increased amount of currency, but credit has been drastically restricted. The time to withdraw or withhold credit is not when business is recovering, but when it is overexpanding. The great need now is commercial loans for which ample sound security exists. As General Johnson says, "commercial banking is not functioning."

Public attention has largely been diverted from the banking situation since the first vigorous and bold steps were taken by the new administration. Public confidence has been restored, but slow progress is being made in the rehabilitation of closed banks. One-fourth of the banks which were open on the last day of 1932 are now closed. As the Committee for the Nation recently reported, "purchasing power greater than can be released by the government's three billion dollar public works program is tied up in frozen bank deposits. Closed and restricted banks have been put in straight-jackets of bureaucratic control."

In other words too much stress is being laid upon liquidity when with the present banking confidence solvency can be attained without undue liquidity. The restoration of normal banking facilities is a prelude to satisfactory credit conditions.

Whether the lack of credit can be attributed to the government or to the bankers, there is considerable hope for improvement. President Roosevelt appealed early this week to the American Bankers Association in convention to "make credit facilities adequate for the national recovery" and to make increased loans to industry and commerce. Washington advices indicate that if bankers don't loosen up, the government will set up some credit dispensing machinery of its own. The Federal Reserve is enlarging its open market operations, and other steps are being taken to reestablish credit. The eagle must be fed—increasing credit is the next recovery step!

HOWARD E. EVERETT.

Goodrich Announces Tractor Tire

A low pressure tire and tube designed and constructed to meet requirements of a special service—farming, is announced by the B. F. Goodrich Co., Akron, Ohio.

Made in two sizes, 11.25-24 and 6.00-16, the tread design is a heavy lug type with every other center lug cut out, providing a self-cleaning tread. The tire is made of a special rubber compound to resist abrasion, cuts and snags common to field work.

The tube is equipped with a special valve offset to permit installation through the side of the rim instead of the center.

The tires fit standard wheels, of the rod spoke type with spokes securely riveted to rim and hub.

It has been found that it is sometimes necessary to add weights to tractor wheels equipped with low pressure tires to make the unit perform properly in field work. There is no set rule governing the amount of weight.

In many loose soil conditions, Goodrich experiments with the low pressure tire developed wheels with no weight at all are sufficient. When draw-bar pulls are heavier, 300 pounds on each wheel is generally enough, Goodrich tests disclosed.

Weights, in 150-pound sections can be obtained from wheel and rim dealers and quickly attached to the wheel. A total of three weights may be put on each wheel, increasing the load to 450 pounds, but it is only under extreme conditions that this amount must be used and it will be generally found that 300 pounds is ample.

Rubber tires on the farm tractor make it meet more nearly the requirements of an all-purpose unit and eliminate many of the old limitations. A tractor equipped with low pressure tires can now operate in the field as well as the road, around the barnyard or in any of the buildings.

The Goodrich low pressure tire has been built for this special service, and subjected to actual field conditions before being introduced.

Advantages cited by Goodrich for tractors equipped with its low pressure tires are:

1. Have more power for productive work—reduced rolling resistance.
2. Do more work at higher speeds.
3. Lower ground pressure—does not pack seed bed.
4. Increase possible use of tractor—operate on the highway without wheel change.
5. Be more efficient—less fuel cost.
6. Develop a higher draw bar—makes possible high gear on may operations formerly done in second.
7. Ride easier—do more work with less operator fatigue.
8. Have less depreciation—more protection from vibration.
9. Does not pick up or throw dirt over small crops, operator or machine.
10. Save enough money on gasoline, oil and time to pay for itself and pay dividends.

Alexander Legge, industry leader and president of International Harvester Co., died at the age of 67 of coronary thrombosis on December 3, 1933. Pliny E. Holt died at the age of 62 of a heart attack on November 18, 1934. He was the son of Ben F. Holt, pioneer developer of track laying equipment.

AAA Plans in Doubt

While no announcements have been made from Washington of any contemplated changes in the plans of the Agricultural Adjustment Administration for the next crop year, it is generally acknowledged that changes have received serious consideration. Officials however insist that little change has been necessitated as yet in acreage plans, although the program is sufficiently flexible to make them if desirable.

As regards wheat, it is understood that a plan has been favored which would permit farmers to plant without the 15 per cent reduction, with the understanding that should conditions make it seem advisable the farmers would cut 15 per cent of their acreage, utilizing the cut wheat for hay in feeding stock on the farm.

Wheat is the only crop at this time which seems to be considered as necessitating any change in the existing contracts with farmers. The time for winter wheat planting is near at hand. Prior to the drouth it was announced that announcement would be made late in August, regarding the amount of acreage reduction required. Presumably this is still the official intention.

The Be-Ge Mfg. Co. began in Gilroy, California, in 1934 and John Deere introduced the model "A" tractor. Rear tread width of the John Deere "A" was adjustable to match the spacing of different crops. It was interesting to notice that normal spacing of crops was 42 inches which was considered the average width of a horse. The John Deere model "A" continued in production until 1952. The Goodyear Tire & Rubber Co. began making an 11.25 x 42 inch rubber tire for Oliver Row Crop tractors and this was the largest production size at the time.

Some believed that the natural drought reduced grain and livestock production more than the government programs, and was more effective. Weeds were killed in cereal crops by spraying with sulphuric acid. Reports in March indicated at least 2000 strikes were in progress. The 1934 exhibit by International Harvester Co. showed a "Radio-Operated" tractor. A film demonstrating equipment was popular if it was one of the new "talkies," but farmers were already about two years behind in replacing their farm equipment.

Drouths of the Past

The 1934 drouth is popularly believed to have been the most severe ever to have visited the agricultural areas of the United States, but was it? Abnormal weather always seems most severe when it is running its course. Unfortunately western weather records do not go back to 1860, nor are there many still living who are competent to make comparisons between that year and this. Perhaps that drouth was the most disastrous up to 1934. There were many of the intervening years which produced bigger and better drouths than the average, one in 1889 and others in 1894, 1901 and 1913.

The drouth of 1894 was quite comparable to that of 1934, as it represented an accummulation of several years of less than average rainfall, just as 1934 is the culmination of effects from two previous years. If the comparison of these two drouths is any indication 1935 should be a good year from a moisture standpoint, as 1895 produced three inches more than a normal amount of precipitation.

Weather records, however, show many other years which rivaled and in some instances exceeded 1934 in lack of moisture, although mostly more local than general. Cheyenne, Wyo., had only 5.04 in. in 1876. Omaha's low was 15.49 in. in 1910, and North Platte, same state, had its minimum of 10.7 in. the same year. Bismarck, N. D., set its low level with 11.03 in. 1889, and Helena, Mont., the same year established a minimum of 6.71 in. These low marks range from 25 to 60 per cent of average. Whether some of these low records will be broken by the 1934 catastrophe remains to be seen, as the record for the year has yet to be completed, although the crop damage has been done.

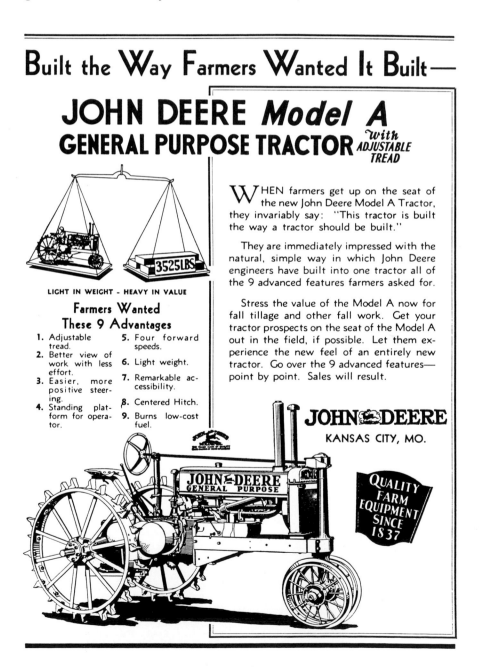
New Black & Decker Drills

A new ½-in. and a new ¼-in. streamline design electric drills have recently been announced by the Black & Decker

The ½-in. B. & D. electric drill

Mfg. Co., Towson, Md. These represent redesigning from the motor to the chuck key. The ½-in. size is a general purpose drill with greater power, better balance and more attractive design, and has an unusually wide range of utility.

The ¼-in. is a companion to the larger unit. Ball bearings on armature shaft

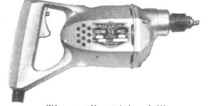

The smaller ¼-in. drill

and spindle give smooth operation and longer life. A larger motor adds one-third more power than previous models of this size and makes it an ideal drill for a farm equipment service shop. The price complete for the smaller unit is $31.

The smaller unit weighs 5¾ lb., and the larger 13½. Bench stands are available for each model.

New Goodrich Tire for Tractor Front Wheels

A new low pressure pneumatic tire designed especially for the front wheels of farm service tractors is announced by the B. F. Goodrich Co., Akron, Ohio. The new tire, made with a heavy circumferential ribbed tread design eliminates the few remaining objections raised on pneumatic farm tractor tires, the manufacturer believes.

Improper steering of all-traction front-wheel tires in soft ground on ridge rows and on side hills has been eliminated by this new design. It allows the tractor to be kept in its true course around turns, and prevents sideslips when working the tractor on the side of hills. All advantages of the metal flange used on steel wheels and the low pressure pneumatic tire for tractors are combined in this design, the manufacturer claims. The new tire is available in 5.25-16, 6.00-9, 7.50-10, 6.00-16 and 7.50-18 sizes.

Eight Millions Unemployed

The total number of unemployed workers in June, 1934, was 7,934,000, according to an estimate of the National Industrial Conference Board. This is an increase of 89,000, or 1.1 per cent from May, 1934, and a decline of 5,269,000, or 39.9 per cent from the total in March, 1933, when unemployment was at its highest point.

The increase in unemployment from May to June was the first increase since January, 1934. Of the total increase, 37,000 occurred in manufacturing and mechanical industries, 60,000 in trade, and 2,000 in extraction of minerals. In transportation there was a decline in unemployment of 34,000.

An Automatic Lock Nut

A new type of lock nut which is automatic in its operation has been announced by the General Automatic Lock Nut Corp., General Motors Building, New York City. The secret of the nut is a locking pin, made of non-corrosive steel, which is engaged by the thread of the bolt and follows the thread. The pin is turned at an angle and establishes a point of impingement against the bolt, the pin biting in and thereby maintaining this point of adjustment continuously against vibration and shock. It is claimed that vibration tends only to keep the nut locked. However, it is pointed out, the nut can be readily removed with a wrench. Sufficient pressure is exerted by the wrench to throw the pin momentarily, releasing the point of impingement and permitting the nut to engage and follow the thread of the bolt in the opposite direction. This process, the company states, can be repeated indefinitely.

Air Tires for Wheelbarrows

The same type of low pressure tires which have been applied to tractors with surprisingly favorable results are now being made available for replacement wheels for wheelbarrows by French & Hecht, Davenport, Iowa, well-known manufacturers of metal wheels. The replacement wheel includes an all-metal wheel upon which is mounted one 16x4-in. low pressure tire, the entire unit selling for $6.25 delivered. At an additional charge of $1, a roller bearing also is furnished.

The wheel replaces the conventional wheel in wheelbarrows in use and adapts it to a wider scope of work. It can be used over lawns, soft ground, etc., without the use or cost of planks, and can with equal facility be used over rough surfaces, such as crossing railroad tracks, etc. The resiliency of the rubber prevents spilling of loads on rough surfaces. It is easier on the operator and permits of use at greater speeds, thus increasing the capacity of the barrow. It is claimed that it will reduce the cost of wheelbarrow operation by one-third, and it is quite possible to perform sufficient extra work in one week to pay the cost of the rubber-tired wheel equipment. Facilities for oiling permit of continued satisfactory use.

New SPEED FACTOR in Farming

*Low Pressure Tires Promise Possibilities for Expediting
Field Operations Without Increasing Expenditure of Power*

New Allis-Chalmers "W"

The *Implement & Hardware Trade Journal* changed its name to *Implement & Tractor*, in 1935 and on April 14, one of the most severe dust storms swept across the Texas and Oklahoma panhandles and into Kansas. The Farm Security Agency began in 1935 and the Rural Electrification Act made low interest loans available as an incentive for companies to extend electrical service to rural areas. In October, Benito Mussolini defied the League of Nations and conquered Ethopia. The Bates Manufacturing Co. took over operation from the Foote Bros. Gear & Machine Co. but failed and closed by 1937. B.C. Oppenheim, president and eldest son of the founder of the New Idea Spreader Co., died on May 22 following an appendicitis operation.

The American Screw Co. of Providence, Rhode Island, introduced a line of screws with the Phillips head. The new cross slot was said to be the first major change in screws in 85 years. John Deere "B" was introduced in 1935 and production continued until 1952. The "B" model was about two-thirds the size of the "A" model. The Fate-Root-Heath Co. renamed the "Plymouth" tractor "Silver King." On May 15, John Lauson Mfg. was forced to liquidate, but is reorganized as the Lauson Co. After 80 years of operation, the Advance-Rumley Corp. was dissolved by stockholders. Oliver tractors in 1935 featured streamlined styling with full engine covering and the industry's first "high compression" gasoline engine. A farm Diesel tractor, the WD-40, was announced by International Harvester. Firestone introduced a new ground gripping tread pattern for their rubber tires. Rubber tires were considered one of the "most important developments in the last 15 years" and some important changes had taken place since 1920.

In June 1935 it was announced that the "Doubtful Experiment Ends! History may eventually accord the Supreme Court credit for the most important contribution to recovery. The NRA decision has definitely ended a doubtful experiment. It has terminated the arrogation by the federal government of authority over commerce which is not interstate. It has restored business and industry to the management of proven able leadership. Initiative and enterprise, characteristically American, can function again as the shadow of regimentation passes."

1936-1941

RECOVERY IS SLOW

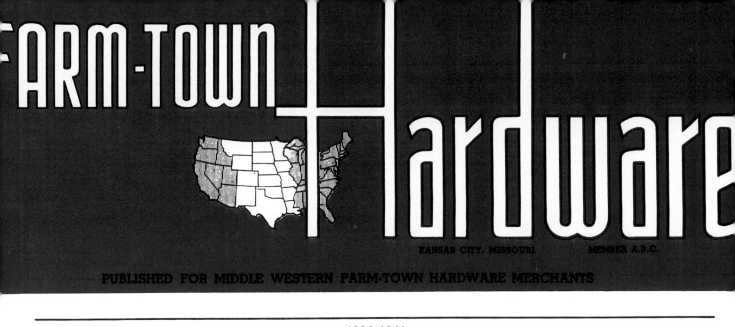

FARM-TOWN Hardware

KANSAS CITY, MISSOURI   MEMBER A.B.C.

PUBLISHED FOR MIDDLE WESTERN FARM-TOWN HARDWARE MERCHANTS

1936-1941

The years of 1936 through 1941 were marked by continuing labor problems.

On May 10, 1936, a controversial documentary film about the Dust Bowl and titled "The Plow that Broke the Plains" was previewed by many of the influential people in Washington, DC. Floods occurred along the Merrimack, Connecticut, Hudson, Delaware, Susquehanna, Potomac, Allegheny and Ohio rivers. Numerous flood control, anti-erosion and other conservation projects were necessary to combat water damage at the same time that wind erosion was causing damage in the Great Plains Dust Bowl. Deere and Company purchased the Caterpillar combine line. More than half of the new tractors sold in 1936 were equipped with rubber tires. It was believed that development of better brakes and higher speeds would allow the tractor to be used on the highway to transport produce and maybe even take the place of the farm truck. The optimistic outlook for 1936 was born out with greatly increased sales by equipment dealers and production by farmers. There was great fear about the oppressive tax burden that would last "the next few years." Net profit for the International Harvester Co. was $29.75 million in 1936. Allis-Chalmers recorded the second best year in their history and doubled earnings of the previous year. Other companies showed similar improvements for 1936.

Modern Tractors Assist Airliners

Parking an airliner at a Kansas City airport by means of a modern tractor. Such power units have become a part of the regular service equipment for keeping planes in perfect condition for their regular trips

Use of Power Units Spreads to Aviation Fields As All Industry Adopts Mechanical "Beasts of Burden" for Handling Excessive Loads — Now a Part of Regular Service Equipment for Planes

Diesels and Joe E. Brown in Thrilling Movie Role

Joe E. Brown, world-famous comedian, is teamed with Diesel track-type tractors in Warner Bros. latest motion picture release, "Earthworm Tractors." This feature of thrills and laughter is now showing in hundreds of theatres, and will soon appear on foreign screens.

"Earthworm Tractors" is the screen version of the series of stories by William Hazlett Upson in a leading weekly magazine. Brown plays the role of the super-salesman hero, Alexander Botts. Opposite him is lovely June Travis as Mabel Johnson. Guy Kibbee is Cyrus Johnson, her father, who is Botts' most stubborn prospect.

The series of hair-raising and impromptu tractor demonstrations that mark the progress of the plot, eventually win the girl and a sizable tractor order from her father. And while Brown enacts his role of "natural born salesman," the tractors are proving themselves able performers, too. All tractors in the picture are Caterpillar Diesels, and during the filming, the Caterpillar Tractor Co. plant at Peoria, Ill., was taken over by Warner Bros. camera men for days.

Director Raymond Enright and screen writers, Richard Macauley, Joe Traub and Hugh Cummings, vied with one another in arranging spectacular and difficult feats for the tractors to perform and the Diesels nobly responded, according to machinery men critics interested in the technical side of the picture. In one sequence, a 16-ton Diesel tractor plunges into a man-made swamp. Driverless, it breasts through the deep mire; wrecks a truck, smashes through a fence, crashes over logs and rocks, and ends its sensational flight by hurdling an automobile. In another scene Super-salesman Botts, who "knows all about tractors," but little about the gear shift, demolishes a railroad station with the attendant smashing of egg cases and milk cans.

The climax comes when Botts takes his prospect along a precipitous mountain trail. Workers are blasting to improve the road. Again the Diesel performs herculean feats amid dynamite charges and avalanches of rock. However, members of the Warner technical staff said the machine emerged undamaged with the exception of scratched paint.

Joe E. Brown attended the world premiere of his picture at Peoria, Ill., July 14. Thirty-five thousand fans gathered around the theatre to welcome him. Newspaper critics at the affair declared "Earthworm Tractors" to be "Brown's best."

New Power Head for Ford V8's

Announcement has been made by the Federal-Mogul Corp., Detroit, of a new power head for Ford V8 engines. This is known as the Federal-Mogul Thermo-Flow Power Head, and combines a number of important operating advantages and economies. It minimizes localized hot spots, permits higher compression ratios and power output, reduces carbon

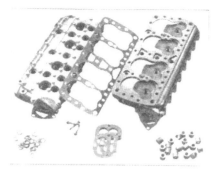

formation and results in materially increased gasoline mileage. Because of its performance and economy features, this new Thermo-Flow Power Head appeals to all users of Ford-powered vehicles, commercial haulers, fleet owners, bus operators, police, fire and ambulance departments, and users of Ford V8 marine power plants.

The new head is the result of several years of intensive engineering research and has been thoroughly tested under a wide variety of operating conditions. On a road test, made by a leading Detroit haulaway company hauling Ford cars from Detroit to Buffalo, this new power head saved 51 gal. of gas on a 1500 mile run. This was a saving of 19 per cent or a saving of $.0051 per mile, figuring the cost of gas at 15 cents a gallon. On a ton-mile basis, this was a saving of 32.5 per cent. Difficult grades were taken in high gear, more power and flexibility were readily apparent in every gear, and ease of handling was repeatedly commented upon.

On closely checked dynamometer tests, this new power head developed 17.3 per cent more power than an aluminum head and 27.3 per cent more power than a cast iron head. It showed 7.4 per cent to 15.8 per cent more miles per gallon, and 4.17 to 14.7 faster acceleration than an aluminum head, under comparable conditions. It is now being offered as original equipment and for replacement on Ford V8 marine conversions made by leading marine engine builders, including Kermath, Lehman, Oscar Smith & Sons, and Scripps, and has figured prominently in many important boat racing victories. Special Thermo-Flow power heads are also offered in certain models by Chris-Craft and by Johnson Motor Co. on its Class A and B Sea Horse outboard motors. Installation of the new head on Ford V8 engines requires no changes, and installation can be made by any competent mechanic in about three hours. All necessary parts are included with every Thermo-Flow power head shipment.

No Fools, No Fun

Flapper Fannie says that you can fool all of the men some of the time and that you can fool some of the men all of the time, but if you'll just leave 'em alone they'll all make fools of themselves most of the time.

Tractor Facts, Please

The Bureau of Agricultural Engineering, Department of Agriculture, Washington, D. C., has undertaken to compile a true and complete history of the tractor. Any information regarding early tractors, either those powered by steam or by internal combustion engines will be welcomed by R. B. Gray, chief of the division. There was a time when the country was "running rife" with aspiring tractor manufacturers and it may be that readers have circulars or catalogs published by them in some unused cranny or shelf nook which might be of real value to the editors of this history. Mr. Gray will suggest a way of handling heavy material upon which the postage would be costly.

Pipe Threading Eliminated by Standard Fittings

"Time out" for cutting pipe to exact lengths, threading, grooving, flaring or screwing up joints in cramped quarters, is no longer necessary, according to an announcement by the S. R. Dresser Mfg. Co., 499 Fisher Ave., Bradford, Pa. With the standard line of Dresser Style 65 fittings just announced, nothing but an ordinary wrench is needed to complete a joint in a few moments.

After inserting the plain-end pipe into the fitting (which comes completely assembled), is is only necessary to tighten two threaded octagonal follower nuts with a few quick turns of the wrench. As this is done, resilient "armored" gaskets at each end of the fitting are compressed tightly around the pipe, forming a positive seal. The resulting joint, Dresser engineers point out, is not only permanently tight but absorbs normal vibration, expansion and contraction movement, and permits deflections of the pipe in the joint. If the pipe is already threaded, it can also be joined in the same way.

The complete line of Style 65 fittings includes: standard and extra-long couplings, ells (both 45° and 90°), and tees, all supplied in standard steel pipe sizes from ½ to 2 in. I. D., inclusive, black or galvanized.

These fittings are recommended by the manufacturer for simplifying joint-making and repair work on both inside and outside piping, for oil, gas, water, air, or other industrial lines. The basic principle is essentially the same as that used in other styles of Dresser coupling

A Timken Fuel Injection Pump for Diesel Engines

Two views of the fuel injection pump which Timken has developed for diesel engines. The obverse (left) is plainly marked, for the benefit of the operator, while the reverse mounting arrangment

Sixty Years' Experience in Making Cowbells Exclusively

J. H. Blum, Jr., one of the nation's leading cowbell makers, performs on a xylophone of various sizes of bells, and they "ring true"

SELDOM classed as a farm implement, but nevertheless a staple item on the shelves of many an agricultural dealer, is that old familiar product, the cowbell. Not so familiar, however, is the unusual story of one of the nation's leading cowbell factories, which sells its product only through independent jobbers and dealers.

On the outskirts of Collinsville, Ill., stands the one-story frame building which houses the Blum Mfg. Co., the only factory in the United States devoted exclusively to the manufacture of cowbells. Like many other so-called little industries characteristic of American enterprise, the management of the firm has been handed down from father to son for several generations.

When C. G. Blum, a tinsmith, began making cowbells nearly 60 years ago, he had to design and build most of the machinery used in their manufacture. Today, with very few changes in design, the same type of machinery is turning out thousands of cowbells which are shipped to every state in the Union as well as Mexico and Canada.

Insisting that a clear, penetrating tone was the prime requisite of a good cowbell, the elder Blum went to W. F. Niedringhaus, then president of the company which later became the Granite City Steel Co., to make sure that the sheet steel used in the bells would "ring true."

When J. Henry Blum, son of the founder, took over the plant, he always placed his orders with G. W. Niedringhaus, son of W. F. Niedringhaus. And now H. J. Blum, Jr., grandson of the founder, buys the same type of steel from Hayward Niedringhaus, president of the Granite City Steel Co., and grandson of W. F. Niedringhaus. To make the coincidence complete, both firms were founded in the same year—1878.

Although most of the cowbells are actually used on cows, according to the younger Blum, a surprisingly large portion of the firm's output is sidetracked to other uses. Football fans use them to increase the noise of the cheering sections; New Year's Eve celebrants find that the clatter of cowbells adds to the festivity, and newly married couples frequently find Blum cowbells, in assorted sizes, securely wired to the rear axles of their autos.

In spite of last summer's severe drouth, the sale of cowbells boomed. The reason: Pastures were dried up and stock had to be turned out into woodland or unfenced pasture.

J. H. Blum, Jr., one of the nation's leading cowbell makers, performs on a xylophone of various sizes of bells, and they "ring true"

Eight sizes of bells, ranging in height from 2½ to 7 inches, are made at the plant. The smallest bells are sometimes used on turkeys, and are frequently given away as advertising novelties.

In making the bells, sheet metal selected for its tone is cut into shapes resembling a double-faced axe blade. With most of the work done by hand, the handle and clapper are fastened on, the steel is bent into the shape of the bell, and the sides are riveted. The finished bell is then copper-brazed. The firm continues to adhere rigidly to a policy adopted many years ago—every bell made in the factory is marketed through wholesale hardware jobbers. Its reputation for producing a cowbell with a real bell-like tone has been responsible for the firm's success.

More Leaded Gasoline Used

A steady rise in the percentage of leaded anti-knock gasoline used by American motorists, with a corresponding drop in the consumption of unleaded regular and third grade fuels, is reported in a nationwide survey of gasoline sales by the Ethyl Gasoline Corp. A parallel trend in compression ratios of automobile engines is noted, indicating that motor car owners are turning more and more to anti-knock fuel as engine efficiencies increase.

Sales of Ethyl or premium gasoline with an anti-knock rating of 76 octane or higher climbed from 558,900,000 gallons in 1924, to 622,700,000 gallons in 1935; and to 783,300,000 gallons in 1936. Engine compression ratio averages for the same years were 5.72, 5.99, 6.19, respectively.

The sharpest rise in Ethyl or premium sales is shown in the Pacific Coast region, comprised by the states of Washington, Oregon, California, Nevada, Idaho, Utah and Arizona, where Ethyl gallonage was 48,500,000 in 1934; 85,000,000 in 1935, and 134,500,000 in 1936. Use of leaded gasoline of the regular grade was reported at 779,800,000 gallons in 1934; 1,017,600,000 gallons in 1935, and 1,126,100,000 gallons in 1936. The percentage of unleaded fuel sold dropped 9 per cent in the same period.

Goodrich Suggestions on Water Inflation of Tires

INFORMATION regarding water inflation of tires, which will prove of value to dealers and their customers, is contained in a bulletin recently published by the B. F. Goodrich Co., Akron, Ohio, entitled "Water Inflation Program." The company recommends the water inflation method of adding necessary weight after a year's study and experimental work. The company finds that the use of water can largely supplant metal wheel weights. In addition it provides normal cushion without the disadvantages of rebound or bounce. The "shock absorber" action reduces the possibility of wheel chatter, increases tractive ability and improves riding qualities.

Water inflation of tractor tires is simplified by the use of a hose adapter which Goodrich has developed. This embodies three principal parts. The water hose connection screws on the hose in the conventional manner. On the other end is the tire valve connection, which screws on the tire valve. Between the two is a bleeder valve by which the built-up air pressure during the inflation process can be released.

Ordinary city water pressures run from 30 to 60 lb. and are usually adequate to fill all sizes of Goodrich Farm Service tires. At these pressures water will be admitted at from 1½ to 2 gal. a minute. Where no water pressure is available, tires may be filled from a tub, barrel or similar container by a gravity flow or siphon action. If air pressure is available, water can be forced quickly into the tire through the use of a pressure filling tank designed by Goodrich.

Only a few simple operations are required for water inflation according to the recommendations of Goodrich: (1) Place the tube valve at top of wheel (do

not jack up vehicle at this point); (2) remove valve core and allow air to escape from the tire; (3) attach hose adapted to valve to hold valve in place; (4) jack up tractor wheel so that tire is not deflected and maximum amount of liquid can be admitted; (5) Attach hose to adapter; (6) fill tire with water, occasionally bleeding built-up air pressure as tire fills (this requires shutting off the water supply during air bleeding); (7) when

tire is filled to desired level, remove adapter; (8) replace valve core securely in valve stem; (9) remove jack and inflate tire with air to recommended pressure; (10) due to the reduced volume of air where water is used, any small air loss through seepage greatly reduced air pressure. Therefore, make sure that there is no valve leakage by properly tightening both the valve inside and the valve cap, and check pressures frequently and regularly.

In the smaller sizes of tires liquid can be substituted for approximately one wheel weight, in the medium tire sizes for approximately two wheel weights and in the larger sizes for approximately three wheel weights.

Complete information regarding calcium chloride anti freeze, which the company recommends as the most practical and economical, are contained in the Goodrich bulletin.

Ariens Plant Completed

A new plant, recently completed, doubles the capacity of Ariens Co, Brillion, Wis., manufacturer of Ariens tillers. It is steel, concrete and glass construction throughout, with new machinery and equipment installed to take care of the increased business. Ariens tillers are used by market gardeners, nurserymen, seedsmen, florists, etc.

The 1937 Model W Parsons Whirlwind is more heavily constructed than its predecessors, yet its complete bearing equipment provides more capacity per unit of power.

Unique Demonstration for New Briggs & Stratton Unit

Demonstrating how battery charging for radio and electric light and gasoline motor power are available in isolated regions from small gasoline-operated generating plants, Briggs & Stratton Corp., Milwaukee, Wis., is sending a fleet of special demonstrator cars on a

This Briggs & Stratton demonstration car, showing the Power-Charger plant in rear compartment, has attracted much attention on recent visits to fairs and corn husking contests.

tour of the country, staging a unique exhibit showing possibilities of the new units. Mounted in a compartment in the rear of an automobile, one of the company's electric-starting 6-volt, 200-watt Power-Charger units lights up a string of 12 electric lamps during the demonstration. Sales representatives accompanying the cars describe to dealers and farm audiences how radio batteries may be charged at home, and how electric light and gasoline motor power is available from the small plant.

The demonstration cars visited several mid-western corn-husking contests during early November, carrying the message of battery charging and motor power directly to the farmer.

Cash buying indicates that the farmer wasn't so badly off, after all.

In 1937, the Farm Security Administration was formed to provide low interest loans.

The J.I. Case Co. purchased Rock Island Plow Co. of Rock Island, Illinois, which made plows, listers, planters, grain drills, cultivators, potato diggers, hay loaders, gasoline engines and tractors. John Deere introduced Model "L" and Model "G" tractors in 1937. The "L" and the similar "LA" model were among the first small garden tractors and production continued until 1946. The "G" model was more powerful than the "A" and production continued until 1953. Deere & Co. purchased the Killefer Mfg. Corp. of Los Angeles, California, makers of heavy duty farming and road building equipment. Montgomery Wards began to sell a small garden tractor built by the

Simplicity Manufacturing Co. and Graham-Paige Motors Corp. built 250 advance promotion tractors for Sears, Roebuck & Co. The Duplex Machinery Co. of Battle Creek, Michigan, produced models in 1937 and 1938, but the 1938 models were also sold by Co-Operative Mfg. Co. also of Battle Creek. Minneapolis-Moline introduced the all new "Z" tractor. The Henry Knapheide Wagon Co. of Quincy, Illinois, manufacturer of wagons since 1848, changed its name to the Knapheide Mfg. Co. Knapheide continued to make wagons and their already popular truck bodies. An outbreak of "sleeping sickness" in the states of Nebraska and Iowa killed a large number of horses and caused an increase in tractor sales in the affected areas.

A SENSATIONAL NEW LINE OF SAFETY STEEL HOUSE TRAILERS

BUILT TO THE STANDARDS OF THE AUTOMOTIVE INDUSTRY

SOLD BY SEASONED AUTOMOBILE MEN

Entered as second class matter under the act of March 2, 1879, and published fortnightly by the Implement Trade Journal Co., 601 Graphic Arts Bldg., Kansas City, Mo., Vol. LI, No. 21. Subscription $1.75 per year.

Cushman Announces "Auto Glide"

The Cushman Motor Works, Lincoln, Neb., has announced the "Auto-Glide" as a contender for "the world's newest and cheapest motor-powered transportation." Filling the gap between bicycle and automobile, the Auto-Glide is claimed to afford 140 miles per gallon of gasoline. At 35 miles per hour it doubles the speed of the bicycle and nearly equals the comfort and convenience of an auto, costing but little more than walking. In the company's opinion the Auto-Glide will appeal to office men, girls and workmen, school boys and girls, employes making deliveries, and folks out for a pleasure spin.

The "Auto-Glide" is powered with the Cushman "Husky" 4-cycle engine, light in weight and low in operating cost. The shipping weight of the engine is 210 pounds—net weight 170 pounds. The wheels are of the steel disk type with tapered roller bearings, 12 by 3½ in., have non-skid balloon tires with removable inner tubes. The seat is 10 by 12 in., deep, soft, upholstered with a flat top comfort cushion.

The Auto-Glide is easy to operate and anyone who has ever ridden a bicycle can immediately handle it. The low weight affords automatic balance. A turn of the hand changes the speed. A push of the left foot throws out the clutch—the right puts on the brake. The Auto-Glide rides and drives on dirt, gravel or paved roads. Its low and top speed of 35 miles an hour assures control and safety.

Leitzke Repair Block

Rein Leitzke, Beaver Dam, Wis., is manufacturing a repair block which is a time and labor saver when removing and replacing sections and guard plates; straightening sickles, sickle heads and other bars and rods; replacing and riveting crank arm shafts and other stub shafts; holding chain when removing links, etc. Sections are sheared off by placing sickle bar, with the points of the sections down, on the shearing edge, giving the section a snappy blow with a hammer, causing rivets to shear off automatically when driven through the holes shown. The wrist pin jig has an offset part way down which holds the pin without marring the threads when riveting it

onto the crank wheel. Ledger plates are removed by placing the guard on the jig and driving the rivet through

hole shown. The guard is placed upside down on the stud when a new plate is being riveted.

Ford Motor Co. to Build Tractor

In an announcement April 27, Henry Ford stated that the Ford Motor Co. would begin at once the mass production of a new farm tractor and that a large number of men would be reemployed for that purpose. The new tractor, known to have been under way for some time, incorporates a plowing system invented by Harry Ferguson, Belfast, Ireland. This system employs a hydraulic lift which keeps the implement being pulled at a constant level as the tractor moves forward.

Commemorating
75 Years
of Faithful, Friendly Service to the Farmer

GEORGE STEPHENS
One of Moline's Earliest Leaders

W. C. MAC FARLANE
President and General Manager

1865 TO 1940

In this modern era of swiftly moving events — of sudden changes and vigorous competition, only those things which possess inherent merit and which qualify as CONSTRUCTIVE can hope to endure.

Testimony to this obvious truth is the progressive service which for 75 years MINNEAPOLIS-MOLINE has given to the farmers of the world and to the general development of agriculture. . . . In the men who guide the destinies of this Company and in the engineers who design Minneapolis-Moline modern machinery, agriculture and industry have both found friend, helper and constant harbinger of the NEW.

To improve methods of farm operation and to synchronize the machinery needs of the farmer with the demands of modern markets — these major purposes have animated the history of Minneapolis-Moline.

Glimpsing ahead from this, our 75th Anniversary Year, the future looms with potential means of still further lessening manual effort for the farmer, of reducing his production costs, and of injecting greater comfort into ALL machines necessary to his business. . . . To the successful unfoldment of these means Minneapolis-Moline dedicates itself.

MINNEAPOLIS-MOLINE POWER IMPLEMENT COMPANY

Now THE *ONLY* EVERY PURPOSE TRACTOR

CHANGED IN 30 MINUTES with 2 WRENCHES AND A JACK

HERE'S WHAT MAKES THE *Difference*

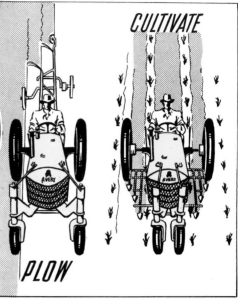

CULTIVATE

PLOW

Here's why Farmers want the Avery RO-TRAK

1. Adjustable front as well as rear wheels make it a two-wheel-in-the furrow plow tractor or an any-width-row cultivating tractor.

2. It has ALL the advantages of a plow tractor when set in plowing position, and ALL the advantages of a row crop unit when set in cultivating position.

3. It can be set for cultivating any crop planted in rows from 14 in. to 60 in. apart. This advantage for farmers is available ONLY in the RO-TRAK!

4. It's AVERY-built — That means extra value, high quality, dependable performance.

Send for Complete Facts about the AVERY RO-TRAK. You KNOW these features make it sell!

AVERY FARM MACHINERY CO.
PEORIA,   **ILLINOIS**

ADJUSTABLE FRONT WHEELS

Both rear and front wheels are adjustable—rear from 56" to 84"; front from 16" to 56".
Plow setting—both front and rear are 56".
For cultivating—whatever width settings best fit the width of rows to be cultivated.
Hundreds of farmers have seen the Avery Ro-Trak. They say—"So practical, so simple, so workable—why wasn't it made long ago?"

PERFORMANCE AND CAPACITY DATA

ENGINE — Heavy Duty 6-cylinder Hercules Engine with special features for the "Ro-Trak"—2½" Crankshaft with 7 Replaceable Bearings.
IGNITION — Automotive type with self-starter—head lights if desired.
SPEED — Throttle governed — Speed from 2 to 18 miles per hour.
BRAKES — Differential for turning—Parking Brake.
TURNING RADIUS—Pivots on either rear wheel in any setting.
SPRING SUSPENDED FRONT END —Weight always divided on front wheels.
CAPACITY—2-plow—Belt Power for 28" Thresher.

For cultivating corn, cotton, tobacco and other crops with rows of similar width, front wheels are set 16", rear wheels 84".

For sugar beets and other 20"-row crops, settings are: front wheels, 40"; rear wheels, 80". There's a correct position for all rows from 14" to 60".

To cultivate soy beans, potatoes, peanuts and other 30"-row crops, front wheels are set 48" apart; rear wheels 60".

AVERY Ro-Trak

After the first A.A.A. was ruled unconstitutional by the Supreme Court, a similar Second Agricultural Adjustment Act was inacted in 1938. Allis-Chalmers purchased the Brenneis Mfg. Co., manufacturer of deep tillage tools, of Oxnard, California, and introduced the "B" model. The Avery Farm Machinery Co. introduced the "Ro-Track" tractor for 1938 with a Hercules 6 cylinder engine. The front end could be adjusted so that the wheels were close together for row crops or spread apart in the wide axle configuration. Production of the "Ro-Track" was soon stopped because of the war. John Deere "A" and "B" models were modernized using sheet metal styling, the Minneapolis-Moline Special Deluxe tractor was produced with full cab and the Oliver "70" tractors with new sheet metal were even more streamlined than before. Harry Ferguson demonstrated the three-point hitch and struck a verbal deal with Henry Ford for its use.

Also new in 1938, the National Refining Co. brought out a new motor oil in sealed one quart cans that sold for 20 cents plus tax. Gasoline powered reel lawn mowers were selling for about $85 to $120. A bait prepared from bran, molasses, epsom salt and water was found to be more effective for killing grasshoppers than the more conventional arsenic bran bait. The 1938 census indicated that one out of every four tractors was 10 years old or older. Old tractors were being converted to rubber tires and new tractors were being ordered with the new tires. Rubber tires were being filled with water to increase weight and some operators were even mounting two tires on each drive axle to increase the traction.

Changes in tractors for 1939 were important to many manufacturers. J.I. Case introduced the new "Flambeau Red" Models "D" and "DC" tractors. John Deere incorporated the new style sheet metal on the 1939 "D" model and introduced the "H" model which had the new sheet metal styling. The small "H" model was intended to replace one horse and continued in production until 1947. Ford introduced the "9N" model which was the granddaddy of the three-point hitch. The "9N" model continued in production until 1942. International introduced the Farmall "A", "H" and "M" models, Massey-Harris introduced its first self propelled combine (Model MH-20) and the Simplicity Manufacturing Co. introduced a riding garden tractor. International Harvester also introduced the "B" model late in the year, but it wasn't sold until the following year. A special pneumatic tire with a slanted tread surface was designed by the B.F. Goodrich Co. for the tail wheel of plows and the Marquette Mfg. Co. introduced the "Carbon Arc" torch equipment and method of electric welding.

Germany increased their exports of farm equipment in 1938 and 1939 especially to the countries of Australia, Belgium, Brazil, Chile, Denmark, France, Greece, Italy, Yugoslavia, Poland, Romania, South Africa and Turkey. The export of U.S. made farm equipment was declining during the same period. The start of the "European War" in September of 1939 immediately affected farming and the farm equipment industry in America. Export of goods from the U.S. to Europe slowed nearly to a stop by September 1939.

A Handbook on Television

A television handbook which will prove of great value as a manual for television set builders and service men has been announced by the Norman W. Henley Publishing Co., 2 W. 45th St., New York City. The book is entitled "Look and Listen" and the author is M. B. Sleeper, television engineer, member of the Institute of Radio Engineers and author of many books on radio. The book contains more than 100 specially made illustrations and opens flat through modern spiral flexible binding. It treats of television transmission equipment, construction, installation, operation and servicing of sight and sound receivers. It is designed to answer such questions as— What is it? . . . Why does it work? . . . and How does it work? The price is $1.00.

Ford Announces New Tractor

FORMAL announcement of the new Ford tractor was made by Henry Ford at a preview and demonstration on the Henry Ford estate at Dearborn, Mich., June 29.

The tractor is a four-wheel unit, with the front tread adjustable, from 48 to 76 inches, for row crop work. It is powered with a 4-cylinder engine, with 3.18 inch bore and 3.75 inch stroke, having a displacement of 120 cu. in. and said to develop 23 brake hp. at 1400 r.p.m, the recommended plowing speed. The tractor weighs approximately 1700 lb.

List price of the tractor was announced at $585, delivered at Detroit. This includes as standard equipment rubber tires, battery and generator, self-starter, governor, oil bath air cleaner, muffler, power take-off, Ferguson system of hydraulic controls for implements, streamlined radiator grill, fenders, instrument board, ignition lock, throttle control and independent brakes on the rear wheels.

A price of $85, Detroit, was also announced on the 14-inch two-bottom Ferguson unit plow.

Production of both the tractor and implements is getting under way at the Rouge Plant of the Ford Motor Co., and deliveries will start shortly. Sales and distribution will be under the direction of the Ferguson-Sherman Mfg. Co., Dearborn, Mich.

The Ferguson line of implements for the Ford tractor includes a 14-inch two-bottom plow, a 12-inch two-bottom plow with sod or digger bottoms, a 10-inch two-bottom plow, a 16-inch single bottom plow, a general cultivator and a row cultivator. Other implements will follow from time to time.

All implement equipment is designed on the unit principle, the tractor and the implement being rigidly hitched to form an integral unit. Attaching and detaching are easily accomplished. The maneuverability, especially the control of backing, was stressed in the demonstrations.

All the implement equipment incorporates the Ferguson system of hydraulic control. The hydraulic implement control consists of a 4-cylinder pump supplying oil under suitable pressure to the ram cylinder and is controlled by a new type of valve. An operating handle convenient to the operator's right hand affords complete control.

As explained by Harry Ferguson, who developed the principles of the new tractor and implements, in addressing a luncheon gathering preceding the preview and demonstration, the hydraulic control not only governs the raising and lowering of the implement units, but assures their penetration at the desired depth independently of wheel movements or unevenness of ground surfaces. The weight of the drawbar load, it was said, also is distributed evenly to all four wheels, preventing the raising of the front wheels from the ground when obstructions are encountered. As demonstrated, when obstructions were encountered the effect was to give the front wheels a firmer ground contact and to raise the rear wheels sufficiently to allow them to spin and with no effect upon the implement.

The general cultivator is intended for tillage operations commonly performed by spring tooth harrows, field cultivators, etc., or as a substitute for the plow under extremely favorable plowing conditions. The tines or shanks of the general cultivator are provided with a steel spring trip release, which enables them automatically to ride over any obstruction and reenter the ground without stopping the tractor or raising the implement.

The row cultivator shanks can be adjusted to meet any row width and can be equipped with any type of shovel equipment desired. For cultivation purposes the front tread of the tractor is changed to the desired row width, the same as the rear tread, and the steering is said to be unaffected. The front adjustment is accomplished in 4-inch steps through a telescoping axle beam and reversible wheel disk. Adjustment of the rear tread is also in 4-inch steps, through a reversible wheel disk and reversible tire rim.

The front wheels are steel disks fitted with 19x4 single rib pneumatic tires on drop center rims, the tires carrying 28 pounds pressure. The rear wheels, also steel disks, are fitted with 32x8 traction tread pneumatic tires on drop center rims, with air pressure of 12 pounds.

The engine power is controlled through a variable speed governor, with a working range of 1200 to 2200 r.p.m. The governor is controlled from the steering column. Ignition is through a battery operated distributor. The engine is started through a conventional type of automobile starter with a finger operated switch on the dash. Lubrication is through a pressure system, and an oil filter with removable cartridge is provided. The air cleaner is an oil bath type with dust receptacle removable for cleaning.

Three forward speeds are provided by a sliding gear transmission in which all shafts run on tapered roller bearings. The speeds are 2½, 3¼ and 6 miles an hour with a reverse of 2¾ miles.

The power take-off shaft extends from the rear of the axle housing and has a standard spline end for fitting to drives of power driven equipment. Internal expanding brakes, 14x2 inches, are provided with individual pedal control for each wheel.

The Ford tractor has a wheelbase of 70 inches, an overall length of 115 inches, overall width of 64 inches, and overall height of 52 inches. The ground clearance is 13 inches under the center and 21 inches under the axles. The minimum turning circle is 15 feet, with use of brakes.

Extra equipment available includes a belt pulley, lighting system and steel wheels with lugs. The pulley is carried by a self-contained drive unit attachable to the rear of the tractor. The pulley has a diameter of 8 inches, with a face of 6½ inches. It is driven at a speed of 727 r.p.m., affording a belt speed of 2800 feet per minute at an engine speed of 2000 r.p.m.

How CLEAN OIL smooths the way for tractor sales

OUTLET

Dirt, dust, carbon, sludge, tiny pieces of metal—together they cause more motor damage, more customer dissatisfaction than all other factors combined.

Purolator keeps the oil clean. Purolator protects against dirty oil and the costly repairs that so often follow.

When you sell Purolator-protected farm equipment—you sell the plus value of clean oil: more dependable, more economical operation. You're making the strongest possible bid for continued confidence!

Purolator-Protection for Diesels. Because of Purolator's outstanding success in the gasoline motor field, leading Diesel manufacturers provide Purolator-protection for both lubricating and fuel oils. Purolator definitely prevents clogging in the injection system. Purolator filtration is the greatest defense against Diesel's arch-enemy—*dirt.*

Whenever you deal with Diesels, make sure of Purolator-protection.

Motor Improvements, Inc., Newark, New Jersey.

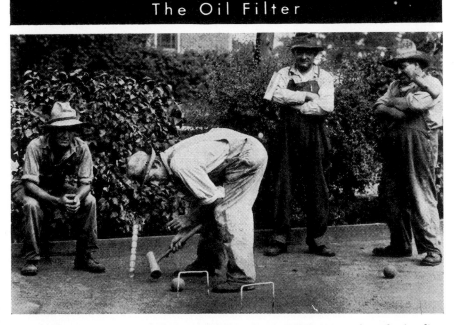

Rural folks take their croquet games seriously. Some of them approach professionalism in being able to go clear 'round without stopping. Normal rainfall and green lawns always boost the sale of sets in a game that has been popular for forty years.

In 1940, a hybrid corn that greatly increased yield was planted and harvested throughout the Corn Belt. This hybrid corn was developed by Donald Jones at the Connecticut State Experimental Station in 1925. Food production was considered a critical consideration of the "European War." Both Britian and Germany imported much of their food even during peace-time and neither were prepared, agriculturally, to wage war. It was noted that "should the armed forces of the war and war industries require it, our mechanized farms could release more manpower than in any previous period, without endangering in any way our food supply. Indeed, a large body of rural manpower could be released today, while food production because of machines on the farm could be enormously increased." Late in 1940, the young men of America were registered "for a year of selective military training." These men would be called in the following years to serve in the armed services and many would indeed be called from the nation's farms, farm equipment dealerships, manufacturing plants and service shops. Fall production of farm equipment in 1940 was only slightly affected by the defense program requirements. Tractor and combine sales in 1940 were only slightly fewer than the record set in 1937 and 19 out of each 20 new tractors sold were equipped with rubber tires.

Allis-Chalmers introduced the Model "C" tractor in 1940 and J.I. Case introduced the "VC" line of small tractors. The "Planet Jr." made by S.L. Allen & Co., the Bolens "HiWheel" tractor, the R.D. Eaglesfield "Unitractor", and the Ohio Cultivator Company's "Black Hawk" were among the one and two wheeled garden tractors that became more popular.

The standard size for the square drive of "midget" sockets was changed from 9/32 inch to ¼ inch and in March, 1940,

the Daisy Mfg. Co. introduced the "Red Rider Carbine Model No. 111." Later in the year, Daisy introduced the "Superman Krypto-Raygun" that projected a picture on the wall each time the trigger was pulled.

Seventy-five percent of the outstanding stock of Stover Mfg. & Engine Co. was purchased by Kalter, Aaronson & Associates and A. Schaap & Sons of New York in January 1941. The Hart-Carter Co. of Peoria, Illinois, purchased the Lauson Co., the gas engine company of New Holstein, Wisconsin, in July and Automatic Products Co. of Milwaukee, Wisconsin, purchased the Bolens Mfg. Co. of Port Washington, then changed the name to Bolens Products Co. in August.

This type of stove has many virtues but it can hardly compare with the modern oil, coal, or gas heaters.

Modern space heaters are not only efficient, but are beautifully styled and add a decorative note to the modern rural home.

This is a full-page advertisement for the Daisy 1000-Shot Red Ryder Cowboy Carbine. The text reads:

READY
THE SENSATIONAL NEW DAISY
1000-SHOT
RED RYDER
cowboy
CARBINE

Licensed by Stephen Slesinger, Inc., N.Y.

RED RYDER — *featuring*

PACKED IN THIS BIG HANDSOME CARTON

16 INCH LEATHER THONG KNOTTED TO CARBINE RING

Red Ryder's OFFICIAL name, brand on stock ties-up with all movie, newspaper, magazine publicity!

FRED HARMAN

Cash In On "RED RYDER"

The HOTTEST Name in BOY-LAND TODAY!

FEATURING:

Carbine Ring! Red Ryder Carbine is the first air rifle having a genuine Western Style Carbine Ring anchored in jacket!

1000-Shot Repeater! The first 1000-shot repeating Carbine in air rifle history!

Golden Bands! The first and only Daisy with Golden Bands . . on muzzle and hand-hold . . . symbolizing "The Golden West!"

Red Ryder Branded Stock! Red Ryder's official signature, picture, and horse "Thunder" are all branded into Carbine Stock!

Lightning Loader! The only 1000-Shot Daisy with Lightning-Loader Invention—best sales feature ever put on an air rifle!

Longer Gun Barrel! Red Ryder Carbine barrel is 3 inches longer than Daisy's original 500-shot Lightning-Loader Carbine!

Full Length Hand-Hold! Long, super-husky, semi-curved authentic Carbine Hand-Hold.

Carbine Style Cocking-Lever! Authoritative Carbine LEVER as used on Western Carbines.

Introduced only last April, Daisy's new 1000-shot RED RYDER CARBINE quickly became the fastest-selling air rifle in 54 years—AND will be *your* biggest seller this Fall BECAUSE—in addition to having the greatest array of mechanical sales features ever seen on any Daisy—RED RYDER CARBINE enjoys these *three* powerful, *nation-wide* publicity promotions·

1. **Red Ryder Newspaper Strip!** America's favorite western cowboy comic strip appears in more than 550 daily and Sunday newspapers *every* week . . . is eagerly read by *millions* of air-rifle age boys. Red Ryder and his CARBINE are already *nationally-known*.

2. **Red Ryder Moving Picture!** "The Adventures of Red Ryder," new movie serial produced by Republic Pictures Corporation, will be seen on the screens of America's theaters by *millions* of kids this Fall and Winter!

3. **Biggest Daisy National Magazine Ad Campaign in 54 years!** All leading comic magazines and juvenile publications starting September 1 will carry big, full-page *RED RYDER CARBINE ADS* all Fall and Winter in Daisy's biggest, most thrilling advertising campaign! This nation-wide RED RYDER publicity *in the movies, in the newspapers, and in national magazines*, plus the most beautiful, authentic, and brilliantly-engineered Daisy *ever produced*, assures you of *bigger* Fall and Winter sales on ALL Daisy Air Rifles *than ever before*. Cash in! Ask your Jobber or write direct—but *hurry!*

Retails at $2.95

The Popular **500 SHOT** *LIGHTNING-LOADER CARBINE*

Daisy's biggest seller in 1939—the original 500-Shot Carbine, featuring Lightning-Loader Invention. Adjustable Double Notch Rear Sight. **$2.50**

Selling BIG in 1940.

Double Barrel 100-Shot Repeater. Retails at . . . **$5.00**

Buzz Barton Special. Telescopic type Sights. Retails at . . **$2.25**

50-Shot Pump, Repeater, Retails at **$4.50**

500-Shot Repeater— nickeled, Retails at . . **$1.95**

60-Shot Buck Jones Special Retails at **$3.50**

Daisy Single Shot Retails at **$1.50**

CHECK YOUR STOCK!

Look over your air rifle stock NOW. Write your Jobber—or Direct—for new Display Rack, Catalogs, Posters, Literature.

FLASH! Daisy's new *OFFICIAL SUPERMAN KRYPTO-RAYGUN* Picture Projectors, Target Buster Game, Targeteer Pistol Outfits, etc.,—now ready! Write!

SELL DAISY BULL'S EYE SHOT

The only *nationally advertised* shot for all Air Rifles. Now made, guaranteed by Daisy!

BULLS EYE

DAISY AIR RIFLES

Ask your jobber . . . or write direct to

DAISY MANUFACTURING COMPANY, 269 UNION STREET, PLYMOUTH, MICHIGAN, U.S.A.

1936-1941

303

The Army Boys Named It
"THE JEEP"

POSSIBILITIES of new speeds and greater mobility in military operations were revealed at the recent maneuvers at Camp Ripley, Minnesota, in which the 35th Division of the United States Army participated with units from the National Guards of the West North Central States.

A new wheeled tractor—the tracklayers have always predominated in Army equipment—was given its first public showing and thrilled the officers and privates with its spectacular performances. It pulled 6-inch howitzers over almost impossible terrain at remarkable speeds, pulled the same equipment at 40 miles an hour on level roads or up 40 per cent grades, crashed through trees 4 and 5 inches in diameter, virtually climbed the trunks of trees 28 inches in circumference to heights of 70 inches at which point the front tractor weight was sufficient to break the tree, forded 40-inch deep streams, spanned ditches, etc.

The tractor was designated merely as a military high-speed prime mover, but the Army boys were quick to christen it "the Jeep."

The tractor is merely an amplification and adaptation of an ordinary farm tractor. It was designed largely by Minneapolis-Moline engineers and is built in the Minneapolis-Moline tractor factory in Minneapolis. Several of the tractors have been under test for some time at the Army proving grounds near Aberdeen, Md., and others have been used by the Iowa National Guard. In a recent test, two of the Jeeps pulling heavy loads of wagons, guns and other equipment kept their places in a convoy which covered nearly 500 miles at an average speed of more than 28 miles an hour.

The Jeep is powered by a standard Minneapolis-Moline Model CE 6-cylinder engine, which has a 4½-inch bore and 5-inch stroke and develops approximately 75 hp. at 1275 r.p.m. The engine has a compression ratio of 5.6 to 1, a piston displacement of 425.5 cu. in. and operates on 70 octane gasoline.

The tractor has a four-wheel drive and a 7500-lb. drawbar pull. The transmission has five forward speeds and one reverse. The front wheels in the standard four-wheel design are equipped with 8.25x20 low pressure pneumatic tires, while the tractor has 11.25x36 dual low pressure tires on the rear wheels. The Jeep is provided with both air brakes and electric brakes for controlling the towed load. The air brakes are operated with a foot pedal, and a hand brake below the steering wheel enables the operator to brake or lock either wheel as desired. The individual braking also permits the tractor to pull itself out when one wheel becomes mired and spins.

An unusual equipment is the front bumper which incorporates a roller, 12 inches in diameter. This roller acts as a bumper and helps in getting over obstructions. The drawbar at the rear is equipped with an air lift, which makes it possible to back into a load and lift or lower the load from the operator's seat. Both front and rear ends are equipped with pintle hooks. The front end drive and the engine base pan are protected by a skid plate of 5/16-inch thickness.

The Jeep has a minimum width of 96 inches, a maximum height of 66 inches and an overall length of 180 inches. The wheel base is 98½ inches, the approximate weight is 13,000 pounds and the turning radius without brakes is 20 feet. It can ford streams as deep as 40 inches. Its center of gravity is 2x33 inches above ground and its angle of repose sideways is 58 degrees.

The rolling bumper has many advantages. It enables the front end of the tractor to roll or climb up a sizeable tree, while the tractor still gets ground traction through the rear wheels. Then when the tractor weight becomes sufficient to break the tree, the rolling bumper enables the tractor to ride roughshod along the fallen trunk. When the tractor hits the bottom of a steep bank, the rolling action of the bumper enables the tractor to roll up the embankment to the top. The roller also offers the same advantages when descending a sharp bank.

"Our company started developing this prime mover in 1937," said W. C. Mac-Farlane, president of Minneapolis-Moline. "It is an amplification and modification of our regular standard agricultural tractors. We are all tooled up, and by the addition of certain auxiliaries that are in standard production with reputable sources of supply, we could produce these vehicles in quantity in about 60 days' time. Our original idea was to endeavor to make available articles that were in standard production in time of peace which could rapidly be converted for army use in an emergency."

The Jeep at Camp Ripley showed the obvious advantage of being able to move guns into places where an ordinary military truck could not take them; also of pulling army wagons and other equipment over terrain too rough for trucks to negotiate.

What part it may play in the future remains for the future to determine. It has possibilities of "getting there first with the biggest guns" or paving the way for tanks and the heaviest military equipment. It certainly should prove an effective military unit in the hands of thousands of young men from farms, whose adeptness in handling farm tractors would enable them to qualify quickly as expert Jeep operators, or "Jeepers" as they probably would be known.

Views showing "The Jeep" in action. Fig. 1—Pulling an army wagon down a steep incline. Fig. 3—Pulling wagon over rough terrain covered with underbrush. Fig. 2—The Jeep with fender equipment providing additional transportation. Fig. 4—Lower right: Two views of the Jeep, the one at right indicating tractor without military equipment.

Explaining the Isms

Socialism: If you have two cows, you give one to your neighbor.

Communism: If you have two cows, you give them to the government and the government then gives you some milk.

Fascism: If you have two cows, you keep the cows and give the milk to the government; then the government sells you some milk.

New Dealism: If you have two cows, you shoot one and milk the other; then you pour the milk down the drain.

Nazism: If you have two cows, the government shoots you and keeps the cows.

Capitalism: If you have two cows, you sell one and buy a bull.

———

Utilizing the charm of an alluring Petty drawing as a sales aid, many dealers are looking forward to good summer sales of Burgess Penlights and SnapLites. Topped by a life-like painting of the charming "Patsy" — created by artist George Petty—an attractive pair of displays hold six Penlights or SnapLites, and have proved to be effective sales-stimulators.

ALWAYS POPULAR

One of the big helps for selling used equipment in 1941 was to install new wheels and rubber tires. Later, when rubber was in short supply, even new equipment was again sold with steel wheels. The Office of Price Administration and Civilian Supply suggested maintaining farm equipment properly, because of a possible shortage of replacement parts and equipment in the future. It was thought that welding broken steel parts would prevent shortages in the steel industry. In July, Americans were "giving Uncle Sam $100,000,000 a week to save for them and receiving in return Defense Savings Bonds and Stamps."

Americans were introduced to the war with the bombing of Pearl Harbor, Hawaii, on December 7, 1941. Production was soon directed by the government and most agricultural equipment manufacturers were building the equipment necessary for the United States to conduct war. Peace time production was not to resume until 1945, but 1941 provided farmers with the highest income since 1920.

E. E. Scherer, superintendent of firearms production for the Winchester Repeating Arms Co., looks over the new Carbine which the Army adopted after exhaustive service tests of five different types of short, light rifles. This Winchester Carbine was selected as the one best fitted to the rugged use of military service. Winchester also manufactures the Garand Rifle.

The new Winchester Carbine is seven and one-half inches shorter and almost 4½ pounds lighter than the Army's basic rifle, the Garand. Though gas operated like the Garand, the Winchester Carbine works on a new principle that takes the gas off much closer to the chamber before it cools. This prevents carbonizing of the piston.

History of The Uncle Sam Button

In October, 1940, a letter from Jay O. Lashar, director of advertising, American Chain & Cable Co., was published in *Newsweek*. Mr. Lashar said, in part:

"Here is a sketch of a design for a new button or pin that I prophesy will be very popular with every real American if you picture it in *Newsweek*.

"This sketch illustrating Uncle Sam reads: 'I am proud to be his nephew,' but it should also be produced in the form of a pin reading: 'I am proud to be his niece' so that the ladies can also wear the emblem and thus proudly proclaim that the Great Sam is also their great-uncle."

Following publication of this letter many expressions of interest were received from individuals and organizations all over the country. The public was quick to grasp the national and non-political significance of the proposed button.

The wording originally suggested for the buttons has been changed to "I am proud he is my uncle," making the new button suitable for wearing by both men and women.

Parisian Novelty Co., Chicago, Ill., has been granted the sole license to manufacture these buttons, and all those interested in buying them should correspond with this source.

Mr. Lashar has agreed that every dollar of the commissions or royalties paid to him by the makers and distributors of the buttons will be donated to the American Red Cross Society.

Alcohol Fuels Costly; Use Far Off Says Study

The use of gasoline of alcohol made from farm products is no more practicable today than at the beginning of its consideration thirty years ago, the Committee on Motor Fuels of the American Petroleum Institute concludes in a comprehensive new study, "Power Alcohol, History and Analysis," issued recently.

Finding that even cheaply priced farm products cannot be processed into alcohol for less than five to six times the cost of gasoline. the committee estimates that a mixture containing 10 per cent of alcohol would consequently cost around 3 cents a gallon more than straight gasoline.

"The claim that technical advantages of alcohol-gasoline fuels justify their extra cost is not supported by facts," the committee says. "Use of a 10 per cent mixture would increase the nation's fuel bill by $690,000,000."

Gleaner Baldwin Combine Equipped with V Belts

The Gleaner Harvester Corp., Independence, Mo., manufacturer of the Gleaner Baldwin combine, has equipped its 12-foot machine with V-belt drives, and there is no increase in the retail price. The popular V-belt drive, which has been used for several years on the smaller sizes, is now furnished for the first time as regular equipment on a large combine.

The principal high speed drives now equipped with V-belts are: Motor to countershaft, countershaft to cylinder, and countershaft to separator. Gates belts are furnished, also malleable hubs with gray iron sheaves which are bolted to the hub with steel bolts. It is so designed that by reversing the position of the cylinder drive sheave on the countershaft and the cylinder sheave, a correct speed of 900 r.p.m. is attained for threshing kaffir and milo.

The Gleaner Baldwin combine is well-known for its design and its many patented features. The 1941 catalog can be obtained by writing to the manufacturer.

Diagram of the V-belt drives in the Gleaner Baldwin combine.

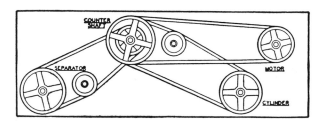

For Defense —Start Saving Today!

You should help as well as the 2,000,000 of our young men in the armed forces. Start today to buy Defense Savings Stamps. Start today doing your share . . .

IN THE SPOTLIGHT

SENSATIONAL NEW BAITS
PAW PAW CHAMPIONS for 1940!

50¢

You've seen baits with a beautiful finish . . . and you've seen baits that were made to sell at low prices . . . but you've never seen such a combination of marvelous finishes and amazingly low prices as are brought together in these outstanding new Paw Paw baits. You would almost say "It can't be done"—but we've done it.

And look at the amazing new features that are embodied in these sensational fish-getters — NEW IRIDESCENT SCALE FINISH and NEW RESINITE WATERPROOF-ING—two great achievements in the bait industry.

Proudly we present these sparkling new baits, at a surprisingly low price that will attract every man who fishes.

... The Baits that will Steal the Show in 1940!

No. 9100 SERIES — Length—2⅝". Weight—⅜ oz.

No. 9100 Series—This small easy-to-cast bait is made for those fishermen who use light tackle. The wood body is light and yet with just enough weight to throw easy and accurately. The shimmering-pearl scale finish shows up mighty well in the water and attracts the big ones from a greater distance. The strong cup-shaped blade adds plenty of wiggling, darting action. 14 patterns.

No. 9100 (Regular, Not Jointed)

No. 9100-Y	Yellow Scale, Black Head and Stripes.
No. 9100-R	Silver Scale, Red Back, Grey Stripes.
No. 9100-W	Gold Scale.
No. 9100-E	Natural Gold Scale.
No. 9100-G	Iridescent Silver Scale, Grey Back.
No. 9100-L	Silver Scale, Red Head and Stripes.
No. 9101	Yellow Perch.
No. 9103	Green, Gold Dots.
No. 9104	White, Red Head.
No. 9105	Rainbow.
No. 9106	Silver Scale.
No. 9107	Natural Pike.
No. 9108	Frog Finish.
No. 9112	Silver Flitters.

No. 9300 SERIES — Length—3¼". Weight—½ oz.

No. 9300 Series—The same design and construction as the 9100 series shown above, but a little larger for heavier tackle. This bait has all the fine qualities of finish and action, and is a real fish-getter. Rises to surface quickly when idle. Also made in jointed style. 14 popular patterns.

No. 9300 (Regular, Not Jointed)

No. 9300-Y	Yellow Scale, Black Head and Stripes.
No. 9300-R	Silver Scale, Red Back, Grey Stripes.
No. 9300-W	Gold Scale.
No. 9300-E	Natural Gold Scale.
No. 9300-G	Iridescent Silver Scale, Grey Back.
No. 9300-L	Silver Scale, Red Head and Stripes.
No. 9301	Yellow Perch.
No. 9303	Green, Gold Dots.
No. 9304	White, Red Head.
No. 9305	Rainbow.
No. 9306	Silver Scale.
No. 9307	Natural Pike.
No. 9308	Frog Finish.
No. 9312	Silver Flitters.

No. 1000 SERIES — Length—4½". Weight—¾ oz.

No. 1000 Series—For the fishermen who like a pike bait, here is a dandy. Beautiful iridescent pearl finish brings out the natural tints and markings, making it irresistible. The cup-shaped blade and fluted head give it a real live-pike wiggle. A real pleasure to show this bait to your customers.

Also made in the jointed style.

No. 1000 Series (Not Jointed)

No. 1001	Yellow Perch.
No. 1004	White, Red Head.
No. 1006	Silver Scale.
No. 1007	Natural Pike.
No. 1021	Silver Scale, Red Head and Stripes.

No. 2500 SERIES — Length—3½". Weight—¾ oz.

No. 2500 Series—This popular bait, with its beautiful iridescent scale finish, lays on its side on the surface, and when retrieved the free-running spinners cause a small ripple like that of an injured minnow. Many fishermen prefer this type of surface bait and their choice is well founded. Made in 6 popular patterns.

No. 2500 Series

No. 2501	Yellow Perch.	No. 2507	Natural Pike.
No. 2504	White, Red Head.	No. 2508	Frog Finish.
No. 2506	Silver Scale.	No. 2521	Silver Scale, Red Head and Stripes.

Write us for information on FREE Display Rack

"WOTTAFROG" for 1940

It's New!

It's Different!

It's a Honey!

Here's a brand new bait for 1940 that is going to create plenty of excitement for anglers—and plenty of sales for the tackle department.

True-to-nature in design and colorations, it is a definite step forward in making an imitation frog that swims, dives and floats with all the real action of a real, live frog.

Legs are hinged to the body, permitting them to move in a kicking, swimming motion. Treble hooks are covered with hair—they look like feet. Body is bright grass-frog green, speckled with black . . . true frog markings.

No. 74 — Length 4". Weight ⅞ oz.
No. 73 — Length 3¼". Weight ⅝ oz.

Two Popular Sizes

LIST PRICE $1.00

Japanese Christmas Lamps are Inferior Lamps

F OR the money spent on inferior Christmas lamps, the American receives mostly an assortment of annoyance, trouble and expense. That isn't guesswork; it isn't a matter of anyone's opinion. A series of tests, described here, were made by the Electrical Testing Laboratories, of New York City, under conditions of the strictest rigor.

The Christmas lamps, both foreign and domestic, used in the tests, were purchased in the open market in widely separated cities. The American-made lamps bought were equally divided among leading manufacturers. The lots were packed carefully and shipped to the laboratory where samples were selected at random for simultaneous tests of 48 American and 48 imported lamps. Tests were conducted at 118 volts. Voltage was maintained within 2/10 of one per cent of the 118 volts set for the test.

Four Times Longer

Complete results are shown in Table 1. The tests show an average life of 46.8 hours for Japanese lamps; 207.4 hours for Mazda lamps. Thus a Mazda lamp at 5 cents has an average life of more than four times that of a two-for-a-nickel Japanese lamp.

This isn't all the story. It is estimated that the average Christmas tree outfit is burned for 50 hours a season. Look at Table 1 once more. Within 51 hours only one Mazda had failed but 35 out of the 48 Japanese lamps were useless. This means that for a single season of operation the customer must buy replacements for 73 per cent of his Japanese lamps; only 2.1 per cent of replacements for American lamps.

Therefore, no more than one failure is likely to occur in the first season's operation of six strings of eight American-made lamps. For Japanese strings, the probability is that there will be six failures in each string—all greatly aggravated in annoyance by the fact that such lamps are almost universally of the series circuit type. When one lamp burns out the entire string is extinguished.

Gross Inconsistency

Another impressive fact brought out by the test figures is the gross inconsistency of the Japanese lamps. Why should a solitary Japanese lamp burn 326 hours (as against 417 hours for the last American), yet seven fail within the first hour and a half? It can only be explained on the basis that all of the seven failures were defective lamps.

TABLE I

Burned Out After	American Lamps	Japanese Lamps
1½ hrs	0	7
20 "	1	15
46 "	1	28
51 "	1	35
75 "	1	43
108½ "	4	46
210½ "	27	47
326½ "	46	48
417½ "	48	
Average Life	207.4 hrs	46.8 hrs
Watts	5.08	4.5
Lumens	66.9	26.8
Lumens Per Watt	13.16	6.56

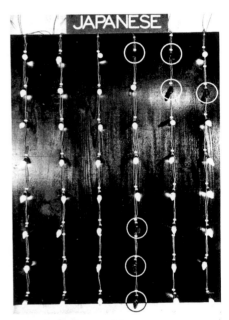

Above: Cut shows test board after 1½ hours. Seven Japanese lamps burned out in that time. The first American lamp did not fail until after 20 hours of operation.

Below: After 108½ hours, all but two Japanese lamps were gone. Meanwhile 44 American lamps, or 91.7 per cent, were still burning. The evidence is quite conclusive.

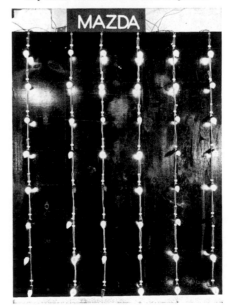

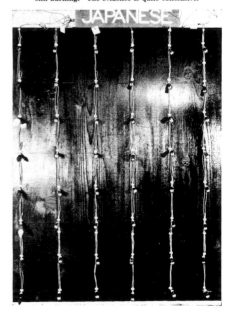

IMPLEMENT & TRACTOR

1942-1945

J.I. Case celebrated the company's centennial in 1942 and the Borg-Warner Corp. purchased all of the factory machinery and equipment except the foundry equipment from the Stover Mfg. & Engine Co., Inc. The machinery was moved to a Borg-Warner factory and was used in the production of munitions and war material. John Deere "D" model was styled like the other models and Ford began production of the "2N" model tractor. The B.F. Avery & Sons Co. of Louisville, Kentucky, purchased the plant at Liberty, Indiana, where the General tractor, a wheel tractor, had been made by the Cleveland Tractor Co. for several years. In February of 1942, the B.F. Avery company announced that it would resume production of the tractor. In March, the Fairbanks, Morse & Co. announced that it had purchased the farm equipment lines formerly manufactured by the Stover Mfg. & Engine Co. of Freeport, Illinois. In April, the New Holland Machine Co. of New Holland, Pennsylvania, purchased manufacturing and selling rights to the Stover limestone pulverizer, tractor saw attachments, saw frames, drag saws, mandrel sets, tank heaters, hog troughs and ensilage cutters. The purchase by both companies included jigs, patterns, patents and the Stover Co. "Good-Will." New Holland also took over Hertzler & Zook Co. of Belleville, Pennsylvania. Massey-Harris Co. started production of military tanks in April at the Racine, Wisconsin, plant purchased earlier from Nash Motors Corp.

The three poster designs shown on this page, and other similar ones, are being used by the USDA Extension Service to push the farm machinery repair campaign to a successful close.

Used Oil May Be Reclaimed By Removing Foreign Elements

Contrary to the general opinion, oil does not wear out. Oil as used in internal combustion engines does deteriorate with use and it becomes diluted to a certain extent with the fuel that blows by the piston rings and it may become acid and may pick up a certain amount of water. With a properly operating filter, carbon and metallic particles will be filtered out so they do not contaminate the oil and expedite bearing wear.

Equipment has been developed which makes it perfectly possible to reclaim oil by removing materials which have caused it to deteriorate. With the objectionable elements removed the oil can be restored to a condition that will be equivalent to its original condition and make it perfectly suitable for lubricating internal combustion engines.

To carry out the reclaiming process the dirty oil is pumped to a heating chamber under automatic float control and absorbent earth is added in suitable proportions. The oil and the absorbent material are then thoroughly mixed by an agitator and heated electrically. Heating and mixing are of sufficient duration to utilize the absorbent material and to drive off any volatile contaminant.

The removal of the volatile matter in the oil is hastened by blowing a continuous stream of fresh air across the surface of the oil which causes any moisture in the oil to be evaporated. This feature permits lower operating temperatures than those which would otherwise be possible. The heaters are under thermostatic control and go off automatically at the proper temperature. Signal lights indicate whether the heat is on or off so that the operator can understand what is going on inside the reclaiming equipment.

The oil and the earth mixture is next dropped to a transfer tank from which it is forced into a two stage filter press by air pressure. In this filter all solid materials including the absorbent earth are removed. The clean oil coming from the machine may then be piped to any suitable storage capacity or may be collected in drums or pails and stored under whatever conditions are desired for the shop where the work is done.

In the tractor shop arrangements can readily be made with the customers of the shop to save all their drain oil and bring it to the tractor shop periodically for reclaiming. This can be purchased, reclaimed, and resold, or reclaimed on a price per gallon basis.

Machines for carrying on the reclaiming operation come in sizes from 3 gal. per hour and a half, up to 12 gal. per hour or hour and a half. The size of the 6 gal. machine which would be the most suitable for the tractor shop occupies a floor space of 23x28 in. and is 68 in. high. The agitator motor for this size is 1/3 hp. The connected electrical load is four kw.

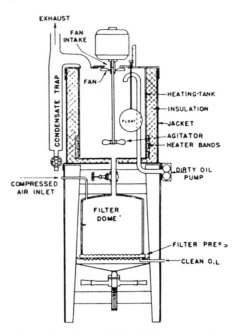

Removal of materials which cause oil to deteriorate is possible with equipment such as that shown above.

It is advantageous to reclaim the same kind of oil as far as possible. Therefore by making arrangements with customers to buy oil and then bring the drain oil back, a certain amount of control can be maintained over the oil that is brought back for reclaiming. The operation of a plan of this kind will build up a remunerative oil business and at the same time cut the oil expense of the customer and assure him of a supply of oil of high quality.

Diesel electric plants, industrial plants, railroads and electric power companies, all of whom use considerable quantities of lubricating oil, have found that the reclaiming method makes a material reduction in the cost of lubricating material. The reclaiming cost is so low that a satisfactory profit can be made by the operation of a machine of this kind at the same time giving the customers high grade oil at a very low gallon cost after reclaiming. The equipment is supplied by the Youngstown Miller Co., 675 Main St., Belleville 9, N. J.

DDT FOR EXPERIMENTATION

A limited quantity of DDT, the war-developed insect killer, has been released to DDT producers for distribution for agricultural and other civilian experimentation, WPB reported recently. Formerly requests for DDT for research work required individual application to WPB.

TIPS FOR SERVICEMEN

Remagnetizing Tractor Magneto

A simple way to remagnetize the magneto on a Fordson tractor which is built into the flywheel, is said to be to use a regular drop cord with a plug on one end to use in a 110-volt A. C. wall socket. The other end of the wires should be fixed so that one wire can be connected to magneto terminal on the flywheel housing and the other wire connected any place on the tractor which will serve as a good ground.

Turn the engine so it will be near the firing point on No. 1 cylinder. Push the plug into the 110 socket and remove insert as fast as possible. Repeat 3 or 4 times. Primary wire running from magneto terminal to coil box must be removed during the operation. A set of magnets can be brought back to a good condition by this method.

Inner Tubes

All manufacturers of inner tubes are now producing butyl type tubes which are far superior to the GR-S (Government Rubber-Styrene type) tubes of synthetic rubber, made exclusively prior to the first of the year. The old type tubes may be identified by the red stripe around the periphery of the tube, while the new butyl tubes have a blue stripe. Dealers should be careful to see that no red-stripe tubes get mixed up with tube shipments that come in. The butyl tubes are said to be able to hold air 20 times as well as natural rubber tubes, which means that unless there is leakage at the valve, they only need to have air added 1/20 as often as formerly.

Conserve Batteries

The battery shortage is still critical, due to the shortage of lead, so that everything that can be done to conserve batteries should be done. We all know that a well charged battery will last longer than one that is allowed to run down frequently. Therefore attention should be given to generators to see that they are charging as they should, and on trucks the voltage regulator should be checked to see that it is controlling the generator as it should.

Where a stock of batteries is kept on hand (if enough batteries can be obtained to build up a stock) the best practice is to have a trickle charger connected with them to keep them fully charged, as a battery will not hold a charge when standing unused for over 30 to 60 days, without losing a substantial part of the charge. Customers expect a new battery to be full of "pep" and will be pretty much disappointed if they buy one that won't turn over their truck or tractor.

"En Raashia we got no shortage of farm implements. We got PLANTY of wimmen!"

This special tool is used for opening up the steel detachable link chain. It is also useful for assembling the chain. The three slots are for three different groups of sizes.

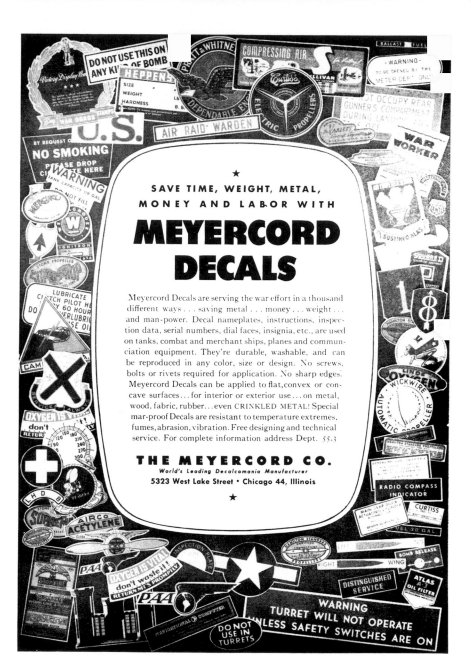
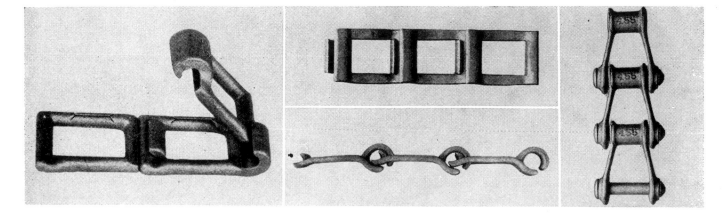

(Left) Ewart cast malleable chain, with one link partially disengaged, showing how the chain comes apart and is assembled. (Center) Two views of steel detachable link belt. The face and edge view show the form of the links. (Right) The smaller sizes of the closed end pintle chain are assembled by riveting over the pin ends, while the larger links are locked by cotter pins.

Thanks for a Million Tons of Scrap

—but Don't Stop Now! Keep the Scrap Moving Till Victory is Won!

ALL AMERICA is watching the magnificent salvage job of the farmers. Already they have rounded up more than a million tons of scrap—yes, close to *a million and a half!*

Right there on all your farms was the War Production Board's toughest salvage problem. Industrial scrap flows in regular channels, but how about the dead metals of Agriculture, scattered all over rural America—the greatest untouched reservoir of all? How could all this *precious* metal come alive and move to the hungry steel mills—*for War?*

Well, the farmers and their friends, the farm equipment dealers, tackled that tough job. They had used this metal in the building of Agriculture—they had laid it aside when it was worn out. And now they have demonstrated that, by George, *they could send it back!*

★ ★ ★

We thank all our farm customers everywhere for rallying to this urgent call of Uncle Sam. We thank the thousands of International Harvester dealers who are giving so much time and work without a penny of profit to themselves because the Government asked Harvester to help get the scrap off the farms.

And we give full recognition to all those who contribute their fine support to the farmers in this harvest of the metal crop—the schools, the churches, the clubs, the farm press and the newspapers, the countless patriots of ten thousand rural communities. Their's is a crusade—with *Victory* as the goal!

KEEP SCRAP MOVING!

A million and a half tons—but don't stop now! Never, while the liberty of your Nation and the lives of your sons are at stake! Just as a man needs food each day, the mills need scrap to build the weapons of Victory.

Keep the scrap moving off your farm, keep the mills at work. Collect your old iron and steel, and rubber, too, and *call up your farm implement dealer.* Salvage cooperation is one of his *extra* services, and will be until peace is here.

INTERNATIONAL HARVESTER COMPANY
180 North Michigan Avenue   Chicago, Illinois

BUY U. S. WAR BONDS AND STAMPS

INTERNATIONAL HARVESTER

1942-1945

With ONE VOICE for VICTORY

SERVICE IN THIS CRISIS IS OUR PATRIOTIC DUTY

All honor to our men and women in the armed forces—and in justice to those who gladly offer their all in the service of our country, WE ALL must back them up on the home front and on the production lines. We of Minneapolis-Moline will do everything in our power to give to those in the armed forces good things to work with and fight with.

Service in a Crisis Made Them Symbols of Democracy

In the heart of South Dakota's beautiful Black Hills, carved from solid granite mountains is this Shrine of Democracy in honor of the leaders who served our country in times of crisis. Their memory, too, is carved in the hearts of all Americans. Today's leaders of our way of life, and all of us, are SO WORKING that VICTORY will be Wrought, the War Won, and the new Peace Preserved, so that no one of us will have served in vain.

Men and Management

THE UNION SPEAKS

America is at war with the Axis Powers. Our land of liberty is in danger. Where yesterday our immediate objective was the constant improvement in our living standards, today that objective, as firm as ever, is secondary to our desire to smash Fascism and preserve Democracy.

Today's war is not only a battle on the military front but also a battle of production. American labor has pledged to do its utmost to increase production. Ships, planes, tanks, guns, and shells must be supplied to the United Nations and a second front immediately established on the European continent engaging the heart of the Axis forces from two sides. A steady stream of supplies must flow across the oceans to our soldiers.

We are proud to announce that our Union has wholeheartedly joined with the Minneapolis-Moline Power Implement Company in forming Labor-Management Committees to devise ways and means to increase production, thereby supplying sinews of war for the United Nations.

With this objective in mind every member of our Union pledges to give the utmost effort of body and brain to establish all-time production records.

We will unite with all forces which have the defeat of the Axis Powers as their goal, we will expose the appeasers wherever we find them and will not permit any interference with the all-out production program. A struck plant is of no more value to the all-out war effort than a bombed plant, so therefore, "strikes as usual" are out. For the duration, we have laid away our strike weapon.

The Labor movement will make all necessary sacrifices to win this war. This war is a people's war, it is our war and we gladly assume our responsibilities to help defeat the Axis Powers.

REMEMBER PEARL HARBOR

ABOVE: MINNEAPOLIS-MOLINE'S LABOR-MANAGEMENT WAR PRODUCTION COMMITTEE MEETS TO ASSURE INCREASED PRODUCTION.

INVEST IN WAR SAVINGS BONDS and STAMPS for FREEDOM'S SAKE!

KEEP 'EM FLYING FOR VICTORY In Our Own AMERICAN WAY!

This advertisement as originally used in newspapers was paid for equally by the company and the locals 1146 and 1138 of the United Electrical, Radio, and Machine Workers of America.

THE MANAGEMENT SPEAKS

WE ARE AT WAR! and all of us here are fighting for our country on the PRODUCTION LINES, while many of our loved ones and friends are on the FIGHTING FRONT. THIS—we of MINNEAPOLIS-MOLINE realize without stint . . . We'll know no peace until VICTORY is ours.

We of Minneapolis-Moline did prepare for war in time of peace. VICTORY must be WROUGHT! Or what difference does it make what we now think or what we may plan for the future? The management and man-power of Minneapolis-Moline realize that economic co-operation in this struggle for our way of life and future freedom means the recognition of the interdependence of all forces working toward (1) the Winning of the War (2) the Preserving of the peace when VICTORY is ours.

Along with our ARMED FORCES, **Labor, Industry** and **Agriculture** are the forces vitally concerned. Without co-operation of labor, industry and agriculture are stopped; without maximum agricultural production, labor and industry are vitally affected. Therefore, co-operation is essential so that victory may be wrought . . . and the only kind of peace acceptable to free men is established.

Minneapolis-Moline and its man power are truly one in the production of machinery for our armed forces, for the farm and for other industries vital to the war effort. All have steadily worked to maintain a spirit of justice and good fellowship. This mutual co-operation has meant much in the production of the "VITAL TO VICTORY" goods we make.

All at Minneapolis-Moline have sought to maintain a healthy attitude of "Give and Take" —realizing that these things are our essential jobs — (1) to build first quality products speedily for our armed forces who will lead us to victory, (2) make first quality tractors and implements for farmers who will produce the food for freedom, and (3) produce first quality machines for other industries, who, in turn, will also produce essential war goods.

MINNEAPOLIS-MOLINE POWER IMPLEMENT COMPANY AND EMPLOYEES

FOR VICTORY BUY UNITED STATES WAR BONDS STAMPS

100%—That's the record of MM employees when it comes to buying United States War Savings Bonds and Stamps every PAY DAY—We're on the Payroll Savings Plan 100%. Every cent and dollar invested in War Savings Bonds and Stamps WILL help the boys in the front lines do a better job for all of us . . . Assure an earlier victory and make our future more secure.

JOIN THE RED CROSS

JOIN THE RED CROSS

Salvage All Scrap for the Big Scrap Now

Our War Industries need metals of all kinds—zinc, copper, lead, steel, iron, aluminum, tin and also rubber, paper, rags, etc. Clean up your homes and places of business and sell or give to a good cause all scrap materials for the sake of Victory and our own Way of Life. Now is the time to do it.

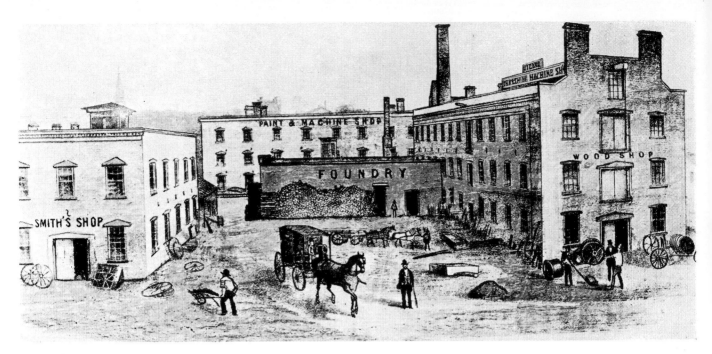

Artist's drawing of Jerome I. Case's factory, Racine, Wis., where he manufactured threshing machines. This is the way the factory appeared in 1857.

CASE REACHES CENTURY MARK

This pioneer company now features seventy-seven distinct lines and looks forward to future service to agriculture

NINETEEN FORTY-TWO for the J. I. Case Company marks the completion of a century of continuous operation and proud achievement—a century epoch in the history of farm equipment—a century of leadership in mechanical grain separation — a century of pioneering such notable developments as the steam traction engine, the gas tractor, the combine—a century in which it has accumulated associations which carry its traditions back to the beginnings of the steel plow, the baler and other farm machines which compose its seventy-seven distinct lines of today.

Nineteen forty-two likewise finds this strong and aggressive corporation, with its traditions, heritage and experience rooted a hundred years into the past, fully established for progress and leadership in a new century of greater change, greater de-

velopment and greater opportunity.

The J. I. Case century record is one of which every American can be proud, for it typifies the opportunity that is America. It had its inception in individual initiative, prospered because of individual enterprise, yet in every one of its hundred years it has given more than it has received to its country and to its customers. The company in its centennial year is truly emblematic of the industry which will enable America to produce the food that "will win the war and write the peace."

THE early history of the company is the biography of the individual who founded the institution—Jerome Increase Case —a name legendary to most men in the industry today, but the name of a young man who

started on his own in a new country in his early twenties, not endowed with material wealth but well equipped with zeal and determination, practical imagination, a knowledge of agriculture, salesmanship and mechanics; a man who lived to become a captain of industry, banker, public servant and outstanding citizen and benefactor.

Jerome I. Case was born in Williamstown, Oswego County, N. Y., December 11, 1819, the youngest of a large family of children. His father was Caleb Case, a practical farmer who had become interested in the crude threshing equipment of his time. His mother was Deborah Jackson Case, a member of the family which produced President Andrew Jackson. The most efficient thresher of that day was the Groundhog, and it consisted merely of a box carrying a spike cylinder through

In 1911 this Case 30-60 tractor was introduced. Ponderous and crude, judged by modern standards, it was a step forward at that time.

which the grain was run and which was powered by a one-horse treadmill. Straw, grain, chaff and all dropped to the ground beneath the machine. Straw was forked to one side, and the chaff was separated from the grain by the wind just as it had been for thousands of years before.

One of these outfits was owned by Caleb Case and to young Jerome was given the job of attending its operation. The lad was fascinated with its work, but he still recognized its limitations as well as its possibilities. The long hours gave him opportunity to think and imagine. Farmers of that neighborhood along Lake Ontario were beginning to grow less wheat. They were raising more fruit and milking more cows. Perhaps some day they would no longer need Groundhogs.

Stories were trickling back east of the great wheat acreages seeded by western homesteaders; of the abundant yields along the Great Lakes and in the Upper Mississippi Valley; of the difficulties of harvesting and threshing in a new country where family farm labor was inadequate and no other was to be had. Young Case sensed his opportunity, bought six Groundhogs on credit and started west.

Jerome I. Case
1819—1891

THE J. I. Case company started when the 22-year-old lad and his six Groundhogs arrived in Rochester, Wis., a small town near Racine, in 1842. Wisconsin was still a territory, the western prairies had not been opened to settlement and Missouri, Arkansas and Louisiana were the only states west of the Mississippi.

Five of the six machines were sold during the 1842 threshing season, and Case retained the sixth for custom work and for experimental purposes. Already he had developed ideas for im-

provement. During the winter of 1842-43 he built a new machine which he tried out in the following threshing season. Then he rented a small shop in Racine where he built several units which sold readily in 1844. In these, the first separators to bear the Case name, the cylinder was provided with cleaning devices and the straw was delivered separately from the grain.

The new Case thresher was a definite step forward, yet each year the young manufacturer was able to add some improvement. Farmers were buying them as fast as he could build them. Larger manufacturing facilities were needed. Five years after starting his venture on credit, still in his twenties, he was the owner of a three-story brick shop, 30 by 80 ft.

A few years later, acquiring a patent on a vibratory apparatus designed as an attachment for a thresher, Mr. Case built it into his machine and the new Case thresher established a standard type of threshing which still prevails. Another notable Case development was announced in 1869, the Case Eclipse thresher, which had no apron and which many considered the most improved and advanced machine of its time.

In 1880, Case again pioneered

First Case gas tractor. Built in 1892.

Above: First steel threshers sold in 1904. Below: The "Groundhog," an early type thresher.

an outstanding thresher development, the agitator thresher known as the Case Ironside Agitator, which attained great popularity among the nation's custom threshermen. J. I. Case himself considered it the best machine he had ever produced.

The first Case all-steel thresher in 1904 is sufficiently recent to be remembered by the older men in the trade. To many this seemed too revolutionary, but the years were to follow when every manufacturer was producing steel units and the wooden thresher was to become a natural for a museum relic.

Interesting though it might be to trace thresher development step by step, space restricts references to more than a few high spots, most of which were originated by Case and the others provided through patent purchases or licenses.

The first cleaning device was added in 1844. Drum and iron cylinders were offered in 1856. The Case "covered" thresher appeared in 1863. An attached stacker and a flax attachment were available in 1870. Separate carriers from the cylinder for straw and grain were added in 1871, timothy and flax sieves in 1876, clover attachment in 1879, shake shoes in 1880 and elevator type baggers, augers and a wagon elevator in 1885.

These were followed by an automatic grain weigher and bagger in 1888, automatic swinging stacker in 1889, special clover concaves with corrugated teeth in 1902, single crank self-feeder in 1893, band cutters in 1894, two separate straw racks in 1897, pea and bean thresher in 1900, recleaner in 1902, peanut separator in 1907, feeder for headed grain in 1909, gearless windstacker with vertical fan in 1911, steel recleaner in 1912 and many subsequent improvements which are too well known to require enumeration.

IT was but natural that a company which had given its entire life to the development of stationary threshing equipment should become a leader in producing such units as the combine for movable threshing. The history of the combine dates back to the 1880s when it was developed to meet special conditions in the Pacific Coast grain areas. These were too large and expensive except for the most expansive acreages, and for years it was commonly believed that combining could have no place in the more humid Middle West, where shock curing had been considered essential. The shortage of men for harvest work during World War I caused some of the larger growers to experiment with combines and a few machines crossed eastward over the Rockies.

Case was among the leaders in sensing the opportunity for a smaller machine, and the Case combine announced in 1923 played a leading role in converting the winter wheat belt to combine harvesting. Through the years since, the Case combine line has grown in popularity and the threshing prestige long enjoyed by Case threshers continues through its combines. Case combines are now made in a variety of sizes to meet the needs of grain growers.

THE first Case threshers were operated by one-horse treadmills. Later came the sweep, considerable of an improvement but still a crude type of power provided by a horse plodding along a circular course for hours at a time. Then a second horse was added, later a second team and eventually three and four teams were used. The efficient performance of a thresher was dependent upon the variations in animal power. When the nags slowed down, the cylinder choked. The lack of adequate power was restricting thresher progress. J. I. Case was building sweep horsepowers, but he was among the first to realize they were not the answer to the thresher power problem.

Industry was using steam. Steam appealed to Mr. Case who knew that more performance could be built into a thresher if there was a reserve of power sufficient to handle grain under any condition of threshing. As early as 1869 Case steam engines were being belted to Case threshers. The Case portable steam engine and thresher combination was shown at the Centennial Exposition in Philadelphia in 1876, where each unit was given a medal. Again at the Paris World's Fair in 1878 these units received medal awards.

The first Case steam traction engine was built in 1878, regarded by many as the last word in a farm power unit. The old Number One of this engine is now on display in the Ford Agricultural Museum at Detroit, and was shown at the Ford exhibit at the Century of Progress in Chicago in 1931 and 1932.

This engine had plenty of power to operate any thresher and plenty of traction to move the old trains of custom threshing equipment from one farm to another—but it couldn't be steered. A team of horses ahead of the engine provided the guiding. But in 1884 it became a farm locomotive in every sense of the word. Steering chains attached to the front axle were wound around a shaft, which was turned by a hand wheel by means of a worm gear. Thousands of these steam units went into the grain fields the world over in the 1880s, the 1890s and in large numbers as late as 1919 and 1920.

Some were coal burners, some burned coal and wood and others burned straw. There were fuel preferences then as now, though usually dictated by necessity. These old steam engines were consistent gold medal winners. In addition to Philadelphia and Paris, they won many awards during the twelve years of the Winnipeg Contests in both belt and plowing competition. One of the boasts of the Case organization is that during a long period of years the company led in sales of both threshers and steam engines.

YET, despite its enviable success with steam engines, the Case company was looking ahead to a day when some other type of power might be more desirable. Steam provided ample power for belt work, but its speed for drawbar operations was little better than that of horses. Fuel had to be hauled to the engine and a tank wagon was a necessary and expensive complementary equipment. The engines were too heavy, too expensive for any except custom operations or bonanza farmers, and the big wheat ranches were breaking down into smaller farm units. An experienced steam engineer also was required.

Otto had developed his internal combustion engine in 1876 and these were rapidly being adapted to stationary power uses. Producer gas engines were being imported from Europe for power plants and factories. Case engineers realized their possibilities and in 1892 an experimental gas tractor emerged from the company's experimental department. Case believes it to have been the first gas tractor ever built. Anyhow it anteceded by several years the first commercial tractors, in which the buyer had the option of either a steam or gas power plant, and was eleven years ahead of the first production of strictly gas tractors as we know them today.

What became of the tractor is not known and the Case archives contain no reference to its final disposition. But D. P. Davies, the company's veteran engineer, now vice-president in charge of engineering and who has been with the company since 1886, was then a young draftsman and did a great deal of work on this unit. Some of his recollections follow:

"The engine was of the four-cycle type with two cylinders; a working stroke was obtained with every revolution of the crankshaft. Very little was known at this early date regarding either carburization or ignition and there was not a single manufacturer of carburetors or apparatus pertaining to ignition.

"The chassis of the tractor, with the exception of the main frame, consisted largely of parts such as were in use at that time on our steam tractor. One forward speed was provided, the reverse being accomplished by sliding a key which could be shifted into neutral position or into forward or reverse speeds. One of the gears driven by the key was operated directly from the pinion of the crankshaft, and the other by means of an idler interposed between it and the pinon on the crankshaft. This was to secure the opposite motion for reversing the tractor

"Because of the lack of proper carburetor and ignition it was decided to drop the tractor at that time."

Even though Case may have been too far ahead at the time with its experimental work, it later was to become one of the pioneers of the tractor industry. An efficient Case gasoline farm tractor was available as early as 1911. In 1912 the Case 20-40 with a 2-cylinder opposed engine was in the field. Then a year later the company announced a new 12-20. All these were available for competition at Winnipeg and won gold medals for design, construction, performance and economy.

Many will remember Case's first 4-cylinder vertical tractor which appeared at the eight national tractor demonstrations in 1916, a three-wheeler which carried a 10-20 rating and was the smallest tractor built by the company up to that time.

More significant and prophetic of future tractor design, however, was the 9-18, a four-wheel unit, which appeared a year or two later with a channel frame and a little later on with a cast steel frame. The 15-27 Case, one of the most successful of the immediate post-war period, appeared in 1919 and also had a cast frame. During the early 1920s the company introduced such improvements as 3-bearing crankshafts, force feed lubrication, removable cylinder sleeves, etc., which accorded the tractors increased power and higher horsepower ratings.

Outstanding Case tractors of the 1930 decade were the L and C series and the R, a smaller unit announced near the end of the decade. The CC, the all purpose tricycle model of the C series, was one of the most notable of Case's long tractor history, the tractor with its complete line of implement equipment controlled by motor lift having played a leading part in influencing farmers toward complete mechanization.

With a tractor experience dating back fifty years, and with more than thirty years devoted to tractor production, the Case company faces the early years of its new century with one of the most formidable tractor lines in the industry, all attractive in their Flambeau Red raiment. The line starts with the V series, powered to pull one or two plows, available in both standard and all purpose models; the S series of full two-plow capacity, also available in both standard and all purpose types; the D series, recommended for three plows, likewise appearing in both standard and all purpose units; and the LA, successor to the L, in standard type only and with capacity for four or five plow bottoms.

A BUSINESS which expanded so rapidly during its early years soon required more supervision than one man could exercise, and Mr. Case displayed rare ability in selecting competent assistants. The business early became known as the Racine Threshing Machine Works, and in its organization were men later to become well known in the thresher industry, such as Messina B. Erskine, who played a prominent part in the mechanical and machinery departments of the company, and Stephen Bull who started as assistant to Mr. Case and later succeeded him as president.

A partnership, known as J. I. Case & Co., was formed in 1863 with Mr. Case as president, Mr. Bull as vice-president and Mr. Erskine as factory superintendent. In 1880 the partnership was dissolved and the J. I. Case Threshing Machine Co., Inc., was organized with virtually the same officers. In 1886 Mr. Case was president, Stephen Bull was vice-president; Frank K. Bull was secretary; M. B. Erskine was superintendent; Charles M. Erskine was treasurer and Charles H. Lee and Jackson I. Case were directors. The "Threshing Machine" was dropped in 1929, since which the corporate name has been the J. I. Case Company.

Mr. Case continued as president until his death December 11, 1891, when he was succeeded by Stephen Bull. Frank K. Bull became president in 1897, continuing in that position until 1916 when he became chairman of the board and was succeeded as president by Warren J. Davis. Mr. Davis retired in 1924 and was succeeded by L. R. Clausen, who has since headed the company. Theodore Johnson, present secretary, also was appointed in 1924. William L. Clark, vice-president in charge of sales, was appointed to that position April 1, 1931, and P. K. Povlson became vice-president in charge of production Nov. 17, 1941.

The first Case steam engine. It was built in 1869.

Back in 1923 Case prairie combines like this one were placed on the market and revolutionized combine design.

Illustration from an early advertisement which appeared in the Prairie Farmer in 1851.

320

THE well known Case Eagle trademark has an interesting historical background. During the Civil War, Company C of the 8th Wisconsin Regiment was known as the Eau Claire Eagles, and its eagle mascot known as "Old Abe" followed the regiment throughout the entire war. Despite the bird's many harrowing experiences in military life, Old Abe survived and lived for years afterward to participate in many Grand Army encampments and other historic events.

So it was natural that Jerome I. Case in seeking an emblem as a trademark for his growing business should have selected this famous eagle. At first Old Abe was shown perched upon the limb of a tree. Later Mr. Case conceived the idea of perching him upon the top of the world, which was somewhat prophetic of the company's later expansion when the trade mark and the slogan that goes with it, "The sign of mechanical excellence the world over," became a familiar sight in more than a hundred nations throughout the world.

JEROME I. CASE remained in close contact with the business he had established until his death, although after it was incorporated in 1880 he was able to give more attention to some of his other interests. He was always active in public affairs, and in 1856 he was elected mayor of Racine as well as state senator. He was re-elected mayor in 1858 and during the same year became a life member of the Wisconsin Agricultural Society. He has been generally credited with having obtained Abraham Lincoln as a speaker at the Wisconsin State Fair in 1859, when the future emancipator said that "the successful application of steam power to farm work is a desideratum — especially a steam plow."

Mr. Case was noted as a banker. In 1871 he was one of the incorporators of the Manufacturers' National Bank of Racine and became its first president. Currency issued by that bank in its early years bore his signature. Later he assisted in establishing the First National Bank at Burlington, Wis., of which he also became president. Still later he was instrumental in establishing banking houses in Monrovia, Calif., Fargo, N. D., and Crookston, Minn. In 1876 he was appointed by the governor of Wisconsin as commissioner from that state to the Centennial Exposition in Philadelphia. In 1877 he became president of the Racine County Agricultural Society in which capacity he served two years.

In 1876 Mr. Case organized and became president of a company formed to build plows which became known as Case, Whiting & Co., but which had no corporate connection with the J. I.

Case company then or later. Two years later the plow company's name was changed to J. I. Case Plow Works under which it was known for many years. The plow company was acquired by the Massey-Harris Co. in 1928, and an arrangement made at that time accords the present Case company exclusive rights to the J. I. Case name.

Mr. Case's hobbies were farming and horses, and he became owner of Jay-Eye-See, one of the oustanding horses of his day. Mr. Case acquired him in 1880 as a 2-year-old gelding from a noted stable in Lexington, Ky. In 1882 Jay-Eye-See became the world's champion 4-year-old trotter when he made a record of 2:19 on a Chicago track. A year later he became the world's champion 5-year old trotter by doing a mile in 2:10¾ at Providence, R. I., the previous record being 2:18. A year later on the same track he became the world's champion trotter, regardless of age, by

trotting a mile in 2:10. The animal's greatest performance was yet to come, however, for he paced a mile in 2:06½ at Independence, Iowa, in 1892 when he was 14 years old, to become the world's champion double gaited performer.

Mr. Case was married to Lydia A. Bull in 1849. A son, Jackson I. Case, served for a number of years as a director of the company, and at his father's death took over the management of a number of his interests outside the company.

EXPANSION into a field of related products seems an almost inevitable by-product of industrial success. It starts slowly at first and accelerates through the years. One of the early steps in Case expansion was in 1878 when the company acquired all interests of the Sawyer Mfg. Co., Oshkosh, in the manufacture of threshing machines and horsepowers.

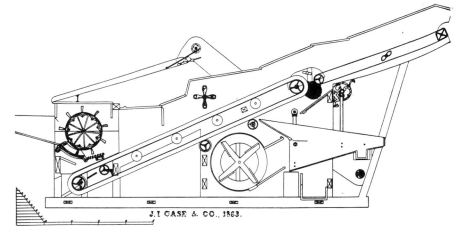

Detailed drawing of apron thresher in 1863. One of the earliest examples of layout on paper prior to construction.

The famous race horse Jay-Eye-See acquired by J. I. Case in 1880. This horse later became a champion trotter.

More than thirty years ago the custom users of Case engines and threshers were asking for additional equipment by which they could lengthen their seasons of custom work. To meet these needs of customers, Case developed a complete line of road machinery which included rollers, dump wagons, graders, scrapers, drags, sprinkling wagons, etc. This line was supplemented in 1911 with the acquisition of the business of the Troy Wagon Co. and the Perfection Road Machinery Co., Galion, Ohio. During the early years of the century, the company also developed corn shellers, saw mills, hay balers, engine gang plows and other power machines.

The first major step in expansion, however, dates back to August, 1919, when Case purchased the Grand Detour Plow Co., Dixon, Ill., an acquisition which was to give the company status as a leading producer of tillage tools. The Grand Detour company was an historic institution dating back to the beginnings of the steel plow production. Major Leonard Andrus was among those identified with the establishment of the first permanent steel plow factory at Grand Detour, Ill., in 1837, and another was John Deere. The business continued in Major Andrus' control for many years after John Deere had transferred his activities to Moline, was later incorporated under the Grand Detour name, and still later moved to Dixon, where the factory is now operated by Case. The site of the original factory is now marked by a memorial, erected by Leonard Andrus III, a grandson of the major.

Another expansion step took place in August, 1928, when Case bought the Emerson-Brantingham Co., Rockford, Ill., and thus became a full line manufacturer of farm equipment. The Emerson-Brantingham business was established in 1852 just ten years after Mr. Case founded his enterprise in Wisconsin. It had originally been known as John H. Manny & Co., taking its name from the founder who had been one of the earlier reaper manufacturers, Mr. Manny died in 1856, following which two of his partners, Waite Talcott and Ralph Emerson, a cousin of the poet, reorganized it as the Talcott-Emerson Co., and with the later passing of Mr. Talcott the name became Emerson Mfg. Co. In 1909 the corporate name again was changed to include that of Charles S. Brantingham, who had headed the company after Mr. Emerson's death. One of the important contributions of this company was the foot-lift riding plow. The former E-B factory is now the Rockford Works of the Case company.

Again in 1937, Case added important lines and factory capacity when it purchased the plant and business of the Rock Island Plow Co., Rock

David Pryce Davies, who was instrumental in building the first gas tractor in 1892.

L. R. Clausen, president of the J. I. Case Co. since 1924.

Island, Ill., which likewise had had a long and honorable history extending back to the decade before the Civil War. The business had its origin as a blacksmith shop building walking plows in 1855, when it was known as Buford & Tate. During the Civil War it passed into the hands of B. D. Buford, becoming B. D. Buford & Co. After a disastrous fire in 1882, the business was reorganized as the Rock Island Plow Co. and it was for many years one of the leading builders of farm equipment. In the former Rock Island factory, the Case company is now building its small tractor, a sugar cane tractor and implements, feed mill and disk cultivator.

About the same time Case purchased another factory at Burlington, Iowa, a one-story structure, 1,076 feet long and affording 250,000 square feet of manufacturing space, where the A-6 combine is now being manufactured, and where there is space for further expansion in the 30-acre plant site.

With its two plants in Racine and its facilities in Dixon, Rockford, Rock Island and Burlington, the company has ample capacity for production, with manufacturing largely centered close to its major market and strategically located for shipment to customers by rail, water or highway. A large part of its production facilities is now devoted to munitions and other armament.

THUS the J. I. Case Company looks ahead from the background of a hundred years of substantial contribution to American agriculture and American national economy. It is set for the present and the immediate future with its recent developments of new silo filler, new sliced hay baler, new grain drills, new combines, new corn pickers, new power grain binders, its Centennial plow and other equipment. Its experimental department works ahead into the future, preparing for the greater transition to agricultural mechanization yet to come.

A hundred years of building quality into its products continues into its new century and the same sound, conservative management which marked its first ten decades still prevails. Having seen a nation once torn asunder in civil strife and plunged into three armed conflicts with foreign foes, its faith in America is abundant. It looks confidently ahead to the day when the present mad international nightmare will have ended and J. I. Case can increase its contributions to a greater American agriculture and a greater American prosperity.

"Do ya reckon it will ever replace the WOMAN and the HOE?"

Dear Pop:

Even an old Rainbow Divisioner like you would pop your eyes at the army we're putting together this time. Let me tell you, they're doing everything to make up just about the best bunch of fighting galoots you ever saw.

And that goes for what they do for us off duty, too! Take this new club-house we got just outside of camp. It's got radios, dance floors, nice soft chairs and everything. And, Pop, you can get something to eat that won't cost you a month's pay!

Now, the army isn't running this. The USO is. And most of the other camps got USO clubs too, because you and a lot of other folks dug down and gave the money to the USO last year.

But, Pop, you know what's happened since then. Guys've been streaming into uniform. Last year there was less than 2 million of us. This year there'll be 4 million. And the USO needs a lot more dough to serve that many men—around 32,000,000 bucks I hear.

Now, Pop, I know you upped with what you could last time. But it would sure be swell if you could dig into the old sock again. Maybe you could get some of the other folks in the neighborhood steamed up, too.

It will mean an awful lot to the fellows in camp all over the country. Sort of show 'em the home-folks are backing them up. And, Pop, an old soldier like you knows that's a mighty nice feeling for a fellow to have. See what you can do, huh, Pop?

Bill

GIVE TO THE
USO

Send your contribution to your local USO Committee or to National Headquarters, USO, Empire State Building, New York, N. Y.

This carload of scrap iron consisting mostly of old and worn out farm implements was gathered from 36 different Centerville (Wash.) farms. It will soon be converted into guns, tanks, bombs, shells and armor plate. Piled high with 110,000 pounds of iron and steel, this car is now on its way to the steel mill in Seattle. This load of scrap, gathered by farmers under the sponsorship of the Centerville Grange, is the first to be shipped from Klickitat County.

A-3 PRIORITY FOR INDUSTRY

Raw materials available on basis of 80 per cent of 1940—150 per cent of 1940 for repair parts

TIRES AVAILABLE FOR TRACTORS

Orders issued by the Office of Production Management during the past ten days provide an A-3 priority preference rating for the farm equipment industry, subject to restrictions as to quantity, and also make rubber tires available for tractors and other farm equipment.

The A-3 order is intended to gear the production of farm equipment for the 1942 year to the announced essential farm production goals established by the Department of Agriculture. It is an attempt, through a preference rating, to guarantee materials for the manufacture of definite quotas of machinery. These quotas are in excess of 1940 production as to machinery and equipment needed to produce certain goods related to the food production program, and are considerably less than last year's output as to other equipment employed in raising commodities already in surplus or not covered in the USDA food-for-freedom program.

Some of the production rates as compared with 1940 are:

Potato planters, 58 per cent.
Grain binders, 75 per cent.
Rice binders, 100 per cent.
Pick-up balers, 353 per cent.
Peanut pickers, 208 per cent.

Steel stock tanks, 52 per cent.
Wooden stock tanks, 351 per cent.
Steel stock pens, 50 per cent.
Metal grain bins, 11 per cent.
Silos, 90 per cent.

Horseshoes and nails, 90 per cent.
Wooden wheelbarrows, 100 per cent.
Steel wheelbarrows, none.
Subsoil plows, 50 per cent.
Windmill pumps, 100 per cent.
Small incubators, 60 per cent.

It is understood that the average of the quotas will total about 80 per cent of the production of 1940.

An additional 50 per cent above 1940 will be made available for repair parts, so that the farm machinery already in use may be counted upon for the expanded food production effort.

The new farm equipment order is effective for the fiscal year Nov. 1, 1941, to Oct. 31, 1942.

The new rating for the industry has been provided in lieu of the original plans to guarantee the quotas by direct allocation of raw materials. The allocation effort is said to be continuing, but its enormity in the light of revisions of the whole war production program has made impossible its completion at this time.

The OPM order governing the distribution of rubber tires specifies "farm tractors or other farm implements, other than automobiles or trucks, for the operation of which rubber tires, casings or tubes are essential" as one of seven groups entitled to purchase new tires. The administration of the order, which becomes effective Monday, January 5, is delegated to the Office of Price Administration, which will establish local boards throughout the country. All applications for tires, casings or tubes for farm equipment must be accompanied by a certificate stating that they will be mounted on tractor or implement.

These "Timkenettes" sold $36,417.10 worth of Defense Savings Stamps in a four-day campaign.

PURPOSES OF PRIORITIES

SHORTAGES began to crop up in materials for farm equipment during mid-summer of 1941. To meet this problem, a plan known as the Farm Machinery and Equipment Rating Plan was announced Aug. 22, 1941. Under it, two preference ratings were provided. The first was an A-1 rating for materials necessary for the production of parts for repair and maintenance of existing farm equipment. That order was to expire Feb. 14, 1942, unless revoked at an earlier date.

The second was a B-1 rating for materials for new machinery. It expired Oct. 31, 1941, and enabled the manufacturer to obtain scarce materials during August, September, and October, 1941, in amounts not exceeding 20 per cent more than the amount used in his production of farm machinery during the same period of 1939 or 1940, whichever was higher.

But as the year rolled on, materials became more critical. The Supply Priorities and Allocations board decided, as a matter of policy, that the expansion of purely civilian production facilities was not to be permitted, if this expansion would put a drain on critical materials badly needed for defense.

One of its early decisions was in connection with farm machinery. It established a policy to cover the production of farm machinery and repairs and attachments. That policy was based on the fact that the food program for 1942, set up by the Department of Agriculture required that machinery on farms be kept in operating condition and that there was need for production of new machinery to replace that which was worn out.

83 Per Cent Limitation

The Division of Civilian Supply drafted a program, after consultation with the Department of Agriculture. That program was approved by SPAB. Its effect is to restrict materials available for new farm machinery to an average of approximately 83 per cent of the materials used for similar purposes in 1940 and to permit the use of materials for repair parts and attachments to about 150 per cent of the 1940 level. Then an A-3 rating is assigned, under which, it is expected that farm machinery manufacturers can secure deliveries on the critical materials necessary to produce these machines, repair parts and attachments — a total of 1,793,647 tons of all critical materials.

To implement and put this policy of SPAB into effect, two orders were issued, both dated Dec. 24, 1941. They both cover the period from Nov. 1, 1941, to Oct. 31, 1942. The first is a limitation order No. L-26. It limits the number of units any producer may manufacture, to a definite percentage of 1940 calendar year production by that producer. To the order, and made a part of it, was attached Schedule A, which listed the percentage figures for 17 groups of products, with a total of about 230 divisions within those groups. (See Jan. 17 issue, IMPLEMENT & TRACTOR.) While the average for the entire schedule is estimated at 83 per cent, the actual percentages on the 230 divisions run from 0 for tubular steel wheelbarrows to 353 per cent for hay press combines.

The percentages applied to repairs and attachments average 150 per cent because it is the policy to permit producers to secure scarce materials for all the repair parts necessary to keep farm machinery now on farms in good working order.

The second order, P-95, is an order assigning an A-3 Preference Rating which a producer may use with his suppliers for scarce materials necessary to produce machinery or repairs within the limits of Order L-26.

Immediate Effect

Now what does all of this mean to you, who must distribute these new machines and service those now on farms? So far as repair parts are concerned, you should be able to secure your reasonable needs. If you can't get them from the manufacturer of the machinery, after you have done your best to do so, then get in touch with your closest WPB field office and ask for help.

The government, through the application of an A-3 rating on materials for repairs and new machines, has recognized the need for these on farms and has recognized that farm machinery is an exceedingly essential product. It is your responsibility to put this new machinery in the proper hands where it is most needed. There won't be enough to go around. We all know that. Farmers have money this winter and farmers always buy when they have money to spend. But there just isn't enough material to go around, and you won't get enough machines of all kinds to fill your orders.

There has been no rationing plan laid out. I believe you can do that job better than anyone else. You know your territory, your customers, and the problems. I have discussed this matter with Governor Townsend, chief of the office of Agricultural Defense Relations, and we both feel that this job can be placed in your hands. How you do it this year may have much to do with how it will be handled in 1943.

Can Prevent Rationing

On tractors, combines and other large tools, and on tools where acute shortages may exist, you might find it wise to talk to your county agent or to the chairman of your County Defense Board before making a delivery. But I believe the very best way, after all, is to talk the matter over with the prospect, show him why it is necessary to restrict deliveries only to those who really need the machines, and ask his cooperation. You can make that method work well and demonstrate through your good work this year that you don't need any arbitrary rationing system next year.

There are many items you sell on which the quota is raised over 1940. Milking machines are 206 per cent of 1940, 350 to 500 lb. cream separators 215 per cent. Coolers up to 179 per cent, two-wheel manure spreaders 107 per cent, and many other items in poultry, dairy and small tool categories are up close to or over 100 per cent. So you should fare very well in volume.

Your greatest responsibility lies in keeping those machine tools in the finest operating condition, so maximum crops can be raised with a minimum of new tools. For years, if not for decades, you have been preaching that the farm implement business belongs to the farm implement dealer. You have made this claim on the basis of the service needed to keep farm implements in working order and your ability to render that service. Now you have not only a chance to do so, but a grave responsibility for doing so. Live up to it!

Finally, may I urgently call to your attention the fact that we are at *War*. It is a *total* war. In a total war, everyone and everything counts and counts tremendously.

Effective Machinery Repairs

One basic rule applies. The more materials we give our fighters—now—the fewer of our sons and brothers and sweethearts and husbands we will give later. As my good friend, Wheeler MacMillen says in his editorial in the February *Farm Journal*—"The metals we save by efficient farm machinery repair may make the bomb that destroys the last enemy ship—the better the boys are equipped, the more of them will come back whole."

Materials are terribly short, especially rubber and metals. Every pound you can save this spring and summer means a real contribution to early victory. Make it a personal proposition.

It is just that. This is a total war. So repair machines whenever you can, instead of selling new ones.

The slogan we see everywhere is "Keep Em Flying." By saving every pound of material you can possibly save, you are helping Uncle Sam's boys, and yours and mine, to "Keep Em Flying."

Discarded tractor = 580 machine guns (.30 cal.)

One 1-horse cultivator can be made into two 60-mm mortars . . . Fifteen two-row tractor cultivators can be used to build one light tank . . . Twelve mowers can come from the assembly lines as one 3-in. anti-aircraft gun . . .

Five hayrakers donated to the war cause emerge from the factories as one armored scout car . . . One hand cornsheller as three 6-in. shells . . . One hand garden planter as four .30-cal. rifles . . .

WASHINGTON ORDERS NEW RESTRICTIONS

War Production Board freezes all track-laying tractors. Dealers must sign statement on every order for binder twine to the effect that the item will be sold for use in harvesting of farm products.

The "Supplemental Limitation Order, No. L-26-a" limited the production of new tractors with rubber tires in March and April, then banned rubber tires on new tractors beginning on May 1st. Industry leaders argued that rubber tires, on the more than one million tractors in use, produced more and reduced consumption of other items such as fuel, broken parts and time. B.F. Goodrich announced the invention of tubeless tires in July and it was believed that eliminating the tube would reduce the rubber shortage. "Temporary Rationing Order A" issued by the U.S.D.A. in September was succeeded by permanent order "B" on November 1. Production of new tractors in 1942 was 40 percent fewer than in 1941. The Liberty Ship *John Deere* was launched ahead of schedule.

Blackout Bulb

Designed for blackout lighting in air raids, the new Wabash Blackout bulb provides downlighting in a soft beam of blue light that is safe for indoor visibility during blackouts. The bulb is lined inside with a pure silver reflector lining that hides all filament glare and projects the light downward. Light leaks are prevented by a black silicate coating that covers the bulb up to the extreme lighting end which is a deep blue. The new bulb consumes 25 watts.

Dried Food Goes in Bricks

Each time Uncle Sam pays out a dollar for food to go to our Allies, almost 17 cents goes for dehydrated potatoes, milk, cabbage, beets, eggs, and a few other foods which lend themselves to dehydration and to satisfactory reconstitution.

Food scientists have been experimenting with dehydrated vegetables, meat, and dairy products for years. It was not, however, until the present war created the demand for U. S. food exports, and also created a dearth of shipping that the "experiments" started to roll off the production line in man-sized quantities.

Dehydration plus compression of food means more space to ship tanks, guns, and planes—also less loss of food through spoilage, breakage, and careless handling.

In 1943, manufacturers agreed to standardize the position of the tractor P.T.O. shaft and drawbar as suggested by the A.S.A.E. In general all tractors except Ford-Ferguson manufactured during 1943 complied with the standard. New regulations by the Post Office established postal "district numbers," to speed main delivery. About 80 percent of the corn grown was a hybrid, so very little was being saved as seed corn.

Origin of the Jeep

Recent reference to Jeeps in *Life* and *Time* and to the fact that no one knew the origin of the term finally impelled W. C. Mac Farlane, president of the Minneapolis-Moline Power Implement Co., to write a letter to the publisher of the two papers identifying the original Jeep and explaining the first use of the name. Mr. Mac Farlane did this in the interest of pure truth and historical accuracy—not for publicity. The fact that any publicity has developed should not be held against Mr. Mac Farlane but rests on the burdened and guilty conscience of Bon D. Grussing, advertising and sales promotion manager for MM, who never misses even the ghost of an opportunity to advertise MM in "pure reading matter" columns. Enough of Bon! Mr. Mac Farlane wrote in part:

"The word Jeep was first given to an Army tractor by the Minnesota National Guardsmen at Camp Ripley, Minnesota, during their encampment the summer of 1940, and is not a contraction of the two words 'General Purpose' (GP), but was taken from the 'Popeye cartoons.'

"You may remember there was a peculiar 'animal?' in the cartoon which was part fowl and part animal, supposed to live on orchids. No one knew what to call it, but it knew all the answers and was referred to as a Jeep.

"The Minneapolis-Moline Power Implement Company, as far back as 1938, were working on the conversion of a farm tractor to an artillery prime mover, and, in collaboration with Adjutant General Walsh of the Minnesota National Guard, some experimental models were placed in the Army maneuvers at Camp Ripley in 1940 as a continuation of the experiments conducted there the previous year.

"This vehicle was a 4-wheel drive or, as referred to in military parlance, a 4x4. It was not a truck, nor was it the conventional type of caterpillar tractor customarily used for hauling guns. When Army trucks became bogged down in swamps or dug into sand, or, when artillery pieces got off the road, this tractor was used to extricate the bogged-down vehicle. It could climb trees, go through swamps and sand; it had a roller bumper on front; a machine gun could be mounted on its cab; it cleared the way for tanks through underbrush—in other words, it was neither a caterpillar, a truck, nor a farm tractor or tank—but it did know all the answers and could pinch hit in an emergency for any of these vehicles, and the boys in the National Guard called it a Jeep for that reason because it had no name, nor could it be definitely classified with other Army vehicles. . . .

"Captain Martin Schiska, veteran of the last World War and an employe of this company, collaborated with our engineers in the development of this vehicle and had charge of its operations at Camp Ripley for the Minnesota National Guard. Captain Schiska, by the way, is now with the United States Army Ordnance—probably a Major by this time.

"The original demonstration of the vehicle was handled by Mr. Tony Tchida, an employe of our company and also a member of the Minnesota National Guard. He has demonstrated our vehicles at various Army camps throughout the United States and could give you first-hand information as to the first time it was nicknamed the Jeep."

The latest version of the original MM Jeep, powered by the model Z engine and driving on all four wheels. It will do anything but speak Sanscrit, and they are working on that.

AC Girl Wins "Miss Victory" Title

In a nation-wide contest sponsored by the Hearst newspapers, Barbara Ann Clark, a machine operator employed on machine gun production at the AC Spark Plug Division of General Motors was selected to bear the title "Miss Victory of the United States." A $1,000 war bond was presented the winner, who on a war bond selling tour visited President Roosevelt and other high officials in Washington and New York.

Qualifications which gave Barbara Ann her title as representative of the typical ideal woman war worker of America follow: She is doing a good job of making machine guns. She abhors "absenteeism" from work, and is not backward about making it known to fellow workers, both inside and outside the plant. She has learned to handle every type of work in her department so she can fill in when other workers are absent. She has an outstanding attendance record.

She has submitted several suggestions for plant improvements in the General Motors Suggestion Plan, several of which have been adopted. She "shares the ride" with three other women war workers. The money she gets from them for the rides she puts into war stamps. She regularly buys war bonds under the payroll deduction plan with 20 per cent of her earnings. She purchases war bonds with all the money she receives from the government because her husband is in the navy. She enlisted 21 marines during a marine recruiting drive in Flint.

She is enrolled as a volunteer blood donor. She has purchased no new clothes since Pearl Harbor, preferring to put her money into war bonds and conserve clothing material for our armed forces. She does her own canning but uses sugar saved from her regular allotments instead of applying for extra canning sugar allowed under the regulations. She is an "all-around" patriot who considers hoarding a crime and believes those not taking part in salvage drives are slackers.

Milkweed Goes To War

Boys who once rooted out milkweeds from their fathers' corn fields may very well owe their lives to the fiber of floss found inside the seed pods of this same rural "pest." The hollow, air-filled tubes which help in seed dispersal have been found to be every bit as effective for packing life jackets and life belts as kapok, the material formerly imported from Java for that purpose. The Department of Agriculture has been requested by WPB to collect 1,500,000 pounds of milkweed floss this year for this purpose. Many state and county highway departments are letting patches of milkweed grow until the pods reach maturity. As pods will be picked before dispersal of seeds, there will be no danger of a set-back in fights against the weed as a pest. School children are being counted on to do most of the pod harvesting.

Everybody Wants A Jeep

Ninety per cent of the inquiries received about surplus war supplies to be sold by the government have been about the jeep. Colonel D. N. Hauseman, director of the readustment division of the War Department told the National Association of Purchasing Agents last fortnight.

Said Colonel Hauseman: "Everybody wants a jeep—the city slicker, the huntsman, the fisherman, and the farmer."

No serviceable jeeps are available for sale at present, but when they do become available, the procurement division of the Treasury Department will sell them.

The Oliver Farm Equipment Co. purchased Ann Arbor Machine Co., manufacturer of balers, of Shelbyville, Illinois.

Postwar planning occupied much of 1944 and optimism for an early victory lead to grand plans. "D-Day" landing of U.S. troops on the Normandy coast occurred June 6. The "G.I. Bill" was authorized on June 22. The Oliver Farm Equipment Co. purchased Cleveland Tractor Co. (Cletrac) of Cleveland, Ohio, and changed the company name to Oliver Corporation of Chicago, Illinois. Rationing of all farm equipment except corn pickers ended at a national level in September, 1944; however, local rationing still existed. Rationing of corn pickers was removed by the end of the year. Production was beginning to match the needs of conducting the war. About one-quarter million wheel type tractors were made for the civilian market in 1944.

In 1945, manufacturers began foundation planning for rebuilding their dealer network that was many times totally eliminated by the war. International Harvester and General Electric were among the companies that began plann-

Zero chests will be added to International's refrigeration line after the war.

ing to enter the refrigeration market with freezers, refrigerators and other items to meet the needs of the farm market. Many government surplus sales provided startup inventory for some new businesses. Synthetic rubber was helping to reduce, but not eliminate, the tire shortage. Rice tires for tractors were introduced by tire companies and flame throwers "of the type used in Pacific warfare" were tested as a method of controlling weeds. It was noted that "in dry seasons, only one application of flame is necessary to keep the weeds down." "Phillips screws, with the recessed, cross-shaped driver slot," were coming into wider use. Returning servicemen were beginning to re-enter the manufacturing plants. Victory in Europe was declared on May 8 after the formal signing of the unconditional surrender by Germany on May 7. The U.S. Engine & Pump Co. and the Challenge Co., both of Batavia, Illinois, merged forming the U.S. Challenge Co. An atomic bomb was dropped on Hiroshima, Japan, on August 6, followed by another over Nagasaki on August 9. Japan surrendered on August 14, then signed the formal surrender of Japan on "V-J Day," September 2, 1945. The postwar buzzword was "modernization" and it included the new round fluorescent lighting fixtures, cement block buildings, glass store fronts, aircraft type hydraulic pumps, hydraulic accumulators, etc.

The Graham-Paige Motors Corp. Farm Equipment Division began making the "Rototiller." Tiller and tine assemblies were made by Rumsey Mfg. Co. of Seneca Falls, New York. Graham-Paige ads noted that "There's a new kind of car a-coming! And Joseph W. Frazer is getting set to build it—at Graham-Paige. This great new car will bear the builder's own name—Frazer!" New Idea, Inc. was bought by the Aviation, Corp., known as Avco. Avco had just previously purchased Crosley Corp., Lycoming Motors, New York Ship Building Corp. and several other companies. Another recently formed company, the Kaiser-Frazer Corp., announced plans to build a small car called the "Kaiser." The Ford Motor Co. waived its option to purchase the land and buildings of the Willow Run bomber plant and the Kaiser-Frazer Corp. leased the facility to build the new Frazer medium-priced car, its line of tractors, farm equipment and the Rototiller.

Joseph W. Frazer, president of Graham-Paige Motors Corp., inspects the first Rototiller to come off the assembly line at Willow Run. Production of the new tillage tool started recently, with 49,449 orders on hand from various sources.

Late in 1945, Food Machinery Corp. (FMC) bought Mechanical Foundries, Inc. of Vernon, California, and International Harvester unveiled the smallest tractor yet produced by the company, the "Cub." The new model was powered by a four-cylinder, 10 horsepower engine and was designed for farmers with about 40 acres. The Cherry Co. introduced blind rivets to the civilian market. Many customers believed that mechanics should be licensed. Warnings abounded concerning the indiscriminate use of Dichlorodiphenytrichlorethane (DDT).

The government was still controlling much of the production and economy of the nation. Some of the controls that were begun because of the "total war," were left in place.

Voltage Regulators

Where voltage regulators are used on the tractor generator, it should be remembered that negative grounded and positive grounded regulators can not be used interchangeably. Either the positive or negative side of the battery may be used as the "hot" side, and it seems to make little if any difference which side is grounded. In the automobile field, most of the General Motors cars ground the negative, while Ford, all the Chrysler cars, Packard, Hudson, Nash, ground the positive. Positive grounded and negative in the same way except for one very important feature. This point of difference is the location of the different alloy contact points on the breaker mechanism. Two different materials are used for the points, platinum iridium for one and tungsten for the other. The grounded side governs the direction of current flow through the regulator, and the points have to be so arranged that direction of current flow will not cause transfer of metal from one point to the other. With the correct location of the alloy points, there will be no electrical erosion of the metal, but if the current flows in the wrong direction, the points will only last a few hours before they stick. If the points stick, the regulator becomes inoperative, with the result that the generator armature will probably burn out. This information is not generally available, so it will be worth while for the service man to make a note of it, or at least remember it.

Does DDT Upset The "Balance of Nature"?

"Nature is grossly out of balance when there are extensive outbreaks of insect pests."

This is one of the comments of Dr. P. N. Annand, U. S. Dept. of Agriculture, chief of the Bureau of Entomology and Plant Quarantine, in regard to frequently expressed and widely published fears that DDT is dangerous because it may "upset the balance of nature."

Dr. Annand emphasizes the pressing need for a great deal more research into just such problems. He says that ever since DDT has been available in quantities that made fairly large scale tests possible, the scientific workers of his bureau in cooperation with other interested agencies, have been making increasingly comprehensive tests of DDT in experiments that would give authoritative information as to the effects of the chemical on beneficial insects, birds, fish, and other wildlife.

He emphasizes that results have been generally encouraging. One of the early fears was that DDT would destroy too many bees, which are useful as honey collectors and even more valuable as pollenizers of many crops, particularly legumes and fruit. Tests indicate that DDT is not as deadly to bees as was feared. In work so far it has appeared less deadly than the arsenical sprays now commonly used, and there is evidence that beekeepers may come around to view DDT as a promising relief from arsenic poisoning of bees.

In regard to large scale use of DDT, which appears to offer for the first time a practical control for some forest insects, Dr. Annand says: "The occurrence of these outbreaks in itself is evidence that the beneficial insects, birds and other predaters have failed in holding the population down, and that a supplement is needed to bring the insect population more nearly in balance with the vegetation on which it feeds."

Dr. Annand pointed out that when a forest area is almost completely killed by such a pest as the spruce budworm, the effect is a disturbance of the "balance of nature" that can be compared to the effects of a forest fire. Wildlife, birds, other insects, and even fish are displaced and destroyed rather completely by the forest fire or by the death of most of the trees. In contrast, fairly large scale tests of DDT indicate that DDT may check a pest that is on the rampage and thus actually restore the balance of nature that would otherwise be destroyed. Birds, wildlife, and beneficial insects are soon able to return to treated areas, although a killing of the trees by the pest, if not controlled, might keep them out for years.

GI's Meet High Prices

GI's who aspire to farming will need more than $3,000 to start farming a 160-acre farm in Kansas, L. B. Pollom, supervisor of agriculture for Kansas State Board of Vocational Education, said recently. The figure for mere essentials would run to $3,215, Mr. Pollom said, in listing the following essential equipment:

Tractor	$1,125
Two-bottom plow	151
Eight-foot tandem disk	160
Two-row cultivator and attachments	194
Three-section smoothing harrow	46
Eight-foot soil packer	86
Twelve-hole grain drill	190
Wagon without box	93
Two-bottom lister and planter	173
Side-delivery hay rake	163
Grain-type box	53
Corn binder	180
Mowing machine	116
Cream Separator	120
Hay loader	185
Manure spreader	180
Total	$3,215

The list does not include the time-saving milking machine, the $50 worth of tools needed to keep machinery in repair in the field, or custom machines, such as the hay baler, corn picker, ensilage cutter or combine, Mr. Pollom added. Mr. Pollom was formerly a parts salesman for Allis-Chalmers.

Home Again... **with some mighty sound ideas!**

Welding equipment in the service shop enables the store to offer the farmer the service of converting his tires from steel to rubber.

McMichael Farm and Home Store

GM DIESELS SERVE WHEREVER AMERICA NEEDS POWER

America's fighting Engineers and Seabees really work miracles. Sand dunes are leveled. Jungles are cleared. Landing strips appear overnight. Staggering loads are moved over land and sea.

Helping them work these miracles are General Motors Diesel engines.

Because these engines are rugged and dependable, they get the toughest kinds of jobs to do.

Because they take so little fuel, they save precious transport space.

Because they have been designed for simplest maintenance, they stay on the job and keep on the go.

War is a tough proving ground for engines. It shows their mettle, reveals their stamina. As they perform their wartime tasks, these GM Diesels are proving the service they will continue to render in the many civilian needs for dependable, economical power after the war.

The Army-Navy "E" for efficiency in war production flies proudly over the GM Diesel plant in Detroit.

**KEEP AMERICA STRONG
BUY WAR BONDS**

1942-1945

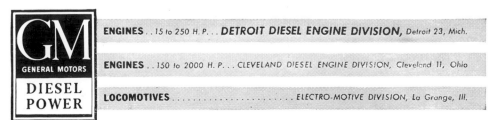

GENERAL MOTORS
GM
DIESEL POWER

ENGINES . . 15 to 250 H.P. . . . **DETROIT DIESEL ENGINE DIVISION,** Detroit 23, Mich.

ENGINES . . 150 to 2000 H.P. . . . CLEVELAND DIESEL ENGINE DIVISION, Cleveland 11, Ohio

LOCOMOTIVES . . . . ELECTRO-MOTIVE DIVISION, La Grange, Ill.

331

1946-1951

NEW FARMING EQUIPMENT
AND METHODS

Now Better Than Ever!

22 NEW FEATURES FOR IMPROVED PERFORMANCE • EASIER MAINTENANCE • LONGER LIFE

NEW 4-speed transmission—four speeds forward, one reverse. A higher speed and greater choice of speeds. Quieter, easier shifting. Easily removable transmission cover plate. Starter works only when shift lever is in neutral.

NEW Swing-back seat. Seat folds back, permitting driver to stand up easily and safely on wide running boards. Extra leg room for greater driving comfort.

NEW Hinged radiator grille. Swings open for quick cleaning of grille and radiator core.

NEW Ford-improved Hydraulic Touch Control of implements. Implements effortlessly raised to transport position, or lowered to operating position with constant control of depth. Quick, easy attachment or detachment.

NEW Well screened vented grille for air intake. Easily removable for cleaning. Air cleaner extension may be readily attached without drilling.

NEW Automotive type steering gear. Less reversible for easier steering. Adjustable for wear.

NEW Duo-servo type brakes. Easy, equalized, positive operation for faster stopping, shorter turning. Pedals for both right and left brakes are on right side; may be operated separately, or together with one foot. Brake adjusting and servicing are easy and simple.

NEW Heavy-duty three-brush generator, with voltage regulator. New long-lived water pump, simple to service. New metallic type carburetor drain and improved throttle linkage.

NEW Full running boards, asbestos shielded on muffler side—make it easier, safer to get on and off, make driving more comfortable.

NEW Heavier, stronger front axle. New disc wheels with standard hub bolt circle. And many other new features.

Watch for the new Ford Tractor—see it at your first opportunity.

**MARKETED AND SERVICED THROUGH A NATIONAL ORGANIZATION
OF DEARBORN DISTRIBUTORS AND FORD TRACTOR DEALERS**

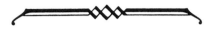

Ford Farming **MEANS LESS WORK . . .
MORE INCOME PER ACRE**

I&T SHOP BOOK

TRACTOR FLATRATES and TRACTOR SHOP MANUAL

1946-1951

The Curtis Mfg. Co. of St. Louis, Missouri, announced a new, improved model of its packaged "air conditioning unit." Earlier, the emphasis was on saving material items, but after the war people wanted comfort and convenience. People who had sacrificed at home and returning servicemen and women all were tired of "doing without." Many who left the farm to serve in the military or to work at war production in the cities would never return to agriculture, except in their memories. It was thought that farm income had begun to level off by 1946. The average net income was about $3,800 in Iowa, $3,000 in Illinois, $3,000 in Wisconsin, $2,300 in Indiana, and $2,000 in Michigan. The highest, Iowa, ranked only fourth among the states in net income per family and Michigan ranked thirtieth. These economic facts caused many to reconsider returning to the "Family Farm." G.I. loans guaranteed by the Veterans Administration was more than $½billion for the first time by March 27, 1946, but farm loans only accounted for about 3 percent of the total. By the end of the year, 20,000 veterans had borrowed a total of $76,173,321 to buy farms and farm equipment. Mechanization of farm work was needed because it was obvious that fewer people were going to produce this and other nations' food. Migration from the farm in south between 1940 and 1945 was 3,203,000 (20%) as compared to 1,876,000 (13%) for the farm population outside the south. Larger equipment was needed by the fewer remaining farmers and construction that had been delayed until after the war all seemed to require the latest equipment and the most skilled people. Farm accidents in the U.S. during 1945 numbered about 1,700,000 or 1.5 times the country's military casualties during the world war that had just ended.

Many dealers before the war stressed sales almost to the exclusion of the repair business; however, most were forced by shortages to emphasize their service shop during the war. Prewar designs for automobiles were satisfactory, as long as they were plentiful. The large number of labor strikes, material shortages and government imposed production limits caused a shortage of new farm equipment. The Office of Price Administration blocked efforts to increase prices and the availability of essential farm equipment led to much frustration for the whole industry from the manufacturer to the farmer. President Truman issued an order to remove price ceilings from everything except dwellings, sugar and rice on November 9, just four days after an election in which Republicans captured control of both the Senate and the House. The election campaign swirled around the slogan "Had Enough?" referring to controls and shortages. Even trade-in allowances for used tires had been set by the government. Repairable tires were worth at least fifty cents and tubes twenty-five cents. Production of farm equipment in 1946 may have exceeded any peace-time year, but the market could not be compared by any pre-war standard. All of the equipment was of obsolete design, at least five years old and most was worn out. Additionally, the South was mechanizing on a scale that couldn't be realized in the remainder of the country. Production of small garden tractors had ranged between 8,788 and 26,759 before 1946, when production increased to more than 114,000.

Crosley Announces Tiny High-Power Engine

A new automobile engine, available for many uses including farm machinery, refrigerator and air conditioning equipment, oil pumps and auxiliary power plants, which is said to be hardly bigger or heavier than a standard typewriter yet powerful enough to give 60 m.p.h. and 50 miles per gal. was announced recently by Powel Crosley, Jr., president of Crosley Motors, Inc., Cincinnati. The engine is said to weigh only 59 lb. but produces 26 hp.

Thin sheet-steel stampings instead of heavy forgings and castings are used in the engine; the cylinder walls are only $\frac{1}{16}$ in. thick. The engine parts are all stamped from thin metal sheet and tubes crimped together, then braced into a single piece by melting pure copper into all the joints by an hour's baking in a hydrogen or gas furnace, it is explained.

Powel Crosley, Jr., president of the company, holds one of the new Crosley engines.

Modern Inventions

We suppose you've heard the one about the absent-minded professor who came home, turned his radio on and seeing nothing, moaned: "My goodness, I'm blind!"

ATOMIC POWER WILL REVOLVE ON BOWER ROLLER BEARINGS

Now that methods of harnessing the power of the atom have advanced from the theoretical to the practical stage, industry may look forward to the time when Atomic Power will revolve on Bower Roller Bearings.

BOWER
ROLLER BEARING CO.
Detroit 14 Michigan

U. S. Must Continue to
DEPEND ON SYNTHETIC RUBBER

Post-war reconversion turmoil would indicate that present shipments of rubber will fall far short of meeting demands this year

IN THE COMPETITION between natural and synthetic rubber, 1946 will not be the year of decision. During the remainder of 1946, at least, the United States rubber industry must continue to rely chiefly on synthetic rubber, for which there is a productive capacity more than ample to meet all quantitative demands. Though shipments of natural rubber will arrive in the United States on an increasing scale, they will fall far short of meeting the demand.

Just what supplies of natural rubber will be available in 1946 is a matter of some speculation. A number of questions have been posed by the disruption of transportation, lack of equipment, shortage of food as well as of consumer goods, displacement of persons, and political turmoil in the Far Eastern countries which for years before the war supplied the bulk of the world's rubber needs.

NEED U. S. EXCHANGE

Among factors favoring speedy resumption of rubber output in these areas are their need for American exchange with which to purchase desired goods, with fairly satisfactory condition of trees generally reported, and the availability of indigenous labor, normally self-sustained in food crops, for native rubber production in Malaya, Sumatra, Dutch Borneo, Siam, and Sarawak and for estate rubber production in Java, Indochina, and Burma. The substantial damage done to buildings and equipment will not be, in most cases, a limiting factor on production, since damage can be repaired as rapidly as production can increase. These favorable factors, however, are heavily outweighed by adverse ones.

Disturbed political conditions are the gravest consideration at the moment. Political troubles virtually bar the shipment of stocks of rubber from Java and Sumatra, and retard the movement of rubber stocks from the interior of these countries and Indochina to ports of shipment.

Likewise, they hold up rehabilitation and production of rubber on estates in these countries.

Unless he can buy wanted merchandise, the native rubber producer feels no urge to produce. Shortage of consumer goods, then, is the next most serious general retarding factor. At present, supplies are not quickly available anywhere in the world. Piece goods and clothing, food items, small tools, and trade goods have to be ordered from the United States, United Kingdom, or elsewhere, newly produced in those countries, and shipped to the Far East.

Channels of distribution and trade have to be reestablished throughout the Far East native rubber areas. This is a responsibility of the governments of the producing countries. To some extent, the interests of these countries in rubber are competitive. To some extent, also, the interests of influential estate owners are competitive with those of native producers in respect to procuring goods for restoring production.

LABOR SHORTAGE

In Malaya, and probably in Sumatra, actual shortage of labor is expected to hamper severely the revival of production on estates. In British Malaya, labor was conscripted by the Japanese for work on the Burma-Siam railway; moreover, it suffered from food scarcity leading to malnutrition and disease. It is estimated that only 30 to 40 per cent of the prewar labor force is available, and ratio of males to females is now reported to be 1 to 2 in contrast with the 2-to-1 prewar ratio. Before estates can approach full production, additional labor will have to be imported from India or China into Malaya, and from Java into Sumatra. Arrangements for

such immigration have not yet been made.

Disruption of transportation, as well as a shortage of transportation equipment, is another serious factor. Motor-trucks for road transport, restoration of service on existing railroads, and reestablishment of service by harbor lighters, river boats, and trading vessels are required to remedy this difficulty in all areas.

New production in liberated areas in 1946 will depend chiefly on what is done to stimulate and facilitate trade with native rubber producers. Their potentialities are large—Malaya, 250,000 tons; Sumatra, another 250,000; Dutch Borneo, 100,000; Siam, 50,000, and Sarawak, perhaps 40,000, or a total of 690,000 tons. With an assured start of some production on estates in British Malaya in 1946, fractional output by native producers would be sufficient to meet the estimate of 1946 arrivals in consuming countries of 197,000 tons from new production in the liberated areas. Already, rubber is being bought by the British in Malaya and by the Dutch in Borneo. Resumption of buying in Siam and Sarawak is under way.

Rubber output is traditionally subject to swift changes, and since present estimates are conservative, there may be a more rapid increase in production than seems probable under the conditions outlined above. Settlement of political troubles would change the picture; it is difficult to believe that, even without settlement, the huge production potentialities of Sumatra can long be completely bottled up, while the world waits impatiently for natural rubber. Under the best of conditions, however, rubber production in the liberated areas is likely to display only a gradual increase.

Enemy Patents Interest Industry

Publications and descriptions of enemy patents which have been taken over by the United States since the end of the war are being published from time to time by the Dept. of Commerce. Many of these are of direct interest to the farm equipment industry and some believed of interest are listed here.

It should be observed that the Dept. of Commerce issues the following words of warning: "The Publication Board, in approving and disseminating reports, hopes that they will be of direct benefit to United States science and industry. Interested parties should realize that some products and processes described may also be the subject of United States patents. Accordingly, it is recommended that the usual patent study be made before pursuing practical applications."

BROWN BOVERIE Diesel engine, model No. BV8M 366. Klockner-Humboldt-Deutz Corp. Drawings and itemized listing (37 items) Exhibit "DD." Off. Pub. Bd., Report, PB 1610. 1945. 92 p. Price: Microfilm, $1.00.

Detail working drawings for the BV8M 366, various parts being applicable to other engines also. There is an index list.

• • •

BUTLER, G. M., and LILLEGREN, A. T. Mahle K.G. (piston plant) near Bad Cannstatt/Stutt. Off. Pub. Bd., Report, PB 503. 1945. 7 p. Price: Photostat, $1.00; Microfilm, 50c.

Description of manufacturing practices at this piston plant which claims to be the largest manufacturer of aluminum alloy pistons for tank, aircraft, submarine and other internal combustion engines. Cast pistons are all made in permanent molds, similar in design for larger sizes; small cast pistons are made in a single-riser mold which conserves metal. Fluxes used are indicated and specifications of aluminum alloys and tool steels and steel for casting molds and cutting tools are given.

• • •

GANZAUGE, MAX T. and BRIGGS, CHARLES W. The German steel casting industry. Off. Pub. Bd., Report PB 1333. 1945. 155 p. Price: Photostat $11.00—Microfilm $2.00.

Detailed survey of steel casting industry based on investigations of representative foundries. Topics treated in general introductory statement are: record of production, classifications of castings, raw materials, melting and molding practices, tapping and pouring methods, gating and risering technique, cleaning practices, heat treatment, welding, inspection and testing, centrifugal casting, permanent molds. Castings produced during war were mostly armament castings or parts of equipment to be used by armed forces or to assist in prosecution of war. Major classifications were: armor castings, tank, tractor and automotive castings, aircraft, submarine and ship castings, projectiles, guns and gun carriages, castings for synthetic gasoline

cracking plants, locomotives and railroads, mining equipment, heavy machinery. The major portion of this report is devoted to descriptions of production at foundries visited. Tables of statistical information and specifications of materials and products are given. Appendix I contains inspection requirements of steel castings for aircraft as set up by Association of Aircraft, Berlin. Appendix II lists composition, heat treatment and properties of various classifications of cast steels. Appendix III presents the results of tests made to study the effect of low Beryllium additions on strength and ductility of cast steel. Appendix IV presents results of study of effects of various heat treatments on cast armor steel.

Appendix V lists specifications for heat resistant steel castings, Bochumer Verein.

• • •

MURPHY, G. Diesel engine research and development in Germany during the war and pre-war period. Off. Pub. Bd., Report, PB 891. 1945. 10 p. Price: Mimeo.—10c.

Report on designs of new Diesel engines and improvements of existing engines, mainly weight reduction of conventional 4-cycle engines. The report covers: (1) M.A.N. double-acting two-cycle engines, (2) Junkers Jumo 205—Opposed piston aircraft Diesel engine, (3) M.A.N. submarine engines.

New Schrader TIRE PRESSURE GAUGES

ARE BACK AGAIN—CHROMIUM PLATED

During the war, Schrader Type Tire Pressure Gauges were *standard equipment* on every U. S. A. Military pneumatic tired vehicle. NOW, these precision engineered Gauges are again available for you *to use* and *to sell.*

Built for accuracy and long wear, every service station and truck owner should have one. Intelligent use will help make your tires last longer.

THE STANDARD "ALL-PURPOSE" *DUAL FOOT* GAUGE

FOR TRUCKS, BUSSES AND GENERAL USE

"Inner Duals" are easily reached for accurate pressure reading. Convenient to carry. Gauge is equipped with the handy "hang-up" ring. Your truck customers will ask *for Schrader Gauges. Use one yourself—the rest will sell themselves.*

RETAIL PRICE
$4.50

THE DELUXE *DUAL FOOT* "TRUTEST SPECIAL"

This portable "master gauge" is designed for checking accuracy of other Tire Gauges and Indicators. Guaranteed for 18 months. Price includes 2 FREE factory recalibrations within an 18 month period.

RETAIL PRICE
$7.00

Schrader
CONTROLS THE AIR

Order your needs from your regular source of supply. And don't forget to stock a few extra gauges for "quick sale" to your truck customers. A profit item!

A. SCHRADER'S SON, *Division of Scovill Manufacturing Company, Incorporated,* BROOKLYN 17, NEW YORK
ORIGINATORS OF THE COMPARATIVE AIR LOSS SYSTEM FOR FLAT TIRE PREVENTION

338   1946-1951

PROFIT
IS A SALEABLE
Commodity

Bath

Kitchen

Garden

Barn

IF YOU CAN SELL A MAN THE IDEA THAT HE IS NOT *SPENDING* BUT *EARNING* MONEY BY INSTALLING A JACUZZI AUTOMATIC HOME WATER SYSTEM, YOU'VE MADE A CUSTOMER OUT OF A PROSPECT.

AND, SURE ENOUGH, THAT IS PRECISELY WHAT HAPPENS. STATISTICS, COMPILED BY DISINTERESTED RESEARCH EXPERTS REVEAL THAT, ON THE TYPICAL FAMILY FARM *"Running Water will pay a return of as much as $20 for every $1 invested in the operation of a home water system."*

Here's How It Works

1. RUNNING WATER IN THE HOME REPRESENTS SAVINGS BECAUSE THE ENDLESS DRUDGERY OF HAND-PUMPING AND WATER-CARRYING CHORES IS FOREVER ELIMINATED . . . TIME FORMERLY SPENT AT THESE DISMAL TASKS MAY THUS BE DEVOTED TO MORE PROFITABLE DUTIES.

2. GARDENS YIELD GREATER QUANTITY AND HIGHER QUALITY WHEN PLENTY OF WATER IS AVAILABLE FOR IRRIGATION.

3. POULTRY, DAIRY STOCK, WORKING ANIMALS, AND LIVE STOCK OF EVERY SORT WORK HARDER AND DO MORE, THUS YIELDING GREATER PROFITS, WHEN A JACUZZI PUMP IS ON THE JOB.

AS AN ADDITIONAL ARGUMENT IN CLINCHING THE SALE YOU MAY, OF COURSE, ENUMERATE THE MULTITUDE OF ADDITIONAL ADVANTAGES OF OWNING A JACUZZI PUMP . . . MODERN COMFORT, MODERN CONVENIENCE, THE LUXURY OF "CITY TYPE" PLUMBING FACILITIES, THE ELIMINATION OF FIRE HAZARDS, ETC., ETC., ETC.

BUT...

There isn't a person anywhere who would not rather make a profitable investment than merely make a purchase. POINT OUT THE PROFIT ANGLE OF OWNING A JACUZZI WATER SYSTEM . AND WRITE UP THE ORDER!

Jacuzzi
The Original Injector Type
PUMPS
& WATER SYSTEMS
FOR DEEP OR SHALLOW WELLS

WE INVITE YOUR INQUIRY RELATIVE TO ESTABLISHING YOUR BUSINESS AS A MEMBER IN OUR NATIONWIDE DEALER-DISTRIBUTOR ORGANIZATION. ADDRESS ALL CORRESPONDENCE TO:

Jacuzzi Bros., Inc. 5327 JACUZZI AVENUE, RICHMOND, CALIFORNIA

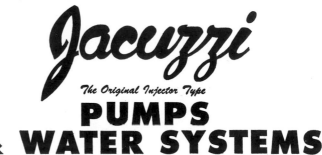

KEEPS HORSEPOWER STRONG

● New tractors, old tractors—they all need Casite f[or] maximum life and power.

In a new engine, Casite assures proper break-in . [. .] carries oil quickly to the tight spots . . . retards fo[r]mation of sludge.

In an older engine, Casite gives a speedy tune-up . [. .] frees sticking valves and rings . . . reduces power-killi[ng] gum deposits . . . improves lubrication.

Sell Casite for new engines, reconditioned engines, o[r] engines. It wins friends and shows a nice profit, to[o.]

For passenger cars, small tractors and small trucks, a pint in the crankcase every oil change and a pint through the air intake every three months. For larger units and diesels, see instructions.

THE CASITE CORPORATION • HASTINGS, MICHIGA[N]

CĀSITE

CĀSI[TE]
with Every Oil Cha[nge]
SLUDGE SOLVEN[T]
FOR EASY STARTIN[G]
MOTOR BREAK-I[N]
MOTOR TUNE [. . .]

"WHY SHOULD I SIGNAL FOR A TURN? EVERYONE KNOWS WHERE I LIVE!"

★ FOR SAFER DRIVING

Install Mitchell Directional Signal Switch

★ In the first four months of 1946 highway accidents injured some 40,000 persons. President Truman recently stated that more people have been permanently injured in automobile accidents than in the two great wars. "The nation cannot afford and will not tolerate this tragic waste of human resources," the president said.

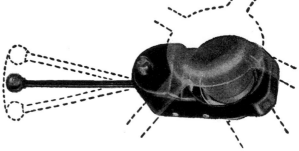

Mitchell clamp-on semi-automatic Directional Signal Switch, designed especially for trucks, busses and tractors.

Promotion of driving safety is a matter of national concern — a constructive force in checking the mounting toll of automobile accidents.

Cars and trucks equipped with the Mitchell semi-automatic Directional Signal Switch unit make a notable contribution to safer vehicular traffic. A flick of a finger on the lever of the unit mounted on the steering post, flashes to both approaching and following traffic either a right or left turn signal. When turn is completed, signal automatically self-cancels.

The Mitchell unit is available in two models. The built-in type is furnished as original equipment on passenger cars; the clamp-on unit may be installed on bus, truck or tractor. Adaptations of these models can be made to fit any vehicle.

UNITED SPECIALTIES COMPANY

UNITED AIR CLEANER DIVISION, CHICAGO 23
MITCHELL DIVISION, PHILADELPHIA 36

Mitchell built-in Directional Signal Switch unit, installed on 1946-1947 cars.

AIR CLEANERS ★ WHEEL GOODS ★ METAL STAMPINGS ★ DOVETAILS ★ IGNITION AND DIRECTIONAL SIGNAL SWITCHES ★ ROLLED SHAPES

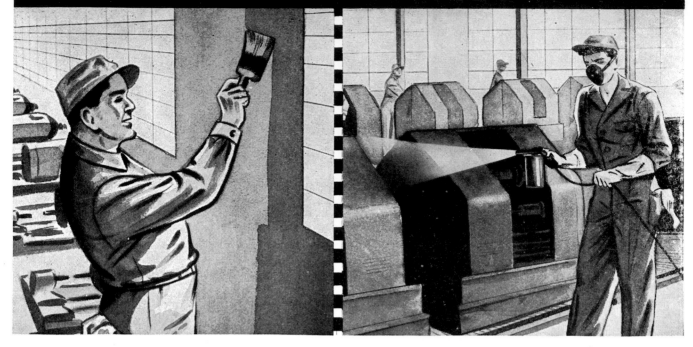
1946-1951

GERMAN TECHNICAL DATA

The Office of the Publication Board of the Dept. of Commerce recently appealed to American industry and science to make suggestions on what data in German industry and science should be selected for microfilming and bringing to this country for American use.

Many tons of documents have been collected in Germany and the government plans to make available at the cost of reproduction whatever data will be of value to American businessmen and industrialists.

IMPLEMENT & TRACTOR'S Policy on COOPERATIVES

The cooperatives are in business, competing directly in a great many lines of enterprise. They make money. Their earnings are profits, regardless of terminology. It is the law of the land that Congress has power "to lay and collect taxes on incomes, from whatever sources derived."

Therefore, cooperatives should pay taxes on their incomes—as corporations if they have adopted the corporate form of organization; as partnership individuals only if they are partnerships.

Cooperative financial statements show that they have ample ability to pay taxes, which is the criterion set up under our tax laws.

There is no other issue involved.

In order to make the most of available opportunities, the Dept. of Commerce has suggested that anyone desiring special information which might be available from Germany send a request to the department. It is further suggested that names of technical experts who might be interested in the work be forwarded to the department. Arrangements can still be made to send industrial personnel on special missions through Germany.

IMPLEMENT & TRACTOR will be glad to forward any suggestions or inquiries from readers to the proper government office.

FARM ELECTRICITY

The electric age is still "in swaddling clothes" so far as the farm is concerned, according to Claude R. Wickard, Rural Electrification Administration. This is true, he says, "even on the majority of the farms already enjoying electric service."

"The water pumps, milking machines, hay hoists, feed grinders and other labor-saving equipment now in use on electrified farms—to say nothing of the household appliances which free the farm wife for more work with income-producing enterprises—have given a tremendous boost to diversified farming. These and other labor-saving electrical devices also cut down production costs of staple crops and improve their quality.

"The surface has barely been scratched in the application of electric power to farm tasks. More and better equipment will be available at lower cost as the Nation's progressive farmers find new and more profitable uses for electric power.

"Rural people who do not yet have electric service are aware of the possibilities and they are becoming very insistent in their demands for service."

"Show Mr. Harvester to Pup-Tent 34, which he will share with a Mrs. Dow and a Mr. Holly."

SURPLUS BARBED WIRE

Flood-stricken farmers along the Susquehanna River in the vicinity of Elmira, N. Y., will get surplus barbed wire to help rebuild their washed-out fences. War Assets Administration announced last week after quick cooperative action by WAA and the Department of Agriculture.

A Department of Agriculture request that a certification of impairment be honored for sufficient barbed wire to rebuild fences recently destroyed over a 30-mile area has been granted, and two and a half carloads of this wire have been earmarked for immediate sale to flood victims from surplus stocks in possession of the New York City regional office of WAA.

With the granting of a special priority for this barbed wire, shipments will be made direct and without delay to assigned dealers designated by the Production and Marketing Administration in Agriculture. Farmers in the designated region will place their orders with the assigned dealers.

To Find Engine Speed

If you want to figure the engine speed on an automobile, for a given car speed, in miles per hour, here is an easy and quick way to do it. Take half of the number of miles per hour and add two zeros, and the result will be the engine speed in revolutions per minute. For example, if a car is going 30 m.p.h., half of that figure is 15. Add two zeros and you have 1500, which is the r.p.m. at that speed. This will be very accurate for cars with a 4.12 rear axle ratio, and fitted with 6.00-16 tires. If the axle ratio varies from 4.12, the engine speed will vary proportionately.

Salsbury Motors Produces Power Package

Automatic operation of the principle of adjustable diameter drive and driven pulleys linked by a V-belt to provide varying drive ratios is the heart of the new, patented Salsbury Power Package, produced

by Salsbury Motors, Inc., 4464 District Blvd., Los Angeles, Calif. For the first time there is made available in the low horsepower field an automatic driving unit in which drive ratios are controlled by the driven unit and clutch engagement is controlled by engine speed, according to the manufacturer.

The engine is Model 600, a single-cylinder, four-cycle, air-cooled type, said to develop 6.5 hp. at 3200 r.p.m. or one-third hp. per cu. in. of displacement, and weighs about 56 lb.

You Can't Go To Town In A Bathtub

The 50th anniversary of the automobile — the machine that revolutionized farm living—is being celebrated this year.

FOR A SUCCINCT appraisal of how rural life in these United States has been altered by the motor vehicle since its introduction in 1896, two anecdotes graphically mirror the motivating forces that put the farmer on rubber-tired wheels even ahead of his city cousin.

The first, from the archives of the U. S. Dept. of Agriculture, reports that when a government investigator asked a farm wife why the family owned an automobile but not a bathtub, the woman promptly replied:

"Why, you can't go to town in a bathtub!"

The second is from "Middletown", by Robert S. and Helen Merrel Lynd. When a life-long resident and shrewd observer of the Middle West learned what the Lynds were trying to discover in their thorough research of life in an Indiana town, he retorted:

"Why on earth do you need to study what's changing this country? I can tell you what's happening in just four letters—A-U-T-O!"

There you have a fairly good explanation of why motor vehicles got rooted more easily into rural life than they did in urban life in the United States.

"It was the cautious, conservative, slow-to-change farmer who accepted the car as a utilitarian vehicle when it was still looked upon as a fad by the city slicker," said David Cohn, in "Combustion on Wheels".

CARS SAVE TIME

This early acceptance in rural areas is enigmatic only to those who fail to recognize that, in the farmer's scale of values, time is the paramount value. If he does not plow when it is time to plow, he will not reap, and no amount of mathematical finagling with account books or juggling of currency values will make the slightest difference in the ultimate effect. So, because he is inherently time-conscious, it was only natural that he should accept the motor vehicle, once he was convinced it was a tool that would save time.

His wife played a part in this acceptance, too. Her reasons were more obvious. To her, the car represented release from drudgery, loneliness, isolation. With a car at hand to take her to town or to the neighbors when the mood for mobility was upon her, she could tolerate a continuation of the old custom of Saturday night baths in the laundry tub—for a little longer, at least.

From such a combination of causes came veritable revolution in rural life wrought by the motor vehicle whose Golden Jubilee is being celebrated in Detroit May 29-June 9 and throughout the United States this year.

To those who have always been urban dwellers, the automobile is interpreted in terms of relaxation and recreation. But with the farmer a motor car becomes a necessary adjunct to his operations. To him it is simply good business.

MORE RURAL CARS

Most Americans may not realize it today, but it is a statistical fact that a quarter of a century ago there actually were proportionately more automobiles rattling along rural roads in these United States than there were cars gliding along our city streets.

In the early nineteen-twenties low-cost cars were a common sight in virtually every agricultural section of the country. On market days and shopping days, whenever roads were passable, every county-seat town in the nation was the mecca of these chugging utilitarian vehicles. Devoid of grace, lacking even a pretense of styling, these were tough and rugged transportation tools. At the wheel was the farmer, his wife at his side. The back seat was invariably crammed with children and produce. The day of trading done, the safari reversed. Arrived at home, the car was often used as auxiliary power for the performance of work. A jacked-up rear wheel with a pulley clamped or bolted on the spokes pumped water for stock, sawed wood, chopped feed or ground corn.

On Sundays the vehicle again sallied forth with its human cargo to church in the morning, to visits with relatives or neighbors in the afternoon. Seasonally, it became a means for expanding mobility for the family, a convenient conveyance for fishing and hunting expeditions, for visits to stock shows in the larger towns, for seeing the circus, for more frequent attendance at motion picture theaters, for attending agriculture college lectures on better farming methods, for participation in extension courses.

RURAL LIBERATOR

The role of the car as rural liberator is revealed in the statistics of 1920 which show that farmers then drove an average of 4,500 miles a year, and that 78 per cent of their mileage was for business purposes. The greatest density of car to population that year was in the eight states of California, Iowa, Nebraska, Kansas, South Dakota, Montana, Minnesota and Wyoming—all predominantly farming or ranching areas.

In that same year—and indeed for almost two decades thereafter—urban Americans automatically said "pleasure cars" when they spoke of automobiles and wanted to make it clear that they did not mean trucks. The city slicker's reluctance to accept the innovation as anything more than a plaything for the well-to-do was a mental quirk that took a long time to die out. Along the

Eastern Seaboard one can still hear that quaint expression "pleasure car".

The strange thing about the American farmer's early acceptance of the innovation is that he was at first more hostile to it than the city cliffdweller. So intense was the rural aversion for the motorized contraption around the turn of the century that in some communities vigilante groups were formed to keep the nuisances at bay. Weary of wrestling with teams of runaway horses, the members of the Farmers' Club of Harlingen, N. J., pledged themselves not to vote for any candidate for public office who owned or operated an automobile. In many sections, farmers armed themselves, with the aroused male citizenry carrying pistols to use on any motorist who failed to stop on signal.

ARRESTED CAR DRIVERS

One West Virginia county (Grand) went so far as to authorize the arrest of all persons found using the automobile. Madison County, in the same state, tempered its antipathy by permitting open possession of a car if the driver slowed down when it met a team on the road and remained motionless until the horses had passed beyond danger.

The farmers' principal objection to the car, of course, was that it frightened their horses. One Michigan trouble-shooter bobbed up with a suggestion for a sort of modern Trojan horse for solving the problem and relieving the border-war tension. This character, one Uriah Smith of Battle Creek, reasoned that since the car frightened the horse, why not make the car resemble the horse? The simple trick could be done, he maintained, by patterning the front end of the car after a horse—that is, build an ersatz horse on the dash. The idea was that when the synthetic horse met the genuine article, the deception would be complete and all would be serene—or so Mr. Smith thought.

Unfortunately, or otherwise, the suggestion never received a tryout, so the feud between the motorist and the farmer was to drag along for several more years. In justice to all concerned, including the horse, it must be admitted that the offending motorist usually found the horse easier to pacify than its owner.

Another far-reaching effect of the automobile's invasion of the back country was the death-knell it sounded for the more remote crossroad trading posts. When new country was first settled, the towns were usually laid out about seven miles apart, so that the average farmer could be within one to two hours' driving distance of the general store and the grist mill.

Even that close proximity, however, was found to be inconvenient during the farmers' rush season, and when country roads were bad in the spring and fall; so, in time, the crossroads store sprang up. Freely sprinkled about in the same pattern were one-room school houses and rustic churches.

The scene gradually changed as the cars multiplied in the countryside. The farmer began to whiz right past the crossroads store and go on to the village, where he found larger stocks of goods for his selection. Today, except in more remote districts, those crossroads stores are only a memory. Many of the schools are abandoned and most of the churches boarded up or torn down.

BOOMED TOWNS

This in turn boomed the medium sized towns, most of which have at least held their own. Even here, however, is a reflection of the transition. Only a generation ago the average country town of any consequence eagerly looked forward to such special occasions as the Fourth of July and the Old Settlers' Reunion. The town was certain to be crowded with visitors drawn by the parade, the oratory, the flying flags, and the fireworks.

Today finds the country town deserted on the Fourth and Main Street has become a ghost street—all the folks have hurried out to picnics at the beach or lake.

The same change has come about in vacations. A vacation was once only an annual event, a hallowed occasion on which the family took a trip by train to the mountains or the seashore. Today just about every Sunday has become a weekend vacation, with endless trips and visits planned with the clocklike regularity of daily life. As for the annual two weeks' vacation, the American family now scrambles into the com-

fortable highway cruiser and goes wandering about the land like a band of Arabs.

Of recent years another trend, set in motion by the automobile, has started to reshape the country scene again. In a return swing of the pendulum, many once remote hamlets are coming back to life again as nerve-taut city dwellers seek out quiet places for rest and relaxation, away from the grind, grime and confusion of life in the city.

In the beginning of the frontier days, no town could expect to prosper unless located along a waterway, where it could be served by steamboat or barge. With the coming of the railroads, the towns began to move inland, into the heart of rich agricultural districts. As the nation developed, certain of these communities grew into congested cities. With 30,000,000 cars spinning along the nation's highways, adding every year to the congestion of these great industrial and trading marts, it was inevitable that there should again be a drift back to the rural areas of people seeking escape from the urban life of narrow confines.

In business transportation, as in pleasure driving, the immense fishnet of highways has gradually swallowed up all but the main-line trunk railroads and their feeder lines. Thousands of miles of railroads have been abandoned as fleets of trucks and motor busses have taken over the business once monopolized by the railways. As a result, thousands of towns and villages once served by trains now depend entirely on rubber tires for their contact with the outside world.

No less than 54,000 towns, or half of all those in the United States, are today entirely independent of both steamship and railway intercourse. The combustion engine on rubber tires has given us a freedom we never knew before.

MACHINERY FOR FARM

All this has been possible because the motor car early won a foothold in the countryside, the traditionally conservative farmers taking the lead in adopting the automobile as the accepted mode of transportation. More and more farms today are being powered by internal combustion engines. Where farming was once a life of sheerest drudgery, power farming has made it a dignified business with machinery today performing the many tasks that had to be done by hand yesterday. For example, in truck gardening a single machine now sets out more celery and other plants in a single day than would have been possible by hand in three weeks a decade ago. Vast fields are plowed, planted, cultivated and harvested with the driver riding the seat of a tractor in every operation.

Putting up hay was once regarded as the hardest job of all on the farm, coming as it did in mid-summer heat and with most of the work being done by hand. No longer is it necessary for the farmer to fork his windrowed hay into cocks, toss it onto a wagon by pitchfork and later perform the same operation in unloading it from the wagon to the stifling heat of the haymow. Instead a baler automatically gathers up the hay, and an elevator lifts the bales into the haymow, no more physical effort being involved in the entire process than merely that of pressing simple levers. In the same mechanized way silos are filled, feed ground and water for house and barn pumped—all by gasoline power or electric motor.

Not only in their daily living, but in more important phases the motor car has broadened the horizons of the men who feed the nation. Primitive medical service was once the scourge of every farmstead. It hasn't been so long since it was a pretty serious matter for a farmer or any member of his family to get sick. Making the rounds by horse and buggy, the country doctor might be delayed by hours on even an emergency call. If the illness turned out to be something like acute appendicitis, the doctor promptly set to work on a kitchen-table operation. Even pneumonia patients were usually cared for at the farm home because of the difficulty in getting them to a distant hospital. As a consequence, the mortality rate from illnesses in the country was much higher than today, now that the motor car has placed the doctor within a few minutes of all but the remotest farm houses. A hospital now is no farther from a patient's home than the few minutes it takes an ambulance to hurry forth on its errand of mercy.

Hard surfaced highways and modern motor power have steadily centralized rural living. Stores are fewer, but they are better. There are fewer, but better schools—also hospitals, churches and places of wholesome entertainment. Daily contacts have multiplied, ending the isolation and loneliness of a generation ago.

PIONEER DAYS PASS

Truly a pioneer phase of American life is passing. There may be nostalgic memories of the little red schoolhouse, the smell of hickory logs on the fire, the gabfests as neighbors gathered about the kitchen fire. But however alluringly disguised by memory, it was a hard life that brought little time for living, little chance to enjoy the fruits of human effort.

Those who have come after the pioneers salute the fine men and women who broke the sod, tilled the field, built their homes—all in the hardest possible way. But today all rejoice that life has been made easier and living enriched, even in the most remote countryside—all because the internal combustion engine, with its mechanical multiplication of the power potential of human and animal muscle, has gone out into the rural areas of the nation on rubber-tired wheels and has made life in the country more attractive than it has ever been before.

Economy Tractor Meets Farm Need

A four-wheel, rubber-tired, 675 lb. powerful economy tractor which sells for less than $500 is now being manufactured by Engineering Products Co., Milwaukee, Wis., in response to what the company felt was a very definite need for this type of tractor.

The Economy Tractor was designed by James E. Turner, president of the company, before his five years' military service as a lieutenant colonel in the Army Air Forces. After his discharge in December, 1945, Mr. Turner organized the Engineering Products Co., and actual sales started in May, 1946.

The tractor is powered by a four cycle, single cylinder, six hp. engine, said to be equivalent to a 60 hp. motor running at automobile speed. It has an automotive type transmission with three speeds forward and reverse, an automotive type differential mounted on Timken roller bearings and an automotive type disk clutch so that it can be driven like an automobile.

Accessories for the tractor, also designed by Mr. Turner, are available. A cultivator, designed primarily for straddle-row work, may also be used for between-the-row cultivation. A ten-in. plow and ten-in. rolling coulter are also available, which will plow five to eight in. deep in first and second gear, depending on the type of soil. A combination bulldozer and snow plow, a spike tooth harrow, a disk harrow, cutter bar and lawn mower are planned for use in connection with the tractor.

Dealer and sales arrangements are now being made to make the tractor available throughout the nation, Mr. Turner states.

Hydraulic Power Units

The George D. Roper Corp., Rockford, Ill., announces a new hydraulic power unit for farm tractors which operates manure loaders, hay loaders, cultivators, postpullers, plows, scrapers, etc.

The compact, self-contained "power package" is designed for universal application on farm tractors, and is known as the "Roper-Pac". The unit is built for either clockwise or counter-clockwise rotation and is reversible in the field. The controls are simple and easy-operating and a valve prevents overloading and overheating.

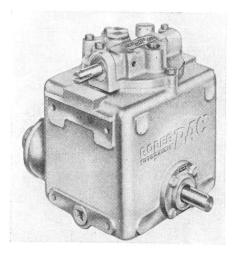

HARVEST BRIGADE

One man is all that is required to operate the harvester for small grain that moves by its own power.

The self-propelled combine introduced just prior to the war led to a great food-saving trek through the grain fields of the West.

ONE of the most practical and spectacular episodes in the age-long history of grain harvesting was the performance of the Massey-Harris Harvest Brigade during the summer of 1944, when 500 red, white and blue self-propelled combines streaked their way across the western half of the United States from the Mexican to the Canadian borders to harvest more than a million acres of grain.

FOOD IMPORTANT

In the midst of the global war, with millions of men in the fighting forces and hundreds of millions of persons throughout the world facing starvation, nothing was of greater importance than food. The billowing grain fields from the Mississippi to the Pacific were promising bounteous quantities of the grain so badly need-ed for the bread of life. But thousands of farmers who ordinarily look after the harvest were in the Armed Services, harvesting equipment was in poor repair and thousands of the combines formerly used in Texas, Oklahoma and Kansas had been dissipated over the map in the "great combine migration" of the year before.

With the foresight inherent from forty years of experience in pioneering combines on the western plains of the United States and Canada's prairie provinces, Massey-Harris sensed the impending emergency. Nothing could have been more tragic than the loss of an abundant crop through inability to harvest. To Joe Tucker, Massey-Harris' vice-president, it was a challenge. Fresh from war-time service on the war production boards of both the United States and Canada, and with the confidence that comes from conviction, he was able to obtain a concession of sufficient material with which the 500 combines could be produced. In return he pledged the harvest of a million acres with a minimum of manpower, and with economy in operation.

Sales of the 14-ft. self-propelled units were restricted to experienced operators each of whom agreed to accept delivery at the far southern limits of the grain belt where cutting could begin at the earliest possible moment and also to harvest a minimum of 2,000 acres.

ORGANIZATION

Backed by an efficient organization, with "Brigadier-General" Joe Tucker in charge, with branch managers as "colonels" and down through the Massey-Harris dealers and service men ranks, with each playing a prominent part in the program, the brigade swung into action in May in the flax fields near Corpus Christi and California's Imperial Valley. The help of national, state and county governmental agencies was enthusiastically extended to the brigade, with the result that practical cutting itineraries were worked out for each operator and many local farmers were given badly needed help. The progress of the brigade made good newspaper copy along the route, radio stations kept farmers thoroughly informed of the brigade's progress through their areas.

WEATHER DELAYS

Rains of torrential proportions necessitated delays in the northward movement to the wheat fields of Oklahoma and the Texas Panhandle. Grain was beaten down, and muddy fields made traction difficult. Then all the grain through the area ripened at once, and wheat on thousands of acres could have been shattered before it was harvested had not the brigade been on the job. The colorful self-propelleds worked northward through Kansas, Colorado and Nebraska in July and were in the northern states in August.

The detachment starting in the Imperial Valley likewise worked northward through California, harvesting varied crops including lettuce and rice and 120-bushel barley. Still another detachment harvested a substantial portion of the crop in the Pacific Northwest.

When the results of the season's harvest were summarized, it was found that 50 combines had harvested more than 2,500 acres; 40 had passed the 3,000-acre mark, nine had exceeded 4,000 acres and two had upward of 5,000 acres to their credit.

From the standpoint of agricultural economics some of the results of the harvest were of unusual significance. For instance, the harvest of more than a million acres had been accomplished with an expenditure of less than a quarter-million man-hours, 245,519 to be exact. This was less than 15 minutes an acre. The average number of acres per ten-hour day was 40.37. Obviously, this conservation of manpower was a distinct war-time advantage. The low manpower requirement was due, of course, to the fact that only one man is required with the self-propelled with its 14 ft. width of cut.

FREED TRACTORS

It is also interesting to note that the use of the 500 self-propelled combines freed an estimated 625 tractors for other uses, such as stubble plowing immediately after harvest or in harvesting additional acres which otherwise would have not been harvested at the proper time.

Another advantage evident from the summarized results was the saving of an extra 500,000 bushels of grain. This represents the difference of a half-bushel an acre in opening a field, which the self-propelled saves, but which is crushed beyond recovery by a tractor and pulled combine.

FURTHER SERVICE

So outstanding was the success of the Massey-Harris Harvest Brigade that it was repeated again in 1945. In that year because of the great performance of the combines in 1944, the company was enabled to increase its production to 750, which were delivered throughout the small grain states and with the 500 from the previous brigade campaign greatly expanded the harvested acreage. Again in 1946, the self-propelleds rendered outstanding service in a "Famine-Fighter" program by which they opened thousands of fields for other harvesting units, saving the losses in-

The M-H self-propelled combine (upper left) at work in a wheat field in the west had a number of illustrious forebears: The No. 9 combine (upper right) was built in 1922 and is said to have been the first engine model. Other forerunners were the No. 10 binder (lower left) and the No. 11 combine (lower right), both of which had wide acceptance.

cidental to their opening operations in addition to handling a substantial acreage in harvesting complete fields.

The two Harvest Brigade campaigns were successful because they were based upon the use of self-propelled units, with their savings in both manpower and engine power requirements. The self-propelleds operated by the brigades were the result of more than 40 years of intimate experience in developing harvesting units to meet the needs of the grain growers in the plains states. At the turn of the century, combines were recognized as efficient machines for the Pacific Coast areas where the grain can remain in the field without shattering until moisture content reaches the absolute minimum. The belief was general that their use in the more humid plains east of the Rockies would be impractical.

Harvesting equipment with efficiency and economy comparable to those used on the western slope was a challenge which Massey-Harris harvester engineers accepted nearly a half-century ago. In 1901 it introduced the Massey-Harris stripper, which merely stripped the heads from the standing grain and which derived its power from the ground wheels. This was followed in 1912 by the company's No. 2 stripper, a traction-powered unit, more on the order of a conventional combine, which was drawn by horses. Both units were moderately successful in harvesting, provided considerable economy over prevailing methods of the binder or header and subsequent separation by threshers.

The No. 2 with some of the features of a conventional combine led to the development of the No. 9 Massey-Harris reaper-thresher in 1922, built in 12 and 15 ft. sizes, which became immediately popular. It is said to have been the first engine-model, and was drawn by either horses or tractors. It became quite popular throughout the Plains states and Massey-Harris became one of the largest producers of combines for that territory.

The Massey-Harris No. 14, a 16-ft. prairie type reaper-thresher was introduced in 1936. This was engine-driven and was tractor-drawn, except for special horse-drawn arrangement provided for shipments to South America. This was followed a year later with the No. 14, built with 6 and 8 ft. cuts and designed for operation with power take-off. Both of these models were unusually popular and enjoyed wide distribution through all the western states.

The Massey-Harris Clipper, first introduced in 1936, was an outstanding development which doubtless will long stand as a landmark in combine design. This was the first "straight through" model, the cut grain and straw being carried without directional change straight through the separation chambers instead of being directed in a right angle turn from the cutting platform. This was first built only in a 6-ft. size, but the cutting width was subsequently extended to 7 ft. the "straight through" design has been widely adopted by the industry.

Another history-making development by Massey-Harris has been the self-propelled combine, in which the engine not only operates the cutting and separating mechanisms but provides power to the wheels, obviating the necessity for using a tractor for traction power. The first Massey-Harris self-propelled was the No. 20, with a 16-ft. cut, made available to the trade in 1939. This was followed in 1940 with a self-propelled Clipper in the 7-ft. size, which found a quick popularity in the soybean areas, as well as on the small grain farms.

The No. 21 self-propelled was produced in 1940-41, just ahead of America's entrance into the World War. It was a 12-ft. unit, but later was enlarged to a 14-ft. cut. This was the model which was used successfully in the Harvest Brigades and which because of its performance in those his-toric harvesting campaigns has become widely popular wherever grain is grown.

R.E.A. headquarters returned to Washington D.C. following its relocation to St. Louis, Missouri, during the war and Goodyear began producing "Zerosafe" freezers in 1946. Farm freezers were thought to be an important step forward for the new electrified farms. The Food Machinery Corp. purchased Bolens Products Co. and the Clinton Machine Co. introduced their new four-stroke, air-cooled engine designed for garden tractors, lawn mowers, generators, pumps and lighting plants. Plexiglas, "the versatile plastic used for bomber noses during the war," was soon used for many peace-time applications. The first national meeting of Flying Farmers, 248 people from 17 states, met in Stillwater, Oklahoma, in mid-summer. Earlier, on February 4, 376 air-minded farmers and ranchers organized the Nebraska Flying Farmers Club at the University Student Union Building in Lincoln. Flying was thought to be the trend of the future, especially for those managing large tracts of farm or range land. Families with incomes of more than $5,000 per year increased from 2 percent in 1938 to 8 percent in 1946.

Predicts Walkie-Talkies For American Farms

Many persons closely associated with radio are predicting that the "walkie-talkie", small portable radio produced in large numbers for the Armed Forces during the war, will make a big hit with the nation's farmers as soon as it can be produced in sufficient quantities at a reasonable cost.

Only one minor change is indicated as probable in the farm walkie-talkie. Storage batteries are more practical than short-lived dry cells, therefore the unit will be heavier. Walkie-talkie on the farm is more likely to be carried by farm equipment than by a person who is on foot. However, the lighter, more portable dry cell set may be kept around for occasional use.

A few of the more than 100 planes flown to Lincoln for the Flying Farmer Day as they were parked at one of the ports used by the flying cowboys.

The spring of 1947 was one of the wettest in history. Corn and oats were planted late or not at all and the hay crop was ruined. The winter wheat crop was, however, the biggest in history. Migratory combines that began to harvest the wheat in Texas continued north through Oklahoma, Kansas, Nebraska, South Dakota and into North Dakota. The new type of big harvesting business had a continuous season from mid spring to early fall following the ripening grain north.

Henry Ford died at the age of 83 in 1947. John Deere began manufacturing tractors at its new plant near Dubuque, Iowa. The first Model "M" was announced in May. Deere also announced the "Roll-O-Matic" knee action tricycle front wheels. Ford began production of the "8N" model, which remained in production until 1952. International Harvester began volume production of the "Cub" tractor which continued through 1979, making it the model that was in constant production longer than any other. Massey-Harris celebrated the company's centennial and late in the year introduced the "Pony."

Model Cub Sparks Sales Drive

Sparking International Harvester's sales drive on its new Farmall Cub, the newest member of the famous Farmall tractor family, is the 1:16 scale plastic model recently developed by Design Fabricators, Inc., Chicago, manufacturing division of McStay Jackson Co., industrial designers. A perfect miniature of the new Cub,

The Cub model is assembled from 21 separate pieces included in each package.

even to the Harvester red color with IH emblem and molded rubber tires, the little tractor has a world of appeal for youngsters, with the added interest created for boys of all ages who like to assemble finished models from component parts. Packaged knocked-down in 21 separate pieces, it is easily assembled by following simple directions printed on the container.

McCulloch Completes West Coast Move

Mass production of gasoline engines in a new plant adjacent to the Los Angeles airport was announced recently by Robert P. McCulloch, president of McCulloch Motors Corp. The company was originally established in Milwaukee, Wis., as McCulloch Aviation, Inc., and began its move to the West Coast in January, 1946. The change in name, indicating a wider scope of products, occurred in October.

1946-1951

Farm Co-operatives attempted to gain a tax advantage for their members. Main line farm machinery dealers, who were threatened by the introduction of larger, more expensive equipment such as tractors, corn pickers, manure spreaders and loaders by the large National Farm Machinery Cooperative, worked to block the proposed tax revisions.

Harry Ferguson, Inc. of Detroit, Michigan, purchased a plant in Cleveland, Ohio, to manufacture the newly designed tractor that incorporated the "Ferguson System" of hydraulic control. Power saws, which used a special cutting chain and were light enough to be operated by one man, were an interesting new product line.

Automotive assets of Graham-Paige Motors Corp. were sold to the Kaiser-Frazer Corp. The Continental Motors Corp. annouced a new 1.5 hp single cylinder engine known as the "AA7," Onan introduced the "CK" model and Ingersoll-Rand introduced their electric impact wrench for mechanics.

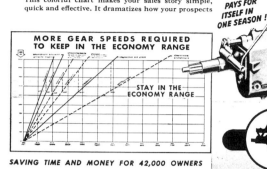

And Now a...
SELF-PROPELLED CORN PICKER

The simple straight-through design of the Clipper combines has been applied to the corn picker. Every operation follows in a straight line from the snapping unit to the shock rolls and on through the elevator to the hauling unit.

Introducing the machine which is in regular production at present that turned in excellent performances harvesting several thousand acres through the Corn Belt last year.

HAVING pioneered the self-propelled combine as far back as 1938, and having been a major contributor to its general adoption as one of the most practical machines for harvesting small grain, it was but natural that Massey-Harris should explore the possibilities of applying self-propulsion to other major farm operations. As a result the self-propelled Massey-Harris two-row corn picker is now in regular production and has given a good account of itself

1946-1951

in harvesting several hundred thousand acres throughout the Corn Belt during the past year.

STRAIGHT-THROUGH DESIGN

The self-propelled corn picker embodies the simple straight-through design which has made the Clipper combines popular. Every operation follows in a straight line, from the snapping unit, to the shocking rolls and on through the elevator to the hauling unit. This straight-through movement obviously greatly reduces the possibilities for jamming, bunching or other similar trouble.

Also outstanding in the design is the substantial frame and balanced distribution of weight. The operator's seat is well located to provide complete vision of the entire operation, and all controls are within easy reach. Maximum safety as well as maximum convenience is provided by this arrangement of seats and controls. At no time is the operator compelled to reach over or near the rolls or any other moving parts, and the picker is undoubtedly as free from personal danger hazard as it could possibly be made. At least 100 per cent safety has been the objective of its two safety-minded designers, Ed and Verne Everett, the well known father-and-son team of Massey-Harris engineers.

OPERATION

In operation, the picker can be started without down rows on any two rows in the field. It will operate under difficult traction conditions and up to a speed of 8 m.p.h. The weight is evenly distributed with two traction wheels carrying most of the load and with just enough weight on the rear wheels to facilitate steering. The picker follows the rows easily, as the design virtually eliminates side draft. Maneuverability is said to be practically as easy in contoured as in straight rows.

The corn picker is powered by the same engine used in the new Model 30 tractor, which is a 4-cylinder L-head unit, with a bore of $3\frac{7}{16}$ in. and stroke of $4\frac{3}{8}$, which operates at 1500 r.p.m. The engine has a 162 cu. in. displacement and a 6 to 1 compression ratio; also is equipped with self-starter.

MORE SPECIFICATIONS

The snapping unit includes two upper and one lower gathering chains. The rolls are spaced 39 in. apart and are adjustable for different sizes of stalks. They will handle rows of 38 to 42 in. spacing. A power lift within easy reach of the operator raises or lowers the gathering points.

Ten 45-in. rolls which are found in the husking unit provide an unusual husking capacity. A total of 130 pegs are installed at the factory and 80 additional pegs are provided. The front wheels are provided with 9 x 24 rubber tires while the rear wheels are similarly equipped with 5.50 x 16 in. tires. The swinging drawbar may be latched in the center or permitted to swing free, the latch being controlled from the operator's platform. The elevator has a sizeable hopper at its lower end for corn storage when elevator is stopped at the turn or when changing wagons. An adjustable chute at the upper end of the elevator permits the operator to distribute the load in the wagon without moving from his driving position.

The picker is said to operate under difficult traction conditions up to 8 m.p.h. The weight is evenly distributed with two traction wheels carrying most of the load and with just enough weight on the rear wheels to facilitate steering. Maneuverability is said to be nearly as easy on contoured as straight rows.

Goodyear announced the use of Nylon in their new "Super" truck tire, later they demonstrated their "LifeGuard" safety tubes. The "LifeGuard" tubes had two air chambers and the inner chamber was formed by an inner 4-ply cord "tire." The inner chamber could deflate slowly through two small holes, but gave plenty of time to slow down in a controlled straight line stop. Firestone began building tires for special heavy duty industrial applications using rubberized wire cord instead of cotton or rayon cords. The B.F. Goodrich Co. developed tubeless tires and began offering them on a limited basis. The United States Rubber Co. announced the return to production of its puncture-sealing inner tube.

"Yes, we finally found a nice little one room place."

Alan Ladd and Dorothy Lamour are stars in a film "Wild Harvest" soon to be released by Paramount which deals with wheat harvest in Kansas. Massey-Harris self-propelleds are the machines used to typify today's harvesters.

AGRICULTURAL DILEMMA

in the

PHILIPPINES

By Manuel Buaken

SOME people think this is a land of shortages. They are unhappy because for a few days or weeks they had to go without meat, and perhaps have had to wait a few months for a new car. These people of the United States are living in a Paradise of abundance . . . they should consider for a moment the plight of another defender of freedom.

My homeland, the Philippines, was a land of plenty, of hospitality and feasting. It is a land of nothing now. When I was there as a member of the United States Army,—and now at this later date, when I have just received a letter from my home province—the picture was and is the same. There is no livestock. Chickens, pigs, horses, cows, carabaos— the Japanese murdered them all. The people have not eaten meat for three years. The carabao was the draft animal of the Philippines; there can be no planting until the stock is replenished or modern farm machinery substituted. The Japanese took all rice, corn, peanuts; they destroyed all producing trees such as coconuts, mangoes, coffee, cacao. They took all sugar supplies. Our people have lived on weeds, on bananas, which grow like weeds, and on green leaves, with occasional catches of fish, where the Japanese had not poisoned the rivers.

A recent letter from my home

No cooking pots, no seed rice, no plows, no able bodied man is the plight of these war ravaged people who also fought for freedom.

town of Santa Cruz, Ilocos Sur, informed me there were no cooking pots of any kind, no seed rice, no plows, and no able-bodied men—and this is typical of every small town of the Philippines, outside of politics-favored Manila.

Brother, can you spare a plow? As a bare minimum for the re-establishment of production, every province of the 48 units of the Philippines needs 10,000 walking plows, 700 peg tooth harrows, 300 grain planters, 500 cultivators, 500 corn shellers, 300 cane mills, 100 rice hullers, 1000 gasoline engines and 50 power units. Add to this list 3,500,000 working bolos—Filipino tool of all work—plus thousands of axes, shovels, crowbars, hatchets, rakes, files, saws, hammers.

The Japanese army in the Philippines was equipped with cheap cotton uniforms, and flimsy shoes, of the poorer quality tennis-wear type. When these wore out, they looted the population of the clothes they wore. So there is in the Philippines no cloth, no buttons, no thread.

The wholly irrational character of Japanese mentality is well shown by the fact that in many cases, they billeted soldiers in wooden-roofed caves, and used well-built homes with hardwood floors as out-houses. People who get enthusiastic about the democratization of Japan should

keep that picture in mind.

The Philippines needs goods. And we can pay well. The Luzon hills are full of gold, much of our minted gold was hidden away from the Japanese, either taken away by submarine, or buried. Our people had plenty of experience, in 300 years of profiteering Spanish occupation, in the caching of gold, so that they knew how to hide our money resources. Our government itself dumped $8,500,000 in coins into Manila Bay during the first days of the Japanese invasion, to prevent its falling into enemy hands. This has now been salvaged for us by the U. S. Army, in accordance with a promise made to us then by General MacArthur. War damage indemnities are bringing us cash; so are United States government loans. Morris Pendleton, of the Los Angeles Chamber of Commerce, who has recently returned from the Philippines, stated in a speech that the credit resources of the Philippines are now exceptionally good, and that importers there can offer security to any one who will ship needed goods to them.

WAR WRECKAGE

We in the Philippines had a double exposure to high power wreckage. Railroads, bus lines, telephone and power installations, modern bridges, modern roads, schools, hospitals, industries—all are gone. We have the support of the U. S. government, as provided by the Millard Tydings bill, to the tune of $480,000,000, for a starter on this long range reconstruction. For this we need American technicians, capital goods, and *goods*.

We need Americans to bid for these government contracts to rebuild our denuded roads, restore blasted bridges, begin construction of the most necessary public edifices, such as the Customs House, Philippine Legislature building, provincial hospitals, water and power dams, rebuild Manila Railroad, etc. We need contractors to handle lumber, iron, cement, bricks, roofing, plastering, plumbing, etc.

Yearly up to the time the Japs came to the Islands, the Philippines bought $4,440,705 worth of drugs and medicines to supply the cabinets of drugstores all over the Island. All these stores were burned to the ground—stocks and buildings alike must be restored. Philippine hospitals must also be completely re-equipped.

In the past, we not only bought our capital goods from you—we ate our daily bread as you serve it up. As Robert A. Smith, writing in *Pic* magazine early in 1941, said: "The bulk of the Filipino population is made up of alert and progressive Christian Filipinos who wear cotton from the mills of Massachusetts and Alabama, smoke cigarettes from Richmond, use electric light bulbs from Schenectady, play radios from Indiana, fasten up the baby with safety pins from Connecticut and put him in the back seat of a car made in Detroit that rolls down a road made of Portland cement, on tires made in Chicago." In other words, we took your thinking and living and made them our own. It is one world! *Brother, can you spare a plow*, some seed, and a cultivator to boot?

AMERICAN INVESTMENT

In the past, far-sighted Americans went to the Philippines, took their capital goods with them, and carved themselves a fortune. Multi-millionaire John Hauserman was of this class. He used to have $454,365.10 average monthly net profit from his Consolidated Benguet Mines. The sum of $2,500 invested in that mine at its organization in 1912, was paid off by 1932 to the tune of 270 per cent. Of course, all that was destroyed, but the boundless resources of the Philippines, and the welcome for Americans, remain.

You will not find obstructionist legal barriers against you. We ask you for fair play, that is all. You will find that taxes are low and the people friendly. Our money is standardized at two pesos for one American dollar, backed by adequate reserves. English is spoken and understood everywhere. American films and American radios have made us at home with your ways.

Our port of Manila was doubled in capacity as a result of necessary war shipping, and its pier facilities are nearly rebuilt now. Commercial banking, with U. S. dollars, has been restored in Manila. Commercial communications are back in operation, so that Manila is the one nerve center alive in Far Eastern commerce. The Manila Stock Exchange has been reopened.

Manila is to be the air capital of the Far East, according to recent announcement by TWA. Pan-American is resuming weekly flights from Manila to the United States.

American veterans will find they can homestead lands in the Philippines on an equal basis with Philippine citizens. Our constitution guarantees that right.

The Philippines needs and wants American business men. The laws which protect us, which have protected our resources from greed and exploitation, are American made. This demonstration of the American spirit of fair play is what made us blood brothers to you on a thousand contested beach-heads, kept us alive in hope in the midst of famine and horror, war and atrocity.

We need two million carabaos—but American inventiveness, aided and abetted by Filipino ingenuity, could adapt some of your farming machinery to conditions of heavy rainfall agriculture.

The Jap is very artistic—he paints a lacquer box very nicely! But he never paints a house! In the Philippines, we paint and impregnate with insecticide and paint again, or the termites eat and eat!

(Left) Here he comes . . . there he goes. Bunting over round bales of hay is done with a motorcycle by Lorry Tendler, supervisor, Allis-Chalmers Tractor Div., proving ground near Elm Grove, Wis. The bales are those made by the new Allis-Chalmers one-man baler (Above) that rolls hay into cylinders instead of the conventional rectangles. The thatched effect of the rools is said to keep the hay dry in the rain except the underside which lies on the wet ground. Tendler discovered that the tedious job of turning the bales by hand can be handled expeditiously by bumping them with the crash bar of a motorcycle. He demonstrates his method (above) by turning over 1045 bales in 38 minutes, it is reported.

The Taft-Hartley Labor Management Relations Act was passed. Half of all farm families had electricity due in part to the "REA." The New Holland Machine Co. was purchased by Sperry Corporation. International Harvester introduced "the most powerful" crawler tractor available; the TD-24.

Implement and Tractor introduced the *I&T Shop Book* service manual to help mechanics fix the large number of tractors that needed repaired. Many of the tractors had been built before 1942 and were well worn by the end of the war. Neither production nor distribution could catch up with demand for quite a while after the ending of hostilities. The *I&T Shop Book* also included the industry's first "Flat Rate Pricing Guide."

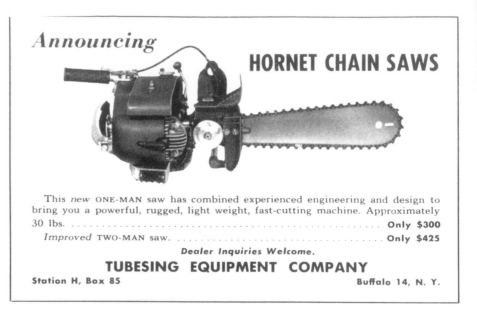

Successful operation of tires with 3½ in. bar heights in the rice and cane fields has led to experimental work in other areas where soils presented unusual traction and clogging problems.

FTC Ruling in Jeep Case

Claims as to the origin of the automotive unit commonly known as "Jeep" were involved in an action instituted by the Federal Trade Commission, May 6, 1943, and ended with a commission ruling, Feb. 27, 1948.

The action was filed against the Willys-Overland Motors Corp., its advertising agency and certain officials of both companies, and was based on certain statements made which, in the opinion of the commission, claimed for the defendant manufacturing company the exclusive credit for the Jeep's origin.

The findings of the commission indicated that these two companies, as well as the American Bantam Car Co. and the Ford Motor Co., each had cooperated with the U. S. Army and its engineers in developing military automotive units which had at various times been known as Jeeps, but that none was entitled to exclusive credit. While the action was in progress, the Minneapolis-Moline Power Implement Co., upon its request, was permitted to intervene in the hearing. The commission's findings also took cognizance of that company's assistance in developing a tractor unit, also known as a Jeep, which units weighed approximately 6½ tons each, and of which 1,018 were sold to the Army and Navy during the war.

In concluding its findings, the commission stated that the acts cited constituted unfair methods of competition in commerce within the intent and meaning of the Federal Trade Commission Act. The commission however made no finding as to the relative rights of the intervenor or the Willys-Overland company to apply the Jeep name to their respective products, such issue not having been raised in the commission's complaint and being essentially a private controversy between the two companies.

SOUTH BEND LATHES

Quality-Built for Quality Work

16″ x 6′ TOOLROOM LATHE $1958

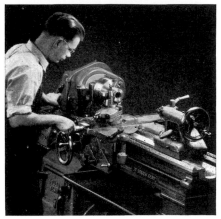

13″ x 6′ TOOLROOM LATHE $1456

9″ x 3′ MODEL A BENCH LATHE $250

Farm equipment can be serviced better when South Bend Lathes are used. They are ideal for repair shops of all sizes. These dependable lathes have accuracy that permits the most exacting operations. They are versatile...a greater variety of jobs can be handled. And they are exceedingly easy to operate.

Your machine shop needs South Bend Lathes. They will enable you to give your customers better service. A greater variety of jobs can be handled... and original factory tolerances can be duplicated. South Bend Lathes are made with 9″, 10″, 13″, 14½″, 16″, and 16/24″ swings—a size for every need. Write for Catalog. State size lathe you are interested in.

PROMPT DELIVERY on all sizes of South Bend Lathes, chucks, tools, and attachments.

TIME PAYMENT PLAN — 25 per cent down, 12 months to pay balance. Moderate finance charge.

BENCH LATHE PRICES

9″ x 3′ Model C, 6 speed	$145.00
9″ x 3′ Model B, 6 speed	195.00
9″ x 3′ Model A, 12 speed	270.00
10″ x 3′—1″ Collet Capacity Quick Change Lathe	881.00

FLOOR LATHE PRICES

13″ x 6′ Quick Change Lathe	$1126.00
14½″ x 5′ Quick Change Lathe	1309.00
14½″ x 5′ Toolroom Lathe	1662.00
16″ x 6′ Quick Change Lathe	1567.00

All prices f. o. b. factory, less electrical equipment. 9″ Bench Lathe prices quoted less bench.

SOUTH BEND LATHE WORKS

BUILDING BETTER LATHES SINCE 1906 • 542 EAST MADISON STREET, SOUTH BEND 22, INDIANA

Allis-Chalmers Introduces a New Design in Tractors

In 1948 Allis-Chalmers introduced the "G" tractor. The unusual style with round tube frame and rear mounted engine continued in production until 1955. The model was welcomed by truck farmers and similar models were still being produced by other manufacturers. Allis-Chalmers also introduced the "WD" tractor in 1948 and, though similar to the "WC", the new model had an integrated hydraulic system. Production was continued until 1953 when it was replaced in the line by the "WD45." Harry Ferguson filed suit against Ford because of patent infringement. Ford was required to pay $9.25 million in 1952. The Soviet Union blockaded the roads to West Berlin on June 24, 1948. The air lift conducted by the United States supplied over 2 million tons of needed material for the next 16 months until the blockade was lifted. Massey-Harris bought the Goble Disc Works of Fowler, California, Minneapolis-Moline purchased B.F. Avery and Sons of Louisville, Kentucky, and New Holland purchased the Dellinger Mfg. Co. of Lancaster, Pennsylvania. Inflation was one of the country's greatest problems and the government continued to fight the rapidly rising costs. Europe needed both food and farm equipment from the farmers and manufacturers in the U.S., who both seized the opportunity to meet Europe's needs and to increase profits. The Food and Agriculture Organization was the first of the permanent new United Nations organizations launched to help feed hungry people around the world. The British pound sterling was reduced in value from $4.03 to $2.80 late in 1949, threatening the loss of foreign markets.

Bonney Forge Offers New Tool

The Bonney Forge & Tool Works, Allentown, Pa., has announced a new tool called the Wobble Drive Extension, which is designed for work in close quarters. This tool enables a mechanic to obtain a small amount of swivel and to work on a bolt or nut with the extension leaning at an angle of approximately 12 deg. in any direction without the use of a universal joint.

The male end is a ⅜ in. wobble drive which will receive any ⅜ in. drive socket. The other end is a female ⅜ in. square drive for attaching a ratchet, hinge handle, speeder or "T" handle.

This tool, on which the patent is pending, is made in two sizes: No. AT4 6 in. long and No. AT5 12 in. long.

Work in close quarters is handled easily by this new tool.

1946-1951

361

Good Pal Offers Counter Display

A new, colorful counter display for its battery ignition unit is being offered dealers by the Good Pal Tractor Battery Ignition Co., Napoleon, Ohio. The unit, featuring a Mallory distributor and ignition coil, is easily installed in place of the magneto on many makes and models of farm tractors, the company reports. Easier starting and greater power with less fuel consumption are claimed for the units, which are completely waterproof and equipped with automatic spark advance and retard. The coil is oil sealed.

Good Pal Mallory units are available for Allis-Chalmers Models B, C, WC and WF; Farmall Models A, AV, B, H and M; McCormick-Deering Models W-4 and OS-4; International 1-4; all Oliver 70s; Case Models D, DC, DO (after serial No. 4607033), S, SC, SI and SO; and John Deere Models A, B, G and D. Similar units for other makes and models will be made available soon.

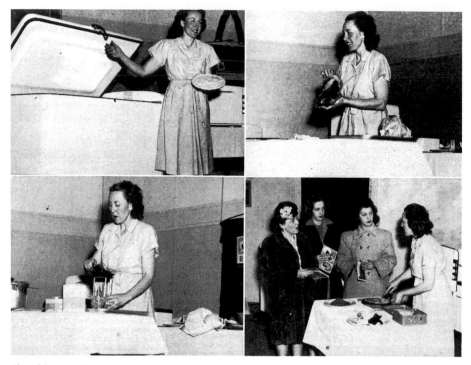

(Above left) Mrs. Belmore removes a frozen pie preparatory to baking it and pauses to explain that the lid of the freezer is counterbalanced—thus eliminating the chance for mashed fingers. *(Above right)* Proper preparation of frozen meats is explained. *(Lower left)* After vegetables that are to be frozen have been blanched in hot water, they are packed for deep freezing. *(Lower right)* Following the demonstration, guests sample the food. Note that Mrs. Belmore is cutting the frozen pie that has been baked.

Conservation took on national importance and soil erosion from wind and rain was fought by farmers and city dwellers. The "Dust Bowl" had to be irrigated to ensure profitable harvests during drought years and the amount of irrigation equipment used constantly increased. Irrigated land in Kansas increased from 248,067 acres in 1949 to 537,566 acres in 1955 and to 1,020,573 acres in 1959. It was also noted that the cheapest available source of nitrogen plant food was the use of anhydrous ammonia fertilizer.

Offers One-Man Chain Saw

A powerful, light weight one-man chain saw which operates with equal efficiency in any cutting position, has been announced by Henry Disston & Sons, Inc., Philadelphia, Pa.

Especially useful for light felling and bucking, for limbing and construction timber cutting, the saw has an 18-in. cutting capacity, and a 2-cycle, air-cooled Mercury gasoline engine delivering 3½ hp. at 4000 r.p.m. Full precision bearing construction insures trouble-free operation at any speed. A specially designed fuel system permits operation at top efficiency even when the saw is used upside down for inverted bucking. A fuel meter assures positive flow of properly mixed fuel to the cylinder, at the same time functioning as a governor to control engine speed.

Like its two-man older brother, the Disston one-man chain saw has a built-in automatic chain lubricator, which has only one moving part—a flexible pickup tube inside the oil reservoir. Starting is made easy by the self-rewinding Magnapull starter, which is built into the engine, and a liberal fuel capacity provides wide working limits on the operator's time.

Control of the new one-man chain saw is concentrated in a well-designed

pistol grip handle. The throttle is at the operator's finger tip, with the fuel mixture control lever nearby. A handy priming valve is provided for easy starting in extreme cold, or when the engine has been run dry. A squeeze on the handle disengages the clutch, and releasing a safety catch re-engages it.

Spring on the Farm—

and NEW IH Equipment for Modern Farming

New Farmall C, equipped with FARMALL TOUCH-CONTROL . . . one of five all-purpose tractors with matched machines for every size farm, and for every crop and soil condition.

Farmall Touch-Control is a *complete* implement control. Hydraulic power in both directions — power to lift implement from the ground and force it into the ground.

← New FARMALL SUPER-A tractor equipped with drill planter and fertilizer attachment. Farmall is a registered trade-mark. *Only* International Harvester builds Farmall Tractors.

Smallest Farmall — the new FARMALL CUB tractor, with cultivator. This is a great combination for fast cultivation.

↑ Speed up the hay harvest with the **new** No. 45 Pickup Baler. Self-feeding, fully automatic. Farmall H (or tractor of equivalent power) handles it nicely. No auxiliary engine needed.

← The new, small No. 4-E hammer mill to be powered by a 3, 5 or 7½ hp. electric motor or the International Cub Tractor. Handles all types of grains and feed.

It's Spring, 1948 . . . and new International Harvester Farm Machines are out in the fields, all over America.

What an array of new IH equipment it is! *Every machine is the leader in its field,* made by International Harvester, pace-setter in farm equipment manufacture. Every machine has been designed and built to make farm mechanization more complete and to bring additional time and labor-saving advantages to the family farm. These machines are as up-to-date as tomorrow. They fit today's *way of farming,* with the emphasis on *soil conservation* and *better land use.* They're made for simple, convenient one-man operation.

Your IH Dealer is the man to see about all that's new in IH Farm Equipment. Every effort will be made to provide you with the machines you need.

INTERNATIONAL HARVESTER COMPANY
180 North Michigan Avenue   Chicago 1, Illinois

INTERNATIONAL HARVESTER IH

Leader in Farm Equipment Progress

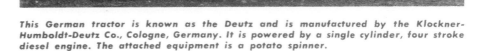

This German tractor is known as the Deutz and is manufactured by the Klockner-Humboldt-Deutz Co., Cologne, Germany. It is powered by a single cylinder, four stroke diesel engine. The attached equipment is a potato spinner.

The Lustron home, built in a factory with the resultant low-cost benefits, is said to be an engineered home far removed from the technique of hammer and nails.

To Distribute Pre-fab Houses Through Implement Dealers

FARM equipment dealers who have the capital and a desire to expand their business operations into other local fields are being accorded priority consideration as local dealers by the Lustron Corp., Columbus, Ohio, manufacturers of porcelain enameled metal houses.

This company has been in production on the new metal homes since the first of the year, having previously acquired and reequipped a sizeable portion of the large plant occupied throughout the war by Curtis-Wright and which was one of the largest producers of naval aircraft. The company is headed by Carl G. Strandlund, for many years identified with the engineering and production departments of the Oliver Corp. and who during the war years was engaged in military production activities. Incidental to his wartime work, he succeeded in perfecting a process for permanently fusing porcelain enamel to steel sheets. This process proved to have many wartime advantages for various units of military equipment, and is now employed in the production of the company's metal homes.

Associated with Mr. Strandlund as the corporation's senior vice-president is J. M. (Joe) Tucker, well-known throughout the farm equipment industry from his previous connections with both Oliver and Massey-Harris.

All Metal Structure

The Lustron house, now being produced by the company, is a 5-room structure, providing slightly more than 1,000 sq. ft. of floor space. All structural parts of the Lustron homes are of metal, with outer and inner walls of heavy gage cold rolled steel having enameled exposed surfaces and with fiber glass insulation between. The roof also is of metal. The houses are furnished complete with plumbing and electrical facilities, Venetian blinds, sink, combination electrical clothes and dish washer, mirrors, asphalt tiled floors, adequate closet space, complete bathroom fixtures including combination tub-shower, etc. Radiant panel heating of modern design is provided which is evenly distributed throughout the entire house. Windows have sturdy all-aluminum sashes, are easily operated by a crank-type handle for opening outward and screens are included. The houses contain living room, dinette, kitchen, master bedroom, second bedroom and a utility room. All interior doors slide into open and closed positions. Built-in facilities also include bookcase, china cabinet with "pass through" to kitchen, vanity cabinet, water heater, kitchen ventilator, etc. The buyer has only to provide his own range and refrigerator to make the housekeeping equipment complete.

". . . and if the roof should ever leak in a pre-fabricated house, just send it back to the factory."

(Above) The dining room as seen from the living room is 9 x 10 ft., has a built-in display cabinet for china and drawer space for crystal and other accessories.

"Once as a joke, I said, 'Would you do my laundry too?' Ten minutes later I had my shirt back—waxed and polished!"

Mass-Produced

Lustron houses are mass-produced with an assembly line efficiency which is outstanding and probably unmatched even in today's highly developed American industrial production methods. The efficiency prevails not only throughout the factory, but extends along the highways and continues through the final erection on the building lot.

The special trailers upon which the unassembled parts are loaded for shipment to the customer start at one end of the plant and are moved in a straight line along the assembly route by a conveyor system. As they progress through the plant they are loaded from both sides at each point of assembly, each part being placed in its special position and in reverse order from which they will be needed for assembly. Most parts are placed in a vertical position on the trailer, and are securely anchored and protected against injury or damage.

Reaching the end of the line, each trailer has been loaded with a complete home package and is ready for an immediate start to its individual building location destination. On the site likewise, construction can be started immediately and the various structural parts and sections are accessible in the exact order needed for assembling.

Also, it is claimed by the company, the houses can be produced at the factory, transported to the home buyer's lot, completely assembled and sold at about three-fourths of the cost for a similar sized home made from any other material. Added to this advantage, the company points out, the structures are proof against fire, vermin, termites, decay, etc.

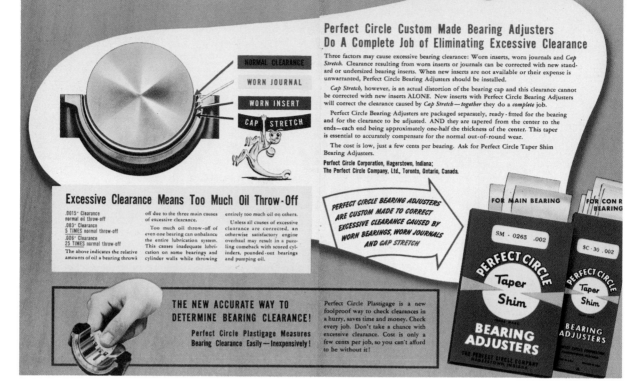

Peterson Announces New Vise-Grip Wrench

The Peterson Mfg. Co., DeWitt, Neb., announces the new model of its Vise-Grip wrench. It features curved jaws and knurled jaw tips and is made in two sizes, with or without wire cutter. According to recent tests, the company claims that the gripping power of the wrench is great enough to support a weight of more than 1½ tons.

The John Deere "R" was the first diesel model offered by the manufacturer known for its two cylinder tractors. The large two cylinder model was introduced in 1949 and continued in production for five years. Oliver introduced tractors with electrical control of the hydraulic system, known as "Hydra-Lectrics." Ferguson entered the U.S. market in 1949 with the TE20 and TO20 models.

The drawbar which provides nine horizontal positions on each of two levels handles virtually all types of pulled implements. This is a drawbar-type disk harrow.

The Korean War caused a labor shortage in some parts of the country during 1950 and 1951. Public Law 78, known as "bracero program permits" permitted Mexican laborers to enter the United States to help combat temporary labor shortage. Farmers, fearing a shortage because of the war, bought an unusually large amount of equipment in the second half of 1950.

The Lowther tractor of 1950 used Chrysler engines and fluid coupling similar to the torque converter used with modern automatic transmissions. The Willys-Overland 'Jeep' was Nebraska tested and implements normally attached to farm tractors were adapted to operation behind the 'Jeep.' The Ford Motor Co. of Detroit, Michigan, purchased the Woods Bros. Thresher Co. of Des Moines, Iowa, and Minneapolis-Moline introduced the "Uni-Harvestor" that was designed to have many self-propelled applications on the farm, including: combine, two-row corn picker, corn picker-sheller, wire-tying baler and forage harvester. L-P Gas was used as an alternate fuel to gasoline, for both new and used tractors.

In 1951, it was necessary to introduce Volume 2 of the *I&T Shop Book* to include repair information for later tractors and the more popular models of small gasoline engines. Prices of many products were frozen and many raw materials were in short supply again. Farmers and dealers were encouraged to salvage scrap metal and to repair usable equipment rather than discard old implements and tractors just because they were not the most modern. Caterpillar Tractor Co. purchased the Trackson Co. of Milwaukee, Wisconsin, that had begun in 1922, New Idea purchased the Horn Mfg. Co. of Fort Dodge, Iowa, and Minneapolis-Moline purchased the B.F. Avery Co.

"My cold war with Junior Higgins suddenly turned hot."

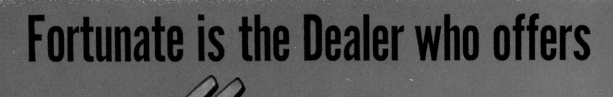

1946-1951

Must Liberty be SCRAPPED?

Not the statue . . . she'd be last to go.
But how about the things she stands for?

How secure is the freedom that has been *given* to us by those generations of Americans who sweated, froze, fought and died all the way from Bunker Hill to Bastogne—to Parallel 38? The foundation for the Bill of Rights and the freedoms it guarantees is only as strong as our determination to safeguard this priceless heritage.

Now, what can *you* do to keep alive the flame of Liberty—to keep it alive and *rekindle* it so that it might burn even stronger through the years ahead?

HERE'S ONE THING *YOU* CAN DO!

Today America is producing double—two assembly lines are going full blast. One line is building the materials necessary for the defense and preservation of our nation. The other is turning out everything from automobiles to arc lamps, from safety pins to box cars so that we might maintain our living standards even while we buckle on the sword for defense of those ideals that have made us a great nation . . . where the flame of freedom still burns bright to serve as a beacon of hope to a large segment of mankind.

All this takes scrap metal . . . *mountains* of scrap metal for steel. To produce a ton of new steel, it takes 1000 pounds of scrap metal. It is estimated that 36 million gross tons of scrap will be needed to meet production demands this year. Next year we will need even more. And scrap reserves are already dangerously low . . . not enough to carry through the winter ahead.

HERE'S HOW *YOU* CAN HELP

Look around your own home, your business, your farm, or where you work. Rusty sheet metal, discarded plumbing, obsolete tools or equipment—any metal that is no longer useful can be sold . . . AND IS URGENTLY NEEDED! Even a few pounds are valuable to industry and worth money to you when turned in to a dealer of scrap metal.

Back America's double production lines! Collect and turn in scrap metal NOW!

Manufacturers of a Complete Line of Modern Machines, Visionlined Tractors, and Power Units for Agriculture . . . and Agriculture is *Basic* to Our Economy.

RUSTING SCRAP METAL IS RED POWDER
TURN IN SCRAP TODAY!

MINNEAPOLIS MOLINE
MODERN MACHINERY

MINNEAPOLIS-MOLINE

Having the merchandise in stock is only half the battle—it must be displayed where a customer can't miss it.

1952-1955

COMPANIES LOOK ABROAD

a trend worth tying to:
MONEY-MAKING GRASS

Grass is becoming a *high value* crop, equal in many cases to grain in return per acre.

In the past ten years farmers' thinking about meadow crops has changed completely. You can remember when little attention was paid to hay quality. Yields were low. It was grown on the poor fields. Few farmers thought about nutritive value.

Now, under farmer-controlled planning in soil conservation districts, grassland farming has come of age. It includes proved seed varieties and special mixtures, fertilizer, better land devoted to grass, timely cutting,

careful handling in curing and storage to save mineral-rich leaves.

Talk with the better farmers of your community. You will find their hay and pasture crops receiving skilled management, and bringing high returns. All this indicates a new appreciation of the value of *good* hay and of the machinery to handle it.

For Allis-Chalmers dealers it is a special opportunity. Their full line of outstanding hay and forage machines means a growing business along with the steady growth of soil conservation and grassland farming.

More and more farmers are choosing the ROTO-BALER because they want their own machinery for high-quality hay. They like smooth, roll-up compression which saves leaves and gives each rolled bale its own thatched roof.

ROTO-BALER is an Allis-Chalmers trademark

Farmers are paying new attention to the right time for mowing—to catch hay when its protein content is high, yet late enough so perennial plants are not weakened.

Allis-Chalmers power-driven rake is engineered for tractor farming. Gear-shift reel enables the farmer to handle a heavy or light crop or to use the rake as a tedder.

Grass silage is on the increase and here is the machine to handle it. The 4-in-1 Forage Harvester also chops wilted hay for mow curing, dry hay, straw, and row crops.

The Allis-Chalmers Forage Blower breaks the bottleneck at the silo, even in heavy crops. Money-making grass needs labor-saving machinery all the way.

ALLIS-CHALMERS
TRACTOR DIVISION · MILWAUKEE 1, U.S.A.

Agricultura
de las Américas

Marca Registrada

1952-1955

Modern agricultural equipment was needed around the world to plant, cultivate, harvest and deliver food to people everywhere. Companies within the United States were able to make much of the needed equipment, but were not the only ones seeking this market. Russia was successful in many areas, especially in communist block countries, but many other countries were not anxious to increase the debt to the United States and several were industrialized sufficiently to provide healthy competition in the world market. Often, American companies were able to buy or build important components such as diesel or hydraulic equipment from the foreign market and thus balance the trade more evenly.

The anti-communist fervor of post WWII America led to unsubstantiated charges by Senator Joseph R. McCarthy that communists had infiltrated the Department of State and the U.S. Army. The U.S. Senate censured McCarthy in 1954.

A new magazine, written in Spanish called *Implementos y Tractores* began in 1952. The magazine helped to sell farm equipment, chemicals and supplies made in the U.S.A. to the farmers and ranchers of South and Central America. Also during 1952, Allis-Chalmers purchased the La Plante-Choate Mfg. Co. of Cedar Rapids, Iowa, International Harvester purchased the Frank G. Hough Co. of Libertyville, Illinois, and the Oliver Farm Equipment Co. purchased A.B. Farquhar of York, Pennsylvania. Ford began production of "NAA" was in production for three years of engine that had overhead valves. The "NAA" was in production for 3 years and the 1953 model was called the "Golden Jubilee" to commemorate the company's 50th anniversary.

IMPLEMENTOS y TRACTORES

"FILLS A VACUUM"

That's what many of the thousands of readers say when expressing their appreciation for this new specialized farm equipment publication.

The reason back of this expression: Never before has technical information on tractors and farm machinery, agricultural techniques and repair of farm equipment been made available in the Spanish language. A publication like IMPLEMENTOS y TRACTORES has been vitally needed to orient and inform importers, distributors and dealers in Latin America. These are the buyers of more than $140 million of U.S. equipment annually—the biggest part of the total export market.

IMPLEMENTOS y TRACTORES was not introduced with the idea of merely improving on a publishing service already provided. It was conceived to provide hitherto unavailable information on every phase of agricultural mechanization, and the use and maintenance of farm machinery. Thus, IMPLEMENTOS y TRACTORES, backed by 70 years of publishing experience in the farm equipment field, provides an editorial service that is unique and necessary—**it's a service not duplicated by any other export advertising medium.**

If you manufacture trombones, hair curlers or sewing machines, IMPLEMENTOS y TRACTORES can't help you increase your export sales. But if your product is farm equipment, replacement parts, maintenance machinery or any item used in connection with agriculture, then IMPLEMENTOS y TRACTORES is the book to carry your message! With great reader impact, it delivers your advertising sales message to every known importer, distributor and dealer in farm equipment in every Latin American country. Your advertising in IMPLEMENTOS y TRACTORES will also influence the top users and big buyers of farm equipment, as well as government officials at policy making levels.

If you're skeptical you may ask: Do readers like this new magazine? Is it successful in capturing their interest? Do they read it? For the answer, read the comments on the opposite page. These are representative of **nearly two-thousand** received thus far . . . and they constitute concrete proof of reader acceptance, preference and appreciation of a new and effective export advertising medium.

Goodyear Introduces New Snow and Ice Tire

A new, improved snow and ice tire which is said to provide superior traction in both loose and hard-packed snow, as well as on icy roads, has been announced by the Goodyear Tire & Rubber Co., Akron, Ohio.

Known as the Suburbanite, the tire is said to be particularly advantageous for access from home to main highway in bad weather, though it can be used effectively by all car drivers who encounter difficulties in snow and ice, the company states.

Tread of the tire is composed of numerous flexible blocks which bend circumferentially under pressure. The blocks are self-cleaning and present sharp edges that dig into snow and provide better grip on ice or slippery pavements. Three circumferential grooves in the tread give added resistance to side slip.

The new tire provides traction on loose and packed snow and ice.

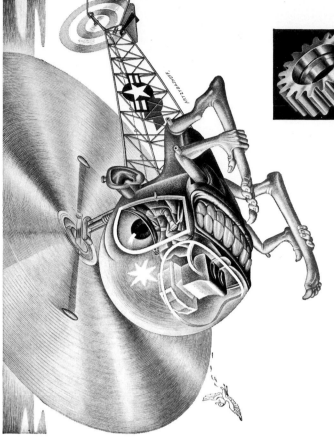

Ever see a helicopter grit its teeth?

When a helicopter's giant blade spins, tiny two-inch gears must carry the load. For teeth with a perfect bite, **Bell Aircraft chose Lycoming's precision production.**

To temper a steel gear to almost incredible hardness . . . to machine it to tolerances of 1/10,000 of an inch . . . Bell Aircraft Corporation called on Lycoming's skill, experience and extensive facilities.

Such super-precision work typifies simply *one* of Lycoming's services to America's leading industries and to its military forces. Long famous for aircraft engines, Lycoming also meets the most exacting and diverse requirements for packaged power, for product development, for high-volume production.

Whatever your problem—however complex your specifications—look to Lycoming!

AIR-COOLED ENGINES FOR AIRCRAFT AND INDUSTRIAL USES, PRECISION-AND-VOLUME-MACHINE PARTS, STEEL-PLATE FABRICATION, GRAY-IRON CASTINGS

LOOK TO **LYCOMING** FOR RESEARCH FOR PRECISION PRODUCTION

AVCO MANUFACTURING CORPORATION

WILLIAMSPORT, PA.
STRATFORD, CONN.

LYCOMING-SPENCER DIVISION
BRIDGEPORT-LYCOMING DIVISION

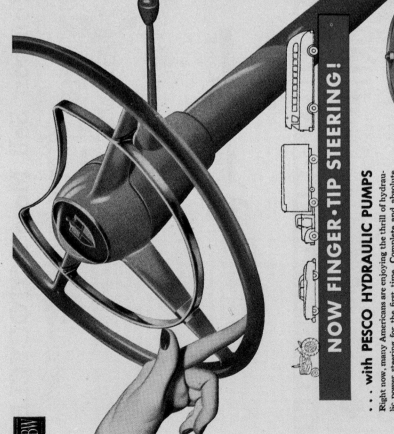

NOW FINGER·TIP STEERING!

. . . with PESCO HYDRAULIC PUMPS

Right now, many Americans are enjoying the thrill of hydraulic power steering for the first time. Complete and absolute control of the vehicle is obtained at all times . . . perfect safety, even on soft shoulders or in case of a blowout. Now, just the touch of a finger to the wheel, at any speed, and the vehicle responds surely and easily.

Less driver fatigue and increased pay loads for truckers are just two of the many benefits of hydraulic power steering. Pesco engineers have worked hand in hand with automotive engineers in the development of a pump for this unit, which is one more in a long series of important developments in the field of pressurized *power and controlled flow.*

Investigate the advantages of hydraulic power steering for the vehicles you manufacture. Perhaps Pesco's experience can be helpful to you. Why not call us?

New Pesco Hydraulic Power Steering unit for passenger cars consists of pump and reservoir.

Pesco engineering products PRODUCTS DIVISION

BORG·WARNER CORPORATION
24700 NORTH MILES ROAD · BEDFORD, OHIO

Here it is!

Wide 18 inch cut
Front "Grass Spray" discharge
2 cycle engine
Only 35 pounds

The LAWN-BOY by RPM

America's Most Modern Lawnmower

R P M Manufacturing Company, the world's largest manufacturer of rotary power mowers, presents America's newest lawn mower . . . the LAWN-BOY by R P M. Here's the *one* mower that offers your customers more new features than any other mower; backed up by the largest, most powerful promotional campaigns that ever introduced any mower. The LAWN-BOY's your guarantee of a trouble-free priced-right lawn mower . . . designed for profit-building sales appeal.

CHECK THESE FEATURES

CONTOUR CUT . . . the feature that checks scalping. The LAWN-BOY's cutting blade is automatically guided by the wheel placement design. Here's the mower that won't scalp.

SIMPLE, uncomplicated . . . the LAWN-BOY's direct drive gives trouble-free simplicity of operation. No belts to adjust, no chains to snap.

GRASS SPRAY . . . here's the feature that assures full cutting power. Grass clippings can't back up in the discharge chute. What's more, this feature eliminates long rows of clippings.

LIGHTWEIGHT . . . all aluminum alloy construction gives a magic, featherweight ease of handling, bound to appeal to *every* customer.

CLOSER TRIM . . . the LAWN-BOY trims as close as 3/8 inch . . . right up against walls, bushes or fences. Saves hours of time spent in trimming or edging. And others!

BALANCED WEIGHT DISTRIBUTION

THE LARGEST MUFFLER AREA ON ANY MOWER

COMPLETELY SHIELDED, EXTRA-SAFE CUTTING BLADE

Styling by Brooks Stevens, ONE OF THE COUNTRY'S OUTSTANDING INDUSTRIAL DESIGNERS

22 inch cut
Front "Grass Spray" discharge
2 cycle engine

21 inch cut
Rear "Grass Spray" discharge
4 cycle engine

WORLD'S LARGEST MANUFACTURER OF ROTARY POWER MOWERS

Available through hardware jobbers and distributors

R P M   M A N U F A C T U R I N G   C O M P A N Y

A Subsidiary of Outboard, Marine & Manufacturing Company
LAMAR, MISSOURI

Also Manufactured in Canada by OUTBOARD, MARINE & MANUFACTURING CO., of Canada, Ltd. Peterborough, Canada

BUY **BIG ORANGE** AND YOU **BUY THE BEST**

4278   4178

.34   1034

**TRACTOR, PLOW & HARROW CLEVISES
TRAILER HITCHES — HITCH CLEVISES,
HITCH AND CLEVIS PINS**

Ask Your Jobber or Write Us

MIDLAND INDUSTRIES, INC.
Cedar Rapids, Ia.

078

78   465   725   120   250

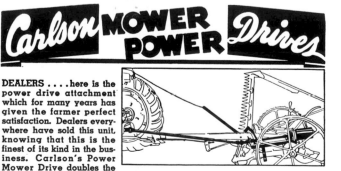

Carlson MOWER POWER Drives

DEALERS . . . . here is the power drive attachment which for many years has given the farmer perfect satisfaction. Dealers everywhere have sold this unit, knowing that this is the finest of its kind in the business. Carlson's Power Mower Drive doubles the cutting capacity of any ground driven mower and saves half the fuel.

**WRITE TODAY
FOR DETAILS**

CARLSON & SONS   BERESFORD SO. DAKOTA

DEARBORN'S DISTRIBUTORS, such as Romy Hammes (Kankakee, Ill.) have introduced over 3000 dealers to the new NAA Ford Tractor.

FORD: A New Tractor for Golden Jubilee

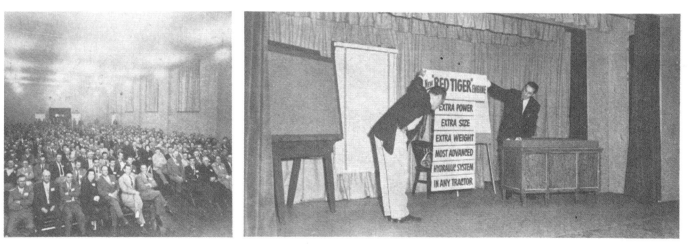

DEALERS GATHER to see and hear elaborate day-long presentation including . . .

PROFESSIONALLY ACTED skit on how to plan sales, make friends and win customers for maximum opportunity in the Golden Jubilee Year.

DEALERS DISCUSS sales prospects after inspection of the NAA.

Valve-in-Head 134 Cubic Inch Engine, Continuous PTO and Live-Action Hydraulic System Featured in New Ford

THE new Golden Jubilee model NAA tractor which is larger, heavier and more powerful than previous Ford tractors is now in production and being marketed through the Dearborn Motors Corp., Birmingham, Mich.

Front and rear tread adjustment range of the new model is the same as on previous models (48 to 76 inches) as are the tire sizes of 4.00 x 19 for the front, and 10 x 28 for the rear. Other general chassis specifications for the NAA compare with the 8N as follows: Wheelbase, NAA 73⅞ inches, 8N 70 inches; length overall, NAA 120¾ inches, 8N 115 inches; width overall (Wheels adjusted to 48 inches), NAA and 8N 64¾ inches; height overall, NAA 57¼ inches, 8N 54½ inches; shipping weight on standard 10 x 28 rear tires, NAA 2510 pounds, 8N 2410 pounds.

Engine

Significant among many of the improvements is the adoption and use of a gasoline carbureted, 134 cubic inch, valve-in-head engine which has a 6.6 compression ratio. The new engine, which is called the "Ford Red Tiger" has a bore and stroke of 3.437 inches by 3.6 inches as compared to the 3 3/16 by 3¾ inches L-head engine as used in the model 8N tractor. The new engine is rated by the manufacturer at 108 pounds feet torque at 1400 engine rpm.

The single piece cylinder block is fitted with centrifugally cast dry type liners, and aluminum alloy pistons of the autothermic type The piston pin is of the full-floating type. Three rings are carried on each pis-

ton; the two compression rings being 3/32 wide, the single oil ring 3/16 wide. Top compression ring of the three ring set is chrome plated.

The crankshaft rotates in three main bearings which are of the precision shell insert type. Similar type bearings are used in the large end of the connecting rods. Main bearing sizes are; Front, 2.500 x 1.500 inches; center, 2.500 x 1.750 inches; rear, 2.5 x 1.640 inches. Connecting rod bearing size is 2.300 x 1.03 inches.

The ball type centrifugal governor which is located under the timing gear cover is mounted directly on the forward end of the engine crankshaft. The minimum speed regulation has been reduced from 800 to 600 rpm and the over-run reduced

approximately 250 rpm to about 200 rpm. The maximum no load engine crankshaft rpm remains unchanged at 2200 rpm.

Other engine features include the use of free valve type exhaust valve rotators, chrome-moly alloy exhaust valve seat inserts, pressurized cooling system, pre-lubricated ball type water pump bearings, weatherproofed battery ignition system, full-flow type oil filter, oil-bath type air cleaner, and a new location for the gear type oil pump.

The engine muffler is now located above and parallel to the engine manifold. This change, according to the manufacturer, minimizes the fire hazard in field or barn, as well as increasing the tractor ground clear-

Ford Golden Jubilee Model NAA tractor is larger, heavier and more powerful than previous models.

The 134 cubic inch valve-in-head "Red Tiger" engine is rated by the manufacturer at 108 pounds feet torque at 1400 engine rpm.

ance. Also, the cooling system thermostat is now installed in the cylinder head instead of the radiator hose as on the 8N model. The hand throttle control has been changed from the notched position type to a cork friction disc type for more accurate control of engine speed.

Power Take-Off

A non-continuous pto which is similar to the superseded model is supplied as standard equipment. A continuous pto attachment is available as optional equipment. The attachment consists of a hydraulically actuated multiple disc clutch which is inserted (in series) between the main drive bevel pinion shaft and transmission main shaft. The separate pump for this clutch is mounted on and driven from the Proof-Meter driveshaft adaptor at the rear of the regular hydraulic pump.

An additional pto can be attached to the engine crankshaft pulley for driving sprayers, hydraulic pumps, and loaders etc. Connection to the front pto can be made without removing the tractor radiator or disconnecting the front axle assembly from the engine.

Hydraulic System

As the name "Live-Action" connotes, the system provides a constant source of hydraulic power while the engine is running, regardless of whether the engine clutch and/or pto is engaged. To obtain the live-action, the vane type pump is mounted on the right rear side of the engine and is driven by a helical gear on the rear of the camshaft.

Featured in the hydraulic system is the "Hy-Trol", a manually operated pump capacity adjustment which permits variation in the cycling speed of the work cylinder. By this means, liquid delivered to the work cylinder can be held at 2.25—2.8 gallons per minute through the wide speed range of 1200 to 2000 engine rpm. Pump has a maximum output of 5 gallons per minute at 2000 rpm and a maximum operating pressure of 2000 psi.

The new hydraulic system, retains the model 8N three-point hitch, touch control, constant draft control, and implement position control. Other improvements in the hydraulic system includes a convenient outlet on top of the lift cover for the installation of a remote cylinder (single or double acting) control valve, and as optional equipment a "Selec-Trol" valve which is manually operated to direct the hydraulic power to either a front or rear-mounted tool.

Unlike the superseded model, the hydraulic system unloading valve, check valve, control valve, pressure relief valve and safety valve of the new model are mounted on the underside of the top cover. Also, the system has a separate (independent) oil reservoir, which is located in the rear axle center housing.

Additional Features

Additional torque capacity of the power drive system is obtained by the use of a larger main drive bevel pinion and ring gear, and differential compensating gears. Compensating gears are rated as having 58 per cent greater load capacity than the gears in the 8N. Also, the reverse gear ratio has been changed, reducing the speed in reverse approximately 30 per cent. Advertised speeds at 2000 engine rpm for the new model with standard 10 x 28 tires are: First, 3.69; second, 4.75; third, 6.54; fourth, 13.64; reverse, 4.27, as compared to 3.23, 4.16, 5.72, 11.92, and 5.31 for the 8N model.

The new model retains the 8N Saginaw worm and recirculating ball nut and sector steering gear, the 9 inch single plate semi-centrifugal clutch, the wheel axle shaft internal expanding brakes which have bonded linings, and the 6-volt electrical system.

Equipment

Standard equipment includes a full-view instrument cluster which consists of a Proof Meter, ammeter, temperature gauge, and oil pressure gauge. Optional equipment for the new model includes a headlight and tail light kit, front bumper, continuous type pto, belt pulley assembly, ASAE pto conversion unit, hydraulic system "Selec-Trol" valve, remote double acting hydraulic cylinder and control valve, 3-point hitch stabilizing arms, 6.00 x 16 or 5.50 x 16 front wheels and tires, and tire pump and gauge.

With very few exceptions, Dearborn implements sold for use with the former 8N model will fit the Golden Jubilee model. For these few exceptions, adapter kits, sold separately, are available.

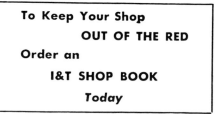

TELEVISION: A NEW MEDIUM

Krause Plow Corp. believed to **
to show dealers new products ar

Here's a typical industry problem:
More than 200 dealers to be introduced simultaneously to more than a dozen new product developments since your last sales convention; limited time; limited space and the need for making your presentation fresh, interesting, different, but effective.

Here's a unique solution:

Call in your local television station manager and plan a TV show over a "closed circuit". This means that only your dealers will be the audience. That's just what Krause Plow Corp., Hutchinson, Kan., did May 4 and 5 for its big sales conference.

Virgil Smith, sales promotion manager for Krause, originally got the idea from a big appliance manufacturer who had successfully tried the system. Station KTVH, channel 12, Hutchinson, was just getting underway for telecasting and hadn't yet taken to the air.

The TV station manager jumped at the chance offered him by Smith's proposal for giving his equipment and crews a real workout instead of the practice "dry runs" that had been going on in the studios.

Hal Lowther, Krause Plow's sales manager, worked with Smith and the TV station manager in planning the overall program and the script for continuity. Lowther played the part of commentator and coordinator for the show which took less than two hours to present 15 major units.

Dealers sat in a darkened auditorium adjacent to, but completely blocked off from, the area where the equipment was placed and from which the telecast was made. Dealers sat in groups around television sets provided by local merchants. There were seven sets to serve almost 200 dealers present. The sets were placed at different points in the large auditorium of

KRAUSE PLOW CORP., Hutchinson, Kan., assembled over 200 of its dealers in town's big 4-H building for a sales training course.

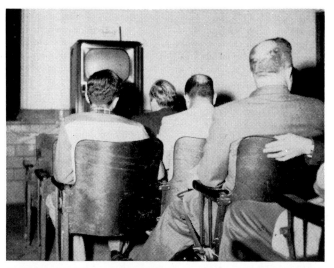

DEALERS GROUPED around individual sets loaned by town merchants to watch presentation of 15 units in less than 2 hours.

OR SALES TRAINING

rst in farm equipment industry to use "closed circuit" technique
resent selling features and methods

the 4-H building on the fairgrounds at Hutchinson. Thus, dealers were able to concentrate on the TV screen and see much more clearly, and magnified many times, fine points that could never be observed if they were actually in a crowd "bunched" around the equipment, as usually is the case at a new product showing.

In a separate part of the big 4-H building, station KTVH had its complete crew of cameramen, technicians and engineers. There were two television cameras, a battery of lights and a control station. Krause's product, design and testing engineers went over the features of each unit of equipment. They used pointers and the TV cameras got close-ups of the areas be-

NERVE SYSTEM of show is mass of wires, tubes and relays controlled by engineers who keep the show on the air. ▶

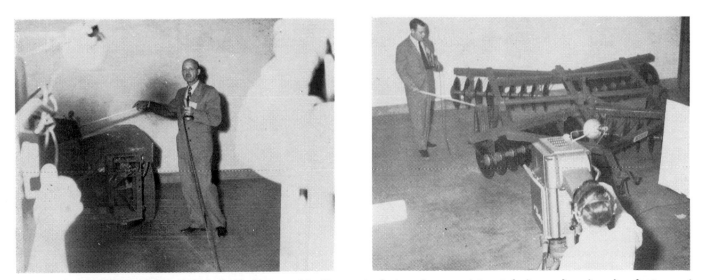

BEHIND THE TELEVISION SCREEN—Part I—Good showmanship: Krause product engineers point out design and engineering features of new products which are shot close up and magnified on screen.

VIRGIL SMITH, sales promotion manager of Krause Plow Corp., got the idea for a TV show from a big appliance maker.

HAL LOWTHER, sales manager of Krause Plow Corp., worked with Smith and the TV station manager in developing the program.

ing pointed to or parts described by the engineers. Hal Lowther introduced each engineer and smoothed the presentation by asking questions and summarizing the sales story presented.

The entire showing gave the dealers an impression of a professional troupe long experienced in such techniques of sales training—obviously it was enthusiasm and personal competence in their respective fields because neither the Krause nor the KTVH team had ever worked together before or had ever attempted such a complete sales training story.

Dealers questioned following the showing indicated that they had been comfortable throughout the one hour and 50 minutes of the presentation and had been able to concentrate on one point at a time. They further indicated that such comfort and concentration are usually impossible when as many dealers have to trot all over a large space and move from one piece of equipment to another and "bunch" and "crane" in an attempt to see and follow the information being talked about by an instructor.

Dealers also commented favorably on the fact that there was no wasted time or motion. The TV cameras swept smoothly from one product to another and from point to point in the product lectures. There was no waiting while everyone moved from one product to another before the instructor could begin his presenta-

tion. The two-camera team was able to handle switching from one area to another with the greatest of ease. Dealers in the auditorium, following the presentation on the screens, were conscious only of a continuous flow of information and each point was highlighted by actually watching a mechanism work, or seeing how it was made. Such detail would be impossible in a typical on-the-spot look and talk type of presentation.

The identical presentation without TV would have required at least 25 per cent larger space and would have consumed at least twice as much time.

Television, in its debut as a sales training medium in the farm equipment field, proved its merits:

1. You can instruct an unlimited number of dealers or salesmen simultaneously in new products, engineering and design features, operating instructions, servicing information, assembly techniques and selling techniques.

2. You can take advantage of a screen projection to show small details of design and engineering which are enlarged many times and completely dominate the screen.

3. You can hold the attention of a large group on one point at a time.

4. You can present more products in less space and in less time than you can by conventional means.

BEHIND THE TELEVISION SCREEN—Part II—The Technical Aspects: In addition to a well planned program and lively script, there must be an abundance of technical skills and competent . . .

1. Camermen and sound technicians   **2. Lighting engineers and directors**   **3. Control engineers**

THE CHAMP'S CHOICE
AT INDIANAPOLIS !

TROY RUTTMAN—1952 winner of the Indianapolis Speedway Classic—set a scorching average pace of 128.922 m.p.h. for 500 miles, to smash all records for this famous track. He rode with Mobiloil !

THE SUPER
DETERGENT OIL

★

THE OIL THAT
DRASTICALLY CUTS
ENGINE WEAR

★

THE OIL THAT
SAVES OIL

★

THE WORLD'S
LARGEST
SELLER

1903 · 1953
50TH
ANNIVERSARY

Mobiloil
Why Sell Anything Less?

ON THE SPEEDWAY or on the farm— wherever engines must operate at their best—Mobiloil* is *first choice* for prize-winning performance.

This world-famous motor oil is heavy-duty—designed to give top protection under the most severe operating conditions. And Mobiloil's new super-detergent action reduces engine wear . . . improves fuel economy . . . keeps farm equipment on the job longer, with fewer repairs or overhauls. Why sell less than the world's fastest seller?

Endorsed *by leading builders of farm machinery*

Advertised *in 20 leading farm magazines*

*Under API Classification, recommended "For Services ML, MM, MS, DG."

SOCONY-VACUUM OIL CO., INC., 26 Broadway, N. Y. 4, N. Y., and Affiliates: MAGNOLIA PETROLEUM CO., Dallas; GENERAL PETROLEUM CORP., Los Angeles

1952-1955

New sheet metal styling and a new 282 cubic inch engine highlight International Harvester's new "R" line of trucks which includes 168 basic chassis models with gross vehicle weight ratings from 4,200 to 90,000 pounds. The new line is said to contain 307 new features.

Front end sheet metal has been restyled for greater air intake and more efficient cooling. The cabs include a new instrument cluster, ignition-key starter switch, new mountings for a more stable ride, and green-tinted, non-glare safety glass which is available as optional equipment.

The new line of light, medium and heavy-duty trucks is identified with an "IH" emblem which replaces the old "Triple-Diamond".

The 168 basic models in 296 wheelbases, the 29 engines and a wide selection of transmissions, auxiliary transmissions, axles and axle ratios make possible a high degree of "Truck-to-job" specialization. Improvements which have been added to the old line of engines provide powerplants which cover the range from 100 to 356 horsepower. Included in the line are engines which burn LP-Gas and diesel fuel, as well as gasoline.

The completely new valve-in-head "Black Diamond 282" engine

powers some of the new models. It has a compression ratio of 6.5 to 1 and, according to International Harvester, develops 130 hp at 3,400 rpm. Featured in the new engine, are a short-stroke crankshaft, high-lift camshaft, exhaust valve rotators, and chrome plated top compression rings. The engine can be equipped to burn either gasoline or LP-Gas.

In addition to the many standard chassis models which are available in the light, medium and heavy-duty classifications, are a number of chassis series for specialized work. Among the special chassis models are the "School-master", for school bus operation; the "Road-liner" for tractor, semi-trailer operations; the "Loadstar" for hauling heavy, compact loads over rough terrain; "Cab-Forward" models for operations requiring the maximum in load-space and maneuverability; and other models for both on and off the highway operation.

International model R-110 pickup truck. Nine new pickup models, featuring 6½, 8 and 9-ft. all-steel bodies, are available in the R-110, R-120 and R-130 Series. This and International's other new light-duty models are powered by a 100-hp valve-in-head engine. International R-line trucks are newly identified by the "IH" emblem.

The two-year old Allied Farm Equipment Manufacturers Association adopted an insignia to identify its members' products and to pledge the integrity of the member short-line manufacturers.

The hostilities begun by North Korean troops invading South Korea just three years earlier, were ended by a truce signed on July 27, 1953. The name of *Implementos y Tractores* magazine was changed to *Agricultura de las Americas*. Also in 1953, the *I & T Shop*

Manuals were divided into several smaller books, so that repair instructions for newer tractors could be included more easily. Demand was increasing for service tools, repair parts, and qualified service technicians. Modern equipment made it necessary to update service skills constantly as indicated by ignition point gaps that could no longer be set with a "thin dime" by unskilled operators.

Allis-Chalmers purchased the Buda (engine) Co. of Harvey, Illinois, New

Idea purchased the Ezee-Flow Corp. and Oliver Farm Equipment Co. purchased Be-Ge Mfg. Co. of Gilroy, California. International Harvester purchased the manufacturing rights for two-wheeled, rubber-tired tractors from Heil Company of Milwaukee, Wisconsin. Efforts to make attaching and removing equipment from tractors easier and quicker included International Harvester's "Fast-Hitch" and the Allis-Chalmers "Snap Coupler" systems. The last Case threshing machine was made in 1953.

Sparking the Stars

OF THE MOTORAMA !

...and Original Equipment on Nearly as Many
New Cars as All Other Makes Combined

CADILLAC EL DORADO
The 40th year Cadillacs
have been AC-equipped.

BUICK SKYLARK
The 46th year that Buicks
have been AC-equipped.

OLDSMOBILE STARFIRE
The 43rd year Oldsmobiles
have been AC-equipped

PONTIAC STARCHIEF
The 28th year Pontiacs
have been AC-equipped.

CHEVROLET CORVETTE
The 38th year Chevrolets
have been AC-equipped.

AC
SPARK PLUGS

PATENTED
CORALOX
INSULATOR

AC SPARK PLUG DIVISION   **GM**   GENERAL MOTORS CORPORATION

In 1954, the Supreme Court ruled that public schools must be racially integrated, Caterpillar introduced the D-9 model and "I & T Shop Service" introduced a new book named the *Small Engine Service Manual* describing repair procedures for small gasoline engines. The British National Institute of Agricultural Engineering built an experimental hydrostatic tractor that was said to be the "World's First Gearless Tractor." The tractor used a Fordson Major Diesel engine to drive the pump and a five cylinder radial hydraulic motor in each rear of the rear wheels. Massey-Harris-Ferguson Limited was formed by an "amalgamation" of the Massey Harris Co. and Harry Ferguson, Inc.

Allis-Chalmers purchased the Gleaner Harvester Corp. of Independence, Missouri, in 1955. Future tractors were predicted to be "atom powered, with delta planetary gear system, nucleonic controls and either remote controlled or air conditioned." The Chamberlain tractor built in Australia used the British Perkins 57 hp four-cylinder, diesel engine and Allis-Chalmers began making the "D270" model in England using a Perkins 24.5 hp three-cylinder, diesel engine.

Positive Crankcase Ventilation On Engines

Willys Motors, Inc., 1150 N. Cove Blvd., Toledo, Ohio, announces that positive crankcase ventilation is now being installed on four and six-cylinder Power Giant industrial engines. Positive crankcase ventilation makes internal air circulation independent of the engine's physical motion relative to the external atmosphere. The Willys type of ventilation is said to automatically regulate internal air flow to engine speed. Air is pulled in through the air filter to remove harmful dust and dirt that would normally cause oil sludge. It passes into the oil filter tube and circulates within the engine block removing oil thinning or acid forming condensates. According to the manufacturer, this type of ventilation helps lower maintenance costs and provides longer engine life.

Announces New Tubeless Tire

B. F. Goodrich Co., Akron, Ohio, has announced a new tubeless tire, the Safetyliner, that incorporates many of the safety features of the Life-Saver tubeless tire. The new tire will be standard equipment on many 1955 automobiles, in addition to being available for replacement use.

An inner liner that is part of the tire itself is said to change dangerous bruise blow-outs to slow-outs, permitting the driver to bring his car to a safe stop. According to the manufacturer, the tread design of the Safetyliner provides maximum non-skid and traction qualities while squeal is virtually eliminated. The new tire is available in both black and white sidewalls in all popular passenger car sizes.

Snafu

A colonel was transferred to a new command. On reaching his depot, he found stacks of old documents accumulated in the archives of his predecessors, so he wired to headquarters for permission to burn them up. The answer came back: "Yes, but make copies first."

Good Advice

Daughter: "What should I do if the brakes give way?"

Father: "Steer for something cheap."

WHAT'S AHEAD--
IN TRACTOR DESIGN

In an exclusive interview with I&T's technical editorial director one of the industry's best known design engineers points out the direction tractor and farm equipment engineering will take over the course of the next several years.

A. W. Lavers, Lavers Engineering Co.

Mr. Lavers, what can farmers expect in basic design changes in tractors over the course of the next several years?

In the next three years changes from 1955 models will be confined, in the main, to increased working speeds and general appearance. But, you can expect really big changes in the next five years; particularly in transmissions designed for easier shifting, lower cost, quieter operation, with more speeds and perhaps using hydraulic and automatic principles.

What other big things are happening on the farm equipment design front?

In the next five years you will see streamlined implements, simplified and designed to perform better in unison with the tractor. They'll be easier to mount and adjust to varying conditions.

In the next eight years, perhaps even less, you'll see more engine economy from improved fuels including 100 octane gasoline. We also expect improved ways of adjusting wheels for row crops which might involve some departure from the tricycle type. Also, there will be more design and production of the "common power plant" type of harvesting equipment utilizing the basic concepts contained in the present offerings of Minneapolis-Moline and Ferguson.

Are any steps being taken design wise and production wise to lower the initial cost of farm equipment?

Yes, definitely. The common power plant just referred to is one of the directions in which steps are being taken. Machines will be developed for faster attachment and detachment of the power plant that will also serve as a versatile tractor. This will certainly reduce the cost of harvesting equipment and, in many cases, make self propelled units, each with its own power plant, obsolete. Other steps involve parts elimination and simplification.

Specifically what direction is tractor design taking with regard to improved means for transmitting engine torque to the rear wheels or to the tracks?

New transmission and final drive assembly design eliminates a lot of friction and permits delivery of more power to wheels or tracks. Transmis-

---

THE BIG CHANGES THAT WILL BE EVIDENT BY 1965

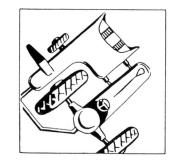

Transmissions and Final Drives . . .
Transmissions will have more speeds and will be shiftable with tractor in motion. They will probably operate automatically and may utilize hydraulics.

With higher octane gasolines on the way the delivered power per gallon (and perhaps per dollar) should be sufficiently economical to preclude the dominance of alternate fuels.

A Power Plant with Many Machines . . .
You can expect extension of the basic concept contained in Minneapolis-Moline and Ferguson offerings. You'll see more harvesters quickly interchangeable with a master power unit.

This will reduce the investment in harvesters and make many of the self-propelleds obsolete.

sions with fewer gears or more speeds with the same number of gears is an important development that will accelerate. Under consideration, and in some cases being tested are mechanical planetary transmissions, hydrokinetic transmissions and combinations of gears and fluid elements of the basic types represented by Hydra-Matic and Dynaflow. It is too early for a more specific prognostication.

Specifically what direction is tractor engine design taking?

Specifically in the direction of better use of higher octane gasoline, thereby getting more power and greater economy with the engines of the same size. Also, engines will be better lubricated due in part at least to better oils.

What about fuels of the future? Will diesel supplant gasoline and will LP-Gas usage increase?

We believe gasoline will be greatly improved and we do not believe that gasoline will be supplanted by diesel fuel. This is especially true as higher compression engines become prevalent. LP-Gas usage will increase in areas of favorable price differentials.

Are present day tractors designed to withstand overloads that are unreasonably high and do they last too long—or are they inadequate in either respect?

Load and life design factors vary considerably by manufacturers, but the high, in our opinion, is not excessively high on tractors.

What about such factors in implements and harvesting equipment?

In our opinion many harvesting machines are too heavy and have far too great a life built into them. Considerable cost could be saved by design improvement and a better total machine would result.

Do you see, in the near future, any really big developments that will revolutionize farming practices?

Yes, progress toward basically new methods of tillage, planting and harvesting already show promise.

The placing of fertilizer and the use of new types of fertilizer are coming into the picture.

Harvesting equipment that is more versatile and, at the same time, more efficient and less costly makes this area of equipment subject to a higher obsolescence rate.

We see unlimited opportunity to improve on present designs of farm equipment so that units will do a better job all around: Be easier to operate, safer to operate, more economical to operate, have better appearance and simultaneously cost the farmer less. And, that's going to get more important if we want to foster a faster replacement policy on the part of a large segment of farmers.

FARM EQUIPMENT DESIGN—1965
Are These Too Fanciful?

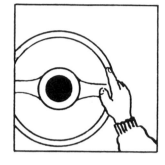

EASIER HANDLING . . .
Power steering, a luxury in passenger cars and rapidly becoming a necessity in heavy trucks, could become standard equipment on, at least, row crop tractor models in the next 5 to 8 years.

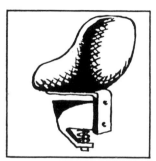

OPERATOR COMFORT . . .
Driver fatigue is a major cause of farm accidents. It also reduces the amount of work accomplished. Scientifically designed seating, and some type of escalator stairs to facilitate mounting and demounting for drivers are needed.

OPERATOR SAFETY . . .
In addition to designing more dynamic stability into tractors, it would seem desirable to add removable devices to protect drivers under some conditions. One would be the "rollover bar" familiar to stock car racing.

IMPROVED FUELS . . .
The biggest development will be production of 100 octane and even higher octane gasoline for farm equipment power plants. Such fuel improvement will have to precede the development of the higher output engines.

Farmers-of-tomorrow think it's lots of fun to ride a tractor "just like Dad's". The little tractors are made on a regular assembly line at the Eska plant, as illustrated on this page.

1—Here is where the Tractorcycle starts to take form. Aluminum castings for the body come from the foundry in halves, must be drilled and punched to ready them for assembly.

2—A conveyor belt carries the bodies along assembly line, past various benches and workmen where additional parts are assembled just the way they are in a real tractor plant.

How a toy riding tractor "just like Dad's" can attract farmers into a farm equipment store is good news for many dealers, especially those who are located a little outside the main trading area in their communities. Use of the Tractorcycle as a traffic builder isn't just a good theory; it's been tried by many dealers, and it really works. Here's the story of the young fry's version of "Dad's tractor", and how it can build good will and increase store traffic for dealers.

When the Eska Co., Dubuque, Iowa, brought out its first Tractorcycle four years ago, it was designed after a well-known row-crop tractor, simply to make it different from the garden variety of tricycles and kiddykars then on the market. The idea of using it as a merchandising device didn't crop up until enterprising farm equipment dealers themselves entered the picture, and discovered that the Tractorcycle was more than just a riding vehicle for children.

The fascination youngsters find in the brightly-painted vehicles made in the image of the Farmall 400 was immediately demonstrated by the repeated success of the "band wagon technique" in selling them. When one Tractorcycle was placed in the hands of a four to eight year old in any city block, it was no time at all until most of the neighborhood parents had been nagged into similar purchases for their children.

Dealers have found many ways to use the tractors in their merchandising programs. One dealer, for instance, cleared a space on his floor and roped off a ring where visiting children could have free rides on half a dozen Tractorcycles provided for the purpose. It didn't take long for parents to discover that there was one place in town where their youngsters could be parked for an hour's royal entertainment, while father and mother went shopping. On extra-busy days like Saturday, his wife or daughter acted as chaperone and traffic regulator.

One dealer discovered that interest in his annual "family night" really zoomed when he announced a Tractorcycle race between customers' youngsters as part of the program. Preliminary elimination contests staged in his store attracted good crowds, too.

Other dealers are enthusiastic about their use in fair exhibits. Free rides for the youngsters give their parents time to look over the dealer's main exhibits.

Of course, any dealer who uses the Tractorcycles for

TRACTORS DO A MERCHANDISING JOB

ride Tractorcycles, and you'll get

ur latest farm equipment displays.

4—Rubber tires are placed on the Tractorcycle's wheels in exactly the same way as tire casings are stretched on automobile wheels. Wheels are packed separately from frame.

3—A moving chain carries the frame, minus wheels, to the paint room ,where it's given a first flat coat and a coat of red enamel. Then it spends 30 minutes in the drying oven.

5—Some idea of the volume of production at the Eska factory may be gained from this assembly of Tractorcycle rear wheels, ready to be shipped. This is a single day's output.

6—Every 25 seconds this moving belt delivers a completed Tractorcycle from the assembly floor to the trucking dock, ready for shipment to dealers in all parts of the country.

their merchandising value will be able to make profits-through-sales from them, besides. The children who occasionally ride the tractors in his store are going to wish they had one at home, to ride whenever they please, and this can mean many sales. Dealers who use the tractors as a "come-on" at local fairs report that parents are apparently in a "holiday mood" during these events, and are more likely to purchase a tractor there than during a sober day's shopping in town.

With much of the "merchandising expense" absorbed by profits from the Tractorcycle's sale, the cost of using the little tractors as traffic-getters becomes nominal. If you've discovered how important it is to get farmers into your store, to get them used to dropping by even without a specific purchase in mind, you'll find the Tractorcycle is a big step in that direction. Making your store "headquarters" for farm families (see I&T, July 17, '54, p. 30) is a tried and proved way of increasing sales. A Tractorcycle "rink" in addition to the usual comforts and conveniences will mark your store as a good place to stop when farm families come to town.

1952-1955

1956-1960

FIERCE COMPETITION
BETWEEN MANUFACTURERS

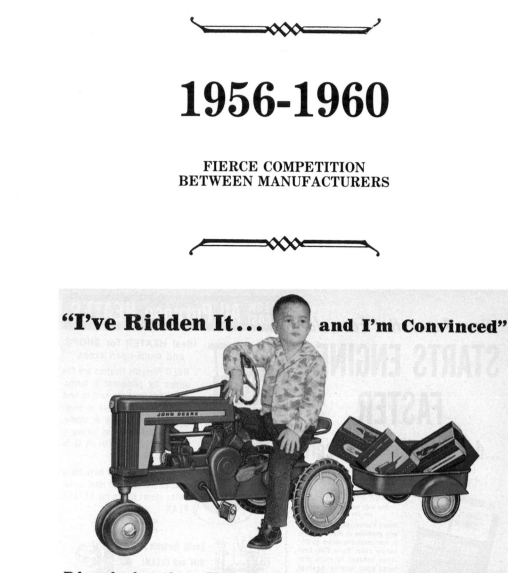

"I've Ridden It... and I'm Convinced"

Discriminating Youngsters Unanimously Endorse
the Distinctive NEW JOHN DEERE TRACTORCYCLE

TOY TRACTOR
Right Up-to-Date

Here's another good-will builder for you . . . the new John Deere Toy Tractor . . . with brand-new features that'll make it even more popular than before. The new John Deere Toy Tractor—basic equipment on every toyland farm—is as up-to-date as *right now* with new, modern three-point hitch and two-tone colors. The three-point hitch really works, too. Plows and other drawn tools can actually be raised and lowered by means of a lever to the right of the driver's seat.

The John Deere Toy Tractor, modeled after the John Deere General-Purpose Tractor, is sturdily built of heavy, die-cast metal to take plenty of rough handling. Wheels are heavy-duty, solid rubber, and the steering wheel actually turns the front wheels. Stock plenty . . . they go fast!

Of course, the young "tractor connoisseur" is a bit prejudiced. His dad's been talking a lot lately about the exclusive new features of the brand-new John Deere Tractors. Fact is, Dad and his John Deere dealer have been putting their heads together on a new tractor. Now that Junior's sold on John Deere . . . looks like Dad's a *sure bet.*

You'll promote your tractor and implement line . . . build good will among customers and prospects by displaying and selling the New John Deere Tractorcycle. You can add a handy new profit, too, especially with Christmas coming up.

Youngsters Love Them

Youngsters will love the new John Deere Tractorcycle because it's "up-to-the-minute" in the latest John Deere Tractor features. To kids, it's . . . "just like Dad's."

The new John Deere Tractorcycle highlights the spanking new John Deere Tractor two-tone color scheme . . . distinctive new front-end trademark . . . "power steering" steering wheel . . . adjustable seat. It's made of heavy cast aluminum, and boasts sturdy chain drive, heavy, solid rubber tires, and two coats of long-lasting, gleaming enamel. Big-capacity two-wheel trailer and miniature John Deere Umbrella are popular "extras."

Be sure and stock Tractorcycles and other John Deere Farm Toys. They'll provide promotional, advertising, and good-will value that you just can't buy elsewhere.

THE **ESKA** COMPANY

100 West Second St., Dubuque, Iowa

FARM IMPLEMENT NEWS

1956-1960

By the mid-fifties, farms had become food manufacturing facilities and farmers were the managers/owners of these not so small businesses. Farmers couldn't control nature, even with modern irrigation and insecticides, but a great deal of business sophistication was necessary to operate the farm as a money making company. Modern farmers had to be convinced that new equipment would make them money and that usually translated "BIGGER." The fierce competition that developed between equipment manufacturers was reminiscent of the earlier "Harvester Wars." International Harvester offered thirty basic tractors "to fit every job, every crop, every fuel" and 158 pieces of equipment to fit their "Fast-Hitch."

Hungarians revolted and overthrew their communist government in the fall of 1956, but the Soviet army intervened after about 150,000 Hungarians had escaped to other countries. Deere & Co. purchased Heinrich Lanz, A.G., one of the largest manufacturers of farm equipment in West Germany. Many American companies were developing or acquiring very good manufacturing facilities in Europe or Canada. A torque converter was offered as an option on the Shepard "SD-4" diesel tractor. Crop dryers, some using gas turbine (jet) engines, were the trend on farms as were the 2.5 and 3.5 hp riding mowers to handle chores around the home. Twelve volt electrical systems cured the hard starting problems for the new high performance engines. Many of the new appliances available to the rural living families, such as new clothes washers, were damaged by the often hard water from local wells. This problem led to the popularity of water softeners. Used equipment had always been a problem to the dealer, but much of the equipment traded in during the mid-fifties was really still good and usable, it just wasn't big enough, didn't have enough power and "only had a 4-speed transmission." Some dealers converted these trade-ins to LP Gas, installed power increasing engine kits or added more speed ranges to the transmission just to make the used tractors saleable. The M&W Gear Co. of Anchor, Illinois, introduced a dynamometer priced inexpensive enough for most tractor shops. Soon A&W Tractor Products of Colfax, Illinois, and other companies would also offer dynamometers.

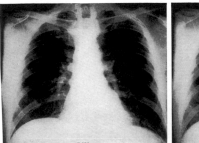

now — *six months from now—?*

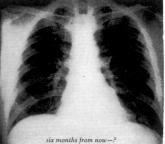

Monmouth

MICRO* AND CLEVITE* 77 BEARINGS...
you can't install greater engineering, precision and performance!

Monmouth bearing qualities can back up your desire to provide the finest of engine repair service. Why? Because they are the product of the world's leading manufacturer of automotive engine bearings . . . and you benefit by the most advanced bearing engineering and bearing manufacturing techniques known today.

You can quickly get Monmouth Micro or Clevite 77 bearings that are precisely right for any car, truck, bus or tractor. Available from N.A.P.A. jobbers coast to coast.

Monmouth
TRADE MARK
ENGINE BEARINGS

Clevite Service
The Cleveland Graphite Bronze Co.
Division of Clevite Corporation, Cleveland, Ohio, U.S.A.

*The words Monmouth, Clevite and Micro are registered trade marks of Clevite Corporation

---

Super POWER
FOR YOUR TRACTOR

McQUAY-NORRIS *Super-Power* CONVERSION ASSEMBLIES

MORE HORSEPOWER • HIGHER COMPRESSION • MORE ECONOMICAL

LARGER BORE

The increased bore in McQuay-Norris Super-Sleeve Assemblies gives added horsepower, which means greater draw-bar pull for that extra rugged job. You'll find that those hills and soft spots won't slow you down and eat up valuable daytime hours. But the larger bore is only one of the features of McQuay-Norris Super-Sleeve Assemblies.

LIGHTWEIGHT ALUMINUM ALLOY PISTON

The pistons are made of special high-grade lightweight aluminum alloy and are equipped with a chrome top ring which actual tests gives the longest wear life under the toughest conditions. The piston pins are special four-heat-treatment steel and are machined from solid bar stock for the greatest resistance to wear and impact. And in addition to all this—

READY TO INSTALL

McQuay-Norris Super-Sleeve Assemblies are ready for installation. They are finished on the O.D. so that no machining or reboring of the block is necessary. Thousands of these sleeves have been installed and are giving users increased horsepower, greater economy and longer life, plus all around satisfactory operation. Your tractor will run better, you'll get more work done and feel better at the end of the day when your tractor is equipped with McQuay-Norris Super-Sleeve Assemblies.

AVAILABLE FOR...

ALLIS-CHALMERS • FERGUSON • McCORMICK-DEERING • FORD • CASE MASSEY-HARRIS • OLIVER-HART-PARR • COCKSHUTT, CO-OP & FARMCREST

McQUAY-NORRIS MANUFACTURING CO., ST. LOUIS -- TORONTO

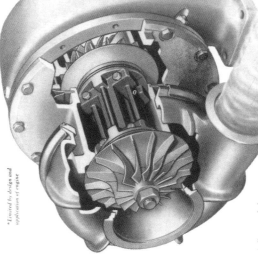
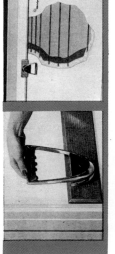
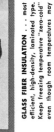
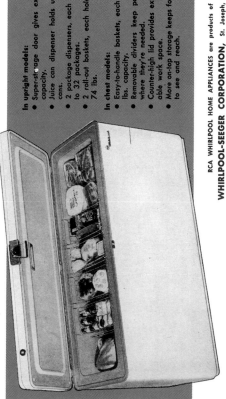

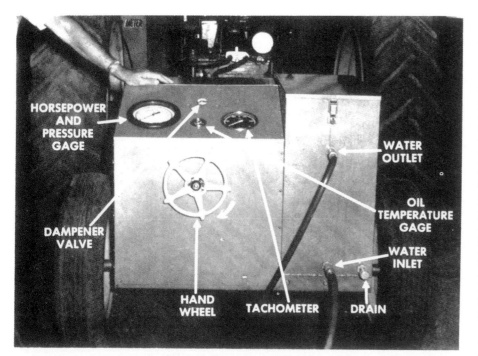

Easy to interpret gages, portability and a high degree of accuracy are features of the new M&W dynamometer. Horsepower readings are made directly on the combination horsepower and pressure gage. Manual adjustment of the dampener valve reduces the gage needle fluctuations.

In 1957, the U.S. Senate approved the "Eisenhower Doctrine" that proposed to send troops to aid any Middle East nation opposing communistic agression. Sputnik I, launched by the Soviet Union, became the first man-made satellite. J.I. Case Co. purchased American Tractor Co. (Terratrac) of Churubusco, Indiana, and the Electric Wheel Co. was purchased by the Firestone Tire & Rubber Co. Electric Wheel had stopped production of tractors about 1930. Tecumseh Products Co. announced on April 13, 1957, that it had taken over operation of the Power Products Co. of Grafton, Wisconsin, and Outboard Marine Corp. announced that it had acquired Cushman Motor Works, Inc. of Lincoln, Nebraska. The Curtiss-Wright Corp. of Utica, Michigan, sold 36 to 3000 hp Mercedes-Benz Diesel engines to the American market, Volvo began selling their 32 and 43 hp tractors in the United States and David Brown developed a tractor specifically for the North American market.

Power steering was still fairly new for automobiles, but was drawing some real attention for farm tractors. Traction was a continuing problem that was often solved by adding a lot of weight to reduce tire slippage, which then caused the tractor to be very difficult to steer.

THE 1957 CLINTON ENGINE LINE
19 models with thousands of variations 1 to 9¾ hp.

THE PANTHER FAMILY
6 models to choose from

Powerful as a Panther on the attack—as quick starting, light and smooth as a Panther on the move. Hundreds of applications. Quick starting every time under all operating conditions. Interchangeable with any engine using 4-inch or 8-inch bolt circle. Horsepower capacities from 1¾ to 2½ horsepower. Lightweight: 15 and 18 lb. models available. Equipped with new mesh type air cleaner... full float type carburetor... high tension fly wheel magneto.

THE GEM FAMILY
6 models to choose from

Phenomenal power from an engine series with a maximum weight of 19 lbs. Models available from 2½ to 3 horsepower. Noted for every time quick-starts in any weather, under all operating conditions; moisture and dust-proof high voltage magneto with fully covered ignition points ... Comes equipped with new dry type air cleaner... float type carburetor... rotating blower housing screen.

THE LONG LIFE FAMILY
5 models to choose from

Deluxe air-cooled engine series for heavier duty operation. 2 to 3.6 horsepower. Models available with 6-to-1 speed reducer. Quick starting because of increased spark. Uses dust and moisture-proof magneto with covered ignition points. Equipped with positive overspeed governor, new dry type air cleaner and full float type carburetor.

THE RED HORSE FAMILY
2 models to choose from

The ultimate in gasoline engine power. Built especially for the heavy jobs where engine power is needed to take hold instantly and keep going until a job is completed. Equipped with deluxe ignition system with breaker points . . . Alnico permanent magneto with waterproof mounted coil . . . Automatic spark advances and retards mechanism for quick starting and full float type carburetor with double-float for best idle and high speed performance. 6 and 9 horsepower models available . . .With or without 6-to-1 speed reducer.

THE 1957 CLINTON CHAINSAW LINE
5 models, direct or reduction drive

CLINTON D-4 SERIES

The saw that's built for the man who wants a production saw having a diaphragm carburetor. Cuts in any position. Quick starting; coolest running; on-and-off switch; finger-tip controls; automatic clutch. Available with 16 to 26-inch Stellite tipped guidebar.

CLINTON D-3 SERIES

Low in price but this saw is built for every day continuous cutting. Quick starting. Coolest running. Finger-tip controls. Positive on-and-off switch. Automatic clutch; diaphragm pump float-type carburetor Available with 16 to 26-inch Stellite tipped guidebars.

CLINTON D-2 SERIES

The lowest price yet . . . built to provide complete cutting satisfaction. Quick starting. Finger-tip controls for both gas and oiling of chain. Positive on-and-off switch. Automatic clutch; diaphragm pump float type carburetor. Available with 16 to 20-inch Stellite tipped guidebars.

CLINTON CS-577 SERIES

America's new king of the woods . . . a heavy duty professional timber saw with the guts for every cutting job. Torsion drive gets all the power to the chain and withstands load shocks . . . try it with all your leverage applied! Diaphragm carburetor permits cutting in any position. Quick starting. Coolest running. Finger-tip controls. Positive on-and-off switch. Available with 14 to 42-inch Stellite tipped guidebar.

CLINTON CS-323A SERIES

A powerful belt driven saw with patented torsion drive to get more power to the cutting chain and to withstand load shock . . . you can really put the leverage on! Index feature permits cutting in any position. Quick starting. Coolest running. Finger-tip controls. Positive on-and-off switch. Float type carburetor. Available with 14 to 30-inch guidebar.

CLINTON MACHINE COMPANY •

Dept. 1-A

Engine Division, Maquoketa, Iowa
Chainsaw Division, Clinton, Michigan

•

World's largest manufacturers of the most complete line of air-cooled gasoline engines. Over 5,500,000 Clinton Gasoline Engines now in use on the farm, in the home and in industry.

How to Make More Money

ALUMINUM BOATS

Dept. I, P. O. Box 2339, LITTLE ROCK, ARKANSAS

● **BIG PROFIT LINE** — Liberal dealer discount, and fast selling line spell good profits.

● **ESTABLISHED LINE** — Introduced just last year, the Resorter Line enjoyed great sales success among quality dealers everywhere. Made by one of the world's leading boat manufacturers.

● **QUALITY LINE** — Tops in appearance, performance, endurance! They last, and last. You don't have to worry about expensive "make-good" service.

Distributors and Jobbers: Resorter Boats are available with your own brand name. Fully-protected regional territories are open for qualified distributors and jobbers.

WRITE FOR FULL DETAILS

1956-1960

New Trends in Piston Ring Arrangement

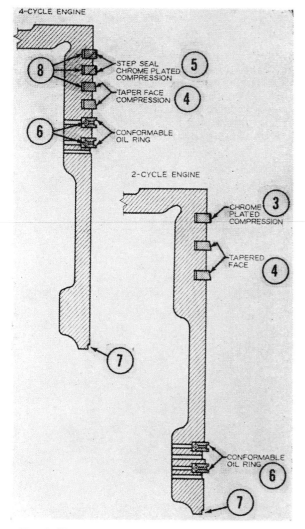

Sketch illustrating some of the trends in piston ring arrangement of modern two- and four-cycle engines, mostly in the automotive field.

Eight distinct trends in piston ring arrangement for modern engines are described and illustrated here

THERE are marked trends in piston ring arrangement for modern engines, according to F. A. Robbins, Koppers Co., Inc., manufacturer of a full line of these products. He mentions eight distinct trends which are tabulated in a recent issue of the SAE Journal as follows:

1. The number of rings is being decreased. No advantage in blowby occurs with more than four compression rings. There is virtually no advantage with more than three. No more than two oil rings are used.

2. Ring locations are changing. Top ring is located lower on the piston so that it remains cooler. Oil rings on four-cycle engines are being located above the piston pin to reduce tendency of piston skirt scuffing.

3. Chromium plated rings increase top ring life and reduce cylinder wear. Several small bore engines have all rings plated sufficiently thick to last the life of the engine.

4. Bottom edge bearing rings are being used in the lower compression ring grooves.

5. Seal joint rings are being used to improve combustion efficiency and reduce oil and ring belt contamination.

6. Oil rings with high unit pressures are using conformable type expanders.

7. Pistons are being fitted for minimum skirt clearance. Bottom edge of skirt is kept sharp for scraping. Cam shaped pistons allow for dynamic loading.

8. Ring land radial clearance is being increased to provide sufficient oil drainage.

1956-1960

REFLECTIONS Perspectives and Personalities

By E. J. BAKER, JR.

Harold G. Wilson, long a representative of the Hyatt and New Departure bearing interests, told with the vividness of a participant of the development of power lawn mowers in the U. S. from 1935 on and of the inspiration for the vertical crankshaft engine that has made the rotary machine what it is now.

When Hal Higgins, the Walnut Creek, Calif., implement historian, learned of the current interest in power mower history inspired by the University of Missouri, he wrote Ransomes, Sims & Jefferies of Ipswich, England, outstanding plow makers and pioneer developers of mechanical lawn mowers.

Then on July 23, last, Higgins, who was in Europe on patent matters, visited Ipswich and was shown some of the pioneer Ransomes mowers.

Briefly, Ransomes, Sims & Jefferies obtained patent rights to what many would say was the first practical mechanical lawn mower—the Budding—built by Ransomes first in 1837.

By 1903, the company was cataloguing and selling a riding 42-inch roller-drive reel-type lawn mower driven by a 6-hp gasoline engine.

Before the gas engine mower was developed, the company catalogued a walking self-propelled steam lawn mower made under Sumner's patent. One may doubt whether many of these were sold, but to justify cataloguing, construction by the inventor must have been followed by some production at the factory.

The illustrations reproduced here were sent to Hal Higgins by Ransomes, Sims & Jefferies, Ltd., and are copyrighted by that company.

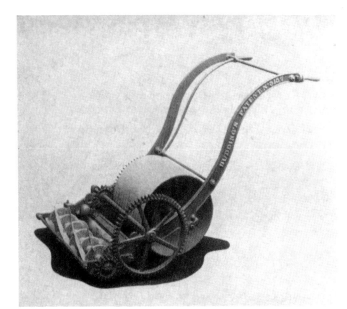

J. R. & A. Ransome started building the Budding lawn mower in 1837 at Ipswich, England. The roller drive idea came from the inventor's study of the manner in which nap was trimmed evenly on upholstery. Some of the largest selling reel-type power mowers today have roller drive. (Photo copyright by Ransomes, Sims & Jefferies, Ltd. From the collection of F. Hal Higgins).

Ransomes first riding motor lawn mower as shown in the company's 1903 catalogue. It had a 42-inch cut and was driven by a 6-hp gasoline engine equipped with a Simms-Bosch magneto. A 36-inch cut walking model was also available with the same engine.

Ransomes, Sims & Jefferies at one time catalogued a steam propelled lawn mower as shown. If steam could be raised in 10 minutes from cold water, as claimed, the engine must have been like those used in naphtha launches in the U. S. A. about this time; certainly not a coal burner. The writer of Reflections has ridden in a naphtha launch in his youth. It was strictly "carriage trade". (Photo copyright by Ransomes, Sims & Jefferies, Ltd., Ipswich, England. From the collection of F. Hal Higgins).

THE BIG FIRST STEP TOWARD TOP PERFORMANCE ON LP-GAS

TRACTORS

THESE ENSIGN CARBURETION UNITS FILL EVERY NEED

Ensign starts quickly, idles smoothly, accelerates instantly, and lugs long and hard at full power. — All this with built-in low cost fuel economy. But that is not all! Look to Ensign quality materials and workmanship for built-in-ability to perform these tasks year after year with absolute minimum field maintenance. Then, later, if and when field service is required, you'll find Ensign has the largest and finest field organization of its kind in the country.

Insist on Ensign — Accept Nothing Less

ENSIGN CARBURETOR COMPANY

1551 E. Orangethorpe Ave. Fullerton, California

Branch Factory; 2330 W. 58th St., Chicago 36, Illinois

Mohawk Offers Riding Tractor

Mohawk Engineering Corp., 3 Weber Ave., North Adams, Mass., is offering a lightweight town and country riding tractor. Incorporating automotive-type differential and steering, the Mity-Mule tractor is powered by a gasoline-driven, air-cooled, four-cycle, 2.75-hp Briggs & Stratton engine. Speed ranges from 3 to 5 mph and one lever controls forward, neutral and reverse. The unit also features power braking, a recoil starter, a remote control throttle and a push-button shut-off.

Mity-Mule is shown with front-mounted reel mower attachment which steers with tractor.

The 215-lb. Mity-Mule has a channel steel frame and is equipped with a full-size contour tractor seat, adjustable foot pedals and a tool box. The front wheels are 10 x 2.75 semi-pneumatic, and the rear wheels (4.00 x 8 pneumatic) are adjustable for width.

The tractor, which is said to haul loads up to one ton, is designed to accommodate quick change drop-pin attachments. Available attachments include: front-mounted reel and rotary mowers; a snow plow and grader; three gang mowers; a roller; spiker-aerators; a sweeper; trailer dump carts and a disk-harrow.

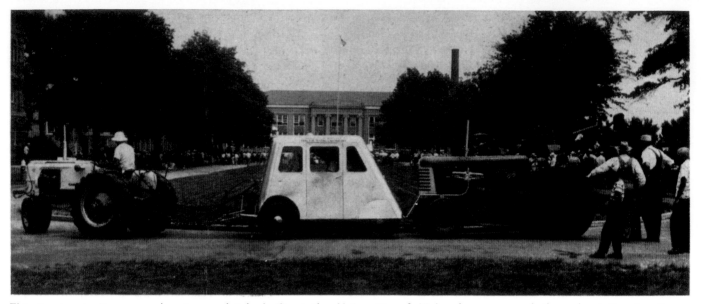

This tractor testing car, almost completely built at the University of Nebraska, was used throughout Tractor Power and Safety Day on the College of Agriculture campus, July 25. Testing engineers driving the laboratory equipment are Keith Jensen, on tractor, and Alvin Brhel and John Crouse, in the test car. PHOTOS: University of Nebraska.

NEBRASKA'S FARM POWER AND SAFETY DAY . . .

emphasized new developments in engines and equipment to reduce work in producing corn

By WALTER PATTERSON, JR.

One of earliest gasoline tractors in U. S. was this 1892 Froelich. Tractor was powered by a 20-hp engine, hit and miss governor and make and break ignition.

MORE than 50 years of progress in the development of tractors and implements was paraded before farmers attending the sixth annual Tractor Power and Safety Day on the University of Nebraska college of agriculture campus, July 25.

The event, an annual open house at the world-renowned Nebraska Tractor Testing Laboratory, attracted nearly 4,500 farmers and farm equipment dealers from a four-state area.

It was not hard to see the advances that have taken place when the parade included an 1892 gasoline tractor and the Ford "Typhoon", which is still in the experimental stage.

The "Typhoon" is an entirely new concept of tractor power. It has a free piston turbine engine and one cylinder with two pistons. The engine generates hot gases to drive the turbine. Factory representatives said that the tractor's advantages include fewer moving parts, and its ability to run on lower volatility fuels.

Farmers also saw two new machines which can help cut corn field operations almost in half. The new machines include a till-planter, used to prepare the soil, fertilize and plant directly in corn stalks, and a corn combine.

In discussing the new corn production machines, H. D. Wittmuss, assistant professor of agricultural engineering at the college, said using the till-planter, a cultivator twice and a combined picker-sheller for harvesting will cut corn growing operations down to four trips over the field.

The one-row corn snapping attachment for a combine has been used experimentally for the past seven years. It attaches to the cutter bar of a combine and strips the ears off the stalks.

Another experimental machine which shows promise for future use is the jet-seeder. The machine uses a jet of air to sow small granular fertilizer, grass seed, legumes and small grains.

Picker-shellers were also on display. The new machines harvest corn when the kernel moisture is between 25 and 28 per cent.

When corn is harvested at such high moisture content, grain drying machines are needed to reduce the moisture for storage, M. L. Mumgaard, Extension engineer at the University, said in explaining new grain drying equipment.

Safety practices took the spotlight when safety specialists at the college presented a demonstration on refueling tractors safely. Propane refueling was given particular emphasis at this year's program.

Also included in the safety part of the program was the dangerous tractor operation demonstration. Four tractor upsets were used to show that a tractor upset can happen to anyone.

The college's tractor testing equipment was used throughout the day in demonstrations. The test car, almost completely built at the University, is used to test approximately 36 tractors yearly.

The German-made Unimog was one of the units tested at the laboratory this year (Test 607, see I&T, July 13, 1957, page 72). It was one of the five foreign models on display at Tractor Day. It resembles a truck-tractor combination.

Other foreign models that were displayed included the French Someca, which also was tested at the laboratory, the German Fendt and

A tractor upset with "Jughead," the dummy tractor operator doing the driving. Four such upsets were seen in the dangerous tractor operation demonstrations.

Lanz and a Russian crawler. The tractors were tested on new concrete and earthen courses which were dedicated at last year's Tractor Power and Safety Day. (See I&T, Aug. 11, 1956) The $30,000 courses are known the world over. Each year thousands of visitors come to see the Tractor Testing Laboratory and methods used for testing.

The need for tractor testing came about in 1920. The first general purpose tractor appeared in 1917. A tractor of this type also was included in the parade of old tractors. Prior to 1920, many manufacturers were selling unsatisfactory tractors, which were abandoned by farmers before they were paid for. Nebraska passed a bill requiring all tractors sold here to be tested. Since then 650 tractor models have been tested at the laboratory. These tests have provided standards for rating tractors and have speeded up improvement the world over.

Engineers pointed out that under identical field conditions the harvesting efficiency of pickers-shellers and picker-huskers is about equal. Last year 2-row picker-shellers harvested up to 235 bu/hr; 2-row pickers harvested up to 185 bu/hr.

MERCE
Diesel

These engines — the world's finest diesels — are now available for sale through the Utica-Bend Division of the Curtiss-Wright Corporation, to manufacturers and users of industrial, marine, construction and materials handling, trucks and busses, railroad, electrical generating, agricultural, oil and gas equipment. . . The Utica-Bend Division will both import and manufacture the Mercedes-Benz diesels, ranging from 36 to 3000 h.p., together with components, spare parts, accessories and fuel injection systems. Mercedes-Benz diesels are noted the world over for their high power to weight ratios and are designed

MODEL OM 636

MODEL OM 321

MODEL NO.	NO. OF CYL.	RATING* HP @ RPM
OM 636	4	36 @ 3000
OM 312	6	79 @ 2400
OM 321	6	96 @ 2600
OM 315	6	123 @ 1800
M 204 B	4	129 @ 1200
†MB 846 A	6	242 @ 1500
†MB 846 Ab	6	320 @ 1500
MB 836 B	6	360 @ 1500
MB 846 Db	6	375 @ 1500
MB 837 A	V-8	451 @ 2000
MB 836Bb	6	513 @ 1500
MB 837 Aa	V-8	564 @ 2000
MB 836 Db	6	666 @ 1600
MB 820 B	V-12	718 @ 1500
MB 820 Bb	V-12	1026 @ 1500
MB 820 Db	V-12	1385 @ 1500

Ab TURBOCHARGED   Db HIGH OUTPUT TURBOCHARGER   Aa

*INDUSTRIAL INTERMITTENT

Other Mercedes-Benz Diesel Engines Available In 12 And 20 Cylinders Up To 3000 Horsepower

AD NO. 29-2

1956-1960

ES-BENZ
ngines

uilt to the highest standards of quality, per-
nce and economy. . . Unusually long engine
nd dependable operation under extreme cli-
conditions are made possible by their exclusive
. . . The sales and service program for Mercedes-
diesels, now being established by Utica-Bend,
les for sales representation in key areas and
es for servicing of engines and distribution of
parts to assure full maintenance of equipment
field. . . For further information, write to:
A-BEND DIVISION, CURTISS-WRIGHT
ORATION, UTICA, MICHIGAN.

MODEL OM 312

WEIGHT LBS.	APPROXIMATE DIMENSIONS		
	LENGTH	WIDTH	HEIGHT
400	2'5''	1'9''	2'6''
808	3'1''	2'2''	3'5''
784	3'1''	2'2''	3'5''
1775	4'6''	2'3''	3'11''
2715	4'9''	2'7''	3'11''
4075	6'2''	2'8''	5'
4298	7'2''	3'1''	5'2''
3560	6'4''	3'5''	4'7''
4266	7'0''	2'10''	5'2''
2970	4'1''	3'5''	3'6''
4080	7'	3'5''	5'4''
3080	4'2''	3'5''	3'8''
4290	7'6''	3'5''	5'4''
5510	8'	4'5''	6'1''
6116	7'11''	4'5''	6'5''
6612	7'11''	4'6''	6'1''

UPERCHARGER †AVAILABLE IN HORIZONTAL CONFIGURATION

MODEL OM 315

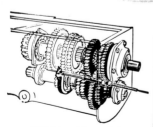

Outboard Marine Acquires Cushman Motor Works

Outboard Marine Corp., Waukegan, Ill., has acquired Cushman Motor Works, Inc., Lincoln, Neb., according to a recent announcement. The transaction involved the exchange of Cushman's total capital

Officials of Outboard Marine and Cushman complete the transaction by which Outboard Marine acquired Cushman. They are (l. to r.) William B. Ammon, Cushman; H. M. Fisher, Outboard Marine; Wayne L. Cooper, Cushman, and Robert Ammon, Cushman president.

stock for 114,000 shares of Outboard Marine stock, the market value of which is approximately $3,000,000.

Cushman will be operated as a wholly-owned subsidiary of Outboard Marine. Present Cushman officers will be retained as officers of the subsidiary. The Cushman plant in Lincoln will continue to make two and three-wheel power-driven vehicles including scooters, light industrial carriers and a three-wheel vehicle designed for the U. S. Post Office Department.

Outboard Marine is a large manufacturer of outboard motors and two-cycle engines. The corporation's products are sold through more than 33,000 retail dealers.

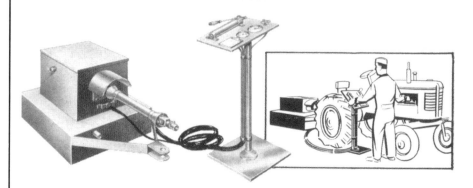

INCREASE YOUR SERVICE VOLUME AND PROFITS
with THE ALL NEW
A & W DYNAMOMETER

- ● **ONE MAN** movable control panel at table height
- ● No pre-warm-up of unit required
- ● Operates at standard power take-off speeds
- ● No lubrication maintenance — no hydraulic fluid to buy
- ● Continuous flow water cooled mechanism

COMPLETE STATIONARY UNIT

$575.00
F.O.B. FACTORY

AS SHOWN

DYNAMOMETER:

There is nothing else on the market like this new, exclusive, trouble-free principle. It is Americas quality dynamometer, actually foolproof and unbelievably accurate. The great advantages of the A & W Dynamometer come from having eliminated the disadvantages of other similar and more expensive types.

The A & W is ready to use the instant you hook it up. No waiting for warm-up. A & W Dynamometers are internally sealed, eliminating malfunctioning or damage to unit . . . and above all, ONE MAN can operate tests. Control panel can be positioned at his finger tips where all adjustments and dial readings are made.

CONTROL PANEL:

A & W remote Control Panel eliminates need for extra man. All controls, gauges and valves are right at hand . . . where you want them. Loading of Dynamometer is done hydraulically. Accurate horse power readings are shown on dials.

Can be purchased with wheels and axle for easy moving, at small extra cost.

Write today, for further information, free folder, discounts.

A&W TRACTOR PRODUCTS
COLFAX 2, ILLINOIS

THE CASE FOR MINIMUM TILLAGE

There is mounting evidence that farmers can save time and operating expenses, and still improve corn yields when plow-plant and minimum tillage operations are executed properly. Presented here are summaries of research findings recently released at three of our land-grant colleges

EXCESSIVE TILLAGE COMPACTS SOIL

By C. L. W. SWANSON, Connecticut

RUNNING over the land by tractors in plowing, disking, harrowing and cultivating for row crops during a growing season can be bad for our soils. This is especially true if done excessively.

Thousands of penetrometer measurements made on soils under cultivation in Connecticut show that tractor and implement traffic produces at least three distinct and varying degrees of compact layers or zones in soils:

(a) disking, harrowing and other seedbed and cultivation treatments after plowing compacts the soil at a three or four-inch depth to form a secondary tillage sole

(b) in plowing, the rear tractor wheel runs in the furrow packing the furrow bottom making a compact plow sole

(c) tractor wheels in row cultivation compress the surface three or four inches of soil, retarding root growth and water intake in the wheel compacted area.

These compaction layers are shown diagrammatically in Figure 1. The differences in the surface of soils in poor tilth (Figure 2) is strikingly shown when compared with Figure 3.

Seedbed and cultivation treatments after plowing loosen the upper three to four inches of the plow layer. In this secondary tillage zone the soils are loose, friable and well aerated.

In the secondary tillage sole below, the soil has been compacted by

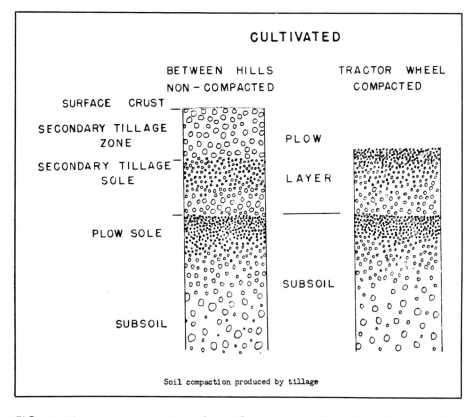

CULTIVATED

BETWEEN HILLS
NON-COMPACTED

TRACTOR WHEEL
COMPACTED

SURFACE CRUST

SECONDARY TILLAGE ZONE

SECONDARY TILLAGE SOLE

PLOW LAYER

PLOW SOLE

SUBSOIL

SUBSOIL

Soil compaction produced by tillage

FIG. I—Many measurements made on Connecticut soils under cultivation show distinct compaction layers. Plow sole in tractor-wheel compacted soils may be about 50 to 75 per cent harder than the underlying subsoil. This plow sole may be broken mechanically with a subsoiler or penetrated with legume roots, which will increase the percolation of soil moisture, often as much as 62 per cent.

seedbed preparation; it has been relatively undisturbed after plowing except for compaction. This may be thought of as residual structure contrasted to that produced in the layer above. Tillage tends not only to compact soils but also to arrange the soil particles in closest packing, resulting in fewer large pores.

In plowing, the rear tractor wheel runs in the furrow, packing the furrow bottom. Compressive action of the plowshare may also pack the soils. Penetrometer measurements show the plow sole to be about 50 to 75 per cent harder than the subsoil below.

One way of breaking up the plow sole is with deep-rooted legumes. In clover plots, for example, percolation was 62 per cent better than in continuous corn. The plow sole may also be broken up mechanically with a subsoiling apparatus. Research on this is in progress.

The diagram shows the effect of tractor wheels in packing soils in row crops after three cultivations. Here the secondary tillage zone is compressed and eliminated by the tractor wheel. Penetration of corn roots was restricted somewhat by this compact layer, but was not of enough consequence to reduce corn yields measurably.

With the advent of the tractor, it is a rather easy matter to till our soils. Because of this, tractors and cultivation equipment have been blamed for the poor structure of many of our soils. It isn't the equipment that is at fault; it is how this equipment is used in soil management.

FIG. 2   FIG. 3

An example of what excessive tillage accomplishes is shown when this crusted, hard surface with run-together structure (Figure 2) is compared with a loose, open, porous surface (Figure 3). The soils in Figure 2 were in corn continuously for six years, compared with those in Figure 3, which was in corn once out of three years in a corn-oats-clover rotation. On soils with a heavier texture than this fine sandy loam, the organic matter supply would no doubt be longer lasting.

In some experimental plots at the Connecticut Station, soils with good tilth (structure) yielded as a six-year average 103 bushels per acre compared with 84 for soils in poor tilth. The high-yielding soils had an average of 13 per cent lower bulk density, required 116 per cent less force to penetrate the soil 3⅜ inches, had 60 per cent greater noncapillary porosity, and had 11 per cent more organic matter. The plots were all fertilized with 5-10-10 equivalent to 1,800 pounds per acre.

This research shows that for highest yields, cultivation should be reduced to a minimum. Soils should be cultivated only for weed elimination or for aeration if the soil surface is crusted.

But before minimum cultivation can be practiced, the soils must be kept high in organic matter and in good structure so that surface crusting and compaction will be minimized. Squeezing the life out of soils by unnecessary cultivation and compaction gives what can be expected —a choked-up soil not responding to management, and reduced yields.

We are entering a new era of thinking relative to tillage and crop growing—till the soil at a minimum, reduce compaction, keep soils high in organic matter and fertilize heavily. All this means new machines, a new respect for our soils and a new outlook on soil management.

SUGGESTIONS FOR REDUCING TILLAGE OPERATIONS IN GROWING CORN

By H. P. BATEMAN and WENDELL BOWERS, Illinois

UNDER proper soil conditions, the seedbed for corn can be prepared by making only one or two trips over the field without affecting yields. This conclusion has been reached after five years of research work in the Department of Agricultural Engineering at the University of Illinois, as well as by the experience of numerous farmers who have reduced seedbed preparation

prior to planting. Here are some suggestions for adapting minimum tillage practices to corn-growing operations.

Advantages of Reduced Tillage

1. *Lower production costs*—Reduced cost is the most obvious and principal advantage of reduced tillage. Each trip over the field you can eliminate means a saving of around a dollar an acre.

2. *Better weed control*—Any seedbed preparation method that leaves the soil between the rows rough and uneven will slow down and reduce weed growth.

3. *Less erosion*—A rough seedbed serves as a better barrier against wind and water erosion. It absorbs more of the rain water, leaving less to run off the surface and carry away valuable topsoil.

4. Reduced soil compaction—Fewer trips over the field during a crop season cause less soil compaction.

5. Reduced labor needs—The extra labor that would ordinarily be used during the early spring or planting season might be eliminated or diverted to other enterprises.

Limitations of Reduced Tillage

1. *Maximum soil pulverizing with the plow is required.* Use a plow or travel speed that pulverizes the soil and completely turns under the trash. It is important to completely turn under weeds and growing forages to prevent their regrowth.

2. *Heavy, sticky, or drouthy soils may retard germination.* One requirement of reduced tillage is a good seedbed in the immediate area of the seed. This is difficult to accomplish in sticky gumbo, hard or dry soils, or fields where heavy growths of legumes are turned under. Under these conditions parts or all of the field will need an extra disking or culti-mulching.

Methods to Reduce Number of Operations

Presently owned equipment can be used in most of the following methods. It would not be wise to invest in special equipment until experience has been gained in reducing operations with equipment units already owned.

1. Eliminate the extra disking operation on fall and early spring plowed fields. Tandem-disk only to kill weeds or to fill in large air pockets. Equipping the planter to bandspray the row will reduce the number of cultivations you will need to make.

2. Pulverize the soil more completely when spring plowing. Instead of using a harrow, pull a section of rotary hoe backwards so that it will not pick up so much trash.

3. Plow the soil at planting time and then plant without disking. Pull a light-tillage machine behind the plow. This method has the best chance of success when the soil breaks up easily, as in corn and soybean fields. This method may take more days to plant as the planting rate is regulated by the rate of plowing.

4. Plow the soil at planting time and prepare a narrow strip for the corn row. This will leave the soil between the rows so that the weeds will be slower to germinate and fewer in number than on a finely prepared seedbed. Plant in the tractor wheel tracks if you can adjust the rear wheel to 40 or 44 inches. On other tractors mount two or four special wheels on the cultivator frame at a 40-inch spacing so that pressure can be applied to till and compact the soil. These special wheels can be made by welding shanks to the hub of a rubber-tired auto wheel. Some companies sell drill planter units that have a special rubber press wheel in front of each planter runner. These units can be attached to a square tool bar.

5. Soil can be till plowed and planted to corn in one trip over the field by using the tractor cultivator at planting time. Adjust the sweeps to till all the area. Only one man and tractor will be needed to plant the corn with this method.

6. Reduce the number of cultivations. A rougher seedbed will reduce the weed growth so that the time of first cultivation can be delayed until the corn is taller than normal. Then, in the taller corn travel time can be faster and larger clods can be handled. New types of sweeps and disk hillers can be used to move soil into the row to smother the weeds. Spraying will reduce further the need to cultivate often.

Planting Rates to Use

If there is doubt about getting complete germination in the rougher seedbeds, increase planting rate 15 to 20 per cent. Or plant a little deeper to insure germination.

Will Reduced Tillage Pay?

The question can be answered next fall if a minimum-tillage area is planted as a "control" next to a normal method of planting. Be sure to weigh out the yield, as it is difficult to observe differences in yield of even ten bushels per acre.

PLOW SOLES AND IMPERVIOUS LAYERS

By CHARLES L. SARTHOU, Oklahoma

DURING the recent Farm Equipment Conference at Oklahoma A&M College, Stillwater, Charles L. Sarthou, department of agronomy, Oklahoma A&M College, said that because of our inability to measure accurately compacted layers of soils that the precise causes of such impervious layers are largely unknown. He indicated that the whole subject of soil mechanics is not as well mapped out as other areas of agriculture and as a consequence there was much justification in his presenting several hypotheses. In discussing the subject of impervious layers or compacted layers, he divided these phenomena into two categories, namely *induced* by some soil movement operation and *genetic*, resulting from natural causes, such as evolution.

In general, he observed that *induced* compacted layers could be broken up mechanically but that, as a general rule, if this operation were performed in a dry period or during a dry cycle that a new compacted layer would be *induced* at the latest level worked mechanically.

Consequently, his observations have led him to the general conclusion that during dry periods a minimum disturbance of the soil mechanically would tend to produce less compacted layers which must be broken up mechanically at a later date. Also, that mechanical disturbance in order to break up a compacted layer in a dry period should be done with the foreknowledge that very likely a new compacted layer would be induced at the depth worked so that planting, root zone, fertilization and irrigation depth would all have to be considered prior to selection of the working depth to break up compacted layers nearer the surface.

FORD'S EXPERIMENTAL TRACTOR

By PAUL L. DUMAS

Automobiles have been glamorized in a series of experimental "Dream Cars" that have drawn crowds to auto shows and copy from feature writers. Even the less esthetic over-the-road-motor truck has been publicized if not glamorized by being run cross country powered with an experimental internal combustion turbine. Some useful technical information has undoubtedly resulted from such combination ventures of the sales promotion and engineering departments but some-

how the prosaic farm tractor was never put into the act. But that was in the past; as of March 14th a farm tractor also became a conversational piece with the unveiling of the Ford Typhoon, first farm machine powered by an experimental free piston type turbine engine.

The free piston engine is an old concept which presently is receiving much engineering attention in the development programs of automobile and engine manufacturers. Its revival is due, at least in part, to

the possibility it may lead the way to a turbine type power plant with a specific fuel consumption comparable to existing conventional piston engines. An additional inherent characteristic is the ability of the gas turbine to operate on a wide variety of hydrocarbon fuels including many agricultural products.

As installed in the Typhoon the 100-hp experimental engine has been de-tuned to 50 hp. The free piston portion of the unit is a single cylinder, two cycle, gas (PV) generator

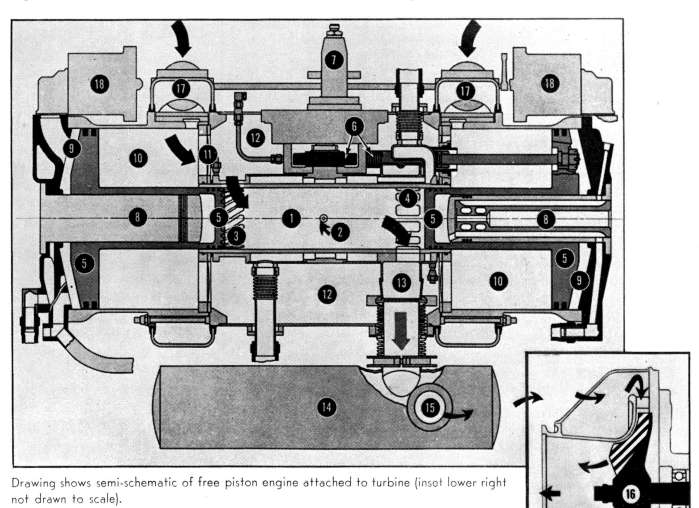

Drawing shows semi-schematic of free piston engine attached to turbine (inset lower right not drawn to scale).

1. Combustion cylinder	7. Fuel injection pump	13. Exhaust tube
2. Fuel injection nozzle	8. Fixed supports	14. Surge tank
3. Intake ports	9. Bounce cylinders	15. Port
4. Exhaust ports	10. Compression cylinders	16. Turbine wheel
5. Free pistons	11. Reed valves	17. Air intakes
6. Rack and pinion	12. Air box	18. Starting cans

The experimental free piston engine mounted in a Ford farm tractor is shown towing a dynamometer. This new concept of farm tractor power is undergoing tests at the proving grounds operated by Ford's Tractor and Implement Division. The free piston turbine has but one cylinder containing two pistons; fuel is ignited by compression temperatures of induced air; it has no crankshaft, spark plugs and no mechanical valves.

which supplies heat energy to a turbine which drives the tractor. In the generator are two pistons of the stepped type, the combustion step being of 3¾-inch diameter and the stroke 4.2 inches. Turbine wheel diameter is 6 inches.

According to the news release the operation is substantially as follows: Referring to the accompanying illustration the "free" pistons (5) are linked mechanically by a rack and pinion to move in and out of the water cooled combustion cylinder (1) simultaneously. The combustion cylinder has a fuel injection nozzle (2), intake ports (3) and exhaust ports (4). The pistons slide on fixed supports (8) and as they move outwardly compress air in the "bounce" cylinders (9). This air acting as a spring is used to return the pistons inwardly toward the center of the combustion cylinder (1).

The engine is started by a vacuum pump that draws air out of the "bounce" cylinders to pull the pistons into the outward position shown. In moving outwardly the pistons sucked into the combustion chamber a charge of fresh air via the butterfly valves (11) and the reed valves (17). Starting cans (18) then convey a metered amount of air to the bounce cylinders forcing the pistons inward to compress the air previously admitted to the combustion cylinder. In moving inward the pistons also compress the air in compression cylinders (10) from whence it is conducted to the air box (12)

via the reed valves (17). Simultaneously when the air being compressed in the combustion chamber reaches ignition temperature fuel is injected at nozzle (2) and combustion takes place automatically.

As the pistons move outward from the pressure of combustion the exhaust ports are uncovered first and a portion of the hot gases enter the surge tank (14) via the exhaust tube (13). Further travel of pistons uncovers the inlet ports (3) and air from the air box (12) enters the combustion cylinder to scavenge the combustion chamber of any remaining exhaust gas while a fresh charge of air for combustion enters from the outside. Simultaneously as the pistons travel outward their larger steps (5) compress air in the bounce cylinders to again provide the rebound to move the pistons inward for the next compression, ignition stroke.

In the meantime the hot exhaust gases diluted by the scavenging air in seeking escape from the surge tank, flow via port (15), into the turbine (16) where their energy is exerted in revolving the turbine impeller which drives the tractor.

Aside from its power plant the Typhoon provides a test chassis for such experimental stage advanced features as a ten forward, two reverse speed transmission, which is shifted without halting by finger tip control, built-in front and rear lighting, a feathering feature for attaching implements and 90-inch wheelbase.

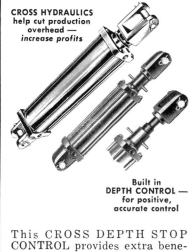

PLASTIC PIPE AND ITS FARM USE

Recent improvements in materials indicate that we can
expect plastic pipe of higher working pressures and temperature
tolerances that will expand its use into fields beyond
water systems, its present dominant application

By DALE LOUNSBURY

Plastic pipe has gained wide acceptance within the past four years in both industry and in agricultural applications. When used within pressure and temperature recommendations, "flexible" plastic pipe offers many advantages over other types of pipe used on today's farms and ranches. Plastic pipe's ease of installation, light weight, low cost and far lower replacement factor are some of the reasons why it is replacing other types of pipe in many urban and rural domestic water systems.

The manufacture of plastic pipe is of relatively recent origin. Its major commercial production began in this country in 1948. Since that date sales have expanded from an estimated $50,000 a year to about $50,000,000 at retail valuation in 1956. According to spokesmen for the plastic pipe industry, 1957 sales should show an increase of about 10 per cent from the 1956 figure. An estimate of about a million domestic water systems employing plastic pipe gives some notion of the extent of market penetration in this particular application.

The use of plastic pipe in this country prior to 1950 was generally unsatisfactory because of poor dimensional stability, temperature and pressure tolerance and a general lack of quality and uniformity. However, development of greatly improved plastic materials and improved extruding techniques have resulted in great improvement in those properties, thus initiating the tremendous growth and development of the plastic pipe industry. National standards of dimensions, performance and material requirements of three of the four major classes of thermoplastic pipe (a plastic which softens when heated and rehardens upon cooling,

Greatest single use of plastic pipe at present is in domestic and rural water systems and in jet well installations.

it will do so repeatedly), have been developed and issued by the United States Dept. of Commerce. The fourth standard, on ABS (acrylonitrile-butadiene-styrene) pipe is expected shortly. The National Sanitation Foundation (NSF) has approved the majority of plastic pipe formulations of 100 per cent virgin material as suitable for transporting drinking water. Such pipe is marked with the NSF seal of approval.

It is generally considered that there are four major types of thermoplastic pipe. They are, namely: polyethylene, ABS (acrylonitrile-butadiene-styrene), PVC (polyvinylchloride) and Butyrate (cellulose acetate). Because space limitations do not permit a general comprehensive discussion of all four major types of plastic pipe, their uses and characteristics are shown in an accompanying table.

Generally the following advantages are claimed for all thermoplastic pipe:

High resistance to corrosion, chemical and electrolytic attack.

Low initial cost and maintenance.

Ease of installation.

Lower labor costs.

Longer length, fewer fittings, low friction loss.

Less collection of deposits.

Light weight.

One disadvantage of plastic pipe is that it is generally unsuitable for use for hot water above 160 deg. F. Its low working pressures and operating temperatures are sometimes

a handicap in its use also. Some types of thermoplastic pipe are prone to damage by rodents by gnawing. However, there is apparently no evidence that it is preferred to other substances.

The principle limiting factors in the use of all thermoplastic pipe are temperature and pressure. It is usually considered that, in general, thermoplastic pipe should not be used if liquids being transported exceed 140 to 160 deg. F. This temperature varies with the type of pipe concerned. The usable pressure will vary with the diameter, the wall thickness of the pipe, and the temperature and density of the transported liquid, as well as the properties of the pipe used.

Of the four major types of plastic pipe, only polyethylene has been extensively used in agricultural applications. However, the continued improvement of such plastic properties as tensile strength, dimensional stability, temperature and pressure tolerance should encourage the expansion of its use for incorporation in farm equipment. This will enable plastic pipe to rival metal pipe for use in high pressure, large scale sprinkler irrigation systems. Information from plastic materials manufacturers indicates the suitability of plastic materials for such things as spray booms, tanks and fittings on various agricultural equipment handling fertilizers. The dairy farm equipment field is another area where plastic pipe will gain wide acceptance as it is presently superior to many types of similar materials being used. It should particularly lend itself to use in pipe line milking systems. It is also quite possible that improved plastics will find use as radiator hose and hose for hydraulic systems and other similar uses in the not too distant future.

The greatest single use of all plastic pipe at present is its use in farm water systems and jet well installations. Polyethylene flexible plastic pipe has been mainly used in these applications. Owners who have been interviewed believe the material to be entirely suitable in providing domestic water on farms. In fact, according to estimates of the Society of the Plastics Industry, Inc., well drillers and contractors who are members of the National Water Wells Assn. believe that over 70 per cent of all newly installed jet wells

in the United States are using polyethylene pipe.

Plastic pipe comes in three forms, rigid, semi-rigid and flexible. Polyethylene, the only flexible plastic pipe, is offered in sizes ranging from ½ inch to 3 inches in diameter and is available in rolls ranging from 250 to 500 ft., depending on the size of the pipe. It is available in tar wrapped cartons or reels for selling from the floor rack. Sizes from four inches to six inches are available in straight lengths.

The rigid or semi-rigid type of plastic pipe comes in 20 and 30-ft. lengths, and diameters of ½ to 6 inches, depending on the manufacturer. They are handled in bundles and can be stored on regular pipe racks.

Farm equipment dealers will become increasingly aware of plastic pipe as it expands into its several areas of farm application. First, dealers will be asked more frequently by farmers for specific types and diameters for expanding or replacing in their water systems. As such plastic pipe will become an important re-sale item or shelf merchandise. Second, dealers will see more applications of plastic pipe as parts of new farm equipment. Take, for example, its possible use as booms in spray rigs and as hose for conducting air, water or other fluids on major machines. Third, there will be many applications in the growing practice of irrigation where plastic pipe or materials can find a use.

Plastic pipe will become an important resale item for most equipment dealers. Greater pressure tolerance will expand its use in sprinkler irrigation systems.

Branch personnel attending the meeting in Racine saw this diversified equipment in the new 12-Month Sales Line.

NEW CASE LINE DESIGNED FOR YEAR-ROUND SELLING

A NEW, diversified 12-Month Sales Line was introduced by J. I. Case Co. at a recent meeting of branch managers and assistants in Racine. The expanded line is designed to provide machines which are usable and salable in seasons when business normally is slack in some types of strictly agricultural equipment.

In addition to farm equipment units, new products demonstrated before the more than 100 persons attending the meeting included light industrial machines, adapted for use by local contractors and custom op-

The 310 wheel tractor, with factory-installed loader-backhoe combination, handles excavating, trenching, back-filling and other types of earth-handling work.

erators, as well as by the farmer himself. While related to the heavy line of Case industrial tractors and earth-handling equipment, the 12-Month Line items are built with an eye to increased sales and service for farm equipment dealers, with a minimum of added investment and service parts inventory, company officials point out.

The most striking innovation for Case is the 310 wheel tractor, with a factory-installed loader-backhoe combination. Designed primarily for the small contractor, parks, cities,

townships and counties, the unit handles excavating, trenching, back-filling and other earth-handling jobs on farms, and makes off-farm income possible for farmers who purchase it. Dealers are expected to find ready markets in the construction industry and local government projects.

An already introduced item in the 12-Month Line is the Model 310 agricultural crawler, designed to give better traction and flotation in hillside, muddy field and muck-land farming. The crawler is available with dozer blade and a free running winch. The blade is readily removable, making it a practical outfit for frequent shifts from farm work to custom or contract jobs when farm work is not pressing.

For strictly on-farm selling, the line includes two new tractors, a new two-row mounted corn picker and three new plows.

The new Super 400 diesel tractor, in the 4-plow class, is available also with gasoline, LP-Gas and distillate engines. It features a bigger engine with more horsepower, pressurized cooling, eight forward speeds and independent pto. In the 3-4 plow class, the 350 tractor is being introduced as a straight addition to the Case tractor line. The 350 features a longer wheelbase, more weight and traction, a 42-hp engine, a 12-speed transmission and constant running pto.

Pull-type performance in a mounted picker is claimed for the 425 two-row model. Forward design gathering chains are arranged specifically to get more of the down corn which is usually troublesome to mounted pickers. A short turning range in both operating and raised position and automatic raising of the wagon elevator whenever the picker points are raised are other features of the picker.

The three new plows—Model M with three bottoms, Model C with four bottoms and Model A with six bottoms—are of girder beam and brace construction, with mechanical or hydraulic lift control.

Branch personnel attending the meeting will later conduct duplicate meetings for dealers in their territories, making available full information on the new products and on sales helps provided by the company. In addition to the sales opportunities inherent in the new 12-Month Line, Case has recently introduced new flexibility into its dealer franchises, and complete financing facilities have been made available through the Case Credit Corp., a wholly owned subsidiary.

The Super 400 diesel tractor features a bigger engine, more horsepower, independent pto and eight forward speeds.

The 350 tractor, in the 3-4 plow class, has a longer wheelbase, better traction, 12-speed transmission and 42-hp engine.

Ford's new Styleside pick-up features passenger car styling and a load area 25 per cent larger than older models. It is rated at 5,000 lb. GVW.

NEW DESIGN AND HIGHER ENGINE POWER IN FORD'S 1957 TRUCK LINE

Nearly 300 different models are offered in line as variations of four basic power plants

Ford's new 1957 truck line, originally not scheduled for production until 1958, was introduced to Ford dealers this month. Offering nearly 300 different models, the line is said to represent the biggest change in the history of the company, and cost $77,000,000 to design and put into production.

The new line features Styleside pick-up models with a streamlined load area 25 per cent bigger than last year; a series of six tilt-cab transport truck models; a new Ranchero model combining passenger car styling and comfort with pick-up truck utility; great payload capacities in medium and heavy duty models, and an expanded parcel delivery series.

Engine power throughout the line has been raised an average of 8 per cent, and engines have been redesigned for better breathing. A new rotor oil pump gives more effective lubrication, a dry-type air filter is more easily cleaned and heavy duty carburetors and lighter flywheels permit faster acceleration. Hydraulic clutches are standard.

Four basic power plants are of-

fered, with 27 different modifications for special types of service.

Ford's largest selling truck, the F-100 pick-up, has been radically redesigned for the new Styleside models. Cab contours are carried straight through to the rear of the truck. The pick-up is rated at GVW of 5,000 lb. and its load space has a capacity of 56.05 cu. ft. There is a steel floor for durability and a wrap-around rear window is provided as optional equipment.

Four tilt-cab trucks come in six series with GVW's from 18,000 to 30,000 lb., and four wheelbase lengths, from 99 to 153 inches. A wide range of medium and heavy and extra heavy duty trucks is also provided for all work and transportation purposes.

The new Ranchero is built on a 116-inch wheelbase with a capacity of 30.4 cu. ft. and has a double steel floor for added ruggedness. The driver compartment and controls are the same as those of a passenger car. It is designed especially for those who require a vehicle for both work and social purposes. It can carry a payload of 1,190 lb.

THIS SHOP CARRIES ITS OWN

... and makes a profit, too,
with careful development of
adequate tooling, proper pricing,
incentive pay for mechanics
and even customer education on
price structures.

A Special Shop Management Report
by
JACK PURNELL

BOLL poppin' weather and a yard full of mechanical cotton pickers greeted my eyes when I visited the Sherry-Barbee Implement Co. shop at Weslaco, Tex., one day this summer. Customer service to build sales is the story of this implement dealership. An incentive pay plan for shop personnel, coupled with flat rate pricing, encourages mechanics to promote and render this profitable customer service.

A year-'round, well planned service program to the farmer is necessary in this three-season semitropical area of the Lower Rio Grande Valley of Texas, serviced by Sherry-Barbee. Summer is cotton season, followed by fall and winter vegetables with orchard care a 12-month chore, all of which provides a ready market for the dealer's service shop. Any slack in customer repair work is utilized in readying and servicing new equipment for delivery and in the reconditioning of trade-ins.

A study of this firm's 1959 books reflected an average of over 350 different customer billings per month, in addition to cash accounts. The service shop's availability of 21,660 man-hours were utilized 69 per cent, with 14,986 profitable production hours utilized. The I&T SHOP SERVICE manual was used for pricing the majority of this work, with adjustments on some items peculiar to this territory. Time studies, made with I&T's system as a guide, have extended flat rate pricing to these special repair jobs.

To tell further the story of this successful shop, we must examine

Mechanic Homer Gorham checks out a tractor on the Sherry-Barbee diesel test stand.

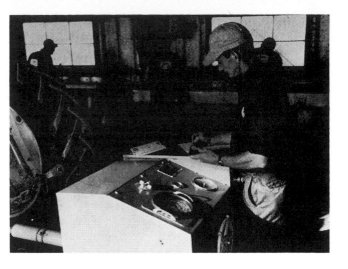

Foreman Gorrell Loran at the dynamometer. Mechanics average $90 per week under flat rate system.

their incentive plan for shop employees, their service attitude to the customer, their intelligent approach to the acquisition of specialized tools, and the close cooperation between equipment sales and service.

The incentive pay plan for shop personnel virtually makes each mechanic an independent businessman in that his earnings reflect his energy and ability, not his age or term of employment. Shop charge to the customer for all work is based on $4.00 per hour, with the mechanic receiving 55 per cent of this charge. He guarantees his work to the customer, and personally stands behind the finished job. Although the shop assures him a minimum earning of $50.00 per week, the records show that the average per mechanic is approximately $90.00 per week.

The shop foreman, although on regular salary, is paid an additional 5 per cent of the shop labor charge to compensate him further for his responsibility of assigning work, pricing jobs, and shop scheduling. All shop personnel receive an annual two weeks' vacation, paid at the rate of their minimum guarantee, and one week's sick leave. In addition, the company meets them half-way on the cost of hospitalization insurance and uniform expense.

In line with management's policy regarding personnel relations, a recent innovation has been a weekly staff meeting held at the shop where the lowest man on the "totem pole" has an opportunity to express his ideas and present his "gripes." Of even more importance to Sherry and Barbee are the monthly department head meetings held at an early

morning breakfast session in a private room at the town's main hotel. Here, key personnel are encouraged to let their hair down and say what they think. This has done much to clear the atmosphere before minor irritations become serious.

Sufficient opportunity to maintain earnings at a high level is assured the mechanical personnel, not only through repair services, but through time spent on readying new equipment for the sales department. Although much of this work is done when repair requests are at an ebb, seasonal fluctuations in sales at times will increase the work load in the shop, making it necessary to add temporary additional personnel for unskilled labor.

The terrific increase in the use of mechanical cotton pickers has been adding considerable mechanical man-hours in readying these new machines for delivery. On new pickers, a flat rate of $100.00 is charged the sales department for preparing the equipment for delivery and operation. This is figured on a basis of 25 man-hours at $4.00 per hour. Local time studies of the operation have shown this to be a profitable

rate for the shop and also fair to the sales department.

Flat rate prices have been established on modification of older picker models to bring them up to the performance advantages of later models. Shop figures for the past year showed 50 per cent of labor time was expended on repair and service jobs, with the remaining man-hours recovered split fifty-fifty between servicing used equipment and new equipment.

Sherry and Barbee feel that the key to sales in their territory is their shop service coupled with their exceptionally well-stocked parts department. Equipment ownership cards are kept on all equipment sold and serviced. This has resulted in more than 50 per cent of their farm customers becoming educated to preventive maintenance servicing, with its obvious profit to both the shop and the farm operator.

Field servicing of diesel tractor equipment is an example of this type of service. Fuel filter exchange for diesel tractor owners is accomplished in the field at a nominal flat rate to the farmer of $2.40 labor plus cost of the two filters.

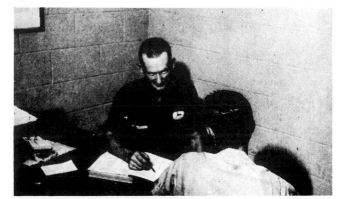

Loran shows a farmer what the price will be on a repair job, a regular S-B practice.

The shop and sales department have six radio-equipped pick-up trucks and cars traveling the territory a major part of the time. This service is a prime sales factor with the farm customer, for it cuts down-time on busy equipment, particularly during grain harvesting and cotton picking seasons, when combines and mechanical cotton pickers are being utilized to the maximum.

Field service has been established on a flat rate basis. Within a five mile range of the shop, a time charge of $5.00 per hour is made for the mechanic and truck ($4.00 labor and $1.00 vehicle.) If a welder-equipped truck is needed, this flat rate is increased $2.00 per hour to cover usage of the welding equipment. Beyond the five-mile range, a mileage charge of 10¢ per mile and $4.00 per hour is made. A flat-bed transport truck is used for bringing equipment in to the shop for major repairs when needed, and charges for this service have also been established on a flat rate.

In line with increasing shop business, a farmer education program through locally conducted service classes is held annually. These short courses are gratis to the farmer and his equipment operators and they acquaint them better with their equipment service needs for top efficient operation. Conducted at the Sherry-Barbee shops by the foreman and shop personnel, these classes make for better relations with the farm customer.

The farmer attending such classes can hardly help but become aware of the many specialized tools and mechanical know-how of Sherry-Barbee mechanics available to them when needed. They can then better understand the cost factor involved in many repair operations in which the flat rate includes not only labor time but mechanic time charges on specialized machines and tools.

Flat rate pricing which includes not only labor time but machine time has encouraged Sherry-Barbee to invest in not only the usual shop equipment, but in many specialized tools, which have been the means of securing farmer customer business and have brought in considerable business from competitive implement dealers' shops.

This is best illustrated in the special tools acquired for servicing of mechanical cotton pickers, such as doffer grinders, picker bar straighteners, and others. Repair jobs utilizing this equipment are flat priced at $6.00 per hour, which includes labor costs and amortization of the tool over a period of five years on a pre-estimate of tool hours operated.

Other shop equipment paying its way and returning its fair share of profit to the firm includes dynamometers, the diesel test stand, electric and acetylene welders, the Allen electrical test stand, hydraulic presses, and a host of small hand and power tools.

A generator overhaul job bears out this statement in that a flat rate of $6.80 plus parts totaling a maximum of $42.72 returns a fair profit to the shop and parts department. This also brings in additional shop business when the farmer compares this cost with that of a new generator at $80.95.

Sherry-Barbee's shop stands on its own feet, making the firm a yearly profit, and is not carried by the parts department or the sales department. In cost accounting, the shop makes its profit on labor and tools. It pays its own operating expenses and is charged with 6 per cent of the general dealership overhead, based on its income against total sales, because the shop does approximately 6 per cent of total business. Parts sales are not credited to the shop income, yet by efficient operation, it not only pays its own direct operating expenses, but shows a profit after absorbing training costs of its personnel who are sent to factory schools and receive on-the-job training at shop expense.

Kenneth Sherry's attitude toward his customers and business reflects a keen awareness of internal costs and responsibility to the customer. The increase in sales of mechanical farm equipment must be matched by an increase in equipment service facilities at a profit to both the shop operator and the farmer. A close relationship among the three prime operations of an implement dealership — sales, parts and shop service, each carrying its share of the load and each returning a profit — is the secret of assuring the future in this field.

Tommy Anderson works on doffer grinder. On the cover: a picker bar straightener.

Ford Introduces 1958 Tractor Line

Fᴏʀᴅ's Tractor & Implement Div. has introduced 12 new models for 1958. These models are offered in two basic power series: the Workmaster series with an 8 per cent power increase at 32 drawbar horsepower, and the Powermaster series with 44 drawbar horsepower for a 10 per cent increase over last year's models.

The Workmaster name has been given to the Ford tractors with the short stroke 134 cu. inch engine. Five models in the 601 series of utility and all-purpose tractors and one model of the 701 row-crop carry this designation.

The 172 cu. inch engine is used in the six Powermaster tractor models, four of which are series 801 utility and all-purpose models and two of which are series 901 row-crop types.

Standard fuel for the 12 models is gasoline, with factory-installed LP-Gas engines, in both sizes, available as options.

In addition to increased horsepower, the new Ford tractors offer a newly designed grille, brighter headlights, greater fuel tank capacity,

improved instrument panel, new functional styling and two-tone color treatment.

Engine design improvements, according to company officials, include changes such as new cylinder heads for both engines incorporating kidney - shaped combustion chambers, larger manifolds and new carburetors. In the Powermaster tractors, a larger air cleaner and a newly designed muffler, to reduce back-pres-

sure, are standard equipment. Engines in both the Powermaster and the Workmaster models have increased compression ratios of 7.5 to 1. The former compression ratios were 6.75 to 1 in the larger engine, and 6.6 to 1 in the smaller engine.

Two transmissions, a four-speed and a five-speed, are offered. Available for factory-installation with the four-speed transmission is an auxiliary over-and-under transmission option to give 12 forward and three reverse speed ranges.

In tractors equipped with power take-off, two types are available. Models 661, 861 and 961 feature Ford's live pto, permitting the operator, using only a single clutch pedal, to stop the forward motion of the tractor while the pto continues to run, or to stop both the tractor and the pto by completely depressing the clutch. Other pto-equipped models have a standard transmission-driven pto.

Power steering is standard equipment on all row-crop models, and power adjusted rear wheels are standard on two models and available for the third row-crop model. These items are offered as options for utility and all-purpose tractor models.

Three front wheel choices are available in the row-crop models. In addition to a dual-wheel tricycle or single-wheel tricycle front end, an adjustable two-wheel front axle is offered.

COMPARISON OF 1957 AND 1958 FORD TRACTOR MODELS
(Gasoline Engines Standard — LP-Gas Engines Optional)

Model	Cubic Inch Engine	Transmission Speeds Forward	Live PTO	Power Adjusted Rear Wheels	3-Point Implement Linkage	Built-in Hydraulic System
UTILITY MODELS—Four-Wheel, for Farm and Industrial Use:						
1957 (620)	134	4	No	Available	No	No
1958 (621)	134	4	No	Available	No	No
1957 (630)	134	4	No	Available	Yes	Yes
1958 (631)	134	4	No	Available	Yes	Yes
1957 (820)	172	5	No	Available	No	No
1958 (821)	172	4 or 5	No	Available	No	No

NOTE (1957): Power take-off, power adjusted rear wheels available at extra cost.
(1958): Power take-off, power steering, power adjusted rear wheels, over-under transmission available at extra cost.

Model	Cubic Inch Engine	Transmission Speeds Forward	Live PTO	Power Adjusted Rear Wheels	3-Point Implement Linkage	Built-in Hydraulic System
ALL-PURPOSE MODELS—Four-Wheel, for Farm and Industrial Use:						
(Built-in hydraulic system, 3-point implement linkage, ASAE pto, swinging drawbar, Proofmeter, headlights and taillight standard equipment.)						
1957 (640)	134	4	No	Available	Standard	Standard
1958 (641)	134	4	No	Available	Standard	Standard
1957 (650)	134	5	No	Available	Standard	Standard
1958 (651)	134	5	No	Available	Standard	Standard
1957 (660)	134	5	Yes	Available	Standard	Standard
1958 (661)	134	5	Yes	Available	Standard	Standard
1958 (841)	172	4	No	Available	Standard	Standard
1957 (850)	172	5	No	Available	Standard	Standard
1958 (851)	172	5	No	Available	Standard	Standard
1957 (860)	172	5	Yes	Available	Standard	Standard
1958 (861)	172	5	Yes	Available	Standard	Standard

NOTE: Power steering, power adjusted rear wheels, over-under transmission (for Models 641 and 841) available at extra cost.

Model	Cubic Inch Engine	Transmission Speeds Forward	Live PTO	Power Adjusted Rear Wheels	3-Point Implement Linkage	Built-in Hydraulic System
ROW CROP MODELS—Tricycle (Single and Dual Wheel) and Four-Wheel, for Farm Use:						
(Power steering, built-in hydraulic system, 3-point implement linkage, ASAE pto, swinging drawbar, Proofmeter, headlights and taillight, high clearance rear axle, standard equipment.)						
1957 (740)	134	4	No	Available	Standard	Standard
1958 (741)	134	4	No	Available	Standard	Standard
1957 (950)	172	5	No	Standard	Standard	Standard
1958 (951)	172	5	No	Standard	Standard	Standard
1957 (960)	172	5	Yes	Standard	Standard	Standard
1958 (961)	172	5	Yes	Standard	Standard	Standard

NOTE: Three front-wheel options available—Single wheel, dual-wheel tricycle and adjustable front axle two-wheel. Power adjusted rear wheels available at extra cost on Model 741.

During the winter of 1957-1958 Steiger built a very large tractor called "#1" that had a 238 horsepower engine. Implement & Tractor Publications bought *Farm Implement News* in March 1958 and combined it with *Implement & Tractor* in May.

On July 15, 1958, U.S. military troops landed in Lebanon under the Eisenhower Doctrine. The troops were withdrawn by October 25. John Deere captured number one position in farm equipment and tractor sales. The position was solidified with the introduction of an all new line of six-cylinder tractors. International Harvester introduced the Farmall 560 model and New Holland purchased the Smoker Elevator Co. of Smoketown, Pennsylvania. Case tractors were restyled and three models included the new, torque converter equipped with "Case-o-Matic Drive." All Soviet tractors were Diesel powered. In America, Polaroid cameras were delivering printed pictures "in 60 seconds" and two-way radio communication was delivering messages between the boss at the base station and his employees in trucks and cars at remote locations.

Fidel Castro overthrew the Cuban dictator Fulgencio Batista on January 1,1959, after a civil war that lasted three years. Alaska became the 49th state on January 3 and Hawaii became the 50th state on August 21.

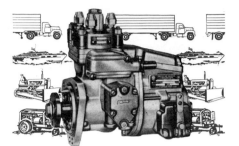

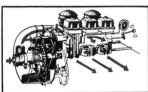

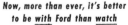

Lawn-Boy Announces 1958 Line of Power Mowers

A versatile, four-wheel lawn-rider vehicle powered by a specially designed 2½-hp engine, plus advances in rotary power mower engine sound-control and cutting performance are featured in the 1958 line of mowers offered by Lawn-Boy, Div., Outboard Marine Corp., Lamar, Mo.

The 85-lb. Loafer can carry a 250-lb. rider up a 35 deg. grade and turns in short radius.

Completely restyled in gold, cream and black, the 1958 line is constructed of lightweight, die-cast aluminum, with a 57 cu. inch twin-chamber muffler for engine sound reduction. The cutting blades of the mowers incorporate aircraft design principles of airfoil grooving to reduce trailing-edge flutter, a major cause of vibration and noise. The blades are also designed to help protect the crankshaft by absorbing shock and impact.

Leading the line is the 85-lb., direct-drive lawn-rider vehicle, the Loafer. Capable of carrying a 250-lb. rider up a 35-deg. grade, it has a specially designed two-cycle engine and can turn easily inside a small radius. Capable of speeds up to 3¾ mph, the Loafer can push or pull a wide variety of equipment such as rollers, aerators, seeders, etc. It has four-inch wide pneumatic tires and an adjustable seat.

Other mowers in the line are the 21 and 18-inch Automower; the 21 and 18-inch Deluxe; the 18-inch Economy and the 18-inch Electric. Electric starting is optional on all models except the Economy, and a leaf-mulching attachment and a windrower are also available optionally.

Owatonna Designs Farm Tractor Maintenance Set

Owatonna Tool Co., 361 Cedar St., Owatonna, Minn., has announced the development of a special universal hydraulic maintenance set for all types of farm tractors. The set contains all basic pullers, attachments and adapters necessary to remove and install gears, bearings, bearing cups, sleeves, pulleys, sheaves, shafts, etc., without damage or distortion of costly parts. The 17½-ton Power-Twin hydraulic unit is said to supply sufficient power to handle 98 per cent of all removing and installing operations, and can be used on other hydraulic equipment as well.

The manufacturer points out that, because the versatile set is adaptable, tractor model changes will not make the tools obsolete. He adds that tool investment can be kept to a minimum by the low cost addition of specific items.

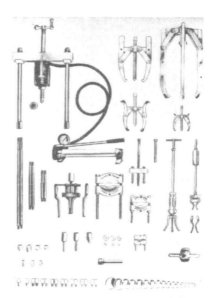

small shots

LET'S PLAY COPS AND ROBBERS!

NO! THAT'S OLD-FASHIONED!

IF YOU WANNA PLAY SOMETHING--

LET'S PLAY AUDITORS AND EMBEZZLERS!

ACilloscope
does the complete job!

No. 1 . . . DETECTS LEAD FOULING. The ACillo-scope is the only electronic analyzer that detects lead fouling, which causes up to 85% of spark plug failures. Moreover, it shows this condition from the start.

No. 2 . . . DETECTS ELECTRODE WEAR. Wear widens gaps. Engines miss at high speed and under extreme loads. The ACilloscope detects this condition, even when it is just starting.

No. 3 . . . DETECTS OIL, GAS FOULING. Deposits from combustion cause many spark plug failures. The ACilloscope points out this condition . . . and here again, it shows it from the start.

SHOWS ALL OTHER CONDITIONS, TOO!

The above conditions are the major causes of spark plug failure. The ACilloscope not only detects them, but also reveals all other conditions . . . improper gaps, cracked insulators, reverse polarity and flash-over. Only ACilloscope assures you of a complete analysis job.

Presto! it's in use!

It takes only four simple connections. You connect red "T" connector to the distributor center tower. You connect black trigger lead to the most convenient spark plug lead at the distributor. Then, you connect battery clips to positive and negative battery posts. Makes no difference which. No chance for error. Allow a few seconds for scope warm-up. When green light shows on screen, have motorist start engine. The ACilloscope is in ACtion!

Give your customers this fast

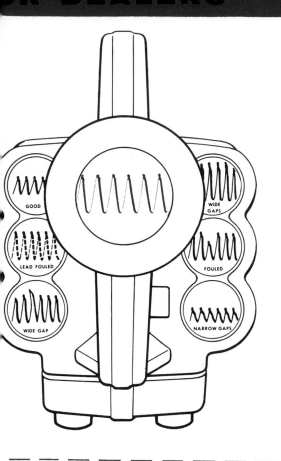

Simple and easy

to read, no training required!

You're looking into the business end of the ACillo-scope, with the key to reading it printed left and right of the central picture. As you can see, you can easily match the pattern you see in the center with the various conditions shown on the key patterns. That's all there is to it. Anyone can read it, with no training at all—servicemen and customers alike.

The ACilloscope features a locked-in circuit, and a conveniently located tuning knob to expand the pattern when studying conditions.

First 25,000

ACilloscope dealers to be listed in national advertising!

Now . . . AC offers you a big national advertising bonus, when you purchase the ACilloscope. The first 25,000 ACilloscope dealers will get their names and addresses listed in a dramatic four-page advertisement scheduled to appear in *The Saturday Evening Post* early next spring. These names will be selected on the basis of the first 25,000 Warranty Cards returned to AC. Thus, your name will be placed before thousands of performance-conscious motorists in your marketing area. You'll be recognized in the most select company in the automotive after-market. AC will send you an advance copy of the ad so you can tie in locally to reap the full benefit of this great national promotion.

ficient spark plug service!

The Pellomatic, a pelleting machine on display at the ninth annual Nebraska Tractor Power and Safety Day recently, drew considerable interest from the crowd. It can produce one or two tons of regular grain feed pellets in one hour, powered by its own motor or by pto.

The Wagner TR14A diesel tractor, largest U. S.-made wheel type unit, made an appearance at the show. It develops 155 hp, weighs 10½ tons.

New Timkens Fine for Wheels

Two new low-cost wheel bearings with a high capacity rating for their size suitable for many farm machine applications have been developed by **The Timken Roller Bearing Co.**, of Canton, Ohio, primarily for use in small economical automobiles, particularly for wheel bearings.

The larger bearing has a bore of 1.0625 inches, an outside diameter of 1.980 inches and a width of .560 inches. The smaller bearing has a bore of .6875 inches, an outside diameter of 1.570 inches and a width of .545 inches.

Application of the two new bearings on the front wheel of a light weight car. The smaller bearing on the left is the outer bearing and on the right is the larger inner bearing.

From Allis-Chalmers:

AN EXPERIMENTAL FUEL CELL TRACTOR

T. G. Kirkland, member of the research team which built and installed the cells in this engineless tractor, discusses a unit of nine cells. Chemical reactions between a mixture of fuel gasses and oxygen—largely propane—and an electrolyte within the cells produce 15 kilowatts of electricity.

Allis Chalmers showed a research tractor powered by a DC electric motor and 1008 fuel cells. The tractor weighed 5270 pounds and was based on D-12 axles. Very little else was like the D-12. International Harvester purchased 98.5 percent of Solar Aircraft, Co. of San Diego, California, manufacturers of turbine engines and Massey-Ferguson Inc. purchased Perkins Engines, Ltd. of Peterborough, England.

Soviet Premier Khrushchev broke off the 1960 Paris summit meeting with President Eisenhower on May 16, after an American U-2 "spy plane" was shot down on May 1. The Oliver Farm Equipment Co. was purchased by the White Motor Co. of Cleveland, Ohio, on October 31, 1960.

New Farmall 560 . . . today's most powerful row-crop tractor!

1961-1968

**BIG TRACTORS GET BIGGER
AND SMALL TRACTORS
GET SMALLER**

Grounds Maintenance

1961-1968

During the 1960's, big tractors got much bigger and small tractors got even smaller, but more important, both extremes were more popular than moderate sizes. Lawn mowers also entered the horsepower race and 2.5 hp wasn't enough to mow the lawn with any more.

DDT and some other compounds were banned from routine agricultural use in the 1960's by the Environmental Protection Agency.

On January 3, 1961, the United States stopped diplomatic relations with Cuba because Castro's new government refused to pay for confiscated U.S. property. The Peace Corps was begun on March 1 by President John F. Kennedy, on April 12, Major Yuri Gagarin, Soviet cosmonaut, was the first man to orbit the earth from April 17 to 20, Cuban troops overwhelmed an invasion force of about 1600 exiles at the Bay of Pigs. In August, the Soviet Union built the Berlin Wall to prevent the escape of East Germans to West Germany and U.S. forces in Vietnam totaled 3,200 on December 11.

The International Harvester HT-340 was destined for the Smithsonian Institution six years later, but the research tractor caused many raised eyebrows in 1961 with no shift lever, no throttle, no brake or clutch pedals, no transmission gears and no cooling water or antifreeze. International Harvester engineers covered an 80 hp T62T gas turbine engine, made by their Solar Aircraft Co., and a hydrostatic transmission with a sleek fiberglass hood and fenders. A high rate of fuel consumption, very hot exhaust temperature and loud noise were some of the serious problems for the HT-340, but engineers were optimistic about finding their solutions.

In 1961, Allis-Chalmers sold riding lawn mowers made by the Simplicity Manufacturing Co. The Clinton Engine Corp. introduced a small, 3.5 hp air cooled engine equipped with Clevite's "piezoelectric" ignition system that has no points, condenser or magneto.

International Harvester HT-340 Research Tractor

"Our intent in developing this tractor," says C. H. Meile, chief engineer of IH Engineering Research, "is to continue our investigation of new types of power systems. This is one of the combinations that is both new and promising. We hear a lot today about 'dream' tractors that will give the operator everything from food warmers to television.

"Our point of view is different. We think operator comfort is important, but we're just as interested in designing a tractor that is easy to operate and that will get its work done in the least amount of time. Tomorrow's farmer, we think, will prefer to enjoy the comforts of home in his home, not in the field. This is the direction we are moving in our research with this tractor."

The low, forward slope of the research unit is possible because of the small size of the gas turbine engine, which is 21 inches long and 13 inches in diameter.

Without its fibreglass covering, the HT-340 looks like this, as it was photographed on an assembly stand. Note rear-mounted fuel tank, front exhaust, air intake on side. IH researchers are, from left, J. R. Cromack, Carl H. Meile, chief research engineer, and Ralph E. Wallace.

Lt. Col. John H. Glenn Jr. became the first American to orbit the earth on February 20, 1962, and the U.S. sent troops to Thailand to prevent a possible invasion. The "Cuban Missile Crisis," including the U.S. blockade of October 22 to 28, forced the Soviet Union to withdraw missiles from Cuba. New Holland purchased Haro-Bed of Fowler, California, and the White Motor Co. of Cleveland, Ohio, purchased Cockshutt Farm Equipment Ltd. of Brantford, Ontario. Cockshutt became a subsidiary of the Oliver Corp., also owned by White.

The Agricultural Stabilization and Conservation Service of the USDA mailed the following instructions to farm equipment manufacturers, wholesalers and retailers, "in the event of surprise enemy attack or in the event the president declares a national emergency."

HOW DOES A TURBOCHARGER WORK?

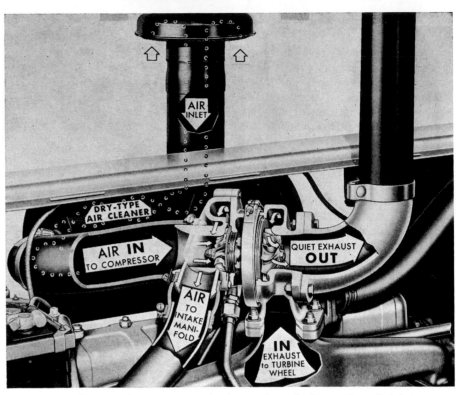

Air at normal atmospheric pressure is drawn through the engine air inlet, passes through a dry type air cleaner, is compressed by the turbocharger, and then passes into the intake manifold. After combustion the exhaust gases pass from the exhaust manifold through the side of the turbocharger powering the compressor and are exhausted to the air.

Instructions for Industry In Case of Attack

Dealers—Retail sales and physical transfers of new equipment must be discontinued and no machines may be moved from the premises even to make delivery of previously closed sales, except to protect goods from fire, flood, or other unusual hazards, or in accordance with direction by state or county ASCS officials.

While the prohibition is in effect, persons seeking to buy new machinery for any purpose are to be referred to the local ASCS office. There will be no prohibition on sale of used machinery unless directives are issued to such effect.

Repair parts may be sold providing the user signs a statement of emergency need and, where practicable, the part to be replaced is surrendered to the dealer.

Dealers may get parts through routine channels, but special orders for unstocked parts require the customer's signed statement of emergency need.

Manufacturers and distributors— The same requirement to discontinue sales of new whole goods and deliveries of already-closed sales applies to both manufacturers and wholesalers. New machinery may be moved only to protect it from hazard and directives from USDA national headquarters or from its regional liaison representative must be complied with.

Parts are to be distributed to dealers "as equitably as possible under existing conditions in accordance with your own policies and procedures."

Special orders from dealers are to receive handling prior to filling regular orders.

According to the booklet, the USDA is responsible for farm equipment distribution programs in case of emergency. "A standby order governing this distribution has been drafted in consultation with representatives of the Farm Equipment Institute and the National Retail Farm Equipment Assn.," it adds. The order will be issued in an emergency to provide the legal basis for actions described in the folder.

Allis-Chalmers Sugar Babe develops estimated 135 dhp, has all-wheel steering, torque converter drive, power shift transmission, power brakes.

HUSKY 600

HUSKY 800

Balmar 60 steering is controlled by reducing speed of planetary gear sets driving wheels on turning side. Ford 172 engine develops 62.5 estimated bare engine hp in gasoline model. Constant mesh 4-speed transmission with auxiliary reversing transmission.

Fordson Super Four features direct coupling between front and rear wheels on each side, front wheel steering, non-interchangeable front axle which cannot be disengaged for road work; 220-cu.-in. engine develops estimated 50 bhp.

John Deere 8010 has hinge steering, develops estimated 150 dhp, available with 3-point hitch for mounted equipment, provides 3-function hydraulic system with 60 gpm pump, has 24-volt electric starting system, and is equipped with four-wheel air brakes.

FWD Wagner WA-4 features hinge steering, engine power ratings of 112 to 250 hp, oscillating axles, planetary gear reduction at each wheel.

A "Hot-line" telephone was installed on August 30, 1963, to improve communication between the Russian and American heads of state in case of a national emergency. President Kennedy was assassinated in Dallas, Texas, on November 22, 1963.

Minneapolis-Moline was purchased by the White Motor Co. in 1963 and New Idea bought the Uni-Tractor line from Minneapolis-Moline.

The Civil Rights Act of July 2, 1964, made racial discrimination illegal, U.S. bombing of North Vietnam began on August 5, 1964, in retaliation for a reported attack on U.S. destroyers and the "War on Poverty" began on August 20. Case announced the 1200 Traction King (4WD) model, Deere moved to new corporate headquarters at the edge of East Moline, Illinois and New Holland purchased Claeys of Zedelgem, Belgium.

Tax Cuts in 1964-1965 began an economic boom, the first U.S. ground combat in Vietnam was authorized on June 28, 1965, and riots in the Watts area of Los Angeles on August 11-16 resulted in 35 deaths.

REFLECTIONS

Perspectives
and
Personalities

By ELMER J. BAKER, JR.
1889-1964

International Harvester 4300 developed 214.23 dhp at Nebraska, features all-wheel steering, oscillating rear axle, heavy-duty 3-point hitch, dual hydraulic systems for power steering and implement control, 6-speed transmission.

M-R-S A-100 features all-wheel 4-option steering, estimated 120 dhp, oscillating front axle, planetary gear reduction at each drive wheel, and 10 forward speeds of 2.2 to 21.19 mph.

Minneapolis-Moline G-706 available with diesel (92 bhp) or LP-gas (95 bhp) engine. Front-wheel drive may be disengaged from tractor seat. Five-speed sliding gear transmission provides speeds of 3.3 to 19.2 mph. Power steering is standard.

Gravely Introduces New Four-Wheel Riding Tractor

Gravely, Tractor Div. Studebaker Corp., Dunbar, W. Va., has announced the new Westchester model lawn and grounds tractor for 1964 which features the use of front wheel drive, a maximum horsepower output of 12 hp, and front-mounted power attachments.

The new model combines a Fiberglas-reinforced plastic body with a number of other features, including Swiftamatic 8 transmission which gives a choice of four speeds forward and four in reverse. The automotive-type transmission has speed capabilities from 1.55 to 4.25 mph. The manufacturer states that the greatest design advantage is the surefooted and super-traction features provided by the front wheel drive. The compact unit has full engine weight concentrated on the front driving wheels.

Attachments for use with the Westchester include a 50-inch rotary mower, 75-inch reel mower, snowblower, snowplow, lawn roller, seederspreader, sprayer and lawn cart. A hydraulic lift is standard equipment.

All controls for the unit and attachments are mounted in a sport cartype console and the semi-bucket seat can be tilted forward to keep it dry. The ignition system, a 12-volt startergenerator combination, is switch-key controlled as with an automobile.

Massey-Ferguson MF-97 offers estimated 90 dhp with LP-gas engine. Powered front wheels permit shorter full-power turns without braking than 2-wheel drive model. Four-wheel drive also permits better weight distribution between axles.

Oliver 1900 (also 1800) front-wheel drive may be disengaged from the tractor seat. Front drive axle on the 1800 is also interchangeable with conventional row crop front axle. Estimated dhp of the 1900 four-wheel drive is 89.

DARK HORSE (Wheel Horse of Course)

Last year we thought this big, powerful, multi-purpose tractor would have only limited use in the home market. We were wrong! Unprecedented sales to homeowners as well as to organizations, institutions and municipalities left dealers' stocks depleted.

Now 10 hp. and in good supply, the biggest of all Wheel Horse tractors presents itself as a real dark horse for the '64 sales race. Like our popular 6, 7 and 8 hp. models, it comes with a complete line of optional quick-attaching tools for year 'round use

For complete information on Wheel Horse and one of the most lucrative franchise arrangements (untarnished by dual or discount distribution), write Wheel Horse Products, Inc. Address . . .

Wheel Horse®

505 W. Ireland Road, South Bend 14, Ind.

MINIMUM TILLAGE EQUIPMENT COMES OF AGE

Iowa State U. held its second Minimum Tillage Field day last month, constituting a sort of "World's Fair" of equipment for this practice. The most significant observation: The basic concept has proved itself to the extent that M-T tools and systems have evolved from improvised, mated-on-farm rigs to factory-engineered outfits.

By Mark Zimmerman

Minimum tillage has been gaining increasing interest for at least 10 years. About 3,500 Iowa, Nebraska, South Dakota and Minnesota farmers proved this trend by attending the May 5 Minimum Tillage Field Day at Sioux Center, Iowa. This crowd, nearly five times larger than last year's Ames attendance, gathered to see both new equipment and techniques.

Iowa State ag engineer Dale Hull presided over the field day by guiding farmers to 18 equipment demonstrations by 11 manufacturers and one local farmer. Thus, equipment ranged from a 7-year-old homemade wheel-track planter to pieces scheduled for the market in 1965. Despite its limited acceptance to date, the broad array of new equipment places minimum tillage at the threshold of widespread adoption.

Field capacity has dramatically increased from the earlier one-row plow-plant and two-row wheel-track plant systems. Predominately 4-row width equipment was displayed. But power requirements have been upped, too. As a result, the light-heavyweight tractors (See I&T, Feb. 7, 1964) were extensively used. With one exception, D-19, 830, 3020, 706, 602 and 1600 tractors powered the minimum tillage equipment.

Most minimum tillage equipment is designed to achieve one funda-

---

The M-T Practice at a Glance

Applying minimum tillage techniques to unplowed ground offers cost-saving, once-over operation. But the over-all system often requires preparatory stalk cutting, disking or rotary tillage. High power requirements and field work postponement until planting time should be considered.

Minimum tillage systems for plowed ground promote better labor distribution with fall and early spring plowing. Large, integral units are often costly and require total practice adoption. Chains of component equipment are usually unwieldy, but allow use of existing equipment and permit tool substitution for other operations.

Minimum Tillage (continued)

Equipment for Unplowed Ground

STRIP-TILL

Howard Rotavator (Harvard, Ill.) offers 2 options: Till 12-inch strips for each row, work 80-inch width ahead of 2-row planter.

TILL-PLANT

Enlarged sweep, 1-inch press wheel, trash guards are key elements of till planter. Fleischer (Columbus, Neb.) and International Harvester both demonstrated till-planting. In one trip, 4-row units till-plant in old corn ridge. Disk hillers cultivate, reshape rows.

MULCH TILLAGE, PLANTING

Case's "Chisel-Planter" incorporates 13-ft. chisel plow, toolbar, 4 planting units. (Set for 1965 market.)

Bush Hog (Selma, Ala.) showed "Variplanter" with subsoiler, plateau-forming disks.

Oliver field cultivator, "Culti-Plant" hitch and planter comprise unit for once-over tillage, planting.

Equipment for Plowed Ground

CULTIVATOR-PLANT

IH cultivator mountings include gangs of shovels, disks, rotary hoe wheels. Planter-mounted spring tine clusters also precede planter disk opener.

Glencoe (Portable Elevator Co., Bloomington, Ill.) fielded this rear mounted field cultivator with dolly-wheel hitch.

Dunham (Richmond, Ind.) "Culti-Mulcher" has 2 spring tooth harrow bars between 2 rollers. It precedes planter in equipment "train."

Emmert (Audubon, Iowa) has two 13-ft. spring tooth harrow bars with depth control which fit beneath Deere planter hitches.

(Other cultivator-plant units shown at Sioux Center were: Case rear-mounted field cultivator, 2-part hitch and planter—the "Culti-Planter." First hitch element is tool bar mounted, second pivots and runs on castor wheels; Deere's No. 9 "Arch-Hitch," which permits use of disk harrow or pull-type field cultivator between tractor and planter.)

STRIP PROCESS

Oliver cluster of 1 sweep, 2 rotary hoe-type wheels prepares row area ahead of each planter opener. Sweep also opens soil for fertilizer placement.

WHEEL-TRACK PLANTING

Deere's No. 10 hitch with 4 hitch wheels replaces planter wheels to provide added firming.

Allis-Chalmers off-set hitch with 2 tires and rear tractor tires providing 4-row packing.

(Other units shown for plowed ground application included Bush Hog's "Varitiller"—which performed equally well on unplowed ground—and a home-made wheel-track planting unit which a local farmer had used for 7 years on 300 acres.)

1961-1968

Fertilizers, herbicides and insecticides can be used in various combinations of liquid and dry forms in minimum tillage. Incorporation devices at the M-T Field Day ranged from the conventional to this Gandy (Owatonna, Minn.) "Ro-Wheel." And some chemicals were merely surface applied.

Seven hitches on display (this one is Case's) permitted tillage tool interchangeability or substitution of added units in place of planter. Farmers looked to these for extending existing equipment use and operation versatility.

Minimum Tillage

mental objective: To *optimize* conditions for crop growth, discourage weed growth and *minimize* field operations. It is then logical to ask, "Why so many types of equipment?"

Six sets of equipment operated on corn stalk ground without prior plowing. Encountering excessive soil moisture and tough corn stalks, the units employing field cultivators for tillage were hampered by stalk plugging. The primary use for these units is in dryland conditions where exposed trash cuts wind and water erosion. However, surface trash in humid areas could retard soil warm-up and seed germination.

An approach seeking to concentrate trash in row middles and plant back in the original rows—till-planting—attracted considerable attention. Features that appealed to farmers are effective trash management and its once-over capability.

Meanwhile, 12 units went through their paces on already plowed ground. This is significant because some farmers continue fall and early spring plowing to aid seedbed prep-

aration and labor distribution. Still other farmers wish to extend use of existing equipment. Special hitches and component attachments permit this assemblage.

Adaption components allow tillage tool substitution to cope with varying conditions. However, maneuverability is often sacrificed with lengthy equipment "trains." Integral machines, on the other hand, usually cost more and forfeit versatility, making minimum tillage adoption an all or none situation.

In brief, no one piece or series of equipment and its corresponding minimum tillage system is equally suited to all farms.

Many functions aren't really eliminated in minimum tillage. Instead—primary and/or secondary tillage; fertilizer, herbicide and insecticide applications; planting and consideration of cultivation (if any) methods are combined for once-over field operation. And at first glance, this appears costly and complicated to farmers.

The involvement of such diversi-

fied knowledge is indicated when one sees the college personnel conducting the field day: An ag engineer, entomologist, botanist and two agronomists were present. In addition, representatives of chemical, seed corn and equipment companies were on hand to answer questions. The *challenge* to both dealer and farmer is to combine these talents.

A rotary device, for chemical soil incorporation, and a high clearance unit, for pre-emergence spraying, also made appearances. Afternoon plowing demonstrations featured tractors pulling up to eight 16-inch bottom plows. An added highlight was the first U. S. showing of the new 4-wheel drive Case 1200 tractor.

Attending Iowa State University personnel and their respective specialties were Dale Hull—equipment; Frank Schaller—fertilizers; Harold Gunderson—insecticides; and E. P. Sylwester — herbicides. The event was sponsored by the Iowa State ag extension service, Sioux County extension council and radio statior WNAX.

Utah State's 'Man and His Bread'

Purpose of "Man and His Bread Museum," recently organized at Utah State University, Logan, is to tell the story of man as a food producer and how he has become the super food producer he is today. Notes the museum's chief, Carl Hugh Jones: "This is the most important story in all human history because developments in agriculture are freeing more and more people from the tasks of planting, caring for and harvesting their own food. Once freed from the burdens of planting and the harvest, the non-agricultural segment of society may turn to production of those things which make possible our high standard of living."

The museum will have exhibits and other programs which show the relationship between food production and civilization. It will have on exhibit examples of agricultural equipment and will demonstrate how they have affected both the farmer and the city dweller.

Present plans include an exhibit area of 7,000 sq. ft. in a new library building on the university campus. The exhibits are expected to be ready for the visiting public soon.

The exhibits will depict the history and development of sources of farm power, of plowing, of harvesting, of threshing and of the development of various varieties of wheat. There will also be special exhibits of pictures and small objects designed to travel from museum to museum and to fairs and special conferences. These special exhibits will cover a great variety of subjects such as sugar beet growing; methods of soil conservation; dairying, ancient and modern; and the history and development of the combine harvester.

While work is being done on these first exhibits, plans are being made for a large museum building which eventually will house more extensive presentations. This will provide room for exhibits on irrigation, dairying, forestry, horticulture and other areas of food production.

There will also be an outdoor museum consisting of demonstration plots, where old varieties of grain will be grown, using the old and primitive methods. Special days will be set aside to demonstrate methods of plowing, preparing the soil, planting, harvesting and threshing.

The museum also has some needs. It needs to improve its library by collecting books and other literature about farming equipment. This collecting has been started by acquiring literature and histories of contemporary manufacturers. There is

Replica of Groundhog thresher, which had only a cylinder and concave, illustrates one step in the development of modern units.

International Harvester gave this ancient McCormick reaper, the start of it all, to the museum.

Reeves steam tractor, partly restored, stands as an example of farm power in use from the 1870's until World War I.

Museum

Ambitious project seeks to show how man has become the food producer he is today. Besides setting up machinery display, the museum is collecting books and periodicals about equipment, and needs some help.

also need for old publications such as the old farm magazines and trade publications (*e.g., the Case Eagle, the John Deere Furrow, Farm Implement News,* and their *Buyer's Guide,* and *I&T* and its *Red Book*).

A special feature of the museum collection will be an indexed picture file of agricultural equipment and methods to be used in preparing the traveling exhibits and in research. There is a real need of photographs for this file. The museum is collecting examples of farming equipment ranging in age from 10,000 years to 10 days.

Any individual or company with material, books, literature, photographs or equipment which might help the museum's collection grow can write to Carl Hugh Jones, museum head, "Man and His Bread Museum," Utah State University, Logan, Utah.

A stationary thresher, built by A. W. Gray's Sons, Middleton Springs, Vt., represents a type midway between the Groundhog thresher and the machine associated with the steam engine.

WHAT IS INTERNATIONAL HARVESTER?

'Fine, old farm equipment institution?'

By Harry O. Bercher, President
International Harvester Co.
(to the New York Society of
Security Analysts, April 21.)

IT SEEMS to me that some important facts about our business are not well known, perhaps because we have failed to give them adequate emphasis.

One such fact is that it has been a long time since farm equipment was the principal element in our business. In the past 10 years, for example, farm equipment sales have rarely exceeded one-third of total sales. Yet in nearly all publications we are classified as a "farm equipment company." That is what is known in Hollywood as "type casting."

We are proud to be in the farm equipment business and to be the world leader in it, but we would like to be known as we actually are.

The major industries in which we are engaged are motor trucks, farm equipment, construction equipment, aerospace products, iron and steel, and cordage. We have more than 1,000 products, not including the many variations on basic machines.

Our major competitors in the products which account for most of our sales are such companies as General Motors, Ford, Chrysler, Caterpillar, Mercedes-Benz, Renault and others at home and overseas. In farm equipment, of course, one would add such fine companies as Deere, Massey-Ferguson, Allis-Chalmers and others. I think that we can safely say that few, if any, companies have a more gilt-edged list of competitors than we have. And we take them all very seriously, indeed.

It is a difficult problem to try to hang a brief label on a company under modern conditions. Perhaps ours should be called an *automotive* company, for more than 80 per cent of the products we sell are self-propelled and self-guided, travel on wheels or movable tracks, and generally get their power from engines which we design and manufacture. That meets the dictionary definition of "automotive."

The essential fact is that we are a diversified capital goods company. Our business, in the broadest sense, is applying mechanical power to do the work our customers want done. Almost everything we make is used by the buyer to earn his own living.

The image of our company has been that of a fine, old business with very high quality products; a financially strong company with an excellent dividend record; but not very exciting, not very aggressive, not going anywhere very fast. In short, there has been the feeling

that we were an "institution," with all the overtones that are implied by that word. Let me make it quite plain that my associates and I are not interested in being custodians of a business museum, however fine. We are more interested in the future of things like gas turbine engines and the advanced metallury of our Solar Division.

(Mr. Bercher noted that in the later 1950s the company "simply didn't earn enough profit" and described corrective programs that had been undertaken in all divisions, some of which were completed and others well under way by the end of 1961.—Editor.)

In our 1961 fiscal year our sales were $1 billion, 612 million. On that, we earned $48.4 million, a return on equity capital of 4.74 per cent and a return per share of common stock of $3.02. At the end of our last fiscal year (1963) our sales were $1 billion, 957 million, our record up to now. Our net income in 1963 was $68.3 million, equivalent to $4.58 a share and to 6.5 per cent of equity capital. During that two-year period, our sales increased 21.4 per cent, our earnings increased 41 per cent, and our rate of return on equity capital improved 37 per cent.

We do not regard these results as evidence of a job completed, but only as evidence of a job well started and as proof that our program is a sound one.

Our company's sales for a full year passed $1 billion for the first time in 1948. So it took the business 117 years to hit the billion-dollar mark. We expect that for 1964 our year's sales will be something in excess of $2 billion. If so, we will have added our second billion in only 16 years, and can start working on the third.

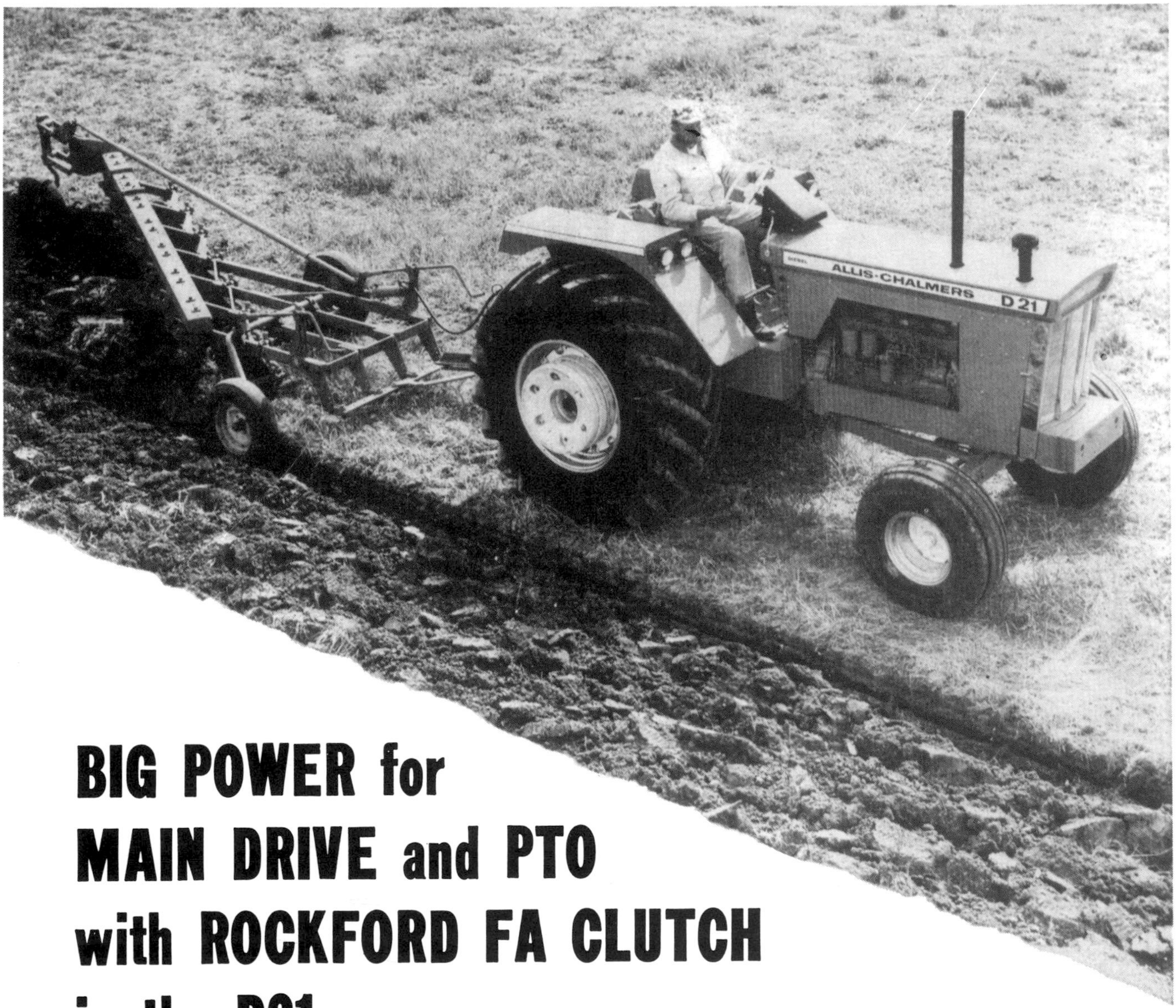

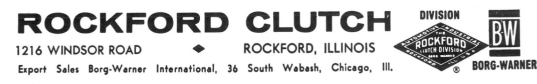

BIG POWER for MAIN DRIVE and PTO with ROCKFORD FA CLUTCH in the D21

Big farming calls for the big power in the Allis-Chalmers D21 farm tractor. Whether its pulling a 7-bottom plow, double discs or other high-performance implements, the D21 power train takes all the "horses" that the big diesel engine puts out and puts them to work. Rockford FA Dual-Drive clutch has patented release lever design that requires low pedal pressure. The low-inertia, high-strength clutch design speeds shifting. FA clutches are available in a wide range of torque capacities. Find out how Rockford Clutches can be incorporated in your equipment designs for dependable performance. Call or write for OEM engineering recommendations without obligation.

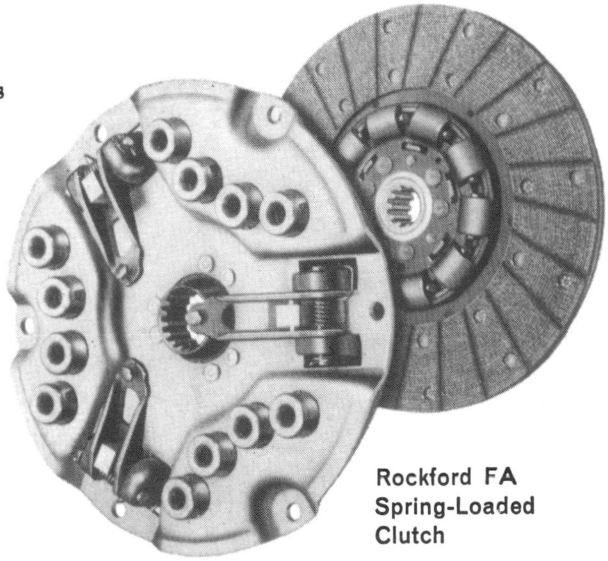

Rockford FA Spring-Loaded Clutch

ROCKFORD CLUTCH DIVISION

1216 WINDSOR ROAD ◆ ROCKFORD, ILLINOIS

Export Sales Borg-Warner International, 36 South Wabash, Chicago, Ill.

AUTOMOTIVE · INDUSTRIAL · THE ROCKFORD CLUTCH DIVISION · BORG-WARNER · ROCKFORD · ILL. U.S.A. ®

B-W BORG-WARNER

Year	Test Milestone	Test No.	Make and Model
1920	Cast iron unit frame	18	Fordson
	Practical power take-off	24	International 15-30
	General purpose tractor and one of the first with storage battery, battery ignition, electric starter & governor	33	Moline Universal D
	Crawler tractor	45	Cletrac W
1922	4-wheel drive	84	Rogers 4-Wheel Drive
1925	Successful all purpose-type	117	Farmall
1926	Distillate fuel-type	128	Hart-Parr
1930	Imported tractor (from Ireland)	173	Fordson
1931	Direct engine drive pto and first wheel-type to pull its own weight	192	Bradley General Purpose
1932	Diesel fuel-type (crawler)	208	Caterpillar
1934	Rubber tires	223	Allis-Chalmers WC

THE LONG LINEUP AT

■ A BUSY season for Prof. L. F. Larsen and his crew of tractor test engineers is shaping up at the University of Nebraska. Tractor No. 940 will likely be put through its paces at the Lincoln track by the end of 1965. (Test No. 876 wound up the 1964 season.)

With 60-plus tractors slated, 1965 will see the second highest number of tests for a year since the Nebraska Tests began.

Sixty-nine tractors were assigned test spots in 1920, but due to rescheduled and withdrawn tractors, only 62 official ratings were made. So in terms of complete, published results, we could see a new record this year.

In 40 years of full-scale testing (discounting the war years), an average of 22 tractors a year have been tested. Other years with extra heavy test schedules are 1958 (47), 1959 (49) and 1960 (50).

The 1000th tractor should be tested in 1967, which is, appropriately, Nebraska's centennial year. Hence,

plans are being made to present an historical program on tractors at the 1967 Nebraska Tractor Day. Such notables as the first tractor tested, John Deere's Waterloo Boy; a Farmall regular; Fordson and an early rubber tired model may be seen on display.

Other important engineering advances in tractors, according to their appearance at Nebraska, are listed in the accompanying table.

Prolific manufacturers are responsible for the busy 1965 test season.

Year	Test Milestone	Test No.	Make and Model	Year	Test Milestone	Test No.	Make and Model
	Last kerosene fuel-type	229	McCormick-Deering W-12		Turbo-charger; largest crawler tractor	584	Caterpillar D-9
1935	Diesel fuel-type	246	McCormick-Deering WD-40	1956	Best fuel economy for gasoline, LP-gas and diesel fuel-types	598, 590, 594	John Deere 620, 520, 720
1936	High compression gasoline engine	249	M-M Twin City KTA HC		Last distillate fuel-type	606	John Deere 720
		252	Oliver Hart-Parr Row Crop 70 HC	1958	Torque converter with lockout	679	Case 811-B
1938	Hydraulic lift	296	Graham-Bradley	1959	Full power shift transmission	701	Ford 881
1940	Three point hitch and hydraulic control system	339	Ford-Ferguson 9N	1960	Hydrostatic power steering	759	John Deere 4010
1948	Torque converter (crawler)	397	Allis-Chalmers HD-19	1962	Turbo-charger	811	Allis-Chalmers D-19
1949	LP-gas fuel-type	411	M-M U Standard LPG		Largest wheel tractor	815	International 4300
1954	Hydraulic power assist steering	528	John Deere 70		Largest 2-wheel drive tractor	828	John Deere 5010
1955	Partial range power shift transmission	532	Farmall 400	1963	Alternator	855	Allis-Chalmers D-21
	Super charger (crawler)	550	Allis-Chalmers HD-21 AC	1964	Vacuum advance ignition distributor	874-5	Oliver 1650 and 1850

NEBRASKA'S TRACK

By Mark Zimmerman

In the last few months, there have been introductions of an entirely new line, a revised line and various other new or modified models. Testing is required when changes in basic power train design affect a tractor's pull, speed, horsepower or fuel economy.

The previous high for a given make is the 20 Case tractors tested in 1960. The new Ford line could challenge this record with its many models and their engine and transmission options. And Massey-Fergu-son will probably be next with about 10 tractor tests. If combined, the offerings from both companies could account for half the season's tests.

Nebraska Tractor Tests are conducted from March through November. By crowding abilities and facilities to the limit, engineers can test about two a week for a maximum of about 78 tractors a year. Bad weather generally shortens the season, but pto testing in a laboratory environment is always possible. This and careful scheduling enable the engineers to approach their maximum limit.

As the season progresses, watch for additions to the list of tractor test highlights. (One can also observe the effects on tractor design and performance trends reported in I&T, Nov. 21, 1964, *The Farm Tractor: A Case Study in Mechanical Evolution.*)

Incidentally, the 1966 I&T *Red Book* will be a little bigger than usual. It will carry complete reports of tractors tested this year. □

Supercharging the Diese

What superchargers do, how they work—gear-driven or turbine powered.

■ POWER output of a diesel engine depends upon the rate at which it can burn fuel. In most cases, the limiting factor on burning rate is the weight of air available in the cylinder.

In a *naturally aspirated* diesel engine, the partial vacuum produced by the travel of the pistons is the only means for getting air into the cylinders. As engine speed increases, the time available for the natural flow of air decreases. So the weight of air available, in relation to the total amount needed, decreases.

One approach to avoiding the drop-off in engine performance as speed increases is to provide some means for forcing air into the combustion chamber. This process is known as *supercharging*.

Early superchargers were mechanically driven by the output shaft of the engine. In many applications, the potential increase in output of the engine was offset by the power consumed by the supercharger, and by the complexities of the mechanism required to drive the supercharger at high speed.

Turbo-Superchargers

A more recent approach is that of using the exhaust gases from the engine to drive a turbine-like fan wheel. Connected to the same shaft is a blower wheel which literally pumps the air into the engine.

The harder the engine works the more air it needs. Fortunately, more exhaust gases are also available to turn the turbine faster, so that it can provide the needed extra air. Thus the turbo-supercharger tends to be self-regulating. There's no need for operator control of the device.

Because the supercharger is driven by exhaust gases, the energy it absorbs reduces exhaust gas temperature by 120 to 150° F. at rated load. In addition, the need for a muffler is reduced, and in some cases eliminated.

The extra air which the super-

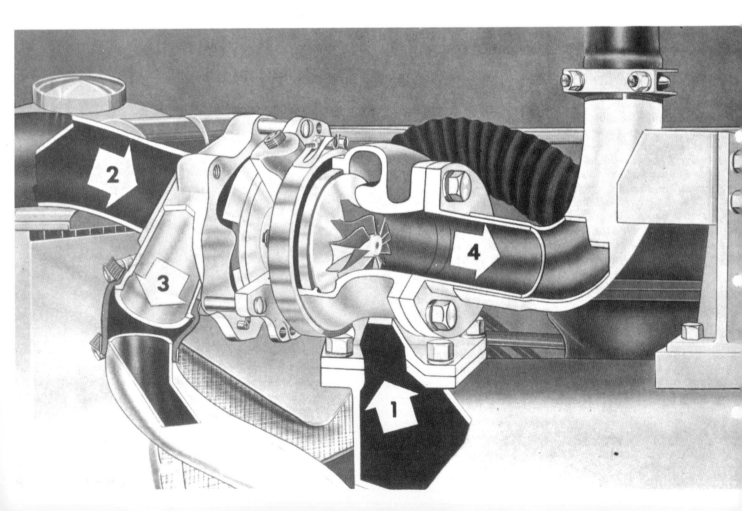

Engine

By Melvin E. Long

Aftermarket units installed on Case 830, Deere 4020. (Photos: M&W Gear.)

charger makes available produces better scavenging, or cleaning out of the combustion chamber after each power stroke. Thus, the critical combustion chamber parts — head, valves, injectors, and pistons—actually tend to run cooler in a supercharged than in a naturally aspirated engine.

The excess of air supplied to each cylinder also serves other useful functions. The more complete combustion which it permits clears up the exhaust smoke; more important, the better combustion also increases fuel economy.

Another benefit unique to the turbo-supercharger is its ability to compensate for the effect of altitude on engine performance. Naturally aspirated engines suffer at high altitude because there is less air pressure available to force air into the intake manifold. In the turbo-superchargers, however, the decreased atmospheric pressure permits the exhaust gases to drive the supercharger at a *higher* speed, thus, it compresses more air and forces it into the cylinders. In general, turbo-superchargers can maintain almost sea-level engine performance up to altitudes of about 7,000 ft.

Gear-Driven Superchargers

A new unit called a differential diesel engine has been developed by F. Perkins Ltd. of Great Britain and Perkins Engines Inc., its U. S. affiliate. This arrangement includes the main power transmission, in addition to the diesel engine and the supercharger. The system is designed to produce a relative constant power output by increasing torque available as engine speed decreases. In the usual engine-transmission arrangement, torque falls as speed declines.

Basically, the system includes a planetary-type differential gear between the engine and the output shaft. The differential gear also drives the mechanical supercharger.

Relatively simple design of turbo-supercharger: Gases from exhaust manifold enter (1) and drive turbine wheel at high speed. Clean air from air cleaner enters (2), is compressed by high-speed rotating blades of compressor. Compressed fresh air flows from outlet (3) to regular intake manifold. Exhaust gases exit (4) to exhaust stack. Energy withdrawn by turbine cools exhaust gas. Blades tend to break up exhaust noises, destroy sparks. Muffler usually is needed. (Photo: M&W Gear.)

Factory-installed unit on Allis-Chalmers D-19. (Photo: A-C.)

When engine speed drops, the planetary-gear arrangement reacts automatically to increase the speed of the supercharger in relation to engine speed. Thus, air intake—and torque output—is increased.

A torque converter eliminates the need for a mechanical clutch. This arrangement permits automatic, two-pedal control of a vehicle as can be obtained from a conventional automatic transmission. But it promises to be a less costly arrangement than the conventional engine and automatic transmission.

Aftermarket

Turbo-superchargers are available for field installation on several current and late-model diesel farm tractors. Price to the farmer, including installation, is approximately $500. For this he can expect to receive a significant increase in tractor power output and an increase in fuel economy.

To avoid undue overloads on the tractor power-transmitting components, the user should be encouraged to use the extra power output in the form of higher speed of field operation. The alternative of pulling larger implements at the same speed produces undesirable excess loading in many of the tractor parts.

Experience to date indicates that premature failures caused by overloading tractor parts has not been a significant problem.

In some cases, the air cleaner must also be replaced with one of larger capacity to handle the increased volume of air being used by the engine.

No particular special service or maintenance is required by these add-on turbochargers. Lubrication is provide by a connection to the engine crankcase lubricating system. Since the speed of the rotating shaft may be as high as 80,000 rpm, it's essential that the bearing always have an adequate supply of oil. To help ensure this, the time between engine oil changes should be reduced about 25 per cent from that originally recommended for the tractor. Conscientious servicing of the oil supply also tends to provide an extra margin of protection for the engine parts, which are more highly loaded as a result of the use of the turbocharger. □

Differential-diesel arrangement includes mechanically driven supercharger which is driven faster — to increase air flow to engine — as engine speed falls. (Photo, drawing: Perkins Diesel.)

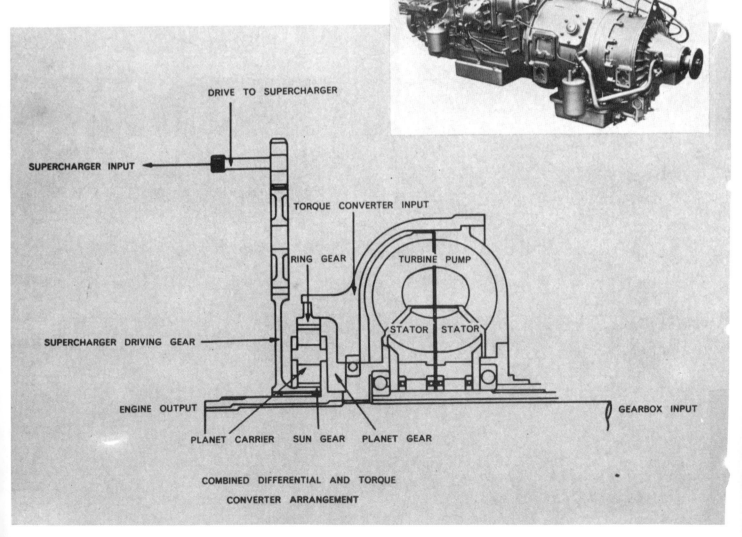

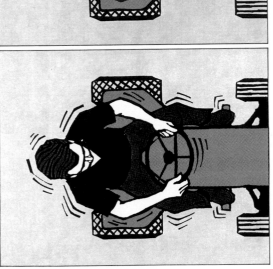

Briggs & Stratton has produced a new 12-hp engine SYNCHRO-BALANCED for smooth operation!

Because nothing "shakes up" a tractor buyer more than vibration...

MOST RESPECTED NAME IN POWER

4 CYCLE GASOLINE ENGINES® BRIGGS & STRATTON

In a remarkable engineering advancement, Briggs & Stratton has reduced vibration 60% vertically and 85% horizontally in an all-new 30.2 cubic inch 12-hp Synchro-Balanced engine designed for use with compact tractors. Vibration normally transmitted through the steering wheel, seat and footrests is cut to a minimum for smooth engine performance... greater operator comfort and efficiency... longer tractor and equipment life.

How Synchro-Balanced design combats engine vibration

Counterweights at each end of the crankshaft are geared to rotate in a direction opposite from the crankshaft counterweights. In the horizontal position (A), the synchronized counterweights cancel the unbalanced weight of the crankshaft. In the vertical position (B), the crankshaft counterweights combine with the unbalanced weight of the crankshaft to balance the weight of the piston.

Give your product this outstanding competitive advantage. For full details on the new Synchro-Balanced engine, write:

BRIGGS & STRATTON CORPORATION, MILWAUKEE, WISCONSIN 53201

Hydra-Drive Hydra-Brake

...two of many Sales-Closing Features

of COLT COMPACT TRACTORS

HYDRA-DRIVE ... instant operating response with the Colt hydraulic power drive system. Hydra-Drive control operates with push/pull fingertip ease. No drive belts, no clutches, no shaft drives. It's an eye-opening demonstration, at full throttle, to go forwards and backwards instantly without clutching or killing the engine!

HYDRA-BRAKE ... instant stop with patented brake in hydraulic drive system. No brake drums in wheels—no linings to wear out!

Yes! These are but two reasons why the all-new, 1966 COLT will have a powerful impact on the compact tractor market. Take the big leap into the big league with a COLT ... sold only by independent distributors and dealers. For details about specs, all-season attachments, prices— write—wire—call:

COLT MANUFACTURING COMPANY, INC. 133 SOUTH FIRST STREET WINNECONNE, WISCONSIN 54986

For More Details Circle (76) on Reply Card

The first mechanized tomato harvest was possible because of special hybrid tomatoes and newly developed equipment. Allis Chalmers purchased the Simplicity Manufacturing Co. of Port Washington, Wisconsin, in October, 1965.

The first U.S. landing on the moon by unmanned spacecraft occurred on May 30, 1966, Medicare became effective on July 1 and France withdrew from NATO on July 1.

Grounds Maintenance magazine was begun in 1966 to serve the growing number of professional landscape contractors and grounds keepers. Versatile introduced a big 4WD tractor and Allis-Chalmers introduced the first 6-row corn head for 20-inch rows. Some of the new diesel engines were using the 9.5 mm diameter "pencil nozzles" and hydraulic motors were beginning to be used as an auxilliary drive for the front wheels of some tractors. Rickel, Inc. introduced the Big-A bulk fertilizer spreader. The Big-A was equipped with three large low-pressure tires that could be driven over wet, muddy fields without compacting or rutting the ground. Toro Mfg. Corp. of Minneapolis, Minnesota, bought the assets of Quick Mfg., Inc. of Springfield, Ohio, effective August 1, 1966. New Idea announced a new name, Avco New Idea Farm Equipment Division, and a new corporate symbol.

Israel launched a surprise attack against neighboring Arab nations, beginning on June 5 and ending on June 10, 1967. J.I. Case Co. was purchased by Kern County Land Co. The Kern County Land Co. was soon purchased by Tenneco. The Dickey-John Corp. of Chatham, Illinois, and Puzey Bros., Inc. of Fairmont, Illinois, both sold electronic monitoring systems that notified the operator on the tractor if a planter stopped dropping seeds. Oliver offered a high speed plow that could be drawn at 7 mph and at Iowa State engineers were developing a machine that made 660 pound round hay bales.

Office mechanization has paid off. Bookkeeping machine installed 3 years ago cut 50-man payroll job down from over a day to 2 hours, also does other work. Copying machine has simplified much paper handling, too.

The all-weather bulk fertilizer spreader can be used under any field conditions.

North Koreans seized the USS Pueblo and crew on January 23, 1968, the Rev. Martin Luther King Jr. was assassinated in Memphis, Tennessee, on April 14 and presidential candidate Senator Robert F. Kennedy was shot in Los Angeles, Caliifornia, on June 5. About 650,000 troops from the Soviet Union and Warsaw Pact nations invaded Czechoslovakia and ousted the liberal communist government. In 1968, J.I. Case Co. purchased Drott Mfg. Co. of Wausau, Wisconsin, and Davis Mfg. Co. of Wichita, Kansas. International Harvester showed its truck of the future, the "Turbostar." Improvements in diesel engines and the fuel "crunch" that followed, stopped further development of the gas turbine engine for truck and tractor use.

SNOWMOBILES

Kawasaki Announces New Touring Sportcycles

Kawasaki Aircraft Co., Ltd., 208 S. LaSalle St., Chicago, Ill., has introduced a new line of sportcycles with models through 650 cc.

The oil pump is synchronized with both the engine rpm and the throttle grip opening.

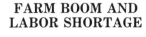

Farmers tell engineers about
TRACTOR REPOWERING

IMPLEMENT & TRACTOR

1969-1974

A farm boom from 1970 to 1973 was caused by two smaller than normal worldwide harvests. Famine occurred in parts of Africa and Asia, and increased consumption of the already scarce grain. The Soviet Union and other countries began buying grain from the world markets, commodity prices more than tripled and farm income more than doubled.

The U.S. began withdrawing some of its more than 541,000 troops from Vietnam in 1969, President Nixon ordered U.S. troops into Cambodia on April 30, 1970, the cease fire began on January 28, 1973 and the American troop withdrawal was complete by March 29. All U.S. combat operations were officially ended in Southeast Asia by congressional order on August 15, 1973.

In 1969, the Case eagle trademark was retired and Case began turning their attention to the industrial market. The Steiger Tractor Co. was incorporated and moved to Fargo, North Dakota, and, in July, astronaut Neil Armstrong was the first man to walk on the moon.

Case, John Deere and International Harvester offered 140 hp tractors in 1969. Cabs and air conditioning became almost standard equipment on the larger and more popular size tractors. International Harvester pushed to sign up automobile dealers who didn't have a light duty truck line, especially Pontiac and Buick dealers. Inflation of 61 percent between 1967 and 1975 raised the production costs of International light trucks so much that it was difficult to compete with GMC and Ford. Japanese small tractors were beginning to be imported to the U.S.

NEW TRACTORS:

Deere builds a speedy unit, buys two giants

RATED AT 94 pto hp, the new model 4000 weighs 7,700 lb.; it's designed to work in 4th gear or higher, pulling many implements at 5 mph or faster.

A hybrid of 4020 and 3020 models, now pegged at 94 and 70 pto hp, the 4000 is powered by the 4020's 6-cylinder, 404-cu.-inch diesel engine. Thus it produces 4020 power at the pto, but is intended to pull 3020-size implements 1 to 1½ mph faster. This means that disk harrows, planters and rotary hoes can be pulled at speeds up to 5, 7 and 8 mph respectively, conditions permitting.

Deere reports pricing the 4000 to "compare favorably with competitive 76 to 80 hp tractors." The speed-based design means more horsepower per amount of weight, and a lower cost per horsepower. Many features are regular 4000 equipment, permitting standardized assembly line operations and reduced production costs.

Besides basic design, the 4000 has many specially matched components, including tires, that are speed-based. Modern implement design, such as high speed plow bottoms and shock-absorbing features, permit faster working speeds. To get intended benefits, farmers must avoid too low a gear, bigger implements and lugging the tractor down. This poses to dealers a new approach to selling tractor power.

Two big ones

The WA-17 and WA-14 tractors deliver 220 and 178 drawbar hp (in 4th gear) respectively. They carry six-cylinder, 855 cu. in. Cummins diesel engines developing 280 (turbocharged) and 225 flywheel horsepower. A 2-range, sliding gear transmission gives 10 forward and two reverse speeds. The range selector is air-assisted; the 4-wheel brakes are air-actuated. The mechanical foot-operated clutch is said to permit on-the-go shifting by "double clutching."

Both models articulate 80 degrees (40 degrees from center) and oscillate 30 degrees (15 degrees from horizontal). A flexible coupling joins the two units or bogies of each tractor; two double-acting cylinders, a 30-gpm pump and 40 gal. reservoir are for power steering. The wheel hubs of front and rear axles have planetary reductions. An engine-driven 15 gpm pump with 2-spool valve, and a pto-driven 35 gpm pump with 2- or 3-spool valve are the hydraulic system options. Single and dual tires are available in several sizes.

The tractors are produced by FWD Wagner to Deere specifications, as arranged by a recently formed Outside Manufactured Products Div., and sold, warranted and serviced exclusively by Deere. □

The Ford 9000:
130 PTO
Horsepower

Know your strong points.

Delco-Remy contact sets.

Each contact point is made of a tungsten wafer cut from a rod, not a flat sheet, to keep it working longer. And hydrogen brazed to the lever arm to keep the shakes away.

More than that, we use a steel breaker spring that's heat treated to keep it taut. And a molded rubbing block that's distortion-resistant. So you don't have to bother with readjustment after wear-in.

We even put our contact sets through 21 different torture tests. All to make sure they're strong enough to put in . . . and forget.

And you can get them with no trouble at all. Right now. From leading farm equipment dealers, or from United Delco.

Delco-Remy
Division of General Motors · Anderson, Indiana

'Total Merchandising Works'

Regardless of what you sell, it takes a full promotional effort to get the job done. This farm equipment dealer's success with snowmobile selling is proof.

By Tom Holloway

■ "TOTAL MERCHANDIS- ING is one of the keys in selling snowmobiles," reports Eric Zellmer, president of Watertown Tractor & Equipment Co., Watertown, Wis. A Ford dealer, Zellmer handles the Polaris and Chapparal lines of snowmobiles. He also sells snowmobile accessories, including a $45 snowmobile suit, helmets, trailers for hauling the vehicles, sleds, racing and hop-up equipment for speed enthusiasts, and scores of other items to "total merchandise" one of the nation's fastest-growing sports.

Machines on display at Zellmer's range from a small, low-priced Chapparal at $595, to a huge, powerful Polaris priced at $1,650. Manufacturers of both vehicles offer a wide variety of models between the high and low figures.

"With any sale of $900 or more, I thrown in a free $45 snowmobile suit and helmet," Zellmer says. "This is part of our total merchandising."

Zellmer also sells automobile tires and garden tractors. He said these items help attract urban residents to his store as well as farmers.

Watertown Tractor & Equipment Co. started selling snowmobiles in 1969. The first year's volume on snowmobiles was $32,000. In 1970, the volume rose to $55,000 and 1971 sales were $68,000.

"Although we did not reach our estimated goal for 1972, we did show an increase over the previous year. We sold $89,800 worth of snowmobiles. We also think we have another good year coming up," he said.

The young Wisconsin dealer sees a good future for snowmobiles and the recreational vehicle market in general. "The manufacturers are building better, more reliable, machines today than they were a few years ago. The manufacturers are also looking for better quality dealers," Zellmer explained. "A lot of the poor quality dealers have been forced out in the last few years. Snowmobiles are a high capital item. In 1969, there were 11 snowmobile dealers in this area. Today there are six dealers, including myself.

"Sometime in the future, I think we will have split the snowmobiles, mini-bikes and garden equipment from the farm equipment to make an independent operation. However, I don't think this is going to take place for some time," Zellmer said.

Zellmer said his snowmobile sales are strong because of service. Each machine is given an average of four hours of pre-delivery service. Zellmer charges a flat rate of $32 per machine for pre-delivery.

"After 10 to 20 hours of operation, we ask the customer to bring the machine in for a check-over," Zellmer said. We recommend that the customer spend from one to two hours with the serviceman. The mechanic makes all of the necessary adjustments and works with the customer to explain the operation of the snowmobile."

Because of the improved quality of the new machines, Zellmer's 1971 warranty costs decreased 40 per cent even though his sales volume showed considerable increase. This was also true of his 1972 warranty costs. "The quality of the machines these days is 100 per cent better than it was two years ago. the manufacturer has gone a long way in improving the machines."

Snowmobile parts can be a problem for one dealer. "It seems we can never stock enough parts," Zellmer related. "The Polaris parts man was in once and said our parts inventory was the largest of any Polaris dealer he had seen in his recent calls. We normally stock about $3,700 in parts, but during

As part of his total merchandising plan, Eric Zellmer gives a snowmobile suit and helmet away to anyone who buys a machine which costs $900 or more.

the busy season our parts inventory may peak as high as $5,000.

"This year," Zellmer said, "I doubt if we have any parts problem at all. I was told that for the first time, parts will be produced first for the 72-73 season before the new models of snowmobiles are. This means we will have replacement parts on hand before we even have the new machines to show. This is a complete turnabout over the past years."

The dealer who plans to make a profitable business out of snowmo- biles should plan to purchase spe- cial tools for service and recondi- tioning. Timing tools are an especially important item. "We looked into the possibility of ac- quiring a dynamometer for snow- mobiles but there was a question of whether we could afford it. So we still don't have the dynamometer," Zellmer said.

Price-cutting was a problem for Watertown area snowmobile deal- ers a few years ago. At that time, the market was glutted with many different makes; some of them were poorly made. "Now that nearly half of the dealers have gone out of business and the quality of the ma- chines has improved, the price- cutting situation has taken care of itself," Zellmer explained.

Zellmer has also initiated a sea- sonal open house at the dealership for snowmobile owners to view slides on snowmobile operation and learn about his service capa- bilities.

Zellmer's five salesmen, all farm equipment men with the exception of the tire salesman, are paid a straight 25 per cent commission on the gross margin for snowmobile sales. Zellmer is also doing a good business in reconditioned used ma- chines. They sell for $350 to $475 each.

The trade-in value of a used snowmobile is based largely on the condition of the metal and rubber track which propels the machine over the snow. "The replacement cost for one of these tracks is about $150," Zellmer said.

"This is a total family sport. That's what makes it so great," Zellmer said. "Some families have two or three machines. I know of one family with four snowmobiles. The head of this family is the type of guy who doesn't hunt, golf, or bowl. But, he sure loves snowmo- biling."

One state's new snowmobile laws

A COMPREHENSIVE BILL governing the operation of snow- mobiles in Wisconsin recently has been passed by that state's legislature. The bill was made up of 26 small pieces of pending legislation and out- lines safety requirements, sets out prohibited riding territories and authorizes the use of state funds for the development of trails and local law enforcement teams for the snowmobilers.

The bill includes: (1) a minimum operating age of 16, with youths from 12 to 16 being able to get a safety certificate to drive a machine while in the presence of an adult; (2) a registration fee increase of from $6 to $9 for a three year period; (3) a new system of distributing ⅔ of the registra- tion fees to the counties for development and maintenance of trails and the remaining third toward establishing a snowmobile recreation council to develop new snowmobile programs and enforce the existing snow- mobile laws; (4) a maximum noise level for the machines of 83 decibels by July 1, 1972 and 78 decibels by July 1, 1975.

The legislation also sets speed limits in densely populated areas.

Both ardent snowmobile enthusiasts and skeptical property owners who consider the machines "obnoxious intrusions" rallied to support the bill which contains favorable clauses for each group.

Scaled Down Industrial Tractor

Powered by a 17 hp, 3-cylinder diesel, this industrial-type tractor, manufactured by Kubota, Ltd., is now being marketed in the United States. The L-175, designed in Japan, features a front and rear pto shaft, three-point hitch for category #1 implements and availability of a variety of tire sizes.

The White Farm Equipment Co. was formed by consolidating the Oliver Corporation of Chicago, Illinois, and Minneapolis-Moline of Minneapolis, Minnesota. Ohio National Guardsmen killed four Kent State University students during a war-protest demonstration on May 4, 1970, and riots in Poland during December were stopped by Soviet troops. The Occupational Safety and Health Act of 1970 was soon to make a lasting impact on all jobs.

The 26th Amendment to the U.S. Constitution lowered the voting age to 18 on July 5, 1971, wages and prices were frozen by President Nixon on August 15 to combat inflation and communist China was admitted to the United Nations on October 25. Product reliability was foremost in the consumer's mind and affected all manufacturers.

For farmers who like something different, IH has introduced the model 1468 tractor with a V-8 engine. It is basically the same as the previously introduced model 1466, but has a naturally aspirated, 550-cu.-inch-displacement V-8 diesel instead of a 436-cu.-inch, turbocharged six. PTO horsepower output is estimated to be 133 for both models by the manufacturer.

An unusual feature of the V-8 engine: Fuel is injected into only four cylinders when the tractor is operating under no load or partly loaded conditions. Under heavy load conditions, fuel is injected into all 8 cylinders. A Bosch in-line fuel injection pump is used. Engine bore and stroke is 4.50 x 4.312 inches; compression ratio is 17:1 and engine governed maximum speed is 2400 rpm.

A couple of new developments: A Honda dealership was opened in Owensboro, Ky. last year with good results so far; in the Evansville shop, one mechanic has developed a reputation for rebuilding Model T Fords.

1969-1974

To get more power at a less expensive price, some tractors were repowered like the John Deere 5020 shown with a Cummins V-903. The high cost of another tractor and operator provided the reason to increase the power of currently owned tractors. Larger than normal implements were necessary to make use of the additional power. Sometimes it was possible to buy new large equipment, but often, the proper size was made from two or more old, smaller implements. Soon the manufacturers began offering the larger sized tractors. New regulations by the Occupational Safety and Health Act (OSHA) concerning roll over protective systems (ROPS) went into effect in 1972.

J.I. Case bought David Brown of England in 1972.

Steiger began building International Harvester 4WD models in 1973. Vice President Spiro Agnew resigned on October 10, 1973, and President Nixon resigned August 9 the following year.

A world wide recession, combined with inflation, began in 1974 when the Organization of Petroleum Exporting Countries (OPEC) raised the cost of crude-oil 500 percent from the 1973 price.

A chemical formulation known as WD-40, in aerosol spray cans or bulk, is proving effective in displacing moisture and putting a stop to rust and corrosion problems while providing a perfect light lubricant, for farm machinery and equipment. WD-40 has been found effective in protecting equipment that is used with various fertilizers.

Protector of Farm Machinery and Equipment

Model 315

BUSH HOG'S® BIG, TOUGH RELIABLE 15-FOOT FLEXIBLE ROTARY CUTTER NOT ONLY CLIPS PASTURES, CLEARS HEAVY UNDERBRUSH, CHOPS AND MULCHES STALKS, BUT BUSH HOG'S REPUTATION AND COMPETITIVE EDGE CUTS THROUGH CUSTOMER RESISTANCE.

LEARNING TO COPE WITH OSHA

An insurance safety specialist is telling some dealers that if their operations do not meet the requirements of the Occupational Safety and Health Act of 1970, they could face some heavy fines. What have you done to prepare yourself for the day the man from OSHA comes in your door?

by Charlie Cape

■ "WHEN TWO TRAINS approach each other at a crossing, they shall both come to a full stop, and neither shall start up until the other has gone."

That is a Kansas law which has been on the books for many years. Like so many other laws, it was written hurriedly at a time when it was deemed necessary to the public well being.

Jerry Juergens, divisional supervisor for Federated Insurance Co.'s loss control division, says the Occupational Safety and Health Act of 1970 (OSHA) has generated some regulations that could be considered similar to the old Kansas statute.

"The classic one we're seeing, and several contractors have been cited (fined) for it, is in the General Environmental Control section of the industrial regulations of OSHA," Juergens said. "There is a regulation that says if you use ice to cool your drinking water, the ice cannot come in contact with the water.

"This regulation was made back in the days when they had natural ice and, admittedly, it was unsanitary. But today, the ice is every bit as sanitary as the water. This thing should go by the boards and will have to."

Many people have questioned the inclusion of many of the OSHA regulations, calling them greatly outdated and contrary to today's working atmosphere.

OSHA was born

"My opinion on this," Juergens said, "is that the drafters were necessarily in a great hurry to get some regulations on the books. So, they went to the American National Standards Institute

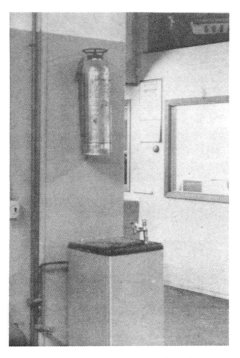

(ANSI) and the National Fire Protection Assn. (NFPA) and apparently used everything available.

"These regulations, codes and standards were devised over the years by well-meaning, knowledgeable people. A consensus was reached that these were standards of good practice. They were never meant to be laws."

Juergens explained that his insurance people used these standards as a guide in their inspections over the years.

Juergens did add, "We tried to use judgment in our own inspections, however. Some of these codes were arrived at through compromise. For example, you have to have one 2A fire extinguisher for every 3000 square feet. The reason it is not for every 2000 square feet is simple—after much discussion, it was decided that the code would call for one extinguisher for every

Which one of these fire extinguishers does not meet OSHA standards? If you said the one on the right, you were correct. That extinguisher does not have a red background and is too high above the floor. The other extinguisher is properly marked, painted red and the top is not more than five feet off the floor.

3000 square feet. This was meant as a guideline, though, not a law."

Through wording and the varying interpretations of the new law, there have been reports of different inspectors doing different things in their inspections. Juergens said one of his field representatives visited an automobile dealer who had been inspected by an OSHA compliance officer.

"This dealer had 20 citations; the whole wall was plastered full of them. All were very specific, right down to the frayed cord on the pop cooler. But nothing was said about the spray painting set-up, which was non-standard," Juergens said. "This seemed unusual, because for the most part the painting operations of auto dealers were really being hit."

Inspectors vary

"It's the individual inspectors. It's going to take a lot of time for them to correlate their ideas and get their thinking together so they all cite the same things," Juergens said.

Congressman Keith Sebelius of Kansas explains regulations in the law in a newsletter he sends to many of the people in his district. "A citizen writing to the Dept. of Labor more than likely receives a reply that the regulations are tentative and subject to change," he says, when writing for copies of the original ANSI and NFPA standards.

This grinder, says the insurance man, meets OSHA standards because it has a safety shield covering the right amount of space and the tool rest is in place. Many shop grinders are currently being replaced or modified.

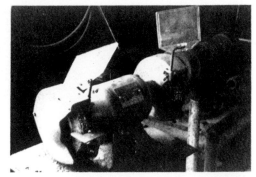

William R. Noble, the Washington counsel for the National Farm & Power Equipment Dealers Assn., points out that the law was passed to help insure the safety of the employees. But due to its wording, construction and questionable clauses, Noble calls OSHA "the most frightening piece of legislation we have ever had to face."

Ken Kamholz, manager of the loss control department of Federated Insurance Co., feels that the law is basically good. He says that one of the better sides of this law is the fact that it makes the employer more safety-minded and this in turn makes his employees safety-minded.

Accidents still occur

"The fact that we have laws will not stop accidents. Machines and humans have accidents, but if the human becomes more safety-conscious, there is a possibility that the number of accidents will decrease," said Kamholz.

"When you take a look at the situation as a whole," Kamholz said, "many of the injuries are the result of an unsafe act instead of an unsafe condition. In the farm equipment dealership, for example, we (Federated) pay claims for many

Insurors have long told policyholders: 'Mark circuits in the fuse or breaker box so the area each circuit controls is known.' Now they're adding: 'It's the law.'

back injuries, but within OSHA, the only section which even comes close to requiring proper training in any work procedures is in the administering of first aid.

"There is nothing in the law which requires the employer to teach his employees the proper way to operate machines and safely do the work required of the job. Only

This lifting hook on a chain hoist would not pass an OSHA inspection because it does not have the required safety catch which would prevent the hook from slipping out of what it is fastened to.

It is a good idea, Juergens says, to have all outlets and electrical equipment wired with the three-prong plug to allow for proper grounding. At left, this has not been done; at right, a washer has had the three-prong plug added, but this would not pass inspection because of its condition.

The railing with the steps (left) is in compliance, but there should be a guard rail for the storage area over this store room. At right, there is a guard rail with toeboard, but the rail is too high and no midrail is provided. 'Some inspectors,' says Juergens, 'are following the law to the letter and could well write a citation for either condition.'

in the construction section of the law is anything said about supervision or making employees aware of some hazards present in their work."

Infractions Viewed

Juergens went to visit three midwestern farm implement dealerships with I&T so that some of the possible infractions of the OSHA law could be pointed out and discussed first hand.

In one of the dealerships, the "Safety and Health Protection on the Job" poster was not posted. Juergens said the initial fine for this infraction would be $50 if the OSHA inspector were to call.

In another dealership, an open flame heater was found in the paint booth. This, Juergens said, could cause the man from OSHA to cite an imminent danger condition and start proceedings to close down the shop completely, as well as to levy

a large fine against the dealer. He said the heater should be moved outside the booth and proper venting installed.

In two dealerships' paint booths, the venting fans and their motors were contained within the booth. In both cases, the motors were caked with paint due to sucking of the particle-laden air by the fan. Juergens told the dealers that the motors had to be placed outside of the booth to prevent this caking,

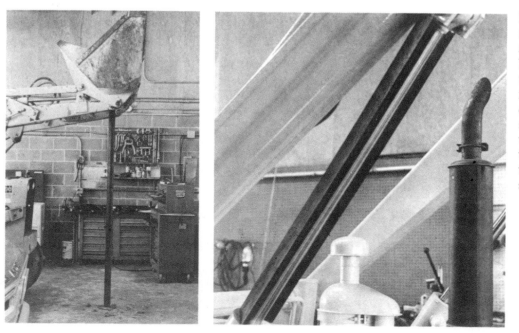

Piece of pipe is used to support bucket of loader and angle iron is placed beside lift cylinder rod to assure safety during repairs. 'This isn't required by law,' Juergens says, 'but an inspector would be impressed by the dealer's extra effort to assure his workers' safety.'

for if enough paint were to accumulate on the motor, overheating would occur, increasing the possibility of fire.

Other things Juergens looked for included open-front toilet seats, hot water in the restrooms, all means of exit marked as such, proper grounding of power tools, railings with midrail and toeboard on storage areas located on elevated surfaces, fire extinguishers properly mounted, safety shields on grinders, first aid kits and many other items.

At one of the dealerships, Juergens was asked what to do if a man from OSHA should come to inspect his dealership. "Be courteous and helpful, show him what he wants to see and don't hesitate to do this. We have been led to believe that, generally, people who give the OSHA inspector a hard time may end up with more problems than they would face with a cooperative attitude."

One dealer asked Juergens, "What about supporting HR 12068?" He was referring to a current proposal which would exempt employers (outside of manufacturing) with 25 or fewer workers from OSHA provisions. The dealer was shocked by his reply:

"We think it's the wrong way to go. These small companies are the ones who need OSHA the most.

"Larger companies in most cases have had safety programs for a long time and have had to do relatively little to comply with the new regulations. Many smaller companies have needed to upgrade safety considerably, and OSHA provides the impetus.

"Rather than spending time, money and effort trying to be excluded from OSHA provisions, try to get some of the non-essentials of the regulations removed; the basics of the law can be of real use in preventing death or serious injury."

Ken Kamholz agrees: "It would be better for dealers and particularly the dealer associations to work at persuading Congress to remove the impractical and insignificant points from OSHA than to try to amend it to apply only to certain groups."

These two features of dealer paint shops could get their owners in trouble, Juergens says. The open flame heater is inside the paint booth; the ventilating fan motor, being inside also, has become caked with paint particles. Both situations could be cited by an OSHA inspector as presenting an imminent danger.

FIFTY YEARS OF THE FARMALL

Announcement of the 50th birthday of IH's great innovation brings to light an old Elmer Baker column about the inspired achievement of Bert R. Benjamin, developer of the Farmall tractor.

International's 1466, latest in the company's farm tractor line, has about 12 times as much work capacity and vastly increased comfort and versatility, but the basic tractor shape is still the same: High-clearance rear drive wheels, engine mounted amidships, small front steering wheels to go between rows.

■ THIS YEAR MARKS the fiftieth anniversary of the first successful attempt at building a genuine row-crop tractor. International Harvester scored a major breakthrough in design with development and introduction of the first Farmall in 1922.

The following year, 26 improved tractors were sent out on a second trial run, and in 1924, IH started regular production of the Farmall tractor at its now-defunct Tractor Works. By October, 1926, production had shifted to the newly equipped Farmall Works in Rock Island, Ill.

"The Farmall has undergone many changes in power and utility since it was introduced," says an IH spokesman, "Though each year has seen important refinements, the essential features have remained the same." These features were the high rear-wheel drive for maximum clearance under the rear axle; the small, narrow front wheels designed to run between row crops; and the means of mounting implements and other attachments to either the front or the rear.

The IH spokesman added: "Through its ability to perform not only the belt, drawbar and pto functions of the standard four-wheeled tractor but also to work in row crops efficiently, the Farmall tractor made possible for the first time the horseless farm. Since the Farmall's introduction, all manufacturers have produced tractors of similar conformation."

When I&T staffers saw the IH release about the Farmall's golden anniversary, memories were jogged somewhat: Hadn't the late Elmer Baker, Jr., once talked about the development of the early Farmall in one of his essays? A few hours of digging unearthed a memorable "Reflections" column titled "Committee Designing" which appeared in I&T Dec. 26, 1959. The Farmall, in Baker's opinion, was the exact opposite of committee designing, and he gave his view as to why it was so successful:

"Then came the challenge to provide farm power cheaper and better than muscles—human, bovine and equine—could develop it. We had a lot of committee designing then. Locomotive boilers were put onto land wheels, steering mechanisms were devised, the differential was invented by some early genius. Traction engines displaced some of the muscles in the heaviest type of farm work.

"Along came Mr. Otto, and we

substituted the internal combustion engine for steam. Farmers began to talk about tractors rather than traction engines. Henry Ford, picking up ideas and personally developing them, took the bulk monkey off the farm tractor's shoulders and produced one light machine with interchangeable parts by progressive assembly. But it still was the old land locomotive refined.

"Then one man, not a committee, got an idea that a tractor that wouldn't displace the horses and mules on a farm was only half the answer to the farm power problem. So he started to work on units that would straddle and cultivate row crops with implements mounted as well as doing the land locomotive work of other tractors.

"That man was Bert R. Benjamin, and he worked for the Harvester company experimental department. He may have had a desk on the old 7th floor at 606 S. Michigan Ave., Chicago, but we always found him somewhere around that old shack close to McCormick Works next to the Illinois and Michigan Ship Canal, then gradually filling up from disuse.

"First he turned out a two-row motor cultivator, with the dual drive wheel mounted on a steering spindle at the rear and a wide spread arch frame in front with double row cultivator gangs hung thereto.

"The machine cultivated corn beautifully, but it would not pull a breaking plow in sod, or an equivalent load, so it was not the answer to the problem of displacing muscles in farm work.

"Ultimately Bert turned the machine around. That put the double spindle-mounted steering wheels up front. He mounted the engine and transmission amidships on a channel frame and designed a rear axle and final drive with enclosed bull gears and pinions that permitted straddling two rows with cultivator clearance (26 inches). Then, in an effort to devise as fine cultivation as was otherwise possible only with a one-row walking

cultivator, used by many to lay by a crop, Bert added interconnections between the steering mechanism and the next-the-row cultivating shovels so that swinging away from the row to avoid plowing up a hill of corn instantaneously swung the shovels conformably also. The Farmall, as it was ultimately called, did beautiful work, but it was not put into production as soon as it was mechanically and functionally proved.

"That step did not come until about 1925, two years after Ford had reduced the price of his Fordson tractor from some $525 to just $395, two days before the opening of the National Tractor Show at Minneapolis. It cost as much as $395 to buy the engines offered by some motor builders to tractor assemblers.

"That tractor show was more of a wake than a sales event. No maker could meet the Fordson price and remain in business. Ford himself lost millions, it was disclosed after his death, but he did not know it and had plenty to lose if he felt like it. He sold about 80 per cent of the new tractors during that World War I postwar depression.

"Harvester met the competition, insofar as they did, by giving away a Little Genius P&O plow with a Titan or a Mogul. Ford had to buy his plows from Oliver, and J. D. Oliver, Sr., was not wont to cut prices for anyone.

"Harvester also as a desperation decision gave the production go-ahead on Bert Benjamin's Farmall. It couldn't put them in a worse position than they were, and it might bail them out.

"The company made quite a lot of the machines the first year and offered them only in Texas, where they could be tested on both cotton and corn. The Farmall swept the field. It just about chased the Fordson out of the Lone Star State.

"McCormicks had now tasted blood, and they went after Ford for keeps, they thought. They did chase him over to Cork, Ireland, which put the Fordson on the British market. None were imported into the U.S.A. for some years, and Ford did not get back in the tractor business in the U.S.A. until he met up with Harry Ferguson at a time when Ford had time, money and inclination to poke sticks again at the McCormicks. He loved to, we've heard from the horse's mouth.

"The Farmall tractor was the antithesis of committee designing. What might be called committee designing in the Harvester company at that time was evident in the International 10-20 gear drive and 15-30. They were assembly-line machines made to automotive standards the way the original Fordson had been. From the fan to the flywheel, they were E. E. Sperry's general ideas ('I can design an engine like that in 14 hours' — his remark to the Reflector once). Clutch to the rear axle was Otto Schoenrock's work, largely. You will find the same style construction in some of the early four cylinder Olivers, as well as others.

"Only overall design, of course, is meant in crediting this work to Sperry and Schoenrock. Many other engineers worked on details and parts, as in fact they did for Bert Benjamin when the Farmall was laid out for production.

"The point is that the 10-20 and 15-30 were simply good tractors — pattern tractors in fact — but they were not revolutionary. They were proved-in-the-past design. The Farmall, however, was revolutionary. It was an inspired inventor's inspired dream, and it became a reality and a revolutionary influence upon power farming." □

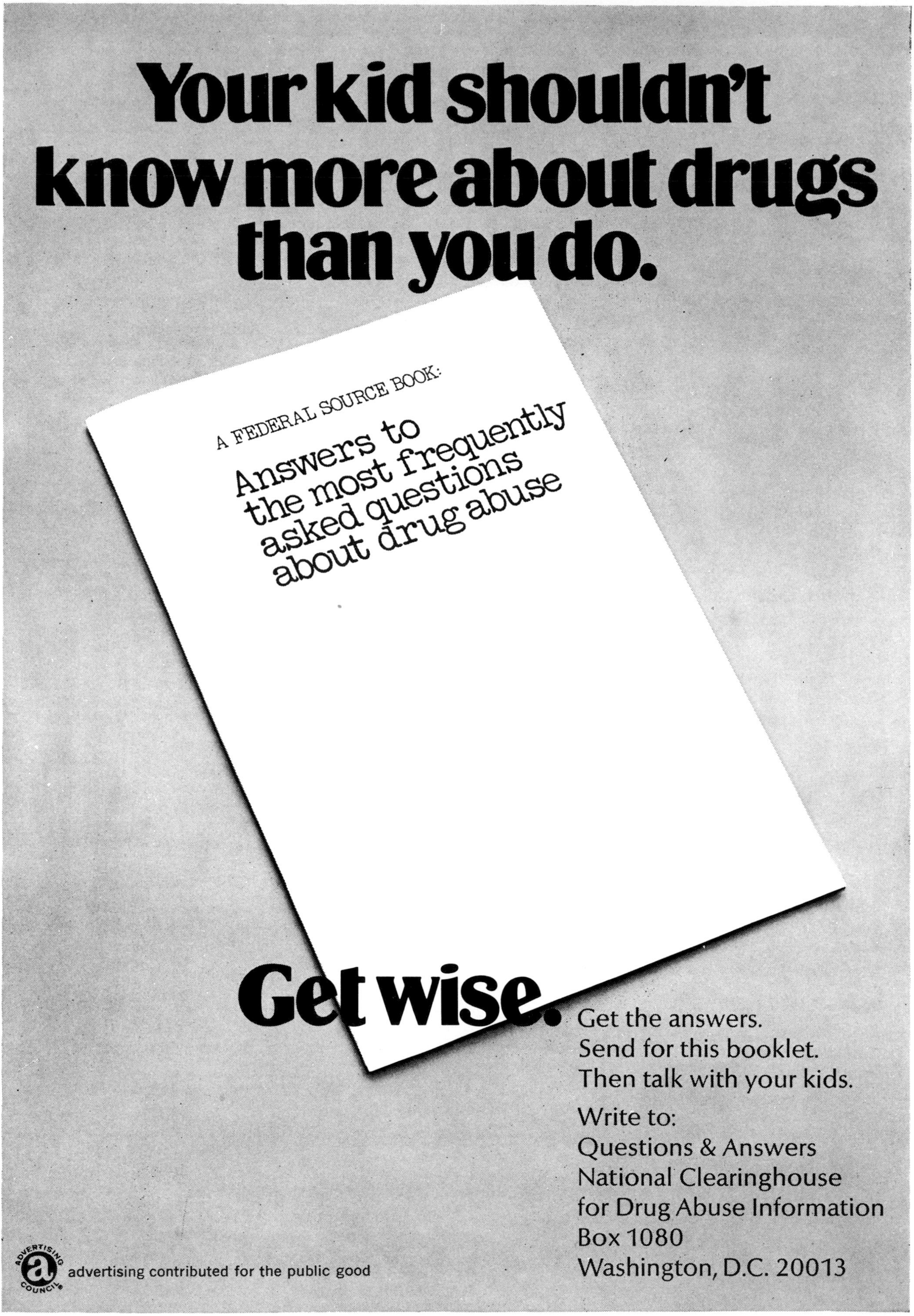

Your kid shouldn't know more about drugs than you do.
A FEDERAL SOURCE BOOK:
Answers to the most frequently asked questions about drug abuse
Get wise.
Get the answers.
Send for this booklet.
Then talk with your kids.
Write to:
Questions & Answers
National Clearinghouse
for Drug Abuse Information
Box 1080
Washington, D.C. 20013
advertising contributed for the public good

WHAT'S NEXT?

Answers: 1974 Demand Very Strong
'Get Used to Chronic 10% Inflation'
Canadians Will Demand Better Services
Watch for Energy-Saving Production Changes
'Construction in the '70's—$1.5 Trillion'

By Bill Fogarty

One answer to the energy shortage comes to us from the British Agricultural Machinery Journal. The Leeside Rover is a peat-powered steam engine capable of performing the same farming operations as a tractor. It is shown here equipped with a four-bottom plow. The inscription on the back of the machine reads "God Speed the Plough."

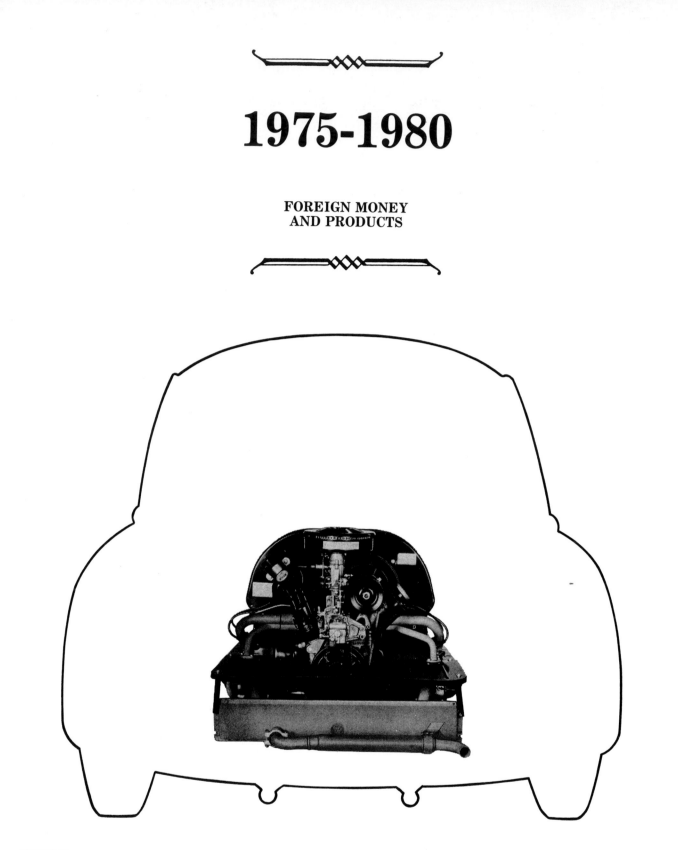

The heart of a bug
is available for transplant.

Implement & Tractor

The business magazine of the farm and industrial equipment industry

red book

Tractor and equipment specifications
Hydraulic components data
Nebraska test reports

1975-1980

Cambodia fell to the communists on April 17, 1975, followed on April 30 by the fall of South Vietnam. Unemployment in the U.S. was 9.2 percent during May, the highest since the "Great Depression." The number of Americans without work would increase from 8.5 million in 1975 to nearly 12 million by 1982. Civil war raged in Lebanon between rightist Christians and leftist Muslims during 1975 and 1976, until the Syrian army occupied the nation. There was also a civil war in Angola during 1975 and 1976, aided by Soviet arms and Cuban troops. By December 20, 1975, more than 130,000 refugees from Vietnam had been resettled in the U.S.

The state of Minnesota passed the Family Farm Security Act in 1976 to help protect farms from foreclosure. Deere filed suit against International Harvester for patent infringement. The suit was about the adjustment for a corn head patented by Lester Schreiner of Iowa and owned by Deere. Deere was

awarded $28 million in 1978, but International Harvester was not required to pay until the early 1980's.

International Harvester introduced the "axial flow" combine in 1977. The U.S. and Panama signed treaties on September 7, 1977, to turn the Canal over to the country of Panama on December 31, 1999. Control of the Canal Zone was turned over to Panama two years later. On November 19, 1977, President Sadat of Egypt became the first Arab national leader to visit Israel. Sadat's visit was part of an attempt to achieve peace in the troubled Middle-East.

The American Agricultural Movement organized "tractorcades" in 1977 and 1978 to various government offices including Washington D.C. The farmers were lobbying for increased price supports. By 1978 foreign investors owned farm land worth $800 million in the United States.

A revolt overthrew the ailing Shah of Iran on January 16, 1979. The previously exiled Muslim leader Ayatollah

Ruhollah Khomeini returned to Iran and created an Islamic state with himself as ruler for life. The population of the United States reached 220 million on February 5. A peace treaty, mediated by President Carter, was signed by Egypt and Israel on March 26. The U.S. Congress gave the President the power to invoke gasoline rationing on October 23. The UAW began a strike at International Harvester on November 1. The strike occurred following a time when tractors had been overproduced and sales were very slow. The result was that the strike lasted about six months.

Iranians captured the U.S. embassy in Teheran on November 4 and held 52 American hostages until January 20, 1981. Soviet troops invaded Afghanistan on December 27.

Grain exports from the U.S. reached a high of 71,000 metric tons during the period of 1979-1980. President Jimmy Carter imposed a grain embargo in 1980 to stop the sale of grain to the Soviet Union.

Only one in twenty Americans lived on a farm in 1980 as opposed to 1940 when one in four was a member of farm family.

A temporary recovery occurred in the general economy during the second half of 1980. The prime interest rate shot up from 15.5 percent to 21.5 percent from October to December and the bottom fell out of the market for big tractors.

The challenge for ag engineers:

Increase our output,
but conserve what we have

HINOMOTO ®

Looks small but works big.

A new concept in agricultural machinery –

Up until now, big farm jobs required the kind of big power found only in large tractors. And small jobs demanded lighter machines, with more versatility for work in confined spaces.

So much for cost-efficiency on the farm.

Now, Hinomoto can solve this sticky problem with its new line of powerful, maneuverable machines. Actually, two tractors in each one.

Today's profit minded distributors will welcome Hinomoto new line of powerful, high-performance, yet remarkably compact tractors. Confident in the open under heavy loads. At home in the garden handling smaller jobs.

Toyosha. The company that put versatility back into tractors and efficiency back on the farm.

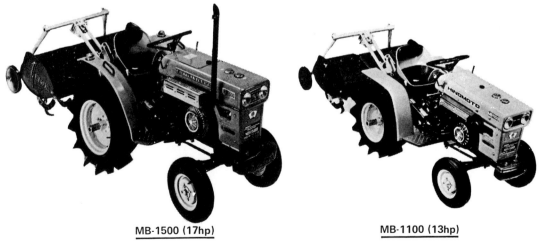

MB-1500 (17hp)   MB-1100 (13hp)

Export Policy

Toyosha is ready to place full trust in dealers within their alloted area if conditions permit.
- wants your positive aid.
- has plans to establish a stock-base and maintenance center.

For distributorship contact:

TOYOSHA CO., LTD.

55 (IT9), Joshoji-16, Kadoma-city, Osaka, 571 Japan
CABLE ADDRESS: TOYOSHA NEYAGAWA
TELEX: 05347763 MBTRAC J

1975-1980

Can the new little tractor from Kubota move mountains?

No. But the husky B6000 can sure move a large customer group into your compact tractor market.

With a rugged 35.20 cubic inch two-cylinder engine, this versatile tractor is perfect for park, landscape, golf course and highway maintenance.

Your regular farmers will find it an economical second power source when a full-sized tractor means more muscle, changeover time or fuel than they want to spend. And your big garden users will appreciate the *real* tractor features and quality of the B6000.

No other lawn and garden tractor offers a hydraulic lifting system like the B6000's (category 0 as well as category 1 implements accommodated), or 4-wheel drive, 8 speeds (6 forward, 2 re-

verse), differential lock, 2 PTO shafts —rear and front. And the reliability and economy of diesel power.

Add to big machine features a broad 40-1/8″ stance. The B6000 puts a solid grip on the ground for pushing and pulling a broad range of quick-hookup implements: front loader, plow, cultivator, disc harrow, mower, snow and earth removal blades.

There's no doubt about it. The new one from Kubota is going to move a lot of people into your showroom. So call our toll free number (800-421-1058) or send the coupon for details on dealerships, parts, service, and delivery across the country. Other models: L175 (17 hp), L225 (24 hp), L260 (26 hp)-all diesels.

IF FUEL SAVINGS ARE IMPORTANT TO YOU IN HARVESTING, READ ON.

When fuel was cheap and plentiful, not much attention was paid to the engines of self-propelled harvesting equipment.

It's a whole new ball game today.

In 1974 fuel costs were more than half the cost of labor in harvesting. And fuel prices have jumped since then. With the way they continue to soar, fuel costs could soon equal labor costs in harvesting grain, beans, corn, forage, and other crops.

A better way. Deutz air-cooled diesel engines can cut fuel costs to the bone. Compared to many water-cooled diesels, users report savings from 20 to 40%. Compared to gas guzzlers, even more. Air-cooling lets a Deutz operate at a more efficient temperature for cleaner, more complete combustion, reducing costs and pollution.

A better way. Deutz offers more than substantial fuel savings. Advantages like no radiators, pumps or hoses to freeze up. No cavitation.

A better way. Made by German craftsmen for the world's largest manufacturer of air-cooled diesels, Deutz engines are acclaimed world-wide for their trouble-free performance—from the Sahara desert to Alaska—from combines to forage harvesters.

The next time you look at self-propelled harvesting equipment, take a long, hard look at the engine. If it's a Deutz air-cooled diesel, you'll be looking at an engine that will put you dollars ahead—more dollars ahead every time fuel prices go up.

Isn't Deutz a better way?

1975-1980

The Leyland 154 is looking for a job.

The 154 is the small tractor with the big performance. It has the pull, the lift, the PTO power and the mobility to do those hundred and one jobs around the farm. At its heart is a 25 hp four cylinder diesel engine, smooth running and easy to start in all kinds of weather. A full line of implements available for the 154 permits its use for plowing, cultivating, planting, spraying, row work, loading, mowing and many other uses. Leyland—six models —from 25 to 100 hp.

For more information or a demonstration, contact the Leyland distributor nearest you now or write NEDA, Dept. ITL-124, P.O. Box 5025, Richmond, Virginia 23220.

Leyland Tractors are distributed through members of the National Equipment Distributors Association

1975-1980

478

WHAT'S A NICE LITTLE TRACTOR LIKE SATOH DOING ON A BIG FARM LIKE THIS?

It's a matter of simple economics. Even on bigger farms there are many jobs the smaller Satoh 25 hp and 17 hp gasoline tractors can do more efficiently than the big horsepower fuel guzzlers. Digging post holes. Plowing or tilling small areas. Hauling small loads. Its versatility is limited only by the number of jobs your customers can think of for it to do.

For more information or a demonstration, contact the Satoh distributor nearest you or write: NEDA, Dept. ITS-104, P.O. Box 5025, Richmond, Virginia 23220.

Satoh

15 Satoh Parts Depots serve you throughout the United States.

NEDA

Satoh tractors are distributed nationally through members of the National Equipment Distributors Association.

Four new Deutz tractors

Deutz Tractor Corp., Chamblee, Ga., has introduced four new air-cooled diesel tractors in the 50 to 70 pto horsepower range. They are the 52-hp D 5206, the 60-hp D 6206, the 68-hp D 6806 and the 71-hp D 7206. Power steering is standard on all models except the smallest, on which it is optional. The D 6206 and the D 6806 are also available with four-wheel drive.

Deutz claims improvements in the tractors' transmission housings, rear axle housings, hydraulic lifts, bevel gear drives, rear axles, and both the regular and synchromesh transmissions.

The Deutz line is coming out in a new color scheme: spring green for the hood and fenders with a brown-green for the chassis and fire-engine red for the wheels, rims and seat.

Five Belarus versions from Russia...and three Ebro models from Spain.

TRACTOR PULLING CONTESTS

ENGINEER'S CORNER

By Kenneth K. Barnes

■ ABOUT fourteen years ago, I naively accepted an invitation to serve as an official at a tractor pulling contest. I cannot remember what my specific duties were, but I do remember arriving at the firm resolve to never again become involved in such an event. Therefore, it was with great interest that I found that the June 1972, annual meeting of the American Society of Agricultural Engineers included a session on safety considerations in tractor pulling contests.

This session included four prepared papers, "Why Have A Pulling Contest and What Draws the Crowds" by Ed Hart, president of the National Tractor Pulling Assn.; "Potential Hazards of Stock and Non-Stock Tractors" by Dale Stevenson, tractor design engineer, International Harvester Co.; "Pulling Hazards as Seen by a Contest Observer" by Jack Jenkins, associate editor, *Farm Quarterly* magazine; and "Can Fun and Safety be Compat-

ible" by R. D. Schnieder, ag engineer and safety specialist, U. of Nebraska.

The speakers recognized that tractor pulling contests exist and are going to continue to exist. They recognized that there are many hazards associated with a tractor pulling contest, that it is human nature to compete in the use of work-a-day equipment, and that as competition sharpens, equipment is modified with more attention to competition and less attention to work.

Thus, we have had chariot races. We have horse races. We have rodeos. In the heyday of the draft animal we had horse pulling contests. We have drag racing, we have long distance automobile racing and we have many variations in between. The tractor pulling contest is thus not a surprising outlet for the competitive instincts of men who own and operate tractors and are proud of their skills.

Tractor pulling contests like many other competitive events involve an element of risk for the participant. This risk can be minimized, but cannot be removed. The tractor pulling contest, how-

ever, can and should be conducted in such a way that the risk for the spectator is removed. This requires well-developed rules and conscientious enforcement of the rules.

There is a broad spectrum of opinion regarding tractor pulling contests. Two quotations illustrate this. Schnieder quotes a safety specialist as saying, "I enjoy a good pulling contest and can see why they draw large crowds. But as a safety specialist, I cannot approve of them. If we were to make them safe, they wouldn't draw large crowds. People do not go to pulling contests to be educated, as there isn't anything educational about them, they are for entertainment. If they don't blow up a few tractors, it hasn't been a good contest."

On the other hand, an engineer from industry offers this interesting explanation of tractor pulling contests. He says, as quoted by Schnieder, "Some people accuse the contest of being conducted and the spectator interest as being generated by the hope that some unfortunate accident will occur, which

Participants in the tractor-pull panel at the ASAE meeting: Rollin Schnieder (U. of Nebraska), Dave Stevenson (International Harvester), Jack Jenkins ('Farm Quarterly'), Ed Hart (National Tractor Pullers Assn.).

1975-1980

A couple of products of the Midway shop: The IH 1206 Turbo is seen in action; the 'Fudmobile' is proudly inscribed, 'Built by Jack Freitag and Wayne Kittleson'.

they might witness. Actually, I don't really think that this is the basic cause of spectators' interest. It stems, I believe, from spectators' interest in viewing a piece of equipment which they are familiar with being operated at speeds and conditions beyond normal capacity of the particular vehicle and by skills way beyond that which are normally inherent in the average operator.''

The ASAE session on the tractor pulling contest was really a plea. It was a plea that recognized that tractor pulling contests will continue and that they can be enjoyable sporting events. The session was, however, a plea for safety in the conduct of tractor pulling contests; safety for the spectators, safety for the officials and safety for the operators. The session was an offer by agricultural engineers with an interest and competence in safety to counsel with those planning tractor pulling events. Agricultural safety specialists would like to work with those tractor pullers who have a sincere interest in conducting a safe and exciting sport.   □

Monsters from Massey

Two-wheel-drive tractors at 160, 190 horsepower

Both the MF 2800 and 2770 tractors are powered by 640-cu.-inch Perkins V-8 diesel engines, the version on the 2800 being turbocharged.

Big baler roundup

If you want to get an idea of how rapidly this approach to haymaking has been establishing itself, take a low-level plane ride around mid-America and see how many fields are dotted with the giant bales. Here's a look at the big balers currently known to be on the market.

Vermeer, generally credited with helping pioneer the concept of larg round bales formed in off-the-ground machines, is currently the onl maker offering three sizes. Upper portion of the bale chamber is forme by a series of flexible belts, and the bale is supported by a series o positively driven platform rollers.

By Melvin E. Long

■BALERS that make large bales—in some cases, round and in other cases, rectangular—are currently one of the fastest growing segments of the farm equipment market.

The concepts of harvesting hay in relatively large bales that must be handled by mechanized equipment were pioneered by a relatively few manufacturers just a few years ago. The concept itself required no technological breakthroughs—just some good, old-fashioned innovative design. The ready acceptance that these machines have received by the customers has prompted several manufacturers, including some full-line companies, to introduce competitive machines.

The potential advantages to the customer include reduced labor requirements, high harvesting capacity, good storage and keeping qualities of the bale even under outside storage conditions, and finally a system that is competitive cost-wise with alternative haying methods.

Even though the balers offered by the different manufacturers may superficially appear to be quite similar, there are significant differences. Here is a roundup of all the machines we found to be currently available at press time. However, at the current rate, additional manufacturers may have introduced their machines by the time you read this article.

In general, big balers can be subdivided into two types: Those that produce a large rectangular bale,

These four models, the Deere 500, the Gehlbale 1500, the Hesston 5600 and the International Harvester 241 Bigroll balers, all include upper forming belts arranged side by side to form the upper portion of the bale chamber. As the bale is formed, it is supported on a single, wide-platform belt.

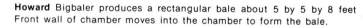

Howard Bigbaler produces a rectangular bale about 5 by 5 by 8 feet Front wall of chamber moves into the chamber to form the bale.

Hawk Bilt 480 forms its bales directly on the ground. A power-driven raddle rolls up the windrow to form a bale. When the bale reaches the desired size, the rear gate is raised by the operator and the bale is discharged from the baler.

Hesston 5400 includes nine belts distributed across the width of the baler; windrow on the ground is rolled up to form bales. Forming grids apply pressure to control bale density.

and those that produce a large round bale. Those that produce a round bale may be further subdivided into two categories: Those that form a bale by rolling the hay on the ground, and those that lift the hay into the machine to form a bale. Further, those that form a bale by rolling the hay on the ground may be subdivided into those that use chains and raddles to form the bale and those that use belts. Those that lift the hay into the machine to form a bale may also be subdivided into machines that use chains and raddles, and machines that use belts or a combination of belts and rollers.

In the chain-and-raddle, ground-rolling baler, two drive chains support a series of crossbars to which rakelike teeth are fastened. As the machine moves forward on the windrow, the power raddle starts forming the bale by folding

over the windrow, much like rolling up a carpet. As the baler proceeds down the windrow, the rolling action continues. The chain and raddle arrangement includes a spring-loaded take up arm that moves to provide additional chain as required to go around the growing bale. When the bale reaches the desired size, it is discharged from the machine when the operator raises the rear gate.

In the belt-type ground-rolling baler, nine belts distributed across the width of the baler are used to form the bale. To assure positive pickup of the hay, metal tines are bolted directly to the belts. In the bale-forming portion of the machine, grids maintain constant pressure on the bale to produce uniform density. These grids are arranged so that they gradually expand as the bale increases in size.

In the chain-and-raddle, off-the-

ground machine, the hay is conveyed into the bale chamber by a series of floor chains equipped with lugs to ensure positive movement of the hay. When the incoming hay reaches the bale forming area, a series of curved leaf springs located between the floor chains provides the initial curling action to start bale formation.

The chains and crossbars of the upper bale-forming elements are arranged so that, as the size of the bale grows, the upper chain assembly expands the bale chamber. It maintains tension on the bale to ensure compaction. When the bale has reached the desired size, it is ejected from the baler by release of the upper bale-forming elements.

In the belt-type off-the-ground machines, the hay is lifted by a pickup loader and fed directly into a baling chamber. The upper

(More...)

Massey-Ferguson 560 and Badger Northland 615 are identical machines manufactured by Vermeer to MF specifications, which include upper bale-forming belts and positively driven platform rollers to support the bale.

New Holland 850 moves hay into bale chamber by means of a four-chain set. Upper bale-forming element is formed by a series of crossbars mounted on two drive chains.

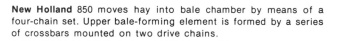

Big balers *(Continued)*

portion of this chamber is formed by a series of flexible belts arranged side by side. These belts are powered to roll up the incoming hay to form the bale. As the bale is formed it is supported on a single platform belt, the width of which is equal to the total width of the upper bale-forming belts. This full width belt helps carry the hay into the machine to form the bale, and helps carry the completed bale from the machine when the upper form-

ing belts are opened.

In the belt-and-roller, off-the-ground machine, the upper belts are arranged in the same general way as those in the belt-type machines that use both upper and lower belts. However, instead of a lower platform belt, the bale, as it is formed, is supported by a series of positively driven platform rollers. Here also, bale density is controlled by the tension applied to the forming belts.

In the large, rectangular baler, hay is compacted by the front chamber wall moving into the chamber to form a medium-density bale. The pickup and feed mechanism does not include any cross ram or knife.

All the balers currently have some means for applying twine to the bale. The large rectangular bales must have twine ties to retain the hay since it is, in effect, folded into layers as it is packaged into the bale. The off-the-ground roll balers apply twine by continuing to roll the bale in the machine but with forward motion stopped so that no more hay is being added to the bale. The ground-rolling balers apply twine as the completed bale is rolled along on the ground. In the off-the-ground roll balers the twine is applied in a spiral wrap but the ends of the twine are not tied. In one type of ground roller, the twine is applied in separate wraps around the bale, while in another type of ground roller, the twine is applied in a continuous spiral wrap. However, in both types of ground-rolling balers, the twine is not tied.

Market notes

●The Farm & Industrial Equipment Institute's Statistics Committee currently does not report retail sales of the big baler group, limiting its surveys to conventional balers, **i.e.,** those which make bales under 200 lbs. A pilot program is currently under way and, if successful, may result in 1976 monthly sales reports.

●Retail sales of conventional balers through the first half of 1975 were down 20 per cent from January-June 1974 to 10,833 units, according to FIEI, but farmers' expressed intentions to buy equipment can be a solid indicator total baler unit volume may be running even or ahead. The Midwest Unit Farm Publications report on its survey of farmers' buying intentions in eight north central states indicated that in 1974, farmers planned to buy 16,950 conventional balers and 6,275 big balers. A year later, the intentions report showed a dramatic shift in preference; the region's farmers said they intended to buy 11,500 conventional units and 11,100 of the big type.

●International Harvester's 241 Bigroll baler, on which production began last February at Memphis, is the Agricultural Equipment Div.'s first U.S.-made machine designed in metric units. □

Kawasaki

The POWER to turn People on.

Kawasaki. Strong merchandising for your products from a top name in high performance engines. Strong engines from the people who understand what heavy duty really means. Strong support from an engineering staff and a distributor/dealer network that only a world-wide company like Kawasaki can provide. Strong warranty protection from a company with quality standards few can match. Add it all up. Kawasaki delivers a lot more than engines. You get the power to turn people on.

Choose 4 cycle engines from 2.3 to 20 hp. Or 2 cycle engines from .8 to 4.3 hp. Kawasaki engineers them all with precision. To give your customers benefits such as fast warm ups, greater fuel economy and longer engine life from a long list of high quality features.

Try us. Let us prove to you that Kawasaki heavy duty engines can give you a competitive edge. Send for complete details today about Kawasaki engines and the power to turn people on.

The Power to turn People on.

650 Valley Park Drive, Shakopee, MN 55379
Phone: (612) 445-6060

Recession and the parttime farmer

By Charlie Cape

■ PARTTIME FARMERS, sundown farmers, city farmers. Whatever they are called, the men who work in the city during the day and farm individually between 50 and 100 acres in the evenings and on weekends make up one group many near-metro dealers have learned to work with and appreciate.

Many of these farmers depend on their ability to obtain good, used equipment to keep their own operations working. Often, they subsidize their farming operations with their monthly pay checks.

Recession and inflation have caused high unemployment and unusual changes in just about everybody's life style. I&T phoned four dealers throughout the country who sell to the parttime farmer group to see how the economic situation has affected their business from this special group of farmers.

George Knecht, Knecht Farm Equipment, Paola, Kan.: "So far this year, sales to our parttime farmer customers are down about 25 per cent. Besides the economy and current layoff problems, there are other things to consider. The energy situation has made it so some people who used to drive down to do business have just quit coming.

"Although we have been able to maintain a decent selection of small used tractors, we are finding that this type of unit is becoming increasingly hard to buy or trade for.

"The number of parttime farmers in our area has stabilized. Between 12 and 18 months ago, the number of people moving their families from the city to a farm was increasing almost daily.

"However, the cost of money, unemployment and the general condition of the economy have put a damper on those who, given the conditions we had two years ago, would be moving to the country now. Land is not changing hands like it did, even though the value of the land is still high. The asking price for the land has not increased appreciably and, in some cases, might even have softened."

Walter May, Walter May Farm Supply, Memphis, Tenn.: "The economic situation hasn't really affected our parttime farmer customers as far as their desire to farm. It has affected their equipment buying habits, however.

"With the price of new tractors going up, many of our fulltime farmer customers are keeping their tractors and selling them on their own rather than trading.

"Generally, the parttime farmer is making his own repairs—this has helped our parts sales but hasn't really affected our service volume. Since the parttime farmer is after quality used equipment and uses it fewer hours than a fulltime farmer would, there seems to be fewer repairs needed.

"Most of our customers, including the parttimers, are paying for their purchases on time.

"The majority of parttime farmers are made up of plumbers, electricians, carpenters, factory men and office men who generally wanted to get their families out of the city. Their farms are located anywhere from 15 to 40 miles outside of Memphis. They are putting in gardens, larger crops, and even buying cattle, more so this year than in several years past.

"Our parttime farmer customers are not working as much overtime at their city jobs. Some used to have to really work to get their crops in, but not this year. If the economic situation has done anything to the parttime farmers in our area, it has given them more time to do their farming chores."

Ned Sharp, Sharp Brothers Implement, Indianapolis, Ind.: "Ten per cent of all our customers are the 'city farmer' type. Last year, that percentage was quite a bit larger.

"This year, I see a lot of them pinching pennies. Their decreased buying is felt all through our dealership. We have a very small supply of used equipment available, therefore, the buying of this equipment by the city farmer has all but vanished.

"The tractors we are getting in on trade are from our fulltime farmer customers who are trading up in horsepower. What we end up with is a farm tractor that is nice, in good shape, but too powerful and expensive for the city farmer to even consider.

"Companies where our city farmer customers work during the day are not allowing as much overtime as they did in the recent past. This is giving the city farmer more time to tend to his land and spend some time with his family. To some of these people, farming has moved or is moving from what was once a hobby to a need.

"Some got very used to the added income they had with the overtime. In that respect, I would say that some of the city farmers are farming more to maintain the level of income and living they have become accustomed to."

Barry Brewer, Ray Bock Equipment Co., Puyallup, Wash.: "About 25 per cent of our customers are 'sundown farmers.' However, our sales are down all across the board, primarily because of the drastic price increases we have had from the manufacturers over the past year.

"Consequently our supply of good used equipment has faded and we have hardly any used equipment at all. Most of our customers and the sundown farmers in particular are keeping what they have and rebuilding this machinery. It's been quite a year in our sales of service and parts. In just the past four months, we have had a 25 per cent increase over the same period last year in our parts sales. Our shop is also keeping tremendously busy.

"The only new equipment we've really been selling is the small import tractor. You just can't buy compatible used implements, so our sales of small implements have been in keeping with the small tractor sales.

"Most of our weekend farmers are raising raspberries and strawberries, so there really isn't any need for harvesting equipment—the harvest is done by hand. They are really hurting in the area where the law prohibits 12-year-old kids from helping with the harvest; this is the age group of kids used by many of the parttimers.

"Some of my customers work in the Tacoma saw mills. The mills have cut back production and some have even shut down because of the lack of housing starts and new construction. Other city employers are just not allowing overtime."

More tractors from overseas

Our 1,150,254th tractor.

57 years ago, Fiat built its first mass-produced tractor.

We've built over one million tractors and now we're thinking about our two millionth.

That's a lot of tractors. But it's only natural when you're Europe's leading manufacturer in the field.

When you serve every category of farming with the most complete tractor line on the market: over 50 different models.

When you build more crawler tractors and tractors with 4-wheel drive than anyone else in the world.

How did we get so far?

Being part of Fiat helped. As agriculture becomes more and more sophisticated, it takes all the resources of a major worldwide corporation to stay in the lead.

But being international isn't the whole answer. In farming, no two countries have exactly the same needs. That's why the men who make the decisions for Fiat agricultural tractors are American in America, African in Africa, Asian in Asia.

And European in Europe.

We've built over one million tractors and now we're thinking about our two millionth.

Fiat Trattori
FIAT

FIAT U.S. REPRESENTATIVE, INC.
Agricultural Machinery Division
200 West Monroe Street - Suite 1607
Chicago - Illinois 60606

The challenge of a hungry world.

I&T 100 YEARS OF LEADERSHIP 1886-1986

Implement &Tractor

The business magazine of the farm and industrial equipment industry | **1981-1986**

The cost of operating a farm continued to increase and the price of grain, livestock and other farm products continued to fall. The overseas markets for grain in the 1970's were appealing to other grain producing countries and the cheaper labor costs in such countries as Argentina, Brazil and Canada affected the long term market. By the 1980's, U.S. grain was purchased as a last resort after the less expensive grain from other countries was already sold. Heavy chemical application to kill weeds began to receive competition from special tillage tools. Production costs were being checked very closely and investments in expensive special equipment was often a way of competing in the cost conscious world market. "No-Till" and "Low-Tillage" methods were developed as less expensive ways of farming. Skid steer loaders, already a popular tool to reduce labor costs in the construction industry, began to be used on farms. The skid steer loader had been developed many years earlier for use on farms.

The recession, which began the previous year, got worse in 1981 and farm prices fell. Unusually good weather resulted in a bumper crop and already huge surpluses became bigger. International Harvester sold Solar to Caterpillar for $505 million in May 1981.

Grain exports for 1982-1983 were down to 54,000 metric tons. The controversial Payment in Kind (PIK) program began on January 11, 1983. International Harvester sold most of its farm equipment assets, including its dealership organization, to Tenneco, Inc., the parent company of J.I. Case, on November 26, 1984. Allis Chalmers was sold to Deutz and Ford purchased New Holland in 1985.

Farming continues to be a challenge and hopefully there will continue to be men and women who are able and willing to meet the demands of the farm equipment industry.

Keeping ALL the Harvest

Automatic spout aiming system by John Deere uses electrohydraulics to improve operating efficiency of forage harvesters.

Angling down from harvester to wagon tongue, the position sensor detects relative movements of harvester and wagon. Signals from the sensor are used by microprocessor to control direction of the discharge spout.

Ford Tractor Operation Goes Truckin'

Ford's new **Cargo series** of medium-duty trucks targeted to the world market will go on sale early this year in Brazil and this fall in North America. Power comes from the modified Ford Tractor engine.

Long Mfg. reaches out to smaller OEM customers

With a new product line and a simple do-it-yourself system for determining radiator size, Long Mfg. is giving smaller manufacturers, or those doing short runs of special machines, a lower-cost method of getting a vital component.

150 million Ford engines ago

and today

The Ford Motor Company is proud of the part it has played in America's growth. Ford's role in putting America on wheels is a legend. Ford also made an important contribution to the industrialization of America.

It began back in 1903 when Henry Ford harnessed the power of one of his cars to a small sawmill on a farm. By 1909 Model T power found countless uses in agriculture and industry. Just a few years later Ford was designing engines specifically for industrial applications.

Today—150 million engines later—Ford supplies many different types of gasoline and diesel engines to meet the diversified power needs of the many different industries it serves.

Over these years, the Ford Motor Company has accumulated vast experience in the industrial engine field. So have Ford's distributors and dealers. Your Ford Power Products Distributor is listed in the Yellow Pages. He will be happy to share his Ford engine experience with you.

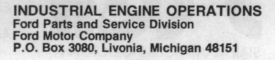

INDUSTRIAL ENGINE OPERATIONS
Ford Parts and Service Division
Ford Motor Company
P.O. Box 3080, Livonia, Michigan 48151

For More Details Circle (49) on Reply Card

Service Department Losses
Got Worse in 1984

Just-in-Time Insights

DEALERS IMPROVED SERVICE SHOP PERFORMANCE

But it's still losing money!

Jay Caylor, veteran parts manager at Brim's Lynden store, uses one of the terminals to get information about the stock of a specific item.

"The reality of parts inventory investment becomes most vivid when dealers are closed or bankrupt."

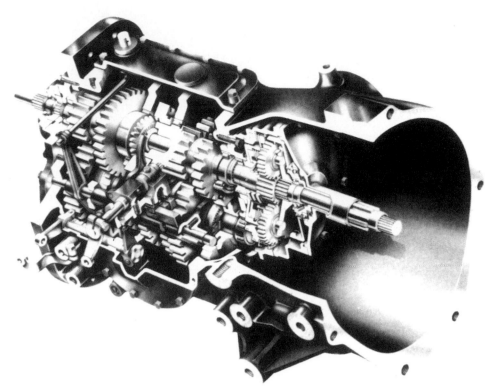

Transmission cutaway shows two-speed gear assembly immediately inside transmission bell-housing. Clutch packs are located directly behind the dual power-shift assembly.

ELECTROHYDRAULIC DUAL POWER-SHIFT CONTROL

Sound waves tie hydraulic cylinder to computer

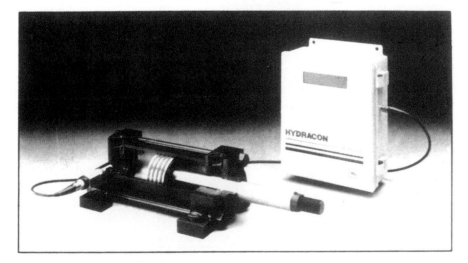

Transducer installed in end cap of hydraulic cylinder generates ultrasound pulses and takes readings to measure position, motion and velocity of piston. The Hydracon system, developed by Hydro-Line, provides accurate readings under varying pressure and temperature conditions and permits closed-loop control of cylinders.

Bill Fogarty

Big Changes in Small Engines

Engineering advances like Honda's overhead valve system for lawnmower engines have added some excitement and new competitors to the small engine market.

Can We Ever Get Back to Square One?

Guidance System Uses Electronics, Hydraulics

Orthman's Tracker Phase II:

Tracker Phase II is mounted ahead of Orthman's Ridge Rider implement. The system automatically steers the implement and the tractor.

Electronic Diesel Fuel Injection Control System

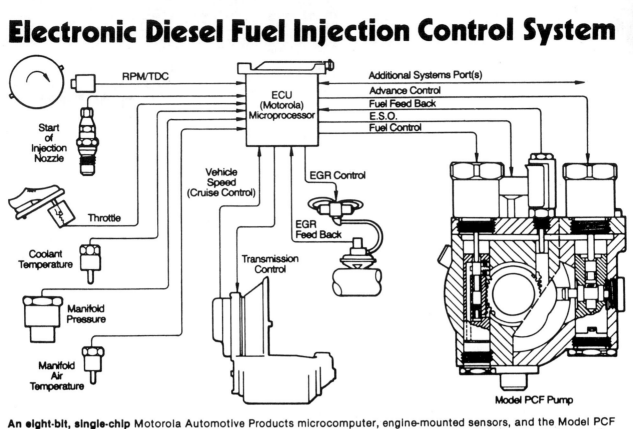

An eight-bit, single-chip Motorola Automotive Products microcomputer, engine-mounted sensors, and the Model PCF diesel fuel injection pump are the major components of the electronic system. Engine mounted sensors analyze engine speed, top dead center, start of injection, water temperature, manifold pressure and exhaust gas recirculation valve position. Start of injection is determined by a special nozzle that incorporates a Hall-effect needle lift sensor. The ECU, based on data received from the sensors, determines the desired advance and fuel delivery levels best suited to engine needs.

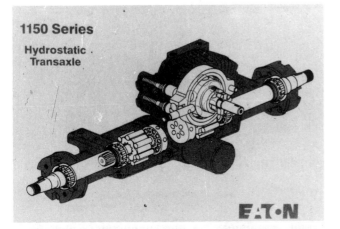

Internal view of Eaton Series 1150 transaxle shows relation of hydrostatic pump, at center, and twin hydraulic motors at inboard end of axles. Configuration and length of output axles can be easily altered to meet needs of the equipment being powered.

A hydrostatic transaxle relying entirely on fluid power rather than gears gives Eaton an innovative driveline system based on off-the-shelf hardware.

When All the Big Guys Change, What Happens to Shortliners?

The tribes gather.

The Tenneco Kids Visit Vegas; Other Players to See Red

DEUTZ SHOWS WIDER NEW LINE

NEW HOLLAND: "NOT BEING BOUGHT, NOT BUYING"

White Farm Gets Aggressive, Tells Dealers to Follow Suit

Seat off an IH unit dating back to the mid '30's has been mounted for use as a utility stool in the kitchen of Mrs. David Watt. The antique backbuster, that could wear the rivets off overall pockets, now shares space with an original oil painting.

To get to the bottom of this repair job Carson H. White has adapted an IH Cub Cadet seat with extra cushion for height, backrest for comfort and tool tray for convenience.

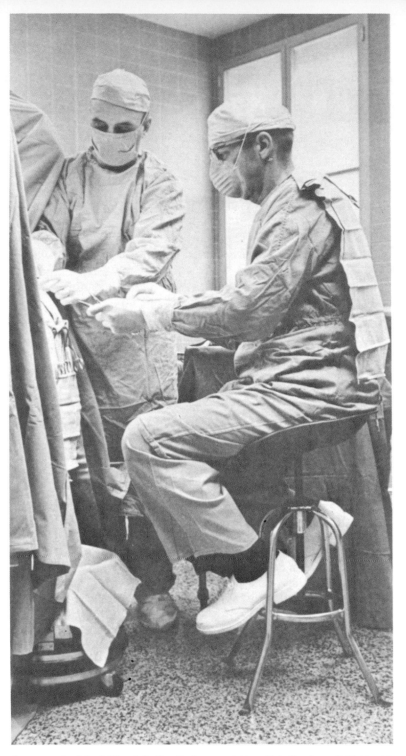

A gynecologist at Hinsdale Sanitarium and Hospital, Dr. Lowell Peterson, does most of his surgery sitting down. With the aid of an IH engineer, Peterson changed a Cub Cadet tractor seat into a comfortable stool. Since gas used in anesthesia is explosive and operating room equipment must be grounded, all paint was removed from the seat.

Urban migration of tractor seats

Salvage yard operators may soon be meeting a new demand for old tractor seats—from city folk. Here are three examples of the metamorphosis—to factory, kitchen, operating room.

INDEX OF ARTICLES

Index of Articles continued...

INDEX OF ARTICLES

LIST OF ILLUSTRATIONS

List of Illustrations continued...

LIST OF ILLUSTRATIONS

List of Illustrations continued...

LIST OF ILLUSTRATIONS

List of Illustrations continued...

LIST OF ILLUSTRATIONS

List of Illustrations continued...

LIST OF ILLUSTRATIONS

List of Illustrations continued...

LIST OF ILLUSTRATIONS

List of Illustrations continued...

LIST OF ILLUSTRATIONS

BIBLIOGRAPHY

Brate, H.R. — *FARM GAS ENGINES*. Cincinnati, OH: The Gas Engine Publishing Co., 1912.

Clark, Neil M. — *John Deere*. Moline, IL: Privately printed by Desaulniers & Co.

JOHN DEERE Tractors 1918•1976. Moline, IL: John Deere & Co., 1976

Dethloff, Henry C. and Irvin M. May, Jr. — *SOUTHWESTERN AGRIGULTURE Pre-Columbian to Modern*. College Station, TX: Texas A&M University Press, 1982.

Jones, Mack M. — *FARM SHOP PRACTICE*. New York: McGraw-Hill Book Company, Inc., 1939.

Horwitz, Elinor Lander — *ON THE LAND American Agriculture from Past to Present*. New York: A Margaret K. McElderry Book, 1980.

Hurt, R. Douglas — *THE DUST BOWL: An Agricultural and Social History*. Chicago, IL: Nelson-Hall, 1981.

Huxley, Bill — *ALLIS-CHALMERS Tractors 1914•1982*. England: Webberley Ltd., 1981.

Marsh, Barbara — *A Corporate Tragedy*. New York: Doubleday & Company, Inc., 1985.

McKinley, Marvin — *WHEELS OF FARM PROGRESS*. St. Joseph, MI: American Society of Agricultural Engineers, 1980.

National Retail Farm Equipment Association — *Farm Equipment Retailer's HANDBOOK*. St. Louis, MO: Farm Equipment Retailing, Inc., 1953.

Pagé, Victor W. — *The MODERN GAS TRACTOR Its Construction, Utility, Operation and Repair*. New York: The Norman W. Henley Publishing Co., 1913.

Putnam, Xeno W. — *THE GASOLINE ENGINE ON THE FARM*. New York: The Norman W. Henley Publishing Co., 1913.

Wendel, C.H. — *Encyclopedia Of American Farm Tractors*. Sarasota, FL: Crestline Publishing.

Wendel, C.H. — *150 Years of International Harvester*. Sarasota, FL: Crestline Publishing.